Process and Device Simulation for MOS-VLSI Circuits

NATO ASI Series

Advanced Science Institutes Series

A Series presenting the results of activities sponsored by the NATO Science Committee, which aims at the dissemination of advanced scientific and technological knowledge, with a view to strengthening links between scientific communities.

The Series is published by an international board of publishers in conjunction with the NATO Scientific Affairs Division

A	Life Sciences	Plenum Publishing Corporation
B	Physics	London and New York
C	Mathematical and Physical Sciences	D. Reidel Publishing Company Dordrecht and Boston
D	Behavioural and Social Sciences	Martinus Nijhoff Publishers Boston/The Hague/Dordrecht/Lancaster
E	Applied Sciences	
F	Computer and Systems Sciences	Springer-Verlag Heidelberg/Berlin/New York
G	Ecological Sciences	

Series E: Applied Sciences – No. 62

Process and Device Simulation for MOS-VLSI Circuits

edited by

Paolo Antognetti
University of Genova
Genova, Italy

Dimitri A. Antoniadis
Massachusetts Institute of Technology
Cambridge, MA, USA

Robert W. Dutton
Stanford University
Stanford, CA, USA

William G. Oldham
University of California
Berkeley, CA, USA

1983 **Martinus Nijhoff Publishers**
A member of the Kluwer Academic Publishers Group
Boston / The Hague / Dordrecht / Lancaster

Published in cooperation with NATO Scientific Affairs Division

Proceedings of the NATO Advanced Study Institute on Process and Device
Simulation for MOS-VLSI Circuits, SOGESTA, Urbino, Italy, July 12 - 23, 1982

Library of Congress Cataloging in Publication Data

NATO Advanced Study Institute on Process and Device
    Simulation for MOS-VLSI Circuits (1982 : Urbino,
    Italy)
    Process and device simulation for MOS-VLSI circuits.

    (NATO ASI series. Series E, Applied sciences ;
no. 62)
    "Proceedings of the NATO Advanced Study Institute on
Process and Device Simulation for MOS-VLSI Circuits,
SOGESTA, Urbino, Italy, July 12-23, 1982"--T.p. verso.
    "Published in cooperation with NATO Scientific Af-
fairs Division."
    1. Integrated circuits--Very large scale integration
--Simulation methods--Congresses. 2. Metal oxide semi-
conductors--Simulation methods--Congresses.
I. Antognetti, Paolo. II. Title. III. Series: NATO
advanced science institutes series. Series E, Applied
sciences ; no. 62.
TK7874.N343 1982   621.381'73   83-4018

ISBN-13: 978-94-009-6844-8   e-ISBN-13: 978-94-009-6842-4
DOI: 10.1007/978-94-009-6842-4

Distributors for the United States and Canada: Kluwer Boston, Inc., 190 Old Derby
Street, Hingham, MA 02043, USA

Distributors for all other countries: Kluwer Academic Publishers Group, Distribution
Center, P.O. Box 322, 3300 AH Dordrecht, The Netherlands

LIST OF LECTURERS

Prof. P. Antognetti
Istituto di Elettrotecnica
V.le Causa, 13
16145 GENOVA
ITALY

Prof. D.A. Antoniadis
MIT - EECS Dept.
Room 13 - 3014
CAMBRIDGE MA. 02139
U.S.A.

Prof. C. Claeys
Katholieke Universiteit Leuven
Departement Elektrotechniek
Kard. Mercierlaan 94
3030 HEVERLEE
BELGIUM

Prof. R. W. Dutton
Stanford University
Stanford Electronics Labs.
STANFORD CA. 94305
U.S.A.

Prof. J. Gibbons
Stanford University
Stanford Electronics Labs.
STANFORD CA. 94305
U.S.A.

Prof. C. R. Helms
Stanford University
Stanford Electronics Labs.
STANFORD CA. 94305
U.S.A.

Dr. L. Mei
Fairchild
4001 Miranda Ave.
PALO ALTO CA. 94304
U.S.A.

Prof. A. Neureuther
University of California
EECS Dept. - Cory Hall
BERKELEY CA. 94720
U.S.A.

Prof. W. G. Oldham
University of California
EECS Dept. - Cory Hall
BERKELEY CA. 94720
U.S.A.

Prof. J. Plummer
Stanford University
Stanford Electronics Labs.
STANFORD CA. 94305
U.S.A.

Dr. H. Ryssel
Fraunhofer Institut IFT
Paul Gerhardt Allee 42
8000 MUNCHEN 60
WEST GERMANY

Dr. K. Salsburg
I.B.M.
9500 Godwin Dr.
MANASSAS, VA. 22110
U.S.A.

Prof. S. Selberherr
Technische Universitat Wien
Gusshausstrasse 27-29
A-1040 WIEN
AUSTRIA

Prof. T. Sigmon
Stanfords University
Stanford Electronics Labs.
STANFORD CA. 94305
U.S.A.

Dr. R. Tielert
SIEMENS ZFE ME 212
Postfach 830952
8000 MUNCHEN 83
WEST GERMANY

Mr. Stan Abbeloos
Bell Telephone MFG CO.
Gasmeterlaan 106
9000 GENT
BELGIUM

Mr. Ginetto Addiego
Univ. of California, Berkeley
Dep. of EECS, ERL
Room 332 Cory Hall
BERKELEY, CALIFORNIA 94720
U.S.A.

Dr. Giuseppe Barbuscia
SGS
Central R & D
Via Olivetti, 2
AGRATE BRIANZA (MI)
ITALY

Dr. J.J.J. Bastiaens
Philips Research Labs
Prof. Holstlaan WAG 1-3-13
EINDHOVEN
THE NETHERLANDS

Dr. Walter Bella
C.S.E.L.T.
Via G. Reiss Romoli 274
10148 TORINO
ITALY

Ing. Delfreo Bianchi
OLIVETTI - DIDAU/QST
Scarmagno A
IVREA (TO)
ITALY

Prof. Giacomo Bisio
Istituto di Elettrotecnica
V.le Causa, 13
16145 GENOVA
ITALY

Dr. B. Broich
BBC Brown, Boveri & Co., Ltd.
Research Center
CH-5405 BADEN
SWITZERLAND

Ing. Antonio Buonomo
Università della Calabria
Dip. Elettrico
ARCAVACATA RENDE (CS)
ITALY

Dr. Alan Butler
Plessey Research (Caswell) Ltd.,
Allen Clark Research Centre
Caswell Towcester
NORTHANTS
ENGLAND

Prof. Gianni Conte
Ist. Elettronica e Telec. Po-
litecnico
C.so Duca degli Abruzzi, 24
TORINO
ITALY

Mr. A.G.R. Evans
Dept. of Electronics
Southampton University
SOUTHAMPTON
ENGLAND

Mr. Alistair Buttar
Electrical Eng. Dept.
Edinburgh University
Mayfield Rd. - EDINBURGH
SCOTLAND

Mr. Christopher Davis
RCA Solid State, MZ 177
U.S. Rte 202
SOMERVILLE NJ 08876
U.S.A.

Mr. Paul Fahey
Stanford University
Stanford Electronics Lab.
AEL 231
STANFORD, CA 94305
U.S.A.

Mr. E.R. Campbell
British Aerospace
P.B. 299
SITE A
STEVENAGE
ENGLAND

Mrs. Kristin M. De Meyer
K.U.L. - ESAT. 91 DEP. ELEKTR.
Kardinaal Mercierlaan 94
B-3030 HEVERLEE
BELGIUM

Mr. John Faricelli
325 Phillips Hall
Cornell University
ITHACA, NY 14853
U.S.A.

Mr. Caquot
CNET
B.P. 42
F 38240 MEYLAN
FRANCE

Ing. Giovanni De Micheli
Via Traiano 13
I-20149 MILANO
ITALY

Ing. C. Fasce
SGS - Central R & D
Via Olivetti, 2
AGRATE BRIANZA (MI)
ITALY

Ing. Daniele Caviglia
Istituto di Elettrotecnica
V.le Causa, 13
GENOVA
ITALY

Dr. Yves Depeursinge
LSRH
Rue Breguet 2
2000 NEUCHATEL
SWITZERLAND

Mr. Gerard Ghibaudo
ENSER, ERA CNRS 659
23 Avenue des Martyrs
38031 GRENOBLE CEDEX
FRANCE

Mr. Peter Claus
Inst. für Hochfrequenztechnik
Techn. Hochschule Darmstadt
Merckstr. 25
D-6100 DARMSTADT
WEST GERMANY

Ing. Ermanno Di Zitti
Istituto di Elettrotecnica
V.le Causa, 13
16145 GENOVA
ITALY

Dr. A.E. Glaccum
British Telecom Research Labs.
Martlesham Heath, Ipswich,
IP57RE
ENGLAND

Bill Cochran
Bell Labs
555 Union Blvd.
ALLENTOWN PA. 18018
U.S.A.

Mr. K. Doganis
ICL AEL 231
Stanford University
STANFORD CA 94305
U.S.A.

Mr. Richard J. Gledhill
Middlesex Polytechnic
Bounds Green Road
LONDON N11 2NQ
BRITAIN

Mr. Collard
I.S.E.N.
3, rue F. Baes
59046 LILLE CEDEX
FRANCE

Dr. J.D. Eades
Robert Gordon's Inst. of
Technology
Schoolhill
ABERDEEN AB9 1FR
SCOTLAND U.K.

Mr. James A. Greenfield
AEL 127 A
Stanford University
STANFORD, CA 94305
U.S.A.

Mr. S.E. Hansen
Stanford University
AEL Bldg. Room 229
STANFORD, CAL. 94305
U.S.A.

Mr. Jan Haraldsen
Okrivn. 27
1349 RYKKINN
NORWAY

Mr. H.B. Harrison
Royal Melbourne Inst. of
Tech. Ltd. (RMIT)

124, La Trobe St.,
MELBOURNE 3000
AUSTRALIA

Mr. Max Hodeau
I.B.M. France - Service 1768
224, Bd. J. Kennedy
91102 CORBEIL-ESSONNES
FRANCE

Mr. I. John
P.B. 252
Six Hills Way
B.A.e.
STEVENAGE HERTS
UNITED KINGDOM

Mr. Michael Roy Kump
Stanford University
AEL 231B
STANFORD, CALIFORNIA 94305
U.S.A.

Mr. Alan Lewis
G.E.C. Hirst Research Centre
Room B 18
East Lane, Wembley
MIDDLESEX
UNITED KINGDOM

Mr. Hon Bor Lo
Dept. of Electrical Eng.
University of Hong Kong
Pokfulam Road
HONG KONG

Ing. Claudio Lombardi
SGS
Central R & D
Via Olivetti, 2
AGRANTE BRIANZA (MI)

Mr. Istvan Madas
SIEMENS AG
Mch B - WIS TE CAD 12
Balanstr. 73
D-8000 MUNCHEN 80
GERMANY

Mr. L. Mader
SIEMENS, ME 31
Postfach 830952
D-8000 MUNCHEN 83
WEST GERMANY

Mr. Americo Marrocco
I.N.R.I.A.
Rocquencourt BP 105
78 153 LE CHESNAY
FRANCE

Prof. Guido Masetti
Università di Ancona
Via Montagnola, 30
60100 ANCONA
ITALY

Mr. Hugh Mc Neillie
General Instrument
Newark road
EASFIELD INDUSTRIAL EST
GLEN ROTHES FIFE
SCOTLAND

Mrs. M. Teresa Mendes
Universidade de Coimbra
Dept. Electrotecnia
F.C.T.U.C.
3000 COIMBRA
PORTUGAL

Mr. Rob Moolenbeek
181 Claybrook Rd,
DOVER MASS 02030
U.S.A.

Mr. Sharad N. Nandgaonkar
Univ. of California, Berkeley
Dep. of EECS, ERL
Room 332 Cory Hall
BERKELEY, CALIFORNIA 94720
U.S.A.

Mr. Theo Naten
Bell Telephone MFG CO.
Gasmeterlaan 106
9000 GHENT
BELGIUM

Mr. Ismail Okter
Istanbul Tech. Univ., Electrical
Faculty
ITU Elektrik Fakultesi
Elektronik ve Yuksek Frekans
Teknigi Kursusu
GUMUSSUYU - ISTANBUL
TURKEY

Mr. Payo-Casares
Thomson-EFCIS D.T.
Av. des Martyrs - B.P. 217
38019 GRENOBLE CEDEX
FRANCE

Dr.ssa Luisa Polignano
SGS - Central R & D
Via Olivetti, 2
AGRATE BRIANZA (MI)
ITALY

Mr. Mark Readdie
G.E.C. Hirst Research Centre
East Lane, Wembley
MIDDLESEX
ENGLAND

Mr. P.M. Savjani
Rutherford Appelton Laboratory
CHILTON, DIDCOT OXON
ENGLAND

Mr. Claude Schlesser
Bell Telephone MFG CO.
Microelectronics Lab.
106, Gasmeterlaan
9000 GENT
BELGIUM

Mr. A. Stolmeijer
Twente University of Technology
Dep. of Electrical Eng.
P.O. Box 217
7500 AE ENSCHEDE
THE NETHERLANDS

Dr. A.J. Walton
Electrical Eng. Dept.
Univ. of Edinburgh
Kings Buildings
EDINBURGH EHG 3JL
SCOTLAND

Mr. Yosi J. Shacham Diamand
Micorelectronic Research Center,
Solid State Institute, Techion
Technion City,
HAIFA
ISRAEL

Mr. Bernard Tremintim
Matra Harris Semiconducteur
BP 942 - La Chantrerie
44075 NANTES CEDEX
FRANCE

Mr. Per Wehlin
Royal Inst. of Technology, Stockholm
Inst. f. Tillampad elektronik
S-10044 STOCKHOLM
SWEDEN

Dr. Martine Simard-Normandin
Northern Telecom Limited
P.O. Box 3511 - Station C
OTTAWA ONTARIO KIY 4H7
CANADA

Mr. Hans Paul Tuinhout
Philips Research Laboratory
WAG-1-4-18
EINDHOVEN
THE NETHERLANDS

Dr. Christoph Weigel
Wacker Chemitronic
Postfach 1140
D-8263 BURGHAUSEN
WEST GERMANY

Dr. Claudine Simson
Northern Telecom Limited
P.O. Box 3511, Station C,
Dept. P 881
OTTAWA, ONTARIO KIY 4H7
CANADA

Ing. Massimo Vanzi
SGS
Central R & D
Via Olivetti, 2
AGRATE BRIANZA (MI)
ITALY

TABLE OF CONTENTS

PREFACE

P. Antognetti

University of Genova, Italy
Director of the NATO ASI

The key importance of VLSI circuits is shown by the national efforts in this field taking place in several countries at different levels (government agencies, private industries, defense departments). As a result of the evolution of IC technology over the past two decades, component complexity has increased from one single to over 400,000 transistor functions per chip. Low cost of such single chip systems is only possible by reducing design cost per function and avoiding cost penalties for design errors. Therefore, computer simulation tools, at all levels of the design process, have become an absolute necessity and a cornerstone in the VLSI era, particularly as experimental investigations are very time-consuming, often too expensive and sometimes not at all feasible.

As minimum device dimensions shrink, the need to understand the fabrication process in a quantitative way becomes critical. Fine patterns, thin oxide layers, polycristalline silicon intercon nections, shallow junctions and threshold implants, each become more sensitive to process variations. Each of these technologies changes toward finer structures requires increased understanding of the process physics. In addition, the tighter requirements for process control make it imperative that sensitivities be under stood and that optimation be used to minimize the effect of statistical fluctuations.

Over the past years, models have been developed which make it possible to simulate IC fabrication process physics. Moreover, the development of computer programs to simulate the complete multistep fabrication process now make it possible to realistically evaluate technologies both for design and process control purposes. Recently process simulators have become available to aid in the design of advanced IC technologies. They provide one--dimensional impurity profiles and oxide thicknesses for any sequence of process steps such as oxidation, ion implantation, diffusion and epitaxial growth.

With the emphasis on decreasing device size as LSI technologies give way to their VLSI successors, there is an increasing need to simulate twodimensional processes in IC manufacture. One aspect of it concerns the studies of the line-edge profiles resulting from a sequence of various lithography, growth and etching processes. Prime examples of the usefulness of such kind of simulation are in understanding projection printing, step coverage in deposition, wafer planarization and linewidth bias and control for composite process sequences. Simulations for individual IC fabrication processes have been combined in a user oriented program recently described in the literature.

Another important aspect of two-dimensional simulation concerns the MOS device design, where the evolution toward VLSI is presenting new challenges. The increasing level of integration requires that the area and power dissipation of integrated circuits be reduced. At the same time, more accurate control must be maintained over device operating characteristics, which may depend on highly two-dimensional structures. The accurate analysis of devices with complex structures requires two-dimensional numerical simulation. Here again recent papers have described user-oriented programs for the two-dimensional simulation of MOS transistors. However, as devices are scaled down, two-dimensional effects of impurity concentration profiles and of surface topology become more significant. Therefore, there is an emerging need to couple two-dimensional device simulators with the two-dimensional process simulators, which are now appearing in the literature.

This book focuses on the physics, modeling and computer simulation aspects of the technological processes and of the MOS devices for VLSI design. The areas of interest are: (a) the description of the technological steps involved in the fabrication and characterization of MOS devices, such as diffusion, oxidation ion implantation, annealing, lithography, deposition and etching; (b) the description of the algorithms used in the simulation of technological processes and of the MOS device; (c) implementation of such algorithms into actual programs for one and two-dimensional simulation of processes and for two-dimensional device simulation; (d) efficient coupling of the different programs described on process and device simulation to obtain an integrated set of design aids.

The content of this book results from the lecture notes handed out at the NATO ASI, held at SOGESTA, Urbino (Italy) from July 12 to 23, 1982. The mornings were dedicated to the formal lectures, while in the afternoon the 65 partecipants had the opportunity to get a direct "hands-on" experience of using the programs via the terminals available at the SOGESTA Computer Center, equipped with a DEC-VAX II/780 computer.

I would like to thank Drs. M. Di Lullo and C. Sinclair of the NATO Scientific Affairs Division for their helpful assistance in organizing the ASI, the lecturers for their timely and accurate presentations and for their careful preparation of the text. I would also like to express my gratitude to DEC for the assistance, to Giovanni (Nanni) De Micheli and R. Riccioni for the organization of the "hands-on" sessions.

DIFFUSION IN SILICON

Dimitri A. Antoniadis

Massachusetts Institute of Technology
Cambridge, MA 02139

1. INTRODUCTION

Solid state diffusion is the physical process responsible for the migration of dopant atoms in the silicon lattice. As such, it is a very important process in the construction of all kinds of electronic devices and circuits on silicon. In the present chapter, the emphasis will be on the presentation of current theories and ideas concerning this physical process. It may be rather surprising that given the significant scientific and technological importance of diffusion in silicon as well as the relative maturity of the field of microelectronics technology, no generally accepted theory exists for diffusion phenomena nor complete and indisputable characterization of diffusivity of the common dopant impurities under several technologically important processing conditions.

2. MECHANISMS OF SOLID STATE DIFFUSION IN SILICON

Various chemical elements may exist in solution in the silicon lattice. The solute element is called substitutional if its atoms occupy regular lattice sites where they substitute solvent atoms. In the case where the solute atoms occupy any of the several void sites in the solvent lattice, the solute is said to be interstitial. Most solutes dissolve both interstitially and substitutionally in silicon, but the ratio of the two solubilities varies over several orders of magnitude from element to element. On the other hand, atoms of Groups IIIA and VA elements are characterized by their ability to form strong covalent bonds with the host atoms in the silicon lattice; as a result of this

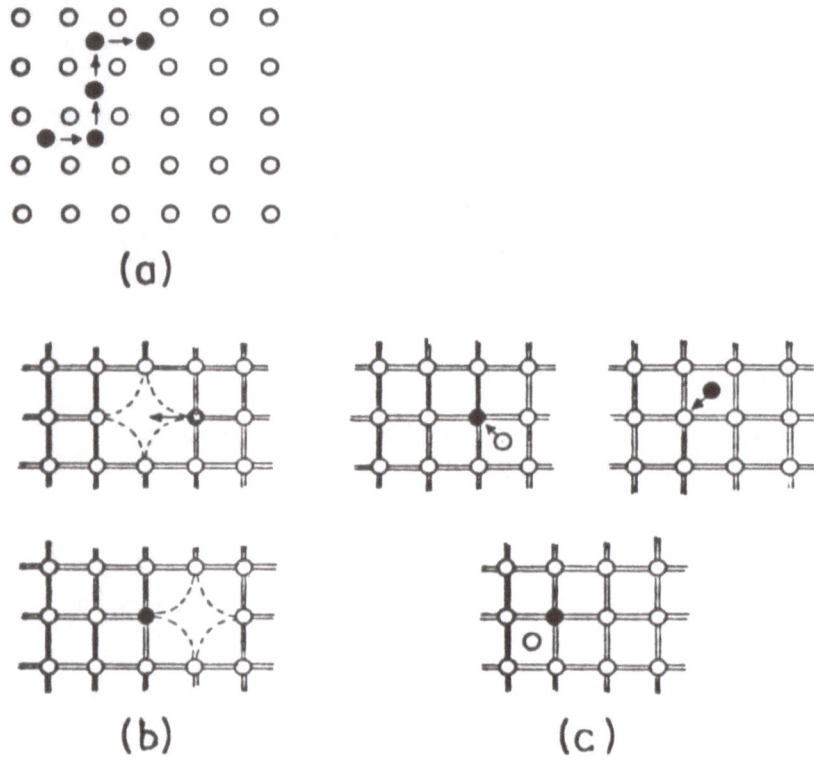

1. Basic diffusion mechanisms in planarized crystal lattice.
   (a) Interstitial diffusion. (b) Substitutional diffusion via
   vacancy mechanism. (c) Substitutional diffusion via interstitialcy mechanism.

property, they occupy almost exclusively substitutional sites. Their low ionization energy in that state makes them ideal dopants for the control of electrical characteristics of the silicon crystal.

Solid state diffusion is the physical process by which solute atoms migrate in the solvent lattice. This process also includes the migration of host atoms in their own lattice in which case it is called self-diffusion. Although phenomenologically solid state diffusion resembles its counterpart in gases and fluids, there exist important differences that stem from the fact that in crystals the moving solute atoms, by necessity, must disturb the order of the lattice and thus they must coordinate their movement with lattice atom movements.

2.1 Types of Diffusion in Lattice

A rather simplistic visualization of the basic diffusion mechanisms in a crystal lattice is provided by Fig. 1. Conceptually, the simplest mechanism is interstitial diffusion where the impurity atoms jump from one interstice to another without requiring permanent displacement of any of the host atoms, as shown in Fig. 1a. However, it is not required that the moving atom be strictly interstitial. The atoms diffusing interstitially may occupy substitutional sites for long intervals of time. In fact, they may well be indigenous atoms diffusing interstitially through the lattice of their own crystal.

Figure 1b illustrates the basic vacancy mechanism for substitutional diffusion by which the solute atoms jump from one lattice site to the neighboring one. This is a point defect mechanism in that it requires a vacancy, i.e., a vacant lattice site, to exist next to the diffusing atom. Thus, diffusion takes place by interchange of position between the lattice defect and the diffusing atom. Under thermal equilibrium, even in the most perfectly grown crystal, intrinsic point defects, i.e., vacancies and/or self-interstitials, exist at concentrations that depend strongly on lattice temperature. It is obvious that the diffusion rate for the substitutional vacancy mechanism would be proportional to the vacancy concentration. This vacancy mechanism is the predominant diffusion mechanism in metals.

Figure 1c illustrates another substitutional diffusion mechanism that also involves a point defect, this time a self-interstitial atom as opposed to a vacancy. As we will see later on, the existence of this so-called interstitialcy mechanism (not to be confused with the interstitial mechanism), together with the vacancy mechanism, can explain several dopant impurity anomalous diffusion phenomena in silicon. In the interstitialcy mechanism,

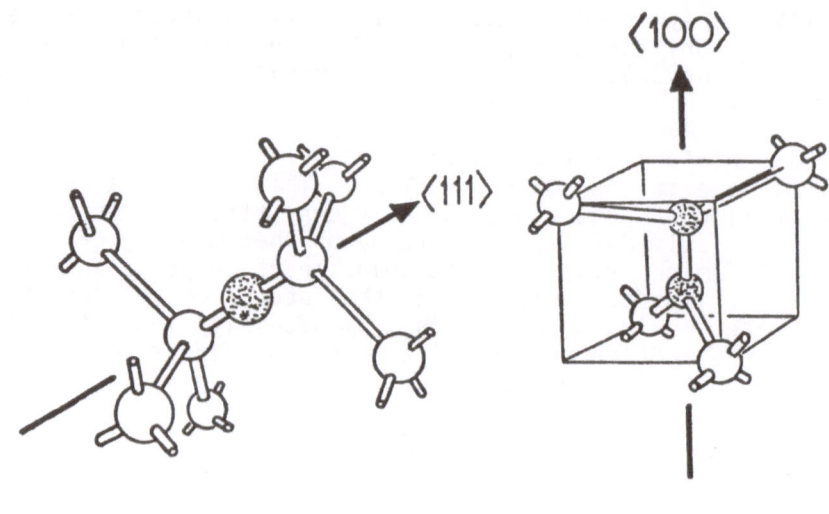

(a) ⟨111⟩ Bond-centered   (b) ⟨100⟩ Split

2. Bonded interstitial in silicon (after Watkins, et al. [19]).

it is simplisticly supposed that the repulsive potential at the saddle point between lattice sites in a normal interstitial path is so high that the interstitial atom chooses to move by pushing one of its neighboring lattice atoms into another interstice while itself taking up the substitutional site. A more accurate picture is that the self-interstitial or the dopant atom in the interstice is actually bonded in a configuration such as those shown in Fig. 2, and that its motion is simply accomplished by reorientation of the bonding orbitals.

It should be noted that combinations of the above diffusion mechanisms may also take place. For instance, the combination of interstitial and substitutional diffusion has been identified for several impurities in germanium and, to a lesser extent, in silicon.

At the time of writing this chapter, there exists good evidence, as described in section 4, that both vacancies and self-interstitials are intrinsic point defects in crystalline silicon, and that both mechanisms of vacancy and interstitialcy diffusion are responsible for substitutional atom migration. However, at the first part of our discussion here we will adopt the classical view that vacancies are the dominant point defects. This is done only in the interest of simplicity of presentation, since for all mechanisms and arguments discussed here for the case of the vacancy diffusion mechanism, similar ones, indeed dual ones, can be devised for the interstitialcy mechanism.

2.2 Diffusive Flux and Diffusivity

The diffusion coefficient or diffusivity $\vec{D}$ of atoms diffusing in a crystal lattice may be defined either macroscopically or microscopically. Macroscopically, $\vec{D}$ relates the gradient of concentration of the diffusing atoms to their flux $\vec{J}$ in an otherwise homogeneous crystal at uniform tremperature according to

$$\vec{J} = -\vec{D} \text{ grad } C \tag{1}$$

This equation is often referred to as Fick's first law. The diffusion coefficient $\vec{D}$ is generally a symmetric second-rank tensor. However, in cubic crystals, such as silicon or germanium, it reduces to a single scalar quantity D, i.e., diffusion in these materials is isotropic.

The microscopic definition of the diffusion coefficient dates back to Einstein and Smoluchowski. According to this definition, D_x, the x-component of the diffusion coefficient $\vec{D}$, is related to the mean square displacement Δx^2 of the diffusing atoms in the x-direction and the mean time interval $\overline{\Delta t}$ during which the displacement takes place according to

$$D_x = \frac{\overline{\Delta x^2}}{2\overline{\Delta t}} \qquad (2)$$

The two definitions can be easily shown to be equivalent.

The macroscopic definition, given by Eq. (1), constitutes the theoretical basis of experimental determination of the diffusivity by the various diffusant detection techniques. On the other hand, the microscopic definition by Eq. (2) is the starting point of the theoretical interpretation and first principle calculation of diffusion coefficients based on physical quantities such as jump-frequencies of atoms, defect concentrations, and lattice parameters.

Experimentally, it is often found that the temperature dependence of the diffusion coefficient obeys an Arrhenius equation

$$D(T) = D_o \exp(-\frac{Q}{kT}) \qquad (3)$$

where k is the Boltzmann constant and T the absolute temperature. In such cases, the diffusivity is characterized by two quantities, the pre-exponential factor D_o and the activation energy Q. As we will see later on, this equation implies a diffusion mechanism controlled by only one point defect species and must therefore be used carefully because often the presence of impurities and impurity gradients in silicon perturb the equilibrium defect concentration, making both D_o and Q functions of impurity concentrations and process conditions.

It has become rather customary to classify impurity species into slow and fast diffusors. Slow diffusors have diffusivities in the range from about 10^0 to 10^2 of the silicon self-diffusivity. On the other hand, fast diffusors have diffusivities several orders of magnitude higher than those of slow diffusors. Typical examples of slow diffusors are the elements of Groups III and V, used as acceptors and donors in silicon. Representative fast diffusors are elements of Groups IA and IB, e.y., H, Li, Cu, Au, etc.

The mechanisms of slow impurity diffusion might be expected to be related to those of Si self-diffusion, classically assumed to be vacancy mechanisms. This assumption is basically an extension of the observation of point defects and of self-diffusion in metals where vacancies are clearly establilshed as the native point defects. However, because of the extremely small intrinsic concentration of point defects in Si (estimated about 10^{16} cm^{-3} at processing temperatures), there is no direct evidence that vacancies are indeed the native defect species. The crucial experiment, where the lattice constant is accurately measured as function of temperature and compared with the crystal thermal expansion to determine whether more vacancies or more interstitials are introduced with increasing temperatures (thus determining the nature of the intrinsic point defects), is still inconclusive for Si.

Fast diffusors have also received considerable attention in the literature, and it has been concluded that they diffuse primarily by interstitial diffusion. Table 1 lists diffusion exponential factors and activation energies for some interstitial diffusors in silicon. Because of the very open nature of the Si crystal, many atoms dissolve interstitially and diffuse likewise. These include most alkali and heavy metals as well as O and C. Oxygen is a good example of a bonded interstitial in Si even at room temperature, but carbon is quite probably substitutional at such low temperatures. A general remark is that interstitial diffusors diffuse much faster than substitutional ones, and they exhibit a lower activation energy of diffusion (compare Tables 1 and 3). This is consistent with the theory presented here since as will be seen later the activation energy for substitutional diffusion contains more than a simple migration enthalpy. In addition to pure interstitial and substitutional diffusors the possibility exists that some species may diffuse by a combination of the two mechanisms. An interesting example of the interstitial-substitutional diffusion in Si is Au which exhibits a low interstitial solubility with fast diffusivity and a higher substitutional solubility with lower diffusivity.

Given its importance in understanding the mechanisms of substitutional diffusion, we will examine the silicon self-diffusion process in some detail.

2.3 Silicon Self-Diffusion

The standard experimental method for self-diffusion determination in solids is based on the use of radiotracers. Controlled amounts of radioactive isotope atoms are introduced into the solid and subsequently diffused at various temperatures. Diffusion coefficients are extracted from measured radiotracer profiles by fitting Fick's second law to the data. Due to the very low concentration of point defects at thermal equilibrium, self-diffusion in the common semiconductors, Si and Ge, is several orders of magnitude slower than in metals. Therefore, the temperature range of useful tracer measurements is limited to a rather narrow range within two to three hundred degrees from the

Table 1

DIFFUSION PREEXPONENTIAL FACTOR AND ACTIVATION ENERGY OF DIFFUSION FOR INTERSTITIAL DIFFUSORS IN SILICON (after Ghandhi [1])

Impurity	Li	S	Fe	Cu	Ag	Au	O	Ni	Zn
D_0 (cm^2/sec)	2.5×10^{-3}	0.92	6.2×10^{-3}	4×10^{-2}	2×10^{-3}	1.1×10^{-3}	0.21		
Activation energy of diffusion (eV)	0.655	2.2	1.60	1.0	1.6	1.12	2.44	$D = 10^{-3}$ cm^2/sec 1100-1360°C	$D = 10^{-6}-10^{-7}$ cm^2/sec 900-1360°C
Temperature range of measurement (°C)	25-1350	1050-1370	1100-1350	800-1100	1100-1350	800-1200	1300		
Type of impurity			Deep-lying usually multiple-charge states						

Table 2

EXPERIMENTAL SELF-DIFFUSION COEFFICIENTS IN INTRINSIC SILICON

D_{so} (cm^2/sec)	Q_s (eV)	Temperature Range °C	Reference	Technique
9000	5.13	1100-1300	Masters and Fairfield [2]	[31]Si and anodic sectioning
1460	5.02	1050-1390	Mayer, Mehier and Maier [3]	[31]Si and sputter sectioning
154	4.65	855-1175	Kalinowski and Seguin [4]	[30]Si SIMS profiling

melting point. In addition, the self-diffusion measurements in silicon are particularly difficult because of the short half-life of the only readily available radiosotope, ^{31}Si. Indirect techniques such as the annealing of dislocation loops are also used for the measurement of silicon self-diffusion on the premise that the observed phenomena are diffusion limited. Figures 3a and 3b are a plot of Si and Ge self-diffusion coefficient as functions of inverse temperature, including selected data from the literature. Table 2 summarizes these data for silicon. There is generally less agreement for silicon than for germanium, which possesses a long-lived radiosotope, ^{71}Ge.

It is commonly assumed that self-diffusion in silicon takes place by means of the vacancy mechanism. We will adopt this mechanism in order to demonstrate the basic principles of point defect diffusion. However, the reader is warned that there are indications that interstitialcy plays a significant role in silicon self-diffusion. All principles developed here for vacancies apply equally well to self-interstitials.

Vacancies (point defects, in general) possess ionization states with energy levels within the silicon band gap and thus their concentration at the various states is affected by the presence of donors or acceptors at concentrations sufficient to perturb the Fermi energy in the crystal. However, in order to present the principles of the theory of silicon self-diffusion, we will first make the simplifying assumption that the diffusion process is dominated by the neutral vacancy species. This would be the case either if the energetics were more favorable for interaction with this vacancy state or if the concentration of neutral vacancies was dominant over other point defects.

At any temperature other than absolute zero there exists a thermal equilibrium concentration of intrinsic point defects in any crystalline latice. This is the result of the minimization of the free energy of the lattice, $G = H - TS$, at thermal equilibrium. For monovacancies in the neutral state, the normalized concentration N_v^x is given by

$$N_v^x = \frac{[V^x]}{n_H} = \exp\left(\frac{\Delta S_f}{k}\right) \exp\left(-\frac{\Delta H_f}{kT}\right) \tag{4}$$

where $[V^x]$ is the absolute concentration of neutral vacancies, n_H is the concentration of lattice sites (5×10^{22} cm^{-3} in silicon), ΔS_f is the change of the thermal entropy of the lattice per vacancy, often called simply "entropy of vacancy formation," and ΔH_f is the change of the enthalpy of the lattice per vacancy also called "enthalpy of formation." Figure 4 illustrates the concept of such a monovacancy in an idealized lattice together with the corresponding free energy of the system (vacancy-lattice) as a

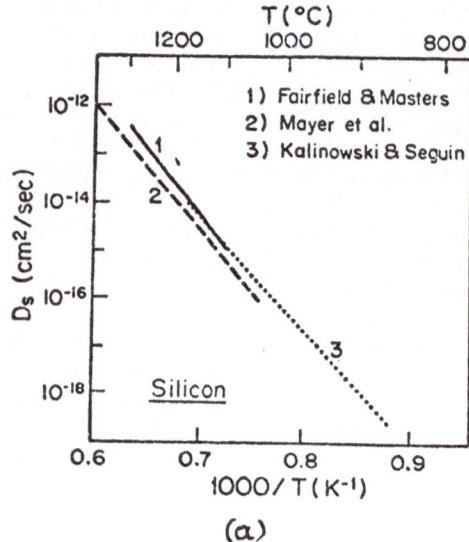

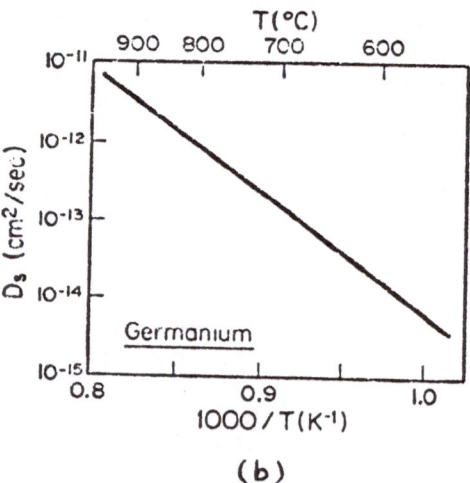

3. Arrhenius plots of (a) silicon self-diffusivity (after Fairfield and Masters [2], Mayer, et al. [3], and Kalinowski and
Seguin [4]) and (b) germanium self-diffusivity (after Widmer
and Gunter-Mohr [5], Valenta and Ramasastry[6], and Letaw, et
al. [7]).

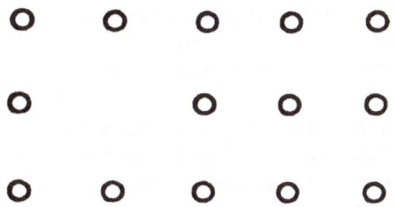

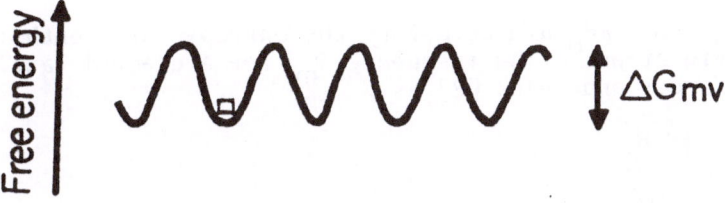

4. Self-diffusion in an idealized lattice by means of a monovacancy mechanism. The free energy of the system as a function
of the migrating vacancy is also shown.

function of the position of the migrating vacancy. Although here it is the lattice atom (tracer of host) that jumps into the vacancy leaving behind it a vacant lattice site, in analogy to the interstitial migration discussed previously it is clearer if one assumes that the vacancy moves as a particle. If ν_{ov} is the jump frequency of an atom in the lattice site adjacent to the vacancy the frequency of this atom's successful jumps over the barrier ΔG_m into the vacancy,

$$\nu_v = \nu_{ov} \ \exp\!\left(\frac{\Delta S_m}{k}\right) \ \exp\!\left(- \frac{\Delta H_m}{kT}\right) \tag{5}$$

where ΔS_m and ΔH_m are as before the entropy and enthalpy of vacancy migration. The frequency ν_{ov} for a diamond lattice is given by Seeger and Cnick [8] as

$$\nu_{ov} = \sqrt{\frac{8\Delta H_m}{3m\alpha^2}} \tag{6}$$

where m is the mass of the diffusing atoms and α is the lattice parameter (α = 5.428 Å, for silicon).

Using the microscopic definition for the diffusion coefficient by Eq. (2) and taking proper account of the projected jump distance along the crystal axis (in silicon $\Delta x = \Delta y = \Delta z = \alpha/4$), as well as the factor of 4 arising from four different ways the vacancy may jump out of its lattice site, one obtains for vacancy diffusion in a diamond lattice,

$$D_v = \frac{1}{2}\!\left(\frac{\alpha}{4}\right)^2 4\nu_v = \frac{1}{8}\alpha^2 \nu_v \tag{7}$$

The self-diffusion coefficient D_{sv} resulting from this monovacancy mechanism is given by

$$D_{sv} = f_v D_v N_v^x \tag{8}$$

where f_v = 1/2 (for diamond lattice) is the so-called correlation coefficient for self-diffusion by monovacancies. This correlation factor arises because a particular atom can exchange position with a given vacancy more than once; this results in a series of correlated exchanges between the atom and the vacancy [17]. Combining now Eqs. (8), (5), and (7), one obtains

$$D_{sv} = \frac{\alpha^2}{8} \ \nu_{ov} f_v \ \exp\!\left(\frac{\Delta S_f + \Delta S_m}{k}\right) \ \exp\!\left(- \frac{\Delta H_f + \Delta H_m}{kT}\right) \tag{9}$$

This expression is of the well-known Arrhenius form

$$D_{sv} = D_{svo} \exp\left(-\frac{Q_{sv}}{kT}\right) \quad , \tag{10a}$$

where

$$D_{svo} = \frac{1}{8} \alpha^2 \nu_{ov} f_v \exp\left(\frac{\Delta S_f + \Delta S_m}{k}\right) \quad , \tag{10b}$$

and

$$Q_{sv} = \Delta H_f + \Delta H_m \quad . \tag{10c}$$

At this point, it is interesting to compare the theory with experiments. Theoretical calculations of ΔH_f and ΔH_m [9] yield the following values for silicon.

$$\Delta H_f = 2.66 \text{ eV (an upper bound)}, \tag{11a}$$

$$\Delta H_m = 1.2 \text{ eV (high temperature)}. \tag{11b}$$

The ΔH_f above is for vacancy formation at surfaces and/or crystallographic defects; Frenkel pair (vacancy-interstitial) formation in the bulk involves at least twice as much energy and is thus an unlikely supply of vacancies. These values give a silicon self-diffusion activation energy of about 3.9 eV as upper bound, which compared to the data in Table 2 is about 0.7 eV lower than the low temperature data [4] and disagrees even more with the high temperature ones [2,3]. As we will see further on, there is some improvement of this situation when the ionized vacancy species are taken into account, but this is not enough to bring totally satisfactory agreement. On the other hand, from the experimental D_{so} and Eqs. (10b) and (6), one obtains, for the entropy of silicon self-diffusion,

$$\Delta S_f + \Delta S_m = 13 \text{ k} \tag{12}$$

It has been pointed out [8] that this large value of entropy indicates that more than one or two atoms participate in the elementary diffusion step. This has led workers in this field [8,10] to formulate the "extended defect" model where it is assumed that the presence of a vacancy or of an interstitial defect leads to deformation of the lattice that extends over several atomic distances from the point defect. Formation of such an extended defect certainly involves higher entropy than a localized point defect but, if it also has a high enthalpy of formation then it may actually have a smaller free energy and thus it may be more probable than localized point defects in the silicon lattice. This is an attractive model, but quantitative arguments supporting it are rather involved, thus falling outside the scope of this discussion.

We turn now to the examination of the implications of the ionization levels of point defects on diffusion. Vacancies (and probably interstitials) possess ionization levels within the band gap of silicon, and therefore at any temperature in thermal equilibrium all species are present in concentrations depending on the position of the Fermi level, E_F. Three ionized states have been identified for vacancies [11]; positive, single negative and double negative. We will denote the normalized concentrations in each state as N_v^+, N_v^-, and N_v^{2-}, respectively. For the neutral state we will use N_v^x. If we accept that silicon self-diffusion, indeed takes place by means of a monovacancy (or monointerstialcy) mechanism and that all states may contribute to it, the total self-diffusivity results from the superposition of the individual diffusivities. Neglecting effects arising from electrostatic potentials and their gradients, D_{sv} is given by

$$D_{sv} = f_v[D_v^x N_v^x(T) + D_v^- N_v^-(T,E_F) + D_v^{2-} N_v^{2-}(T,E_F)$$

$$+ D_v^+ N_v^+(T,E_F)]$$

(13)

From Fermi-Dirac statistics the normalized concentrations of the vacancies at the various ionized states can be closely approximated as follows:

$$N_v^- = N_v^x g_A^- \exp\left(\frac{E_F - E^-}{kT}\right)$$

(14a)

$$N_v^{2-} = N_v^x g_A^{2-} \exp\left(\frac{2E_F - E^{2-} - E^-}{kT}\right)$$

(14b)

$$N_v^+ = N_v^x g_D \exp\left(\frac{E^+ - E_F}{kT}\right)$$

(14c)

g_A^- and g_D are the degeneracy factors of the acceptor and donor states of vacancies, both equal to 2 [17], $g_A^{2-} = 1$ is the degeneracy of the doubly ionized state, E^- and E^{2-} the energy levels of the acceptor states in the silicon band gap, and E^+ the donor state.

The silicon band gap energy $E_g(T)$ at the temperature range above 1000 °K (727 °C) up to the melting point may be approximated by the linear relation

$$E_g(T) = E_{go} - bT \tag{15}$$

where $E_{go} = 1.34$ eV and $b = 4.5 \times 10^{-4}$ eV/°K. Note that this formula is only valid in the above mentioned range. E_{go} has no physical significance and should not be confused with the band gap energy at 0°K, which is $E_g(0°K) = 1.17$ eV. It has been pointed out [8, 11] that the vacancy acceptor levels remain parallel to the conduction band edge E_c while the donor level remains parallel to the valence band edge E_{val} as the silicon temperature varies. Accepted values for these levels, at present, are

$$\Delta E_v^- = E_c - E_v^- = 0.57 \text{ eV}, \quad [9]$$

$$\Delta E_v^{2-} = E_c - E_v^{2-} = 0.11 \text{ eV}, \quad [18] \tag{16}$$

$$\Delta E_v^+ = E_v^+ - E_{val} = 0.13 \text{ eV}, \quad [19]$$

Using Eqs. (15) and (16), it is easy to show that Eq. (14) can be transformed to

$$N_v^- = N_v^x g_A^- \exp\left(\frac{E_F - E_F^i}{kT}\right) \exp\left(\frac{b+k/2}{2k}\right) \exp\left(-\frac{E_{go}/2 - \Delta E_v^-}{kT}\right) \tag{17a}$$

$$N_v^{2-} = N_v^x g_A^{2-} \exp\left(\frac{2(E_F - E_F^i)}{kT}\right) \exp\left(\frac{b+k/2}{k}\right) \exp\left(-\frac{E_{go} - \Delta E_v^{2-} - \Delta E_v^-}{kT}\right) \tag{17b}$$

$$N_v^+ = N_v^x g_D \exp\left(\frac{E_F^i - E_F}{kT}\right) \exp\left(\frac{b-k/2}{2k}\right) \exp\left(-\frac{E_{go}/2 - \Delta E_v^+}{kT}\right) \tag{17c}$$

where E_F^i is the intrinsic Fermi level for silicon given by

$$E_F^i = \frac{E_g}{2} + \frac{3kT}{4} \ln\left(\frac{m_p^*}{m_n^*}\right) \tag{18}$$

and m_p^* and m_n^* are the hole and electron density-of-state effective masses.

In Eq. (17), the Fermi energy, which in a semiconductor is controlled by the dopant impurities, appears only in one exponential term, all the other terms being determined by properties of the intrinsic material. Assuming Boltzmann statistics, if n_i is the intrinsic carrier concentration at the temperature of diffusion, the Fermi energy is related to the intrinsic electron concentration by

$$\frac{n}{n_i} = \exp\left(\frac{E_F - E_F^i}{kT}\right) \tag{19}$$

Equations (17) can therefore be written compactly as

$$N_v^- = N_{vi}^-\left(\frac{n}{n_i}\right)$$

$$N_v^{2-} = N_{vi}^{2-}\left(\frac{n}{ni}\right)^2 \tag{20}$$

$$N_v^+ = N_{vi}^+\left(\frac{n_i}{n}\right)$$

where N_{vi}^-, N_{vi}^{2-}, and N_{vi}^+ are the intrinsic vacancy concentrations at the various charge states, obtained from (17) by setting $E_F = E_F^i$. The term n/n_i is related to the doping concentration N_D and N_A, of donor and acceptor impurities, by the well-known relationship

$$\frac{n}{n_i} = \frac{N_D - N_A}{2n_i} + \sqrt{\left(\frac{N_D - N_A}{2n_i}\right)^2 + 1} \tag{21}$$

It is worthwhile to emphasize at this point that n_i is a strong function of temperature in silicon and, at the common temperatures of diffusion, it is in the range of 5×10^{18} cm^{-3} to 3×10^{19} cm^{-3}, as shown in Fig. 5 [12].

Now, Eq. (13), which gives the self-diffusivity of silicon as a sum of diffusivities contributed by the various charge states of vacancies, can be written as

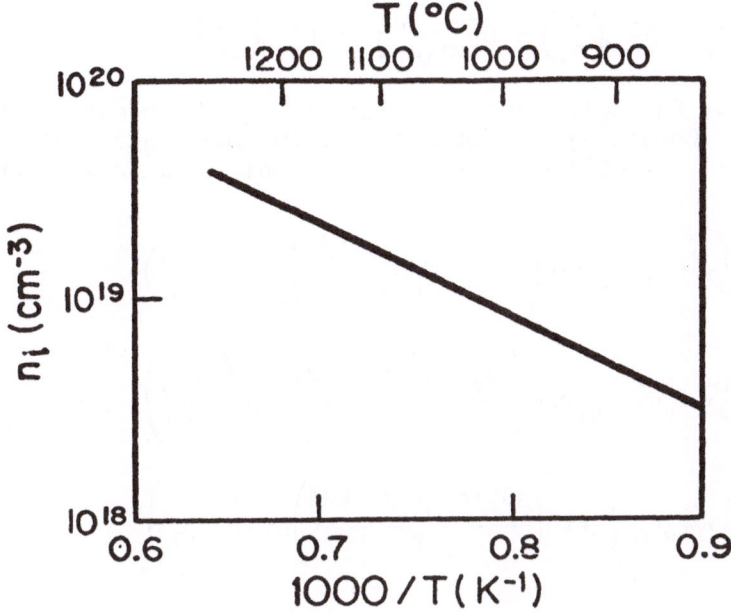

5. The intrinsic carrier concentration in silicon at high temperatures [12].

$$D_{sv} = D_{sv}^{X} + D_{svi}^{-}\left(\frac{n}{n_i}\right) + D_{svi}^{2-}\left(\frac{n}{n_i}\right)^2 + D_{svi}^{+}\left(\frac{n_i}{n}\right) \tag{22}$$

D_{sv}^{X} is given by Eq. (9) and is, of course, independent of the Fermi energy. Combining the form of Eqs. (9) with Eqs. (17), the intrinsic self-diffusivities due to the ionized vacancy states are given by

$$D_{svi}^{-} = \frac{1}{8}\alpha^2 f_v g_A^- \nu_{ov}^- \exp\left(\frac{\Delta S_f + \Delta S_m^- + b/2 + k/4}{k}\right)\exp\left(-\frac{Q_{sv}^-}{kT}\right) \quad ,$$

$$D_{svi}^{2-} = \frac{1}{8}\alpha^2 f_v g_A^{2-} \nu_{ov}^+ \exp\left(\frac{\Delta S_f + \Delta S_m^{2-} + b + k/2}{k}\right)\exp\left(-\frac{Q_{sv}^{2-}}{kT}\right) \quad , \tag{23}$$

$$D_{svi}^{+} = \frac{1}{8}\alpha^2 f_v g_D \nu_{ov}^+ \exp\left(\frac{\Delta S_f + \Delta S_m^+ + b/2 - k/4}{k}\right)\exp\left(-\frac{Q_{sv}^+}{kT}\right) \quad .$$

In summary, the activation energies for silicon self-diffusion by means of vacancies at the four charge states are

$$Q_{sv}^{X} = \Delta H_f + \Delta H_m = 3.9 \text{ eV} \quad ,$$

$$Q_{sv}^{-} = \Delta H_f + \Delta H_m^- + (E_{go}/2 - \Delta E_v^-) = 4.0 \text{ eV} \quad ,$$

$$Q_{sv}^{2-} = \Delta H_f + \Delta H_m^{2-} + (E_{go} - \Delta E_v^{2-} - \Delta E_v^-) = 4.56 \text{ eV} \quad , \tag{24}$$

$$Q_{sv}^{+} = \Delta H_f + \Delta H_m^+ + (E_{go}/2 - \Delta E_v^+) = 4.34 \text{ eV} \quad ,$$

where the enthalpies of migration have been assumed the same for all states. Assuming also equal entropies of migration as well as jump frequency pre-exponential factors ν_o among all states, the ratios of the intrinsic diffusivity pre-exponential factors are

$$D_{svio}^{X} \div D_{svio}^{-} \div D_{svio}^{2-} \div D_{svio}^{+} = 1 \div 35 \div 306 \div 21$$

which combined with eqs. (23) and (24) give the following ratios of intrinsic diffusivities for each vacancy charge state:

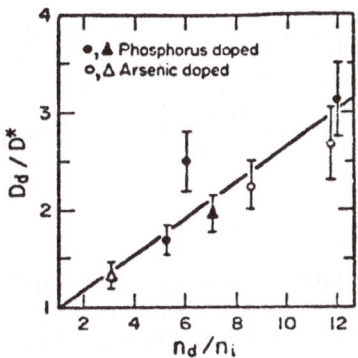

6. Relationship of extrinsic diffusion coefficient to electron
    concentration in silicon near 1090 °C (circles) and 1200 °C
    (diamonds). Solid line is best straight line fit to the data
    (after Ref. 2).

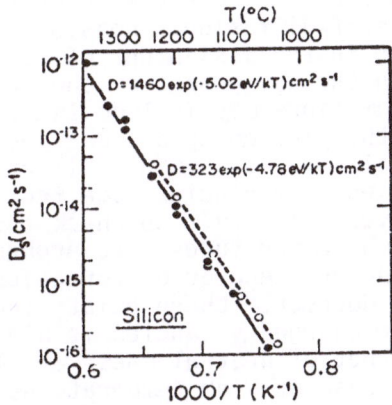

7. Self-diffusion coefficient in intrinsic and boron doped (ap-
    proximately 10^{19} cm^{-3}) silicon (after Ref. 3 and 13, respec-
    tively).

$$D^{x}_{svi} \div D^{-}_{svi} \div D^{2-}_{svi} \div D^{+}_{svi} = 1 \div 35 \exp\left(- \frac{0.1 \text{ eV}}{kT}\right)$$

$$\div 306 \exp\left(- \frac{0.66 \text{ eV}}{kT}\right) \div 21 \exp\left(- \frac{0.44 \text{ eV}}{kT}\right) \qquad (25)$$

It is important at this point to compare the developed theory with experiments. Two general trends are observed: (a) In intrinsic silicon the self-diffusivity activation energy and pre-exponential factor increase with increasing temperature range (Table 2 and Fig. 3a). (b) As n-type or p-type doping is increased, the self-diffusion coefficient in extrinsic silicon increases (Fig. 6 and 7). Based on the purely vacancy-type diffusion mechanism the first trend (a), could be explained by an increased dominance of the diffusivities D_{svi} due to 2- or + charged vacancies since, as can be observed from eqs. (24), they possess activation energies higher than D^{x}_{svi} and D^{-}_{svi}. However, it is easy to see from eq. (25) that D^{-}_{svi} dominates up to the melting temperature of silicon. Thus, we may either conclude that the values of physical quantities or assumptions that have been used above are in error or that the vacancy mechanism is an incomplete picture of self-diffusion. Possible errors in numerical values may exist in the assignment of the charged state levels (eq. 16), or in the assignment of the degeneracy factors. Possible erroneous assumptions may include the equality of formation and migration enthalpies among all charged states, and the equality of ν_0 among all states. Since all of the information about levels and energies of vacancies come from low temperature experiments, it is very difficult to check the consistency of these quantities at high temperatures. It probably seems reasonable to assume that the free energy of formation and ν_0 is the same for all states. However, there exists indication that the migration enthalpy for charged vacancies is higher by 0.1 to 0.2 eV with respect to that of neutral ones at low temperatures [15a]. If this is the case at high temperatures then some reconciliation between the vacancy diffusion theory and trend (a) is possible. However, such an attempt at full reconciliation may be futile while the interstitialcy mechanism is ignored. As will be seen in section 4, the interstitialcy mechanism has been identified as very important for substitutional solute diffusion in silicon at high temperatures. It seems thus unlikely that it is negligible for self-diffusion.

It was mentioned previously that it is observed that silicon self-diffusion is enhanced with increasing n-type or p-type doping [trend (b)]. This is certainly expected on the basis of the vacancy model developed here. As n/n_i or p/n_i increase with doping eq. (22) predicts an increase of self-diffusivity. Then, on the basis of eq. (25) which predicts clear dominance of D^-_{svi}, one would expect the relationship of D_{sv}/D^*_{sv} to be proportional to n/n_i. D^*_{sv} is defined as the intrinsic vacancy mechanism self-diffusivity of silicon, i.e. $D^*_{sv} = D_{sv}(n/n_i = 1)$. However, this is contradicted by the data in Fig. 6 [2], which shows $D_s/D^*_s \approx 0.81 + 0.19(n/n_i)$. This inconsistency also points out that some of the quantities and/or assumptions leading to eq. (35) are in error. Two possibilities exist to explain the data of Fig. 6: (a) based on purely vacancy model, that $D^x_{svi} \gg D^-_{svi}$, or (b) that an interstitialcy mechanism exists in addition to the vacancy one, and it has little or no dependence on n/n_i. At this point it is difficult to resolve this issue without additional experimental evidence.

In conclusion, although the vacancy mechanism captures some of the essential features it appears that it is not a very accurate model of self-diffusion in silicon. Some of the apparent inconsistencies may be due to erroneous parameter values or inaccurate assumptions, but almost certainly a significant fraction is due to the omission of the interstitialcy diffusion mechanism.

2.4 Impurity Diffusion

It is generally accepted that impurities of group IIIA (acceptors) and VA (donors) dissolve substitutionally in Si and thus diffuse by mechanisms similar to those of Si self-diffusion. All of these atom species diffuse in Si faster than Si itself. Since these impurities are mostly electrically active, when present at sufficiently high concentration, they perturb the Fermi energy of the Si host lattice, and thus they perturb the equilibrium point defect concentration at the various charge states. Therefore, on the basis of the previous discussion of Si self-diffusion, the presence of either donor or acceptor atoms in Si at concentrations above the n_i at the temperature of diffusion should result in enhancement of diffusivity of both silicon and of atoms of the same group and in decrease of diffusivity of atoms of opposite type. In interpreting measurements of diffusivities, it is very important to take notice of whether the silicon crystal was intrinsic, i.e., dopant atom concentration lower than $n_i(T)$ or extrinsic during the experiment. In the discussions that follow, this condition will be clearly stated.

Table 3 lists the intrinsic diffusivities of substitutional elements used as dopants in silicon. A general observaion is that they exhibit lower pre-exponential factors and activation energies than silicon self-diffusion (Table 2). The same fact has been reported for diffusion in dilute metallic solutions by Stiegman, et al [14] who first proposed the vacancy mechanism of diffusion. Johnson [15] was able to explain this phenomenon by proposing a model in which solute atoms and vacancies form pairs and migrate without dissociating by a series of cyclic inversions and reorientations: Then the pair binding energy effectively reduces the diffusivity activation energy while the greater number of vacancy jumps required to complete an effective impurity displacement results in a decrease of the pre-exponential factor with respect to self-diffusion. Based on these ideas a quantitative model for substitutional atom diffusion by the vacancy mechanism was proposed by Swalin [16], was later extended by Seeger and Chick [8] and was finally refined by Hu [17]. Fig. 8 plots the hypothesized potential energy of the vacancy as a function of position with respect to the solute atom in terms of coordination site distance. As can be seen it is assumed that there exists a vacancy-solute interaction which extends to several coordination sites. This is consistent with the idea of the extended defect (vacancy) discussed in section 2.3. E_b is the vacancy-solute binding energy and in general it is composed of an elastic and a couloubic term, E_e and E_c respectively. ΔE is the energy difference between third and first coordination site position for the vacancy. This energy is important because in the diamond lattice the vacancy must move to at least a third coordination site before it can complete a full displacement cycle of the solute atom and be still bound to it. Ignoring, for the time being, the various charge states of vacancies, the solute atom diffusivity, D_{Av}, can be written as:

$$ D_{Av} = \frac{1}{8}\alpha^2 f_A \nu_A \left(\frac{1}{4} \frac{C_{Av}}{C_A} \right) \tag{26} $$

where f_A is the correlation factor for an effective displacement of a solute atom [10], ν_A is the solute atom-vacancy exchange frequency, C_{Av} is the concentration of vacancy-solute atom pairs and C_A is the solute atom concentration. The factor of 1/4 arises because only one of possible four first coordination sites can be occupied by the vacancy. According to Hu [10] the correlation factor can be written as

$$ f_A \approx \theta \exp\left(- \frac{\Delta E}{kT} \right)^{\frac{\nu_V}{\nu_A}} \tag{27} $$

where θ is a constant. On the other hand from simple mass-action law, C_{Av} is given by:

Table   3

DIFFUSION PREEXPONENTIAL FACTOR AND ACTIVATION ENERGY FOR SUBSTITUTIONAL DIFFUSORS IN SILICON

Dopant	D_0 (cm^2/sec)	Q (eV)	Reference
P	3.85	3.66	18
As	24.0	4.08	20
Sb	12.9	3.98	21
Bi	1.08	3.85	22
B	5.1	3.70	23
Ga	0.374	3.41	24
In	0.785	3.63	24
Al	1.385	3.39	24

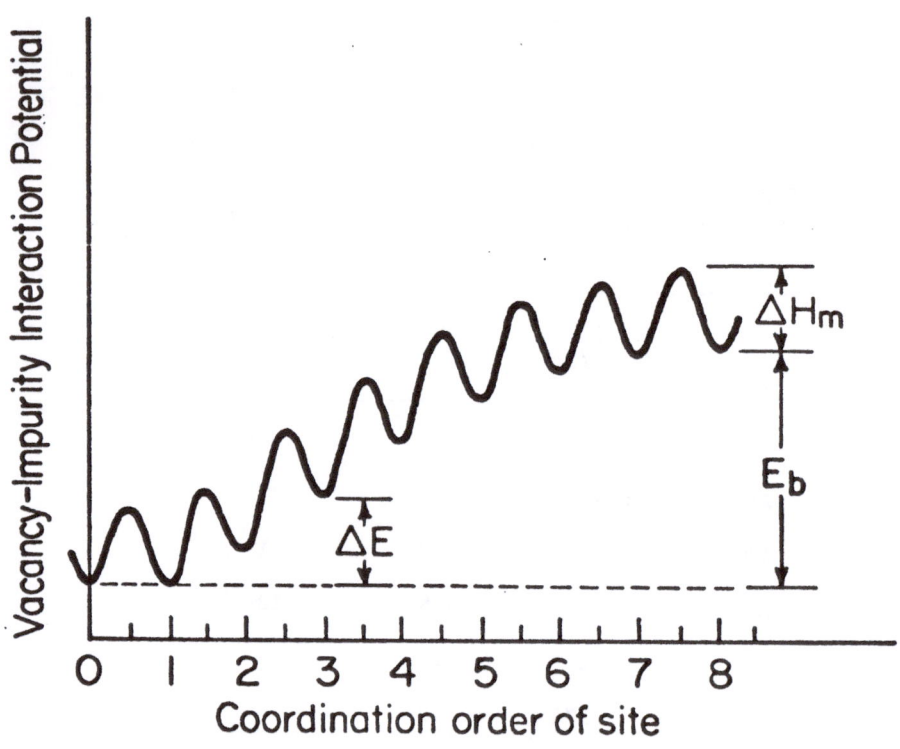

8. Vacancy-solute interaction potential (after Ref. 10).

$$C_{Av} = 4C_A N_v \exp\left(\frac{E_b}{kT}\right) \tag{28}$$

where N_v is the normalized vacancy concentration as already defined and the factor of 4 arises because of the existance of four first coordination sites of the diamond lattice. Combining eqs. (26), (27) and (28) yields:

$$D_{Av} \approx \frac{1}{8}a^2 \theta \nu_v N_v \exp\left(\frac{E_b - \Delta E}{kT}\right) \tag{29}$$

Remembering that the self-diffusion coefficient, ignoring multiple vacancy charge states, is given by

$$D_{sv} = \frac{1}{8}a^2 f_v \nu_v N_v \tag{30}$$

it is obvious that the pre-exponential factor for D_{Av} will be smaller than the same for D_{sv} by approximately θ, and that the activation energy will be lower by $E_b - \Delta E$. The constant θ arises from the lattice geometrical structure part of the contribution to the correlation factor and should normally be of the order of unity. However, a value of θ significantly smaller than unity may arise if the entropy of the vacancy is increased by the presence of the impurity atom. Although the proposed model has the correct qualitative attributes, quantitative comparison between theory and experiment is not possible because of uncertainties in both θ and $E_b - \Delta E$ theoretical calculations. Perhaps even more significantly, the fact that the interstitial mechanism has not been considered contributes appreciable uncertainties.

Extension of the above model to include the various charge states of vacancies is rather straight forward. Using eq. (29) for each of the vacancy charge states, one may write for the impurity diffusion in analogy to self diffusion (neglecting again any effects arising from electrostatic potentials),

$$D_{Av} = D_{Av}^x + D_{Avi}^-\left(\frac{n}{n_i}\right) + D_{Avi}^{2-}\left(\frac{n}{n_i}\right)^2 + D_{Avi}^+\left(\frac{n_i}{n}\right) \tag{31}$$

where

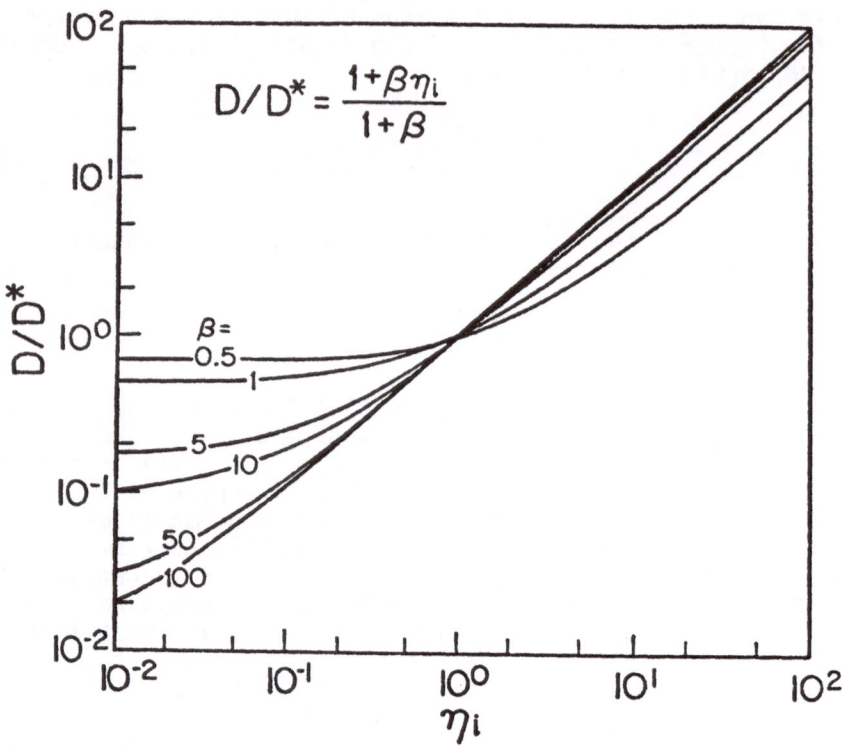

9. Normalized extrinsic diffusivity as a function of normalized carrier concentration in silicon. D^* is defined as the intrinsic diffusivity.

$$D^X_{Av} = \frac{1}{8}\alpha^2\theta\nu_{ov}\exp\left(\frac{\Delta S_f + \Delta S_m}{k}\right)\exp\left(-\frac{Q_v + \Delta H^X_b}{kT}\right)$$

$$D^-_{Avi} = \frac{1}{8}\alpha^2\theta g^-_A\nu^-_{ov}\exp\left(\frac{\Delta S_f + \Delta S^-_m + b/2}{k}\right)\exp\left(-\frac{Q^-_v + \Delta H^-_b}{kT}\right)$$

$$D^{2-}_{Avi} = \frac{1}{8}\alpha^2\theta g^{2-}_A\nu^{2-}_{ov}\exp\left(\frac{\Delta S_f + \Delta S^{2-}_m + b}{k}\right)\exp\left(-\frac{Q^{2-}_v + \Delta H^{2-}_b}{kT}\right) \qquad (32)$$

$$D^+_{Avi} = \frac{1}{8}\alpha^2\theta g_D\nu^+_{ov}\exp\left(\frac{\Delta S_f + \Delta S^+_m + b/2}{k}\right)\exp\left(-\frac{Q^+_v + \Delta H^+_b}{kT}\right)$$

and where ΔH_b for each charge state represents the appropriate $E_b - \Delta E$.

From the form of Eq. (31), it might be expected that, even under intrinsic conditions ($n/n_i = 1$), the diffusivity of substitutional impurities in silicon may not obey a strictly Arrhenius form unless the diffusivity with one of the states of vacancies is clearly dominant. This appears to be the case since, within experimental errors, most of the reliable data reported in the literature do fit an Arrhenius form for all impurities. In order to determine which of the charge states is dominant and to what degree, additional experiments are required under extrinsic conditions. Intuitively, we might expect that, for positively charged donors, V^- and V^{2-} may be more effective for diffusion than V^X. Similarly, for negatively charged acceptors, V^+ may be more effective than V^X. Figure 9 is a plot of D/D^*, the extrinsic diffusivity normalized to the intrinsic, as a function of $n = (n/n_i$ for donors or n_i/n_x for acceptors) for different ratios of β defined as, $\beta \triangleq D^\pm_{Avi}/D^X_{Avi}$. Most impurities do exhibit the monotonic increase of D/D^* for n greater than 1. However, as can be seen from the figure, the region of n in which the most sensitive determination of the ratio $D^\pm_{Avi}/D^X_{vi}$ may be expected is for n much less than 1. Such experiments require that the redistribution of the impurity in question be measured in Si samples uniformly and heavily doped by an impurity of the opposite type, and thus only chemical analysis of the sample can be used. Not many reliable experiments of this kind are reported in the literature. Nevertheless, it does appear that β is of the order of 5 or greater for most impurities. For boron, a value of 3 to 6 at 1000 °C is reported while, for arsenic β appears to be larger than 50. For phosphorus, this simple model is not applicable; it appears that both V^- and V^{2-} species must be taken into account. (See chapter titled "One-Dimensional Simulation of IC Fabrication Processes").

3. MODELING OF DIFFUSIVE MIGRATION OF IMPURITY ATOMS: QUASI-EQUILIBRIUM

In the preceding part of this chapter, we have discussed the basic principles that underlie self-diffusion and impurity diffusion in silicon from the microscopic random jump point of view. In the present and following sections, we concentrate on the macroscopic migration of impurities that result from silicon high temperature processing. Typically, more physical effects than the random thermally activated jumps in a rather homogeneous lattice, as discussed up to this point, must be taken into account. The important ones are the so-called internal electric field resulting from the ionized impurities, the non-uniform point defect concentration resulting from the spacial variation of the Fermi level in the host crystal, point defect supersaturation due to oxidation of Si or radiation damage, point defect pumping from the surface to the bulk and/or lattice strain and dislocation generated by high concentration impurity diffusion and impurity clustering and/or precipitation. All these effects affect significantly the migration of impurities in Si, giving rise to apparent diffusivities that are different from the actual impurity diffusivity. Thus, the conditions of Si processing must be taken diligently into account when modeling diffusive migration, and critical experiments must be scrutinized very carefully before conclusions about diffusivity can be drawn.

In this section, we discuss diffusive migration of impurities under the quasi-equilibrium condition, namely processes during which the equilibrium point defect concentration in silicon is not severely affected by the process. This condition can be met under both intrinsic and extrinsic conditions, where these terms have been previously defined to indicate the carrier concentration with respect to n_i in the host crystal at the diffusion temperature. Closed-form expressions describing the resulting impurity distribution can be derived when diffusion takes place under intrinsic conditions; a relevant fabrication process is low-dose ion implantation and subsequent drive-in. The classical two step process, i.e., chemical predeposition and drive-in under oxidizing conditions is also solved often in closed form. However, solutions to this problem are mostly irrelevant from the practical standpoint because (1) extrinsic conditions giving rise to nonuniform diffusivity often prevail and (2) impurity atoms are lost to the growing oxide. Both effects are rendering the problem unsolvable in closed form.

Under extrinsic conditions, two effects that affect the apparent diffusivity will be discussed here: (a) those arising from gradients of diffusing impurity profiles, often attributed to ion drift due to internal electric fields, and (b) clustering of impurity atoms that effectively slows down the impurity diffusivity. Impurity distributions resulting from such processes may be obtained only by numerical solutions of the relevant equations, as discussed in appropriate chapter of this book. In the present section, we concentrate only on the physical aspects of these effects.

3.1 Extrinsic Diffusion

There are two distinct cases under which diffusion of a particular impurity might be taking place in a crystal under extrinsic conditions. In the most common case, the extrinsicity of the host lattice is due to the high concentration of the migrating impurity alone. Examples of such diffusions are found during predeposition when the surface impurity concentration is at or near solid solubility or during redistribution of high-dose deposited or ion-implanted impurity layers. In the other case, the extrincity of the host lattice is due to another impurity species. An example of such diffusion is found in the redistribution of relatively low concentration of impurities, already existing in the crystal, during the deposition or annealing of heavy-dose impurity layers, as in the case of bipolar transistor emitter diffusion with the base already in place. Another example is the diffusion of an impurity in a crystal already doped uniformly extrinsic. This is often termed "isoconcentration diffusion."

In this section, we discuss only diffusion under extrinsic conditions that do not affect significantly the equilibrium point defect concentration in the host lattice. In other words, it is assumed that the point defect concentration in the various charge states depends only on the Fermi energy. The simplest case of extrinsic diffusion is the "isoconcentration diffusion" since, under this condition, the impurity diffusivity is constant. Of course, depending on the amount and type of the background impurity concentration which determines n/n_i, the diffusivity of the diffusing species is uniformly increased or decreased as can be readily seen from Eq. (31). Thus, Fick's second law may be applied using the correct value of the extrinsic diffusivity. Because of the simplicity in interpreting its results, isoconcentration diffusion experiments are of great value in the study of the fundamentals of diffusion discussed in Section 2.4.

When impurities migrate under non-uniform extrinsic conditions, two effects occur that affect the redistribution: (1) the

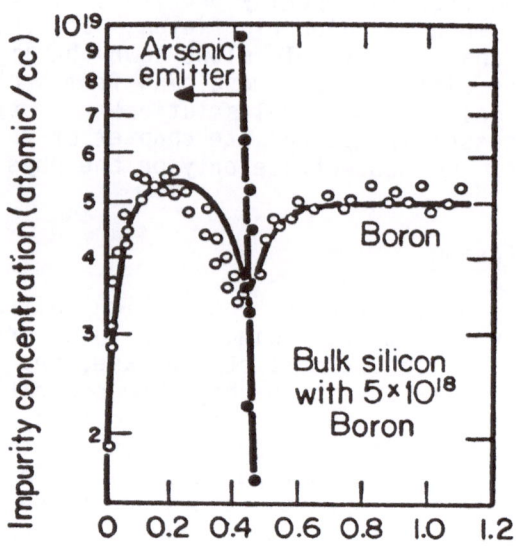

10. Diffusion of high concentration arsenic on top of a uniform boron concentration showing the effect of the arsenic concentration gradient on the boron (after Ziegler, et al. [25]).

diffusivity is a function of concentration and thus varies in space; and (2) there exists an internal electric field due to concentration gradient of the ionized impurities. Historically, the first attempts to explain the enhancement of impurity diffusivity under extrisic conditions dealt with the electric field. The argument for such an enhancement is that fast diffusing electrical carriers pull the ionized impurities by means of the electrostatic field between them. Diffusion under such conditions is common in plasmas where it is called "ambipolar diffusion". Fick's first law is then modified by including a field-drift term as follows:

$$J_x = -D \frac{dC}{dx} + \mu E C \qquad (33)$$

where μ is the impurity mobility obtained from D by Einstein's relation. This equation may be easily manipulated to the form

$$J_x = -D f_e(C) \frac{dc}{dx} \qquad (34)$$

where $f_e(C)$ is the electric field enhancement factor, given by

$$f_e(C) = 1 + \frac{C/2n_i}{\sqrt{(c/2n_i)^2 + 1}} \qquad (35)$$

It can be seen readily that, for $C \ll n_i$, $f_e = 1$ while, for $C \gg n_i$, $f_e = 2$, and thus the apparent impurity diffusivity is effectively doubled. However, this factor of 2 is very small to explain the observed enhancements of diffusivity without invoking the effects of charged point defects discussed earlier.

For most cases of diffusion from heavily doped layers of single impurity species, the question of the electric field effect is mostly academic because the resulting diffusivity enhancement is overwhelmed by that of the concentration dependent part of the diffusivity, and the experimental inaccuracies and parameter (D_{Axi}) uncertainties do not allow detailed analysis. On the other hand, it has been observed that a small dip occurs in a uniform distribution of relatively light concentration of boron when a heavy concentration of arsenic is diffused on top of it. The dip coincides with the diffusing front of the arsenic as shown in Fig. 10. This dip has been offered as a confirmation of the electric field effect on the migration of boron in silicon. Although intuitively appealing, the concept of electric field-drift of ionized impurities is subject to several inconsistencies that arise from the complicated nature of the substitutional atom migration process. For example, the only way that an electric field could affect diffusivity is by increasing the jump rates in one direction while decreasing the jump rates in the opposite direction; presumably it can do that by correspondingly lowering and raising the potential barriers (see Fig. 8) in the two directions thus essentially giving rise to anisotropic migration energy. How exactly to model this effect and how to take into consideration the various charged states of point defects is not immediately obvious. Other consistency problems include the issue of possible field screening, the application of Einstein's relation in heavily degenerate silicon, and the rapid variation of diffusivity with impurity concentration. Therefore, the most natural and foolproof way to treat the problem is to analyze the entire system thermodynamically as Hu has done [10,17]. According to Hu [17], one can define an apparent impurity diffusivity as:

$$D_A = D_{Ai}(1 + \partial \ln \gamma_A / \partial \ln C) \tag{36}$$

where D_{Ai} is the diffusivity that can be calculated from random jump considerations, i.e. the diffusivity that we have been dealing with so far, and γ_A is the impurity activity coefficient given by

$$\gamma_A = \frac{1 + g^{-1} \exp[(E_A - E_F^i)/kT]}{1 + g^{-1} \exp[(E_A - E_F)/kT]} \tag{37}$$

where g is the impurity degeneracy factor and E_A is the impurity ionization level; γ_A is the ratio of unionized impurities at the concentration of interest to that at infinite dilution. It is then easy to show that:

$$(1 + \partial \ln \gamma_A / \partial \ln C) \approx f_e(C) \tag{38}$$

Thus, the impurity diffusive flux can by written in general as:

$$J_x = f_e(C)\left[D_{Av}^x + D_{Avi}^-\left(\frac{n}{n_i}\right) + D_{Avi}^{2-}\left(\frac{n}{n_i}\right)^2 + D_{Avi}^+\left(\frac{n_i}{n}\right)\right]\frac{dC}{dx} \tag{39}$$

Note that the above equation is of the same form as eq. (34) but it has been derived without explicitly invoking the internal electric field. Eq. (39) can also bve used to model the effect of one diffusing impurity on another as in the As-B case of Fig. 10. The coupling then is introduced via the Fermi energy E_F in eq. (37).

3.3 Extrinsic Diffusion with Impurity Clustering

At sufficiently high concentrations, dopant atoms may precipitate or cluster in silicon. Precipitation occurs when the solid
solubility limit is exceeded and is usually associated with large
numbers of impurity atoms forming macroscopic defects within the
host lattice. Clustering involves a small number of impurity
atoms (2 to 4 or somewhat higher) that form bonds with each other
but at the same time maintain bonds with the host lattice. Clusters form before the solid solubility limit is reached. Among
the common dopants, precipitation is often observed in the case
of boron and phosphorus while clustering is mostly attributed to
arsenic. From a device standpoint, the most important effect of
both precipitation and clustering is that the number of electrically active (i.e., substitutional) impurity atoms in silicon is
smaller than the number of atoms introduced during doping. At
the same time, the clustering and/or precipitation effects tend
to reduce the apparent impurity diffusivity, since only a fraction of dopant atoms are substitutional and thus mobile. In
order to describe the redistribution of impurities under these
conditions, the impurity conservation equation including generation and loss must be used. For one-dimensional flow, this
equation may be written as

$$\frac{d}{dt} Q(x_1, x_2) = U(x_1, x_2) - [J_x(x_2) - J_x(x_1)] \tag{40}$$

where

$$Q(x_1, x_2) = \int_{x_1}^{x_2} C(x) \, dx \tag{41}$$

$$U(x_1, x_2) = \int_{x_1}^{x_2} [g(x) - l(x)] \, dx \tag{42}$$

where x_1 and x_2 are the boundaries of the silicon subvolume where
conservation applies and g and l are the generation and loss
rates of atoms at the particular state of the impurity in question. For the precipitate or clustered impurity state $J_x = 0$,
i.e., there is no diffusive flux. Solution of Eq. (40) may be
accomplished only by numerical methods when a model for g and l
exists. Thus, our present discussion will be confined to the
derivation of the apparent diffusivity of impurities in the
presence of clustering. As already mentioned, this is an important effect in the redistribution of arsenic, particularly during
the fabrication of very shallow junctions. Because clustered

atoms are immobile in the lattice, clustering typically leads to an apparent diffusivity that is smaller than the extrinsic impurity diffusivity in the absence of clustering.

For this discussion, the assumed chemical reaction for the clustering of arsenic is according to Tsai et al [26],

$$3As^+ + e^- \underset{K_d}{\overset{K_c}{\rightleftharpoons}} As_3^{+2} \xrightarrow{25°C} As_3 \tag{43}$$

High T

where e^- represents electrons and K_c and K_d the clustering and declustering rate coefficients. It should be stressed at this point that the specific reaction is not very important for the present discussion, and another reaction could have been used as an example as well. The decision as to which reaction takes place during precipitation of arsenic is based only on phenomenological observations, since the reaction cannot be directly observed, and several chemical reactions involving different cluster sizes as well as electrons and/or point defects might be employed. An example of another reaction may be found in Ref. 10.

Defining the concentration of clustered atoms C_c as

$$C_c = C_T - C$$

where C_T is the total concentration and C is the substitutional concentration, the conservation equation (40) for the clustered atoms may be written as

$$\frac{dC_c}{dt} = K_c n C^3 - \frac{1}{3} k_d C_c = 1 - g \tag{44}$$

Where

$$n = C + \frac{2}{3} C_c \tag{45}$$

Further, defining the equilibrium clustering coefficient K_e by

$$K_e = \frac{K_c}{K_d} = \frac{C_c}{3nC^3} \tag{46}$$

eq. (44), under equilibrium conditions (i.e., $dC_c/dt = 0$), gives

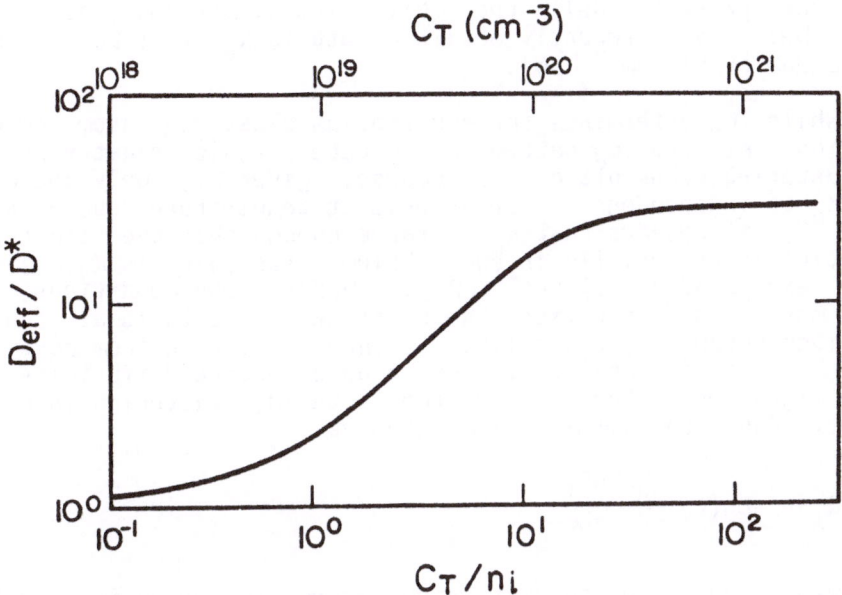

11. Normalized extrinsic diffusivity of arsenic accounting for clustering as a function of normalized total arsenic concentration.

$$\frac{C_T}{C} = 1 + 3nK_e c^2 = \frac{1 + K_e c^3}{1 - 2K_e c^3} \qquad . \tag{47}$$

For the present model, the equilibrium clustering coefficient that best fits recently obtained data is $K_e = 1.26 \times 10^{-70}$ exp[2.062 eV/kT] cm^{-9} [26].

While K_e determines the equilibrium clustered atom concentration, K_c and K_d determine the rate at which clustering and declustering takes place. Of course, given K_e, only one of K_c and K_d is independent. For arsenic at temperatures above about 900 °C, it appears that K_d is large enough that the clustering reaction is essentially in equilibrium. Its value is $K_d = 2.62 \times 10^{10}$ exp(-3.35 eV/kT) sec^{-1} [26]. Under these conditions, eq. (46) may be used to relate the total and substitutional concentrations throughout the silicon. Then, starting from eqs. (34) or (39), it is possible to derive an effective diffusivity for the migration of the arsenic atoms, taking clustering into account. Equation (34) may be written as

$$J_x = -Dh(C) \frac{dC}{dC_T} \frac{dC_T}{dx} \tag{48}$$

Assuming that, for concentrations where clustering is important ($C_T > 5 \times 10^{19}$ cm^{-3}), $D = D^*(n/n_i)$ (a very good approximation for arsenic), eq. (48) yields for the apparent or effective diffusivity of arsenic

$$D_{eff} = D^* \frac{n}{n_i} f_e(c) \frac{dC}{dC_T} \tag{49}$$

Note that the above equation is generally applicable for any clustering model that relates C_T, C, and n in equilibrium. This equation is used to plot D_{eff}/D^* as a function of C_T/n_i, in Fig. 11, for a diffusion temperature of 1000 °C. As can be seen, the effective diffusivity initially increases with C_T above n_i roughly in proportion to C_T/n_i, but its slope starts decreasing above a concentration of 1.5×10^{20} cm^{-3} where the concentration of clustered atoms becomes an appreciable fraction of the total concentration.

4. NONEQUILIBRIUM EFFECTS ON DIFFUSIVITY

Several processes used for device fabrication lead to perturbation of the equilibrium point defect concentration in silicon.

Most notable are ion implantation, oxidation, and high concentration phosphorus diffusion. Generally, a point defect nonequilibrium condition is manifested by enhanced or decreased diffusivity of substitutional impurities even at small concentrations and by the growth of dislocation loops (stacking faults) in silicon. Depending on processing conditions, lattice strain associated with these processes may also give rise to the generation of dislocations. In the present section, we will concentrate only on oxidation and its effect on diffusivity since, among the three processes mentioned, it is the one that most directly sheds some light on silicon diffusion mechanisms, in general. Readers interested in the other two are referred to Refs. 27 and 28 for radiation and ion implantation enhanced diffusion and Ref. 18 as well as the chapter titled "One Dimensional Simulation of IC Fabrication Processes" of this book for heavy phosphorus diffusion.

It has been well established that all of the common impurity dopants B, P, and As exhibit the so-called oxidation-enhanced-diffusion (OED) in silicon at typical oxidation conditions [29-31]. At the same time, depending on the oxidation conditions, oxidation also gives rise to the gorwth of oxidation stacking faults (OSF) [31-33]. Given that OSF have been identified as interstitial type and are unstable in silicon, that is, they shrink during annealing at nonoxidizing conditions, their growth corroborates the hypothesis that oxidation increases the concentration of self-interstitial point defects in silicon. The basic trends exhibited by OED and OSF are: (1) OED and OSF growth are highest in <100>, lower in <110>, and lowest in <111> oriented silicon wafers. (2) OED decreases with increasing oxidation temperature. (3) OSF exhibit both growth and shrinkage, often referred to as retrogrowth, depending on oxidation conditions with the latter occuring at the higher end of temperatures and/or the longer oxidation times. (4) Both OED and OSF growth are related to oxide growth rate in the same fashion, at least for oxidation treatments of rather long duration. (5) Different dopant species exhibit different degrees of OED under the same conditions; in fact Sb exhibits oxidation reduced diffusion (ORD) [34]. (6) At high temperatures and/or long oxidation times ORD may be observed for species that normally exhibit OED.

4.1 Oxidation Perturbed Diffusion

The above observations establish quite clearly the relationship between oxidation-perturbed diffusion of dopant atoms and oxidation stacking fault behavior; they also lead to the conclusion that the following physical mechanisms must be taking place in silicon: a) Dopant atoms diffuse by a dual vacancy-interstitialcy mechanism; b) oxidation of the silicon surface increases the concentration of self-interstitials; and c) oxidation decreases the concentration of vacancies.

In all discussions so far, it has been assumed that one type of point defects, namely vacancy, is responsible for substitutional diffusion in silicon. Including both vacancy and interstitialcy mechanisms (introduced in Section 2), the diffusivity of dopant impuritis, D_A, can be written as

$$D_A = D_{AI}^* \frac{C_I}{C_I^*} + D_{Av} \frac{C_v}{C_v^*}$$ (50)

where D_{Av}^* and D_{AI}^* are the vacancy and interstitialcy motivated equilibrium diffusivities of the impurity. The form of the above relationship is based on the assumption that neither of the equilibrium point defect concentrations, C_I^* or C_v^*, are identically zero. If any of the two intrinsic concentrations was zero then eq. (50) would have to be modified by replacing the corresponding D_{Ax}^*/C_x^* by a finite coefficient since under non-equilibrium conditions diffusion may well be motivated by a non-intrinsic point defect. Defining the fractional interstitialcy diffusion mechanism as $f_I = D_{AI}^*/D_A^*$, where $D_A^* = D_{AI}^* + D_{Av}^*$ is the intrinsic impurity diffusivity, eq. (50) can be written as

$$D_A = D_A^* f_I\left(\frac{C_I}{C_I^*} + (1-f_I) \frac{C_v}{C_v^*}\right)$$ (51)

In the absence of oxidation, the point defect concentrations are at their equilibrium levels and the impurity diffusion is given by D_A^*. During oxidation, excess silicon is produced at the $Si-SiO_2$ interface, part of which flows into the bulk silicon and raises the interstitial concentration above its equilibrium value, thus increasing the impurity diffusivity. The generation rate of the excess silicon at the interface is related to the oxidation rate. The relation is experimentally found to be a power-law dependence of the form [37,38]

$$C_I - C_I^* = K_I\left(\frac{dX}{dt}\right)^n$$ (52)

where dX/dt is the rate of oxide growth and the reaction constant K_I and the power exponent n are related to the point defect generation process at the $Si-SiO_2$ interface. Theories about the origin of the relationship in Eq. (52) and the value of n are discussed in the next section.

From the theory of OSF growth and shrinkage combined with experimental observations it has been shown that the average normalized concentration of self-interstitials in <100> silicon oxidized in dry O_2 can be described by the following equation [31]:

$$\frac{<C_I>}{C_I^*} = 1 + at^{-m} \exp\left[\frac{b}{kT}\right] \tag{53}$$

where m = 0.25, a = 4.2 x 10^{-9} $min^{0.25}$, t is the oxidation time in minutes and b = 2.5 eV. On the other hand independent observations of C_v/C^* are not available; however, as will be seen shortly, this quantity can be easily bound.

Referring now back to eq. (51), the diffusivity D_A is expected to vary with time. We will define then an average diffusivity $<D_A>$ by the equation

$$<D_A> \triangleq \frac{1}{t} \int_0^t D_A(\tau) \, d\tau$$

where $D_A(\tau)$ is the instantaneous diffusivity given by Eq. (51) and t is the total oxidation time. It is easy to show that a constant diffusivity $<D_A>$ produces exactly the same impurity profile as a time varying $D_A(\tau)$ over the same time period t, provided $D_A(\tau)$ is independent of impurity concentration.

Since in any diffusion experiment it is the quantity $<D_A>/D_A^*$ that is determined rather than D_A/D_A^*, and the quantity $<C_I>/C_I^*$ is determined from OSF experiments, it is useful to use these average quantities into eq. (51) and solve for f_I as follows

$$f_I = \frac{<D_A>/D_A^* - <C_v>/C_v^*}{<C_I>/C_I^* - <C_v>/C_v^*} \tag{54}$$

The only quantity not obtainable experimentally is the normalized average vacancy concentration. If the vacancy concentration was insignificant at high temperatures, as pure interstitial defect models would imply, the f_I would be 1 for all elements. This is clearly contradicted in Fig. 12 where $<D_A>/D_A^*$ for phosphorus using data of Antoniadis et al [29], and $<C_I>/C_I^*$ from eq. (53), are plotted as functions of temperature. Thus, in order to

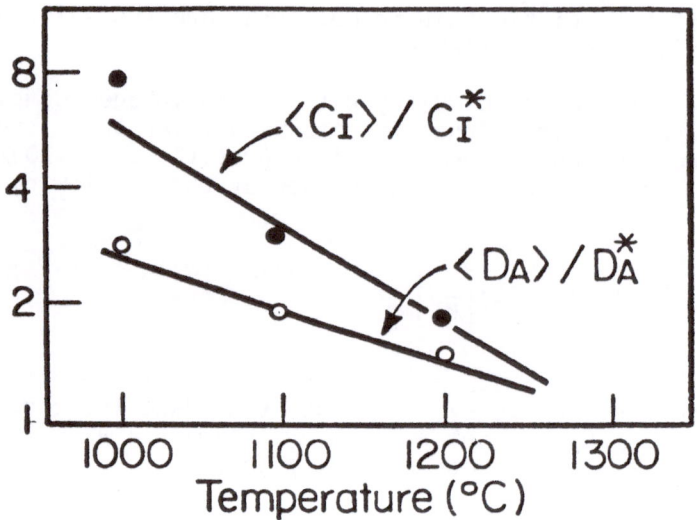

12. Average normalized concentration of self-interstitials and phosphorus diffusion as function of oxidation temperature. Oxidation times: 1000 °C, 6 hrs; 1100 °C, 45 min; 1200 °C, 10 min.

Table   4

FRACTIONAL INTERSTITIALCY DIFFUSION
MECHANISM, f_I, FOR PHOSPHORUS

T °C	$\langle D_A \rangle / D_A^*$	$\langle C_I \rangle C_I^*$	f_I[eq. (54)]	
			$\langle C_v \rangle / C_v^* = 0$	$\langle C_v \rangle / C_v^* = C_I^* / \langle C_I \rangle$
1000	3.2	8.4	0.38	0.37
1100	2.0	3.5	0.57	0.53
1200	1.4	1.9	0.74	0.64

calculate f_I for substitutional elements in an estimate of $\langle C_V \rangle / C_V^*$ must be made. In previous work [38], it had been assumed that $\langle C_V \rangle / C_V^* = 1$. However, from the new evidence provided by ORD of Sb at oxidation conditions where other substitutional dopant atoms experience OED, leads to the conclusion that generally $\langle C_V \rangle / C_V^* < 1$ during oxidation. This can be seen easily from eq. (51) applied for Sb where $\langle D_A \rangle > D_A^* < 1$. Taking into account that $\langle C_I \rangle / C_I^* > 1$, from OSF experiments, eq. (51) leads to

$$f_I + (1-f_I) \frac{\langle C_V \rangle}{C_V^*} < 1$$

which of course yields $\langle C_V \rangle / C_V^* < 1$.

Having now determined that the vacancy concentration is reduced by oxidation we may attempt to construct a model for this reduction. Two mechanisms appear as likely. The first is that a recombination reaction with self-interstitials takes place which would require that self-interstitials and vacancies obey a mass action law as

$$C_I C_V = C_I^* C_V^* \tag{55}$$

The second possible mechanism is that vacancies are consumed by the oxidation reaction which presumably renders the surface a perfect sink for vacancies; thus, near the surface during oxidation $C_V \approx 0$. From the two above mechanisms the variation of $\langle C_V \rangle / C_V^*$ can be bounded as follows

$$0 \leq \frac{\langle C_V \rangle}{C_V^*} \leq \frac{C_I^*}{\langle C_I \rangle} \tag{56}$$

This then allows calculation of bounds for the fractional interstitialcy diffusion, f_I, of different elements from eq. (54). As an example, f_I is given for phosphorus in Table 4. Evidence from recent (unpublished) experiments indicates that f_I of antimony is nearly zero which indicates that it diffuses by a vacancy mechanism only. The fact that f_I for even one impurity species is between zero and one is the best evidence to date that both vacancies and self-interstitials exist under equilibrium conditions in silicon.

For practical applications it is important to obtain a relation between diffusivity and oxidation rate. This is accomplished by combining eqs. (51) and (52) and by adopting eq. (55), although the latter is not yet proven:

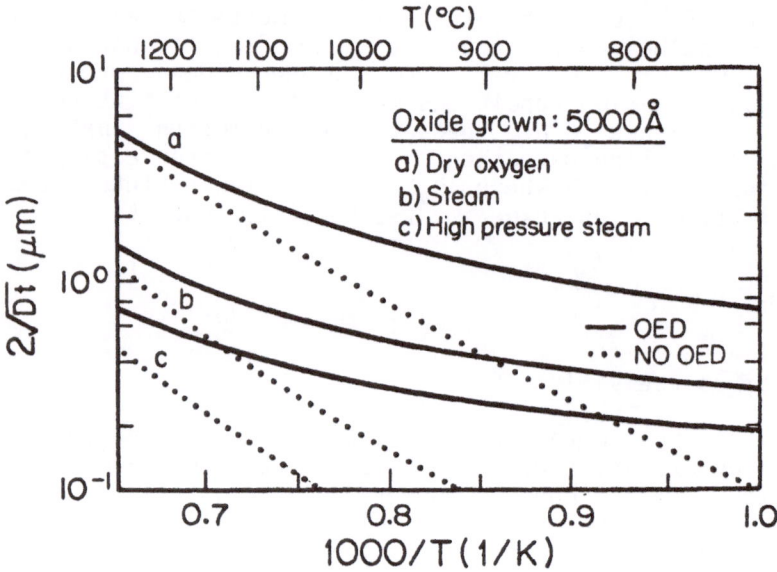

13. Boron diffusion length during oxidation to grow 0.5 μm of oxide in dry oxygen, steam, or high-pressure ambients at different temperatures.

$$D_A = D_A^\star \left[f_I \left(1 + \frac{K_I}{C_I^\star} \dot{X}^n \right) + (1 - f_I)\left(1 + \frac{K_I}{C_I^\star} \dot{X}^n \right)^{-1} \right] \qquad (57)$$

Here $\dot{X}$ is used to denote dX/dt; this quantity is easily obtainable from oxidation experimental data. On the other hand f_I and $K_I/C_I^\star$ must be fitted using both OSF and OED data. It is important to note that both are expected to be dependent on temperature, probably by means of an Arrhenius relationship.

In order to illustrate the effect of OED on the diffusion length of phosphorus, consider a case where 0.5 μm of oxide are to be grown. Equation (57) is used to calculate the impurity diffusion length $2\sqrt{Dt}$ when the oxide is grown in dry oxygen, steam, or high-pressure steam where the oxidation rate is further increased by a factor of 6 at different temperatures. The results plotted in Fig. 13 clearly show the reduction of impurity diffusion length in the two fast oxidizing steam ambients. Also shown in the figure is the diffusion length calculated with only intrinsic diffusivity. At high temperature (>1100 °C), the diffusion length is dominated by intrinsic diffusion and the reduction in diffusion length in the two steam ambients is mainly due to reduction in oxidation time. At low temperatures (<1000 °C), however, extrinsic diffusion dominates and the reduction results from the sublinear oxidation rate dependence of the diffusion enhancement. It is important to also note in Fig. 13 that the reduction of the diffusion length toward decreasing temperatures is only gradual at low temperatures as compared to the rapid decrease when only intrinsic diffusion is considered. This is a result of the negative activation energy of oxidation-enhanced diffusion and emphasizes the importance of OED in impurity profile control at these temperatures.

4.2 Excess Silicon Generation at the Si-SiO$_2$ Interface

As mentioned in the previous section OSF and OED experiments lead to the conclusion that the self-interstitial supersaturation during silicon oxidation is given by eq. (52). Three models have been recently proposed to explain this relationship between supersaturation and oxidation rate. The first, proposed by Lin et al [29], postulates a three-step oxidation mechanism at the interface resulting from the need to provide free volume there, while the second, proposed by Hu [35], postulates a bimolecular annihilation of interstitials in the bulk as well as at the interface. Both models assume that the interstitial generation rate is linearly proportional to oxidation rate and derive the non-linear eq. (52) by balancing the generation rate by respectively postulated non-linear loss mechanisms. The third model proposed by Tan and Goesele [39] attributes the self-interstitial generation to stress at the interfacial layer which is postulated to be relieved by viscoelastic flow of the newly formed SiO_2. The nonlinearity then arises from the power law dependence of this flow on the stress. At the present time knowledge of the atomistic details of the Si/SiO_2 interface is incomplete, and thus it is impossible to verify any one of these models.

REFERENCES

1. S. K. Ghandhi, The Theory and Practice of Micorelectronics, John Wiley and Sons, Inc., 1968.

2. J. M. Fairfield and B. J. Masters, J. Appl. Phys., 38, 3148, 1967.

3. H. J. Mayer, H. Mehrer, and K. Maier, in "Lattice Defects in Semiconductors 1976," Inst. Phys. Conf. Ser., 31, 186, 1977.

4. L. Kalinowski and R. Seguin, Appl. Phys. Lett., 35, 311, 1979.

5. H. Widmer and G. R. Gunther-Mohr, Helv. Phys. Acta, 34, 635, 1961.

6. M. M. Valenta and C. Ramasastry, Phys. Rev., 106, 73, 1957.

7. H. Letaw, L. Slifkin, and W. M. Portnoy, Phys. Rev., 02, 636, 1956.

8. A. Seeger and K. P. Chik, Phys. Stat. Sol., 29, 455, 1968.

9. J. A. Van Vechten, Phys. Rev. B, 10, 1482, 1976.

10. S. M. Hu, Atomic Diffusion in Semiconductors (D. Shaw, ed.), Plenum Press, New York, 1973, Chapter 5.

11. J. A. Van Vechten and C. D. Thurmond, Phys. Rev. B, 14, 3539, 1976.

12. F. J. Morin and J. P. Maita, Phys. Rev., 966, 28, 1954.

13. G. Hettich, H. Mehrer and K. Maier, in Defects and Radiation Effects in Semiconductors,1978 (J. Albany, ed.), Conference Series Number 46, The Institute of Physics, Bristoe and London, 500, 1979.

14. J. Stiegman, W. Schockley, and F. C. Nix, Phys. Rev, 56, 13, 1939.

15. R. P. Johnson, Phys. Rev., 56, 814, 1939.

16. R. A. Swalin, J. Appl. Phys., 29, 670, 1958.

17. S. M. Hu, Phys. Rev., 180, 773, 1969.

46

18. R. B. Fair and J. C. C. Tsai, J. Electrochem. Soc., <u>124</u>, 1107, 1977.

19. G. D. Watkins, J. R. Troxell, and A. P. Chatterjee, in <u>Defects and Radiation Effects in Semiconductors</u>,1978 (J. Albany, ed.), Conference Series Number 46, The Institute of Physics, Bristoe and London, 1979.

20. T. L. Chiu and H. N. Ghosh, <u>IBM J. Res. Dev., Nov.</u>, 472, 1971.

21. H. F. Wolf, in "Semiconductors," <u>Interscience</u>, pp. 153 and 363, 1971.

22. R. N. Ghoshtagore, Phys. Rev. B, <u>3</u>, 397, 1971.

23. M. Okamura, Jap. J. Appl. Phys., <u>8</u>, 1440, 1969.

24. R. N. Ghoshtagore, Phys. Rev. B, <u>3</u>, 2507. 1971.

25. J. F. Ziegler, C. W. Cole and J. E. E. Beglin, Appl. Phys. Lett., <u>21</u>, 177, 1972.

26. M. Y. Tsiu, F. F. Morehead, J. E. E. Beglin and A. E. Michel, J. Appl. Phys., <u>51</u>, 3230, 1980.

27. B. J. Masters, in <u>Defects and Radiation Effects in Semiconductors</u>,1978 (J. Albany, ed.), Conference Series Number 46, The Institute of Physics, Bristoe and London, 545, 1979.

28. W. K. Hofker, H. W. Werner, D. P. Oosthoek, and H. A. M. DeGrefte, Appl. Phys., <u>2</u>, 265, 1973.

29. D. A. Antoniadis, A. M. Lin, and R. W. Dutton, Appl. Phys. Lett., <u>33</u>, 1030, 1978.

30. K. Taniguchi, K. Kurosawa, and M. Kashiwagi, J. Electrochem. Soc., <u>127</u>, 2243, 1980.

31. D. A. Antoniadis, J. Electrochem. Soc., <u>129</u>, 1093, 1982.

32. S. M. Hu, J. Vac. Sci. Technology, <u>14</u>, 17, 1977.

33. S. P. Murarka, J. Appl. Phys., <u>48</u>, 5020, 1978.

34. S. Mizuo and H. Higuchi, Japan. J. Appl. Phys., <u>20</u>, 739. 1981.

35. S. M. Hu, "Oxygen, Oxidation Stacking Faults and Related Phenomena in Silicon," Presented at Mat. Res. Soc. Meeting, Boston, MA, Nov. 17-20, 19xx.

36. W. A. Tiller, J. Electrochem. Soc., 127, 619, 1980.

37. A. M. Lin, R. W. Dutton, D. A. Antoniadis, and W. A. Tiller, J. Electrochem. Soc., 128, 1121, 1981.

38. A. M. Lin, D. A. Antoniadis, and R. W. Dutton, J. Electrochem. Soc., 128, 1131, 1981.

39. T. Y. Tan and U. Goesele, Appl. Phys. Lett., 30, 86, 1981.

40. B. E. Deal, J. Electrochem. Soc., 110, 527, 1963.

THERMAL OXIDATION: KINETICS, CHARGES, PHYSICAL MODELS, AND INTERACTION WITH OTHER PROCESSES IN VLSI DEVICES

James D. Plummer
Integrated Circuits Laboratory
Stanford University

Bruce E. Deal
Research and Development
Laboratory
Fairchild Camera and
Instrument Corporation

INTRODUCTION

Thermally grown silicon dioxide remains one of the cornerstones of modern integrated circuit technology. SiO_2 is commonly used as a mask against dopant diffusion, to passivate active device regions and junctions, to insulate field regions and provide isolation between active components, and as the gate dielectric, is an actual component in MOS devices. As a result, the control and predictability of oxide growth, and the resulting electrical properties, are critical if reproducible device performance is to be achieved.

Despite the fact that thermally grown SiO_2 has been used in the production of discrete devices and integrated circuits for almost twenty years, considerable gaps remain in our understanding of its growth kinetics, the charges associated particularly with the Si/SiO_2 interface, and the interaction of oxidation with other important processes such as dopant segregation and diffusion. The recent resurgence of research activity in laboratories throughout the world in these areas is a direct indication of these gaps in our knowledge.

The motivation for this increased interest in improved physical models of oxidation processes has come from the continued down scaling of active device dimensions. Modern device structures employ lateral dimensions on the order of 2-3 μm and vertical dimensions well below 1 μm. There are no basic physical mechanisms which will prevent a reduction in each of these dimensions by an additional order of magnitude over the next decade. Practically accomplishing this continued scaling, however, depends upon our ability to physically understand and quantitatively model the fabrication techniques which will be used in the construction of such devices.

For devices with relatively large geometries (>5 μm) and loose processing tolerances, relatively simple models suffice for prediction of vertical device structures resulting from a given fabrication sequence. As device dimensions shrink, however, it becomes essential to employ more robust process models and to consider the interaction both laterally and vertically of various processing steps, if accurate simulation of structures is to be obtained. This is important even with today's 2-3 μm device geometries; it will become essential for smaller devices.

Large geometry devices can be successfully modeled as one-dimensional structures. This is true for both process models and electrical models. Devices with lateral dimensions below a few microns, however, require two-dimensional models for accurate simulation. This need has stimulated a large body of work in recent years on twodimensional electrical models of device current-voltage characteristics. One example of this type of work is the 2D program GEMINI. Work of this type has resulted in remarkable advances in our understanding of small geometry device physics.

Progress has not been as rapid, however, in two-dimensional process modeling. While some basic 2D process modeling programs have been developed at Stanford (SUPRA for example) and elsewhere, these programs do not incorporate robust 2D kinetic models which can accurately predict doping profiles and device geometries under a wide range of conditions. This is a direct result of our need for improved physical models of oxidation, ion implantation, diffusion, and CVD. It is quite clear that these processes are not one-dimensional. Recent experimental evidence has clearly indicated that oxidation or impurity diffusion in a localized region of a silicon substrate can substantially affect oxidation or diffusion rates in laterally or vertically adjacent regions of the substrate. There is no clear agreement at the present time on the basic physical mechanisms responsible for such results, although much has been learned in the past several years. It is clear, however, that we must quantitatively understand such phenomena if we are to accurately model small device structures.

A specific goal of much recent work has been to understand and model these two-dimensional effects. Substantial progress in this regard has been made. It appears now that the basic physical phenomena underlying these effects are the roles of point defects-- silicon vacancies and interstitials--in impurity diffusion, thermal oxidation, and other processes. The generation and consumption of these point defects during high temperature fabrication steps appear to be the unifying physical effects which can explain many of the phenomena which have been regarded as anomalous to date. We have used such models to quantitatively understand a variety of process phenomena and have incorporated some of these models in SUPREM II.

More such models are in SUPREM III, and will be described later in this chapter.

It is already clear today that 2D process modeling involves much more than solving 1D equations in two dimensions. Basic physical laws and understanding are lacking at present. Effects such as lateral OED, oxidation under masking Si_3N_4 layers, lateral and vertical effects of high dopant concentrations on diffusion coefficients, and a host of other known experimental effects cannot be explained by a simple extrapolation of known 1D physical laws to 2D structures. A crucial need at the present time is to develop models for bulk point defect (interstitial and vacancy) generation, recombination and diffusion, since it is clear that the local concentrations of these defects determine local diffusion coefficients, oxidation rates, etc. In fact, an alternative statement of the problem we face in 2D process modeling is the need to develop techniques for calculating local (i.e. time and position dependent) process parameters suitable for process simulation. Such process parameters will of necessity be geometry dependent which means that diffusion coefficients and oxidation rate constants, for example will depend on the presence or absence of nearby heavily doped regions or oxidizing interfaces. This will imply a tight coupling between surface geometry and resulting impurity profiles in small devices.

This chapter begins with a review of basic 1D oxidation kinetics and then turns to some of the point defect related and 2D effects outlined above. Not all aspects of thermal oxidation are covered in this chapter, since a companion chapter by Claeys addresses some topics not covered here. Kinetic effects due to Chlorine additions to the ambient (HCl, TCE, etc.), the effect of oxidation on gettering, and the details of OISF growth and retrogrowth are examples which are covered by Claeys. We also do not cover thermal oxidation of other materials important to silicon technology here (polysilicon, silicide films and silicide/polysilicon multilayer structures). Some aspects of the oxidation of these materials are covered by other authors in this book (see Dutton, for example).

BASIC OXIDATION KINETICS

Thermal oxidation is normally carried out in a fused quartz tube in a resistance heated furnace. A schematic drawing of a typical oxidation system is shown in Fig. 1. For dry O_2 oxidation, high purity oxygen is transported into the furnace tube through suitable regulators, valves, traps, filters, and flowmeters. The silicon wafers are generally loaded vertically onto a fused quartz boat where normally up to 200 three- or four-inch diameter wafers may be accommodated. Often, a high density ceramic liner surrounds the quartz tube to minimize ionic contamination or moisture

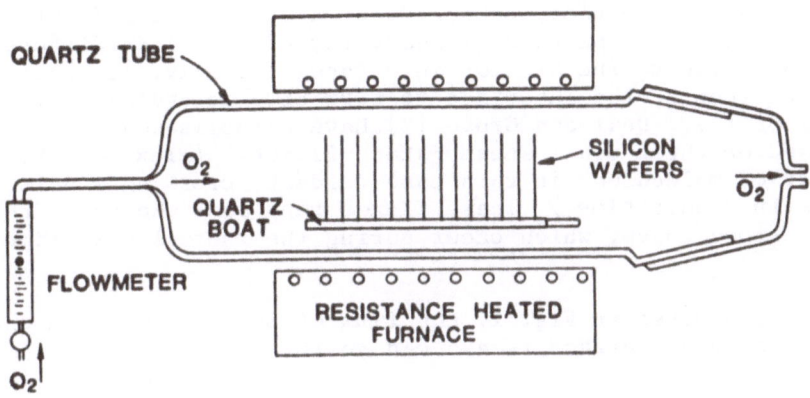

Fig. 1. Schematic drawing of typical oxidation system.

diffusion through the quartz. For a number of years, water oxidation was provided by bubbling O_2 or N_2 through a flask of water maintained at a particular temperature. Thus, a certain wafer vapor pressure could be provided in the oxidizing ambient. However, more recently, pyrogenicsystems have been employed which permit H_2 to react with O_2 at the gas input end of the oxidation tube to provide higher purity water under better conrolled conditions. By adjusting the ratio of H_2 to O_2, the vapor pressure of water can be varied.

The nature of the thermal silicon oxidation process is what makes it unique compared to other deposited dielectric films with respect to silicon device passivation. In this process, silicon reacts with either oxygen or water at elevated temperatures (700° to 1250°C) to form SiO_2. The oxidation reaction may be represented by either of the following two reactions.

$$Si + O_2 \rightarrow SiO_2 \tag{1}$$
$$Si + 2H_2O \rightarrow SiO_2 + 2H_2 \tag{2}$$

It has been shown by special marker experiments [1] that the oxidation species, either oxygen or water, diffuses through the oxide already formed to react with the silicon at the $Si-SiO_2$ interface. As oxidation proceeds, the interface continues to move into the silicon, thus producing a new clean surface. From the densities and molecular weights of silicon and silicon dioxide, it can be shown that for every thickness X_O of oxide formed, 0.45 x_O of silicon is consumed. The exact nature and charge of the diffusing oxidation species have not yet been identified.

As the oxidation of single crystal silicon proceeds, the process may be represented by Fig. 2. As has been shown experimentally

52

[2], the oxidizing species, either O_2 or H_2O, is absorbed into the outer surface of the oxide already formed. It then diffuses through the oxide to the $Si-SiO_2$ interface. Finally, it reacts with the silicon at the interface to form SiO_2 according to reactions (1) or (2). Deal and Grove [3] have established a first-order model based on three equal steady-state fluxes. (Flux is defined as the number of molecules, in this case oxidant, crossing a unit surface area in a unit time.) These fluxes represent the three reactions described above, which occur during the thermal oxidation of silicon.

As is indicated in Fig. 2, the flux of the oxidant from the gas to the oxide outer surface is assumed to be

$$F_1 = h(C^* - C_o) \tag{3}$$

where h is a gas-phase, transport coefficient, C_o is the concentration of the oxidant in the outer oxide surface, and C^* is the equilibrium concentration of the oxidant in the oxide. This

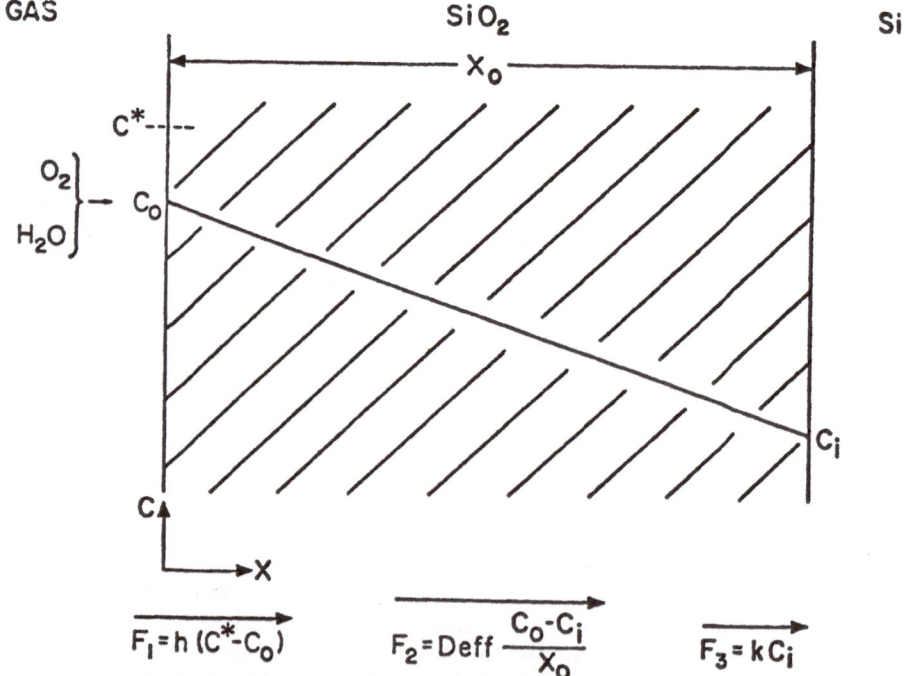

Fig. 2. Boundary and Flux conditions for the Gas-SiO₂-Si system during thermal oxidation.

equilibrium concentration is assumed to be related to the partial pressure of the oxidant by Henry's law:

$$C^* = KP \tag{4}$$

In using Eq. (4), it is assumed that the oxidant is O_2 or H_2O and does not dissociate at the outer surface. Aspects of this assumption will be discussed later.

The second flux, F_2, is that of the oxidant diffusing through the oxide and may be written (using Fick's law) as:

$$F_2 = -D_{eff} \, (dC/dx) = D_{eff} \, (C_o - C_i)/x_o \tag{5}$$

at any point within the oxide layer, where D_{eff} is the effective diffusion coefficient, dC/dx is the concentation gradient of the oxidizing species in the oxide, C_i is the oxidant concentration in the oxide near the oxide-silicon interface, and x_o is the oxide thickness.

The flux corresponding to the oxidation reaction at the oxide-silicon interface may be expressed as

$$F_3 = k_s \, C_i \tag{6}$$

where k_s is the interface reaction rate constant.

If, due to steady state conditions, it is assumed that $F_1 = F_2$ and $F_2 = F_3$, and solving for C_i and C_o between these two relationships, one obtains

$$\frac{C_i}{C^*} = \frac{1}{1 + k_s/h + k_s x_o/D_{eff}} \tag{7}$$

$$\frac{C_o}{C^*} = \frac{1 + k_s X_o/D_{eff}}{1 + k_s/h + k_s x_o/D_{eff}} \tag{8}$$

C_i and C_o can be eliminated and the flux is given by

$$F = F_1 = F_2 = F_3 = \frac{k_s C^*}{1 + k_s/h + k_s x_o/D_{eff}} \tag{9}$$

If N_1 is defined as the number of oxidant molecules incorporated into a unit volume of the oxide, the rate of growth of the oxide layer is described by the differential equation

$$\frac{dx_o}{dt} = \frac{F}{N_1} + \frac{k_s C^*}{1 + k_s/h + k_s x_o/D_{eff}} \qquad (10)$$

If an initial condition of $x_o = x_i$ at $t = 0$ is assumed, where t is the oxidation time, then integration of Eq. (10) gives

$$x_o^2 + Ax_o = B(t + \tau) \qquad (11a)$$

or

$$\frac{x_o^2 - x_i^2}{B} + \frac{x_o - x_i}{B/A} = t \qquad (11b)$$

where

$$A = 2D_{eff}(1/k_s + 1/h) \qquad (12)$$

$$B = 2D_{eff}C^*/N_1 \qquad (13)$$

and

$$\tau = (x_i^2 + Ax_i)/B \qquad (14)$$

Equation (11a) or (11b) is the well-known general relationship. The quantity τ corresponds to a shift in time, which corrects for the presence of an initial oxide of thickness x_i. Further discussion of τ and x_i will appear later. Solving the quadratic Eq. (11) gives oxide thickness as a function of time:

$$x_o = A/2 \left\{ \left[1 + \frac{t + \tau}{A^2/4B} \right]^{1/2} - 1 \right\} \qquad (15)$$

This is the basic equation used in all process modeling programs to calculate SiO_2 thickness. It requires values of B and B/A appropriate to the particular process sequence being simulated. In addition to other assumptions described above, this relationship is purely one dimensional and hence will have to be modified to simulate 2D structures. We will return to this point later.

Figures 3 and 4 show typical experimental thickness vs. time plots for SiO_2 layers grown in dry O_2 and pyrogenic steam ambients respectively (P = 1 atm). Except for the first $\approx$ 200 Å of dry O_2 growth, this data is well modeled by equation (11) or (15).

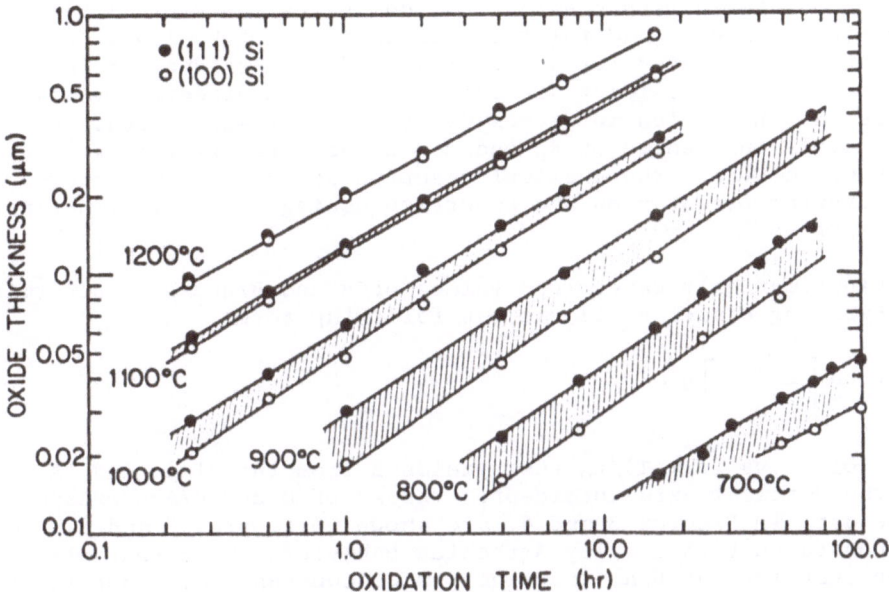

Fig. 3. Oxide thickness vs oxidation time for silicon oxidation in dry oxygen at various temperatures [3].

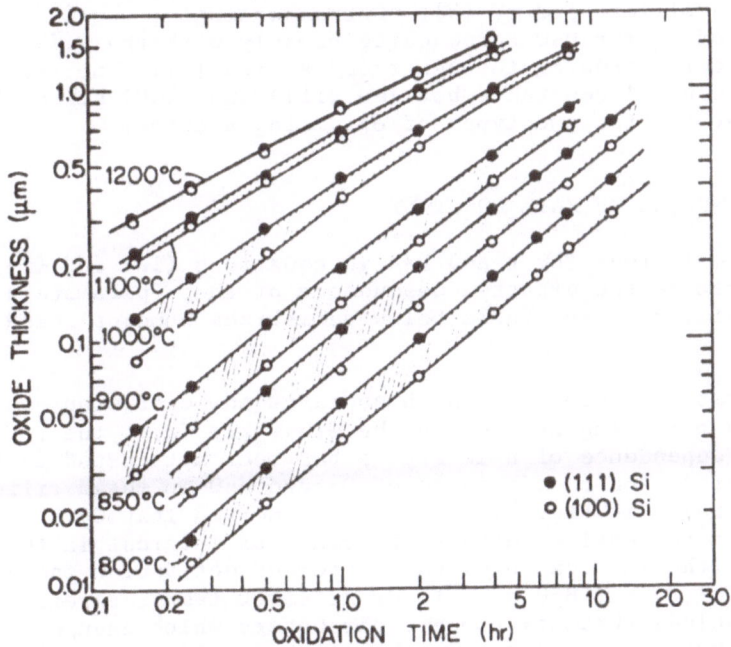

Fig. 4. Oxide thickness vs oxidation time for silicon oxidation in pyrogenic steam (≈640 Torr) at various temperatures [63].

The parabolic rate constant, B, dominates the overall reaction at high temperatures and for thick oxides, while the linear rate constant B/A dominates for short times and thin oxides. B is principally related to oxidant diffusion and can be affected by any process variable which increases O_2 or H_2O diffusion rates. B/A is primarily influenced by k_s and can be affected by process variables which modify the chemical reaction at the Si/SiO_2 interface--crystal orientation and substrate doping level are two important examples.

Experimentally determined values of B and B/A can be determined by re-writing equation (11) in the following form.

$$x_o = \left[\frac{t + \tau}{x_o} \right] B - A \tag{16}$$

A plot of x_o vs $(t + \tau)/x_o$ then yields B from the slope and $-A$ from the vertical axis intercept. Values of B and B/A corresponding to the data in Figures 3 and 4, are shown in Figures 5 and 6. The rate constants both display Arrhenius behavior. Note that all activation energies for B/A, the linear rate constant, are approximately 2.0 eV (46 kcal/mole), which corresponds to the Si-Si bond energy [4]. The activation energy of the dry O_2 parabolic rate constant is 1.23 eV which is reasonably close to that of O_2 diffusion through fused silica, 1.2 eV [5]. Correspondingly, the 0.71 and 0.78 values of E_A for H_2O agree quite closely with the 0.79 value for water diffusing through fused silica [6]. The ratio of the pre-exponential constants between (111) and (100) silicon of 1.7 is consistent for the two types of oxidizing species.

PROCESS DEPENDENCE OF RATE CONSTANTS

The formulations for B and B/A in equations (12) and (13) can be used to deduce the expected dependence of these parameters on process conditions. The table below summarizes some of these expectations.

The parabolic rate constant B contains the diffusion coefficient of the oxidizing species in the SiO_2. In fact, the total temperature dependence of B in Fig. 5 has been attributed to this term. The diffusion coefficients of O_2 and H_2O in fused silica are plotted below in Fig. 7. It can be observed that O_2 shows a higher D over the entire temperature range of interest in thermal oxidation, although because of the different activation energies, the values for O_2 and H_2O are closer at lower temperatures. One can also conclude that any process parameters which change the oxidant diffusion coefficients will change B as well. Such effects have been observed in phosphorus or boron doped SiO_2 layers [7] and in a comparison of D_2O and H_2O oxidation rates [8].

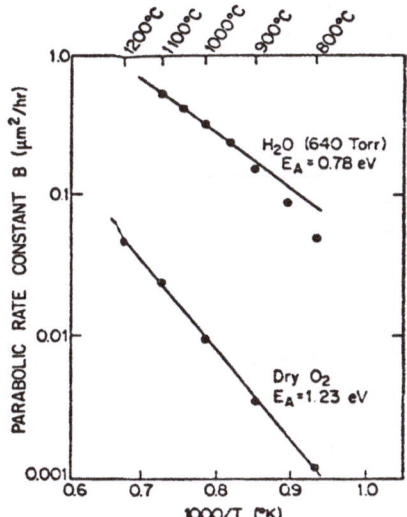

Fig. 5. Dependence of the parabolic rate constant B on temperature for the thermal oxidation of silicon in pyrogenic steam (≈640 Torr) and dry oxygen [63].

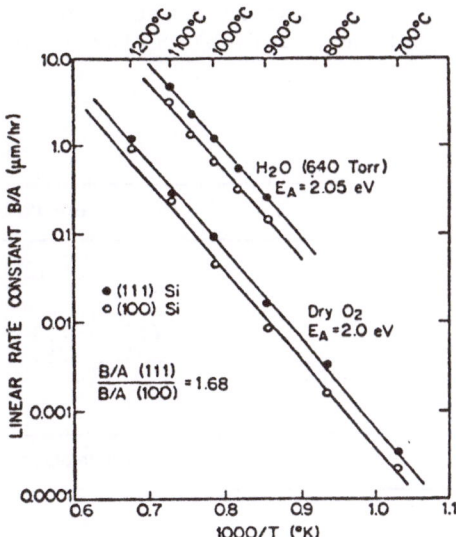

Fig. 6. Dependence of the linear rate constant of B/A on temperature for the thermal oxidation of silicon in pyrogenic steam (≈640 Torr) and dry oxygen [63].

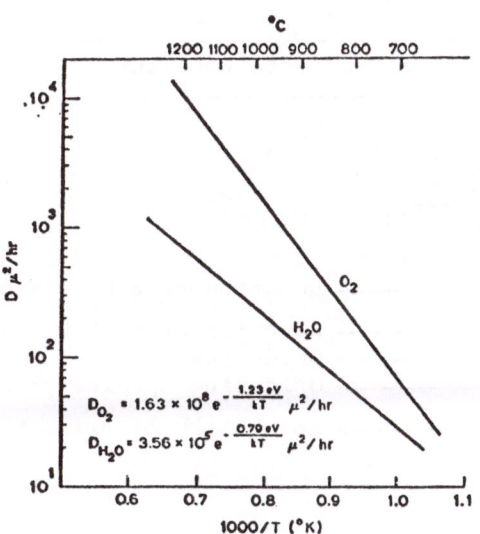

Fig. 7. Diffusion coefficients of O_2 and H_2O in fused silica (SiO_2) vs 1/T [3].

58

One might be tempted to conclude from this figure that O_2 oxidation should be faster than H_2O oxidation; however, the

TABLE 1

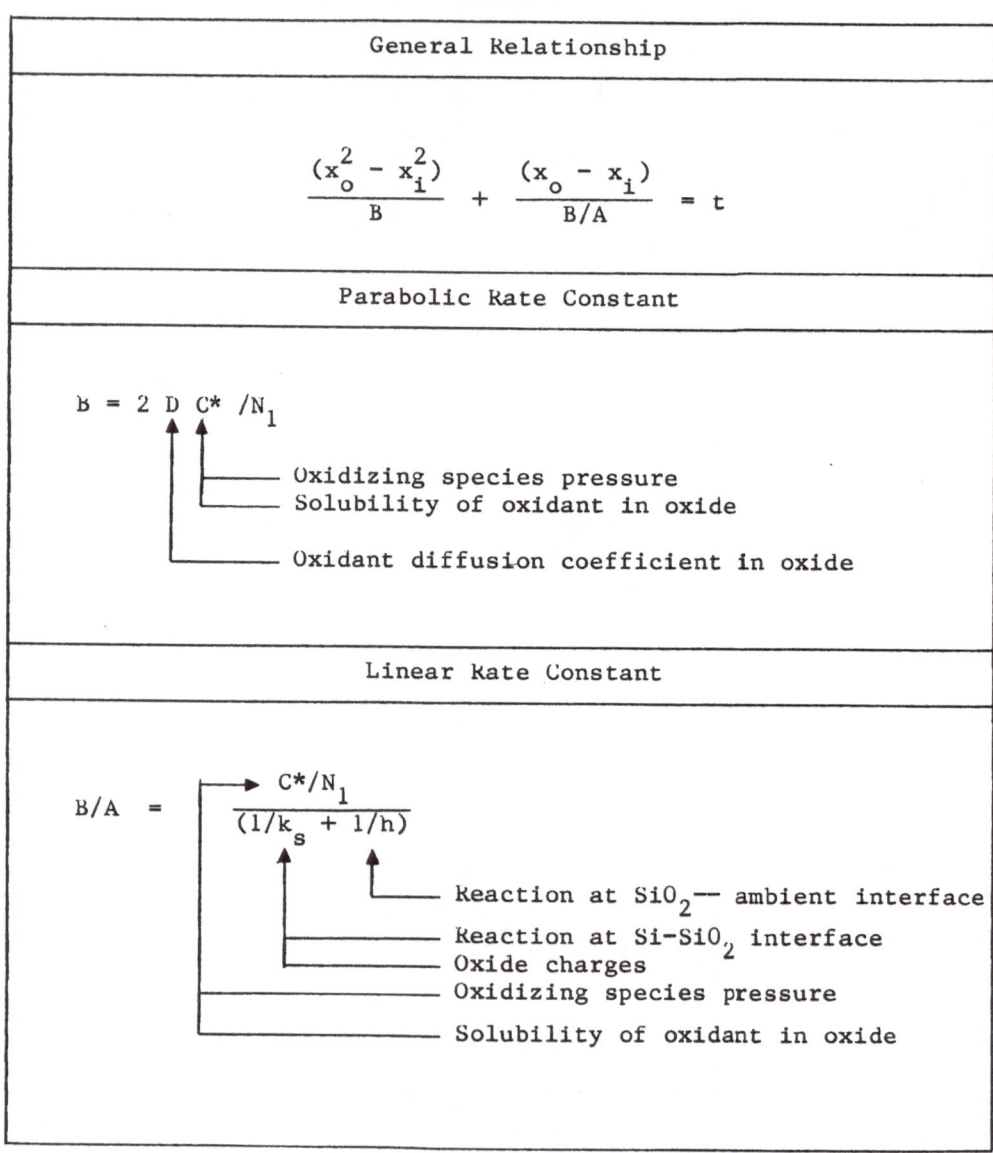

General Relationship

$$\frac{(x_o^2 - x_i^2)}{B} + \frac{(x_o - x_i)}{B/A} = t$$

| Parabolic Rate Constant |

$B = 2 D C^* /N_1$

Oxidizing species pressure
Solubility of oxidant in oxide

Oxidant diffusion coefficient in oxide

| Linear Rate Constant |

$B/A = \dfrac{C^*/N_1}{(1/k_s + 1/h)}$

Reaction at SiO_2 — ambient interface
Reaction at $Si-SiO_2$ interface
Oxide charges
Oxidizing species pressure
Solubility of oxidant in oxide

differences in C* for the two species more than compensate for the
relative diffusion coefficients. C* includes the vapor pressure of
the oxidizing species as well as its solubility in the oxide. This
implies that any change in the partial pressure of O_2 or H_2O
will directly affect the oxidation rate through its effect on B.

The relative solubilities of the two oxidants are approximately
$5.5 \times 10^{16}/cm^3$ and $3.5 \times 10^{19}/cm^3$ for O_2 and H_2O respectively at
1000°C. The temperature dependence of these parameters would be
expected to be small [3] and indeed, all of the T dependence in B
can be accounted for in D_{O_2} and D_{H_2O} as was pointed out
above.

Since C* is proportional to oxidant pressure (Eqn. (4)), B
should be proportional to oxidant pressure. Measurements made at
pressures up to 20 atm. in both O_2 and H_2O confirm this [9,10].
Fig. 8 illustrates this linear dependence of B on P for the O_2
case; similar results are obtained in H_2O.

The linear rate constant B/A also contains C* and hence should
show the same linear pressure dependence as B. This is in fact
observed for H_2O oxidation [9]. However, as Fig. 9 illustrates,
B/A shows a sublinear pressure dependnce in an O_2 ambient at
pressures $\geq$ 1 atm [10]. Similar results have been found for pressures $\leq$ 1 atm [11]. These results have not been explained, but may
imply a pressure (i.e. concentration dependence) of D_{O_2} or diffusion of both atomic and molecular oxygen species.

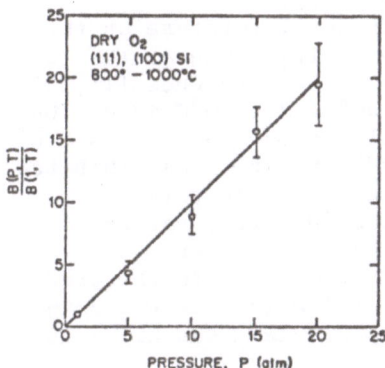

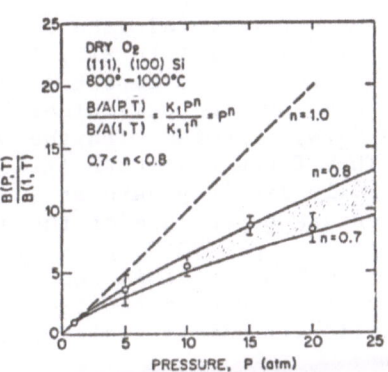

Fig. 8. Normalized parabolic rate
$\frac{B(P,T)}{B(1,T)}$ vs O_2 pressure for (100)
and (111) silicon oxidized in dry
O_2 at 1, 5, 10, 15, and 20 atm.
Temperature range was 800° – 1000°C.
[10].

Fig. 9. Normalized linear
rate constant $\frac{B/A(P,T)}{B/A(1,T)}$ vs O_2
pressure for (100) and (111)
silicon oxidized in dry O_2
at 1, 5, 10, 15, and 20 atm.
Temperature range was 800° –
1000°C [10].

B/A also contains k_s, the Si/SiO_2 interface reaction rate constant and h, the gas/SiO_2 interface reaction rate constant. h is normally assumed to be very large and hence is neglected [3]. k_s contains many hidden features of the oxidtion process and will be discussed in detail later in this chapter. k_s is known experimentally to depend upon substrate crystal orientation and upon substrate doping when the material is extrinsic at the oxidation temperature. In general, any parameter which affects the chemical reaction converting Si to SiO_2, will affect k_s and hence B/A.

THIN OXIDE KINETICS IN DRY O_2

Very thin (< 500 Å) layers of thermally grown SiO_2 are certain to become essential components of VLSI circuits. In fact, some commercially available MOS circuits currently use gate oxides < 400 Å thick. There is a wide body of experimental evidence which indicates that such thin layers are significantly different from thicker layers. Higher dielectric breakdown fields have been reported [12,13]; optical properties are different than thicker SiO_2 layers [14]; and growth kinetics are substantially faster than the linear parabolic model predicts. Present versions of SUPREM model the growth kinetics of such layers by employing approximate empirical correction factors to eqn. 11. SUPREM II arbitrarily multiplies B/A by 10 during the first 200 Å of oxide growth in dry O_2. SUPREM III has a more accurate formulation, based upon recent experimental data, as will be described.

Some experimental data does exist in the literature on the growth kinetics of thin SiO_2 layers [11, 15-17]. One of the first detailed studies was reported by Van der Meulen and Ghez [11,15]. Their work involved oxidations between 700° and 1000°C in dry O_2 and in O_2/N_2 mixtures with the O_2 partial pressure as low as 0.01 atm. They found that both (111) and (100) orientations exhibited essentially linear growth at least down to ~50 Å in thickness. The pressure dependence of the growth rate, however, varied from close to $P^{0.5}$ at 700°C up to close to $P^{1.0}$ at temperatures of 1000°C and higher. This result is in contradiction to the predictions of the linear-parabolic growth law [3] which, as was pointed out earlier, predicts a $P^{1.0}$ pressure dependence under all conditions.

In an effort to explain their experimental results, Van der Meulen and Ghez proposed that the interface reaction is actually more complex than that proposed by Deal and Grove. They suggested that three reactions occur, given by

$$Si-Si + O_2 \rightarrow Si-O-Si + O \tag{17}$$

$$Si-Si + O \rightarrow Si-O-Si \tag{18}$$

$$O_2 \rightleftarrows 2O \tag{19}$$

While the first two reactions are not stoichiometric in SiO_2 as written, they are intended simply to indicate that both molecular and atomic oxygen are assumed capable of reacting with the silicon to form SiO_2. Based upon their kinetic data, Van der Meulen and Ghez determined rate constants for the first reaction (k_1) and for the combination of reactions (18) and (19) (k_3). The corresponding activation energies were found to be 1.91 eV for k_1 and 0.58 eV for k_3. Because of this difference in activation energies, k_1 would be expected to dominate at higher temperatures, and therefore a $P^{1.0}$ pressure dependence and the normal Deal-Grove relationship should apply at these temperatures. At lower temperatures, k_3 should dominate, leading to a $P^{0.5}$ pressure dependence and a growth law with different effective rate constants. Thus they concluded that oxidation by molecular oxygen was dominant at high T while the atomic oxygen reaction becomes increasingly important as T decreases.

Recently Blanc [18] has proposed a kinetic model for thin oxides which in essence uses only Eqs. (18) and (19). That is, he assumed that the reaction involving molecular oxygen [Eq. (17)] does not occur at all. Based upon this, he derives a growth law given by

$$\frac{1}{2} u + \frac{1}{4} (\exp 2u - 1) + bt \tag{20}$$

where $ax = \sinh u$ and a and b are rate constants. For thin oxides, this predicts a linear parabolic growth law similar to [3], but with different effective rate constants. Blanc has compared his theoretical predictions with the data of Irene [19] at temperatures of 780° – 980°C with apparently good agreement. As Blanc points out, however, his model predicts a $P^{0.5}$ pressure dependence for the interface reaction rate; this would seem to be in disagreement with the data of Van der Meulen and Ghez.

Other models have been proposed to account for the enhanced growth rates of thin oxide layers. Grove [20], for example, proposed that molecular oxygen is ionized at the outer SiO_2 surface, forming O_2^-. Coupled diffusion between the O_2^- and holes then effectively increases the O_2^- diffusion rate until an oxide thickness on the order of the extrinsic Debye Length is reached (~150 Å for dry O_2 and ~5 Å for H_2O). This model might also be expected to be at least partially successful in explaining the experimental results of Jorgensen [1] in which externally applied electric fields were found to affect oxidation rates.

A substantial body of experimental data has recently been obtained using an in-situ ellipsometer to monitor the growth in real time of these thin layers [21]. This data covers the temperature

62

range 700 – 1000°C and O_2 pressure from 0.01 – 1 atm. Typical
results are shown in Figures 10 and 11. Figure 10 illustrates the
enhancement in oxidation rate dx/dt which is typically observed for
oxides < 200 Å thick. Figure 11 illustrates the excess oxidation
rate (over and above the linear-parabolic model). Figure 11
suggests that the excess rate is exponential and, in fact, it has
been found possible to model these results with a modification to
the linear parabolic growth law as follows:

$$\frac{dx}{dt} = \frac{B}{2x + A} + K_1 e^{-x/L_1} + K_2\, e^{-x/L_2} \tag{20}$$

where the first term is the usual result and the other two terms
incorporate the thin regime enhancement. L_1 is on the order of 20
Å and L_2 is ≈ 70 Å. SUPREM III incorporates this type of model to
handle the thin regime. At the present time, a physical explanation
for eqn. (20) is lacking.

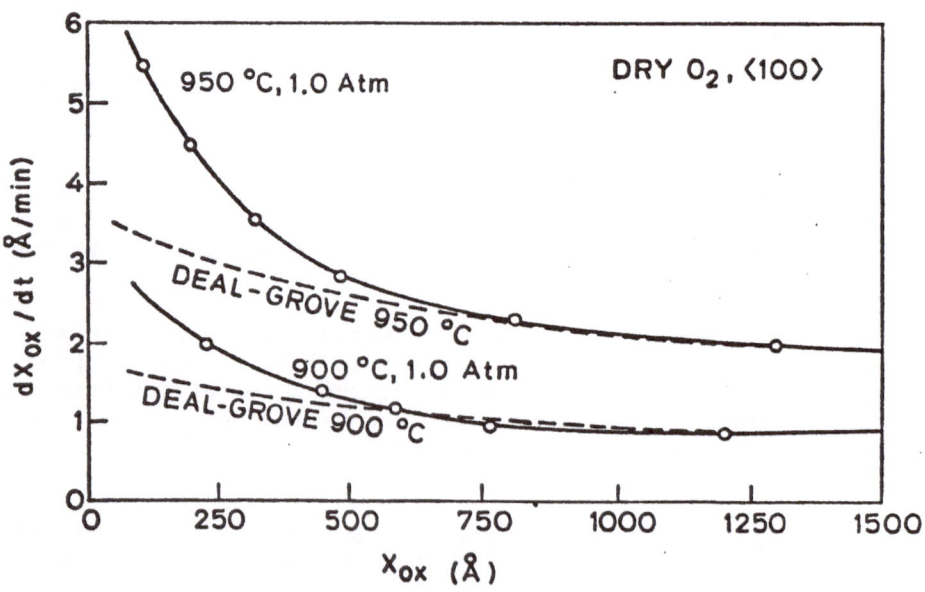

Fig. 10. Rate of growth dx/dt vs. oxide thickness in dry O_2 at
900 and 950°C. Dashed lines represent the theoretical predictions
of the Deal-Grove model [21].

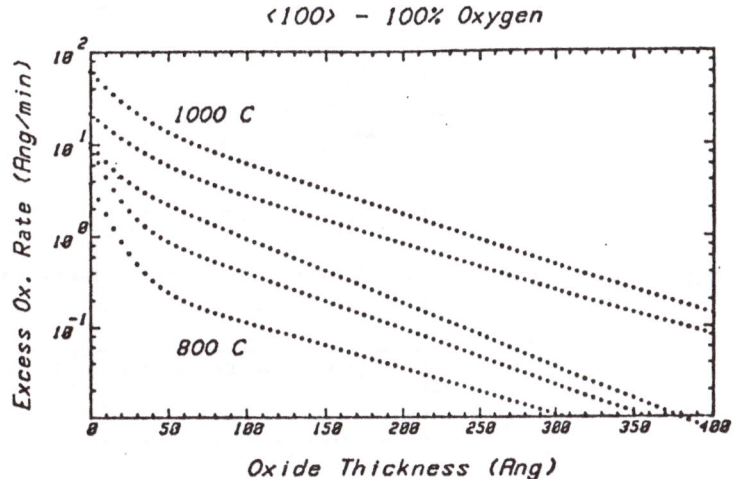

〈100〉 - 100% Oxygen

Fig. 11. Excess oxidation rate (above linear parabolic model), for thin oxides grown in dry O_2 in 100% O_2. T = 800 - 1000°C in 50°C steps [21].

Si/SiO₂ INTERFACE REACTION KINETICS - THE ROLES OF POINT DEFECTS

We saw earlier in this chapter that all of the details of the Si/SiO_2 interface reaction are lumped into the parameter k_s (and hence B/A) in the Deal Grove formulation. Over the past several years, a good deal of effort has been put into examining the interface reaction on an atomic level, with the objective of identifying the physical mechanisms "hidden" in k_s. From this work has emerged an overall picture of the interface reaction processes, which, while incomplete at the present time, nevertheless can be used to understand oxidation processes and the interaction of oxidation with other processes like diffusion. The basic ideas in this picture of k_s are described below.

The processes believed to occur at the interface are shown schematically in Fig. 12. O_2 from the gas phase is incorporated into the surface of the SiO_2 layer and is known to diffuse down to the Si/SiO_2 interface where the reaction converting silicon to SiO_2 takes place. The work of Doremus [22], Jorgensen [1], Rayleigh [23], and Collins and Nakayama [24] presented conflicting evidence concerning the presence or absence of an important charged oxygen ion in the diffusion process. The work of Mills and Kroger [25] clearly showed that a doubly negative oxygen interstitial ion was the dominant species in the high temperature conductivity of SiO_2. The first-order pressure dependence of the parabolic rate

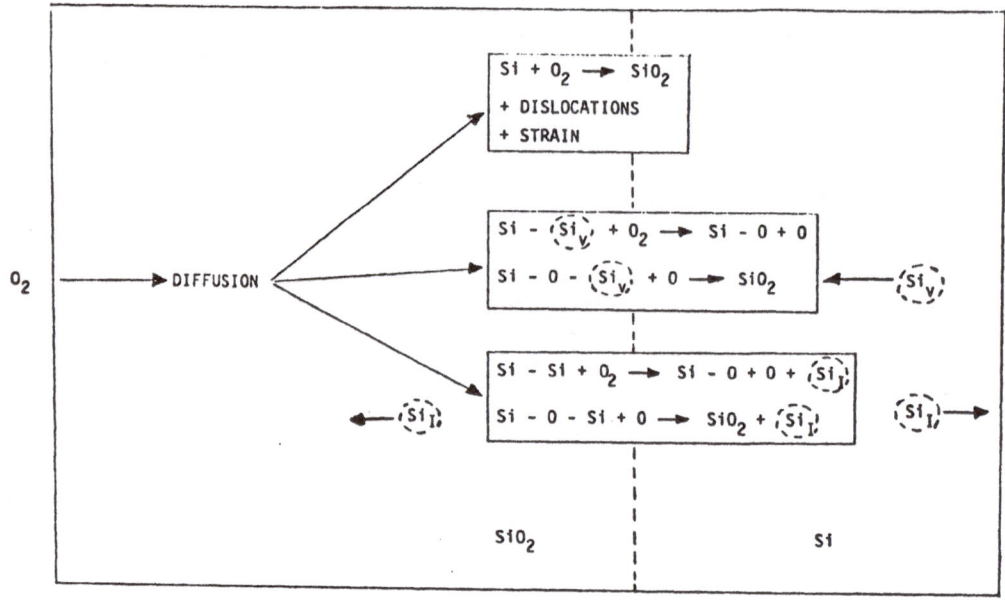

Fig. 12. Detailed model of the interface oxidation reaction including possible sources for the required "free volume" for the reaction $Si + O_2 \rightarrow SiO_2$ to proceed.

constant B in the Deal and Grove model [3] seems to indicate that the dominant diffusing species is molecular oxygen. For the present, the question of the particular oxygen species involved in the diffusion must be regarded as incompletely resolved, and for the purposes of this discussion, we will consider O_2 as the dominant species for thick oxide layers.

When silicon and an oxygen species react in the interface region to form an SiO_2 molecule, a necessary quantity of "free-volume" must also be supplied so that the molecule can fit into the normal SiO_2 network structure [26,27]. From density consideration, the average spacing between Si atoms in SiO_2 is about 1.3 times the average spacing between Si atoms in the Si lattice. This could lead to a 70% strain in the SiO_2 if the lattice continued in a coherent way into the SiO_2 film. If the interface moves into the Si at V cm/sec, the oxide thickens at 2.25 V cm/sec, which means that the "free volume" must flow in at a rate of 1.25 V cm/sec per square centimeter of interface area to produce an unstrained film. If no "free volume" is supplied, then the excess free energy stored as strain in the SiO_2 film is sufficient to strongly retard the oxidation reaction at normal driving forces.

A portion of the required "free volume" for the reaction may be supplied as shown in the top reaction in Fig. 12. In general, we might expect some lattice mismatch to occur at the interface, stored as a cross-grid of dislocation lines (dangling bonds) and some of the mismatch to be stored as strain in the SiO_2 films.

Two alternative means of providing the necessary free volume at the interface are shown in the middle and bottom of Fig. 12 and are based upon point defects in the SiO_2 and silicon layers. In the middle reaction, vacancies (Si_V) from the silicon substrate provide the reaction sites or the required "free volume". In the bottom reaction, silicon atoms are removed from lattice sites to create interstitials (Si_I) and the required reaction sites for the growth of the SiO_2. Note the possible charge states of the point defects have been neglected in the reactions proposed here. Si vacancies are known to exist in at least three charge states, Si_V^+, Si_V^-, and $Si_V^=$, in addition to neutral vacancies; very little is known at the present time about possible Si_I charge states.

It is believed that all three of these mechanisms play a role in the oxidation process. Each of them likely dominates under specific process conditions. For example, the reaction involving Si_V is believed to dominate in the case of heavily doped N^+ substrates when the position of the Fermi level is near the conduction band and the equilibrium concentration of vacancies is dramatically increased. We will consider this specific example later.

The bottom mechanism in Fig. 12 has been postulated by Dobson [28,29] and the excess interstitial flows produced during thermal oxidation may be the mechanism of oxidation enhanced dopant diffusion effect [30,31]. In addition, the interstitials produced during the reaction have been linked to oxidation-induced stacking faults [30,32]. Finally, it might be conjectured at this point that some of the excess interstitials move into the SiO_2 layer. Such a flow would then very probably be closely related to the observed fixed oxide charge N_f that has been discussed as incompletely oxidized silicon atoms [33]. We shall also return to this point later.

The basic concept of the roles of point defects in many related processes has been successfully used to quantitatively model the effects of enhanced oxidation over heavily doped regions [34,35]; to quantitatively model the growth and retrograde growth of oxidation induced stacking faults [30, 32]; to qualitatively explain enhanced diffusion coefficients during thermal oxidation [31]; and to qualitatively explain the dependence of fixed oxide charge (N_f) on process parameters. These and other implications of this type of modeling are indicated in Fig. 13. Such a model thus appears to be of major significance in accurately modeling and understanding many of the fabrication steps used in silicon integrated circuits. It is believed that as device geometries shrink towards submicron

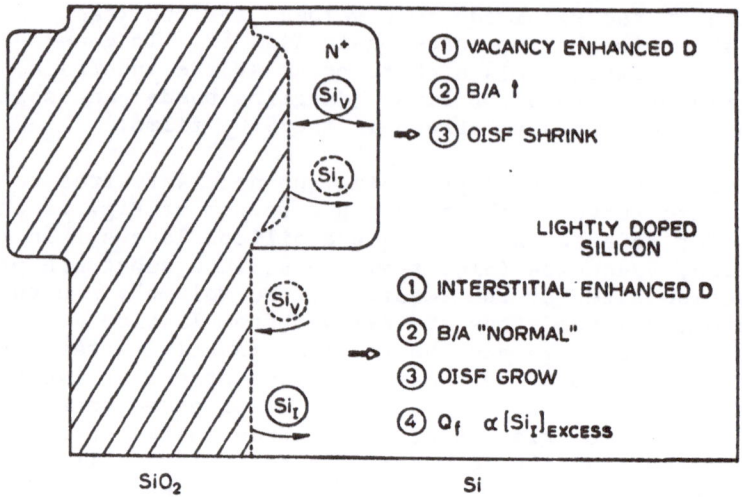

Fig. 13. Qualitative effects of interstitial generation and vacancy consumption during thermal oxidation.

structures, this type of point defect modeling will be essential if we are to accurately model VLSI structures.

Perhaps the most well developed example of the usefulness of the approach illustrated in Fig. 12 is the situation encountered in the oxidation of heavily doped substrates. It is well established that such substrates (particularly N^+ doped) show enhanced oxidation rates over lightly doped substrates. This can be explained, based upon the middle reaction in Fig. 12.

Much basic work on the electrical properties of Si_V (see, for example, [36,37]) has shown that Si_V can exist in several charge states in the silicon lattice in addition to neutral vacancies. Two acceptor levels (Si_V^- and $Si_V^=$) and at least one donor level (Si_V^+) have been identified as illustrated in Fig. 14, taken from [37]. The concentration of neutral vacancies is a function only of temperature [38] with

$$C_{VO} = k \ e^{-E_A/kT}$$

(21)

Current best estimates place $E_A \simeq 2.4$ eV [40]. Charged vacancies have thermal equilibrium concentrations which are a function not only of temperature but also of the Fermi level position. Thus [38],

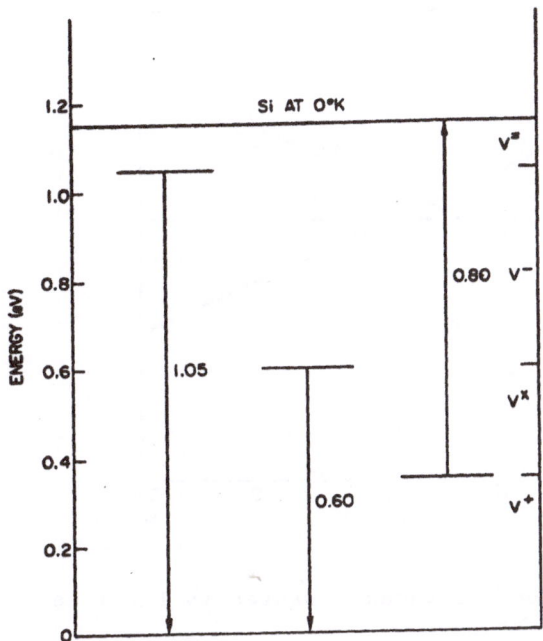

Fig. 14. Vacancy energy levels in silicon at 0°K [37].

$$C_{V^+} = C_{VO} \; e^{\dfrac{E_{V^+} - E_F}{kT}} \qquad (22)$$

with similar expressions holding for the other charge states.

The consequence of these relationships is that not only does the dominant charge state of the Si_V change as E_F and T change, but so does the total number of vacancies. One particular manifestation of this is shown if Fig. 15, where the normalized vacancy concentrations as a function of E_F are shown at 750°C [37]. It is clear from this figure that extrinsic N+ material will have much higher concentrations of Si_V- and Si_V= than intrinsic material and that extrinsic P+ material will have high concentrations of Si_V+. In both cases, the heavily doped substrates will have substantially higher total vacancy concentrations than intrinsic material.

Fig. 12 suggests several parallel paths for the interface reaction. To the extent that these paths may be considered independent, they simply add and we may write [34,35]

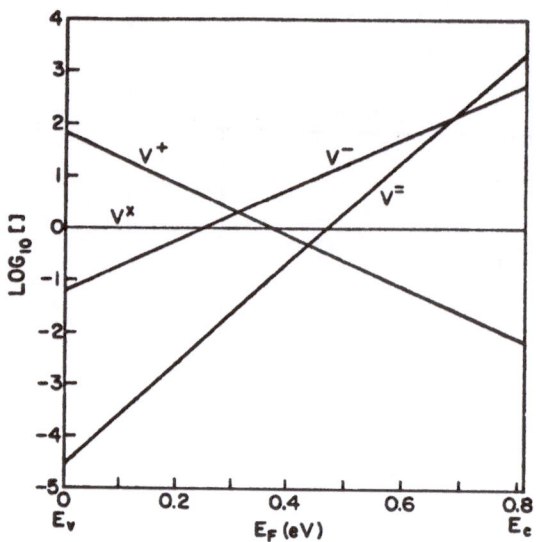

Fig. 15. Concentrations of vacancy states vs Fermi Level at 750°C
[37].

$$\frac{B}{A} = R_1 + K_{C_{V_T}} \qquad (23)$$

where R_1 contains the top and bottom reactions in Fig. 12 and
KC_{V_T} is the middle reaction, postulated to be proportional to
the total vacancy concentration. R_1 exhibits the 2.0 eV activation energy normally associated with lightly doped silicon oxidation
[3].

Based upon the statistics of neutral and charged vacancies,
C_{V_T} may be written

$$C_{V_T} = C_{VO} + \frac{n_i}{n} c_{V+}^i + \frac{n}{n_i} c_{V-}^i + \left(\frac{n}{n_i}\right)^2 c_{V=}^i \qquad (24)$$

where c_{Vr}^i = concentration of vacancies in the rth charge state in
intrinsic material. The n/n_i terms arise because of mass action
effects in extrinsic material [38].

Because of the known energy levels of the charged vacancies
(Fig. 14), equation (24) can be evaluated except for C_{VO}, the

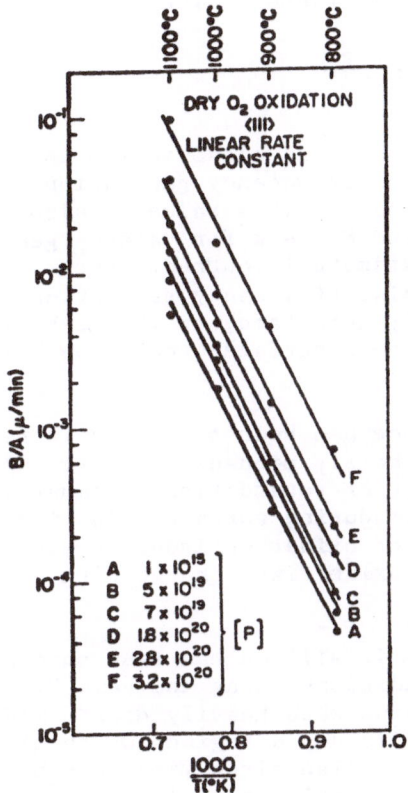

Fig. 16. B/A vs 1/T for samples heavily doped with Phosphorus [34, 35].

thermal equilibrium concentration of neutral vacancies. C_{Vo} can, however, be included in K in equation (23), making it possible to predict B/A for heavily doped substrates with K being the one fitted parameter [34,35].

An example of such an analysis is shown in Fig. 16 where B/A is predicted theoretically for N$^+$ substrates and compared with experimental measurements. The agreement is apparent, lending credence to the overall concept of the role of substrate point defects in the oxidation process. Similar agreement between theory and experiment has been obtained for P-type substrates and for compensated substrates with both P- and N-type dopants [34,35].

The experimentally extracted value of K in equation (23) is

$$K \simeq 2.6 \times 10^3 \, e^{-\dfrac{3.1 \text{ eV}}{kT}} \qquad (25)$$

The 3.1 eV activation energy is close to the total E_A proposed by Van Vechten and Thurmond for vacancy generation and migration (2.4 ± 0.2 eV + 1.2 ± 0.3 eV) [39]. This value is also close to the 3.4 eV proposed by Fairfield and Masters for vacancy generation alone. Depending upon which estimate is correct, then, the vacancy mechanism in Fig. 12 may consist of vacancy generation in the bulk and migration to the Si/SiO$_2$ interface, or the mechanism may be vacancy generation at the interface itself [34,35]. This question remains to be resolved.

Substantial evidence has been presented in recent years to support the existence of the Si$_I$ mechanism in Fig. 12. Such evidence includes the enhancement or retardation of impurity diffusion coefficients in the substrate during surface oxidation (OED or ORD), the growth and retrogrowth of oxidation inducted stacking faults (OISF); and the correlation of oxide fixed charge with oxide growth rate dx/dt.

Impurity diffusion in silicon has been successfully modeled under non-oxidizing conditions using the same Si$_V$ statistics described above in connection with heavily doped oxidation. The same increase in C_{V_T} in heavily doped regions which gives rise to enhanced oxidation rates, also gives rise to enhanced, concentration (E_F) dependent diffusion coefficients [40].

Under oxidizing conditions, additional effects are observed to take place which enhance impurity diffusion even under intrinsic conditions (n = p = n$_i$ at the oxidation temperature). Such effects have been explained by several authors [28,30,31] on the basis of the Si$_I$ mechanism is Fig. 12. Some fraction of the excess silicon atoms present at the Si/SiO$_2$ interface (probably <1%) move into the substrate as the oxidation proceeds. The rate of generation of such Si$_I$ would be expected to be related to dx/dt, the oxide growth rate. In the substrate, the Si$_I$ are believed to enhance impurity diffusion via the interstitialcy mechanism. Excellent agreement with experiment has been found for effective diffusion coefficients of the general form [41]

$$D_{eff} = D_i + K \left(\frac{dx}{dt} \right)^n \qquad (26)$$

where D_i = normal intrinsic diffusion coefficient due to vacancy
mechanisms

$$K\left(\frac{dx}{dt}\right)^n = Si_I \text{ contribution to } D_{eff}.$$

At first glance, one might expect n = 1 in such an expression. It appears, however, that n ≈ 0.4 gives better agreement with experiment [41].

A second area of experimental investigation which has provided indirect evidence of the Si_I mechanism is the work related to OISF growth and retrogrowth [31,42]. Such stacking faults are proposed to grow by precipitation of Si_I which are provided by the bottom reaction in Fig. 12. As a consequence, general relationships of the form [43]

$$\frac{dl}{dt} = K_1\left(\frac{dx}{dt}\right)^n - K_2 \tag{27}$$

where K_2 = shrinkage rate in the absence of oxidation,

$$K_1\left(\frac{dx}{dt}\right)^n = Si_I \text{ contribution to growth rate.}$$

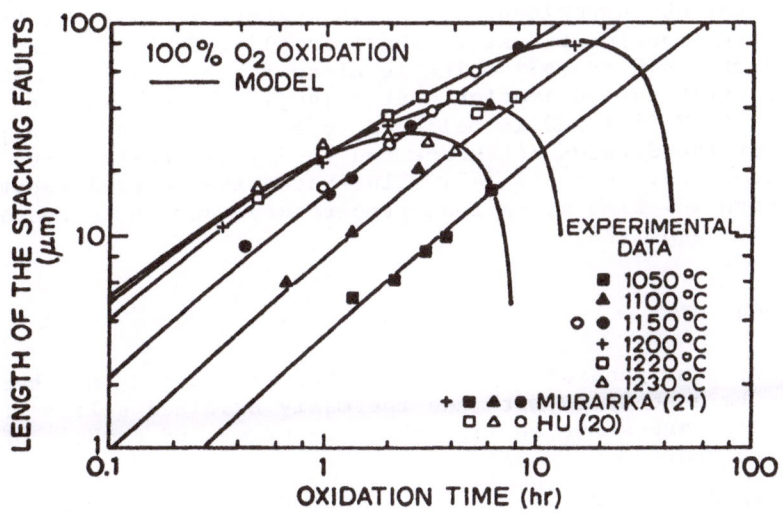

Fig. 17. Experimental data for OISF growth in 100% O_2 oxidation and comparisons with the model [43].

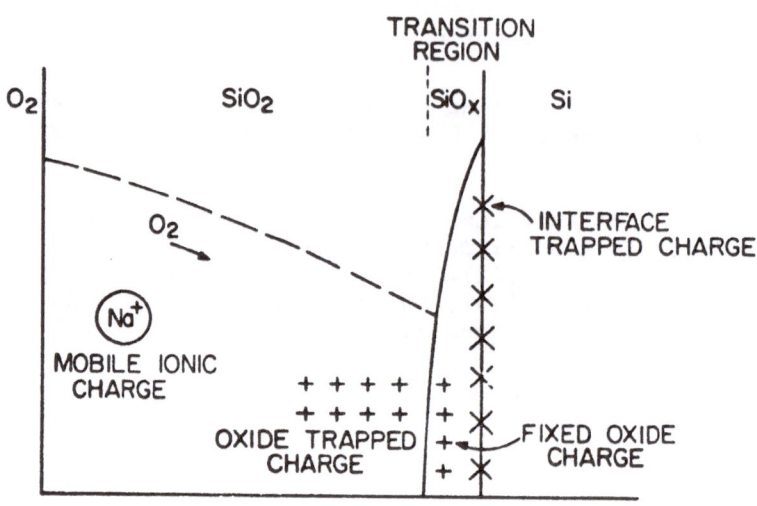

Fig. 18. Names and location of electrical charges associated with the thermally oxidized silicon structure.

have been proposed, and found to agree well with experiment. As was the case in the OED experiments, $n \simeq 0.4$ appears to give the best agreement with experiment. Also experimentally, K_2 has an activation energy of $\simeq 4.8$ eV [43] which is close to the silicon self-diffusion coefficient activation energy [40]. The activation energy of K_1 is $\simeq 2.4$ eV [43] and is believed to be related to the Si_I generation at the Si/SiO_2 interface and to Si_I diffusion away from the interface. Fig. 17 [43] illustrates the general behavior of OISF length vs. time at various temperatures and shows reasonable agreement with eqn. (27).

OXIDE CHARGES

At least four general types of electrical charges have been observed to be associated with the thermally oxidized silicon system [44,45]. These are shown in Fig. 18. The names and symbols of these charges are:

(1) fixed oxide charge Q_f, N_f

(2) interface trapped charge Q_{it}, N_{it}

(3) mobile ionic charge Q_m, N_m

(4) oxide trapped charge Q_{ot}, N_{ot}

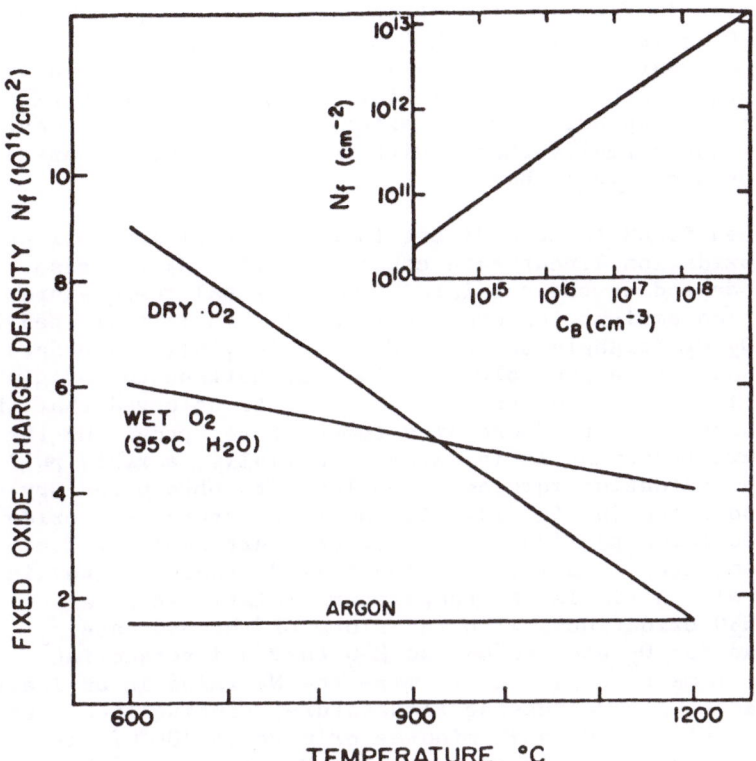

Fig. 19. Dependence of fixed oxide charge for (111) silicon on final oxidation and/or annealing temperature. The insert indicates the maximum allowable N_f to prevent inversion of P type silicon surfaces of varying doping concentrations.

By definition,

Q = effective net charge per unit area at the Si-SiO$_2$ interface

N = Q/q = net number of charges per unit area at the Si-SiO$_2$ interface

Interface trapped charge is also expressed in terms of the number of charges per unit area per unit energy of the silicon band gap and, for this definition, D_{it} should be used. It can be noted that at least two of the charges, N_f and N_{it}, are closely related to the Si-SiO$_2$ interface region and undoubtedly are directly related to the oxidation process.

As indicated in Fig. 18, the fixed oxide charge N_f (previously designated Q_{ss}) is located in the oxide in a narrow region (less than 50 Å) next to the $Si-SiO_2$ interface. It is generally positively charged with densities in the range 10^{10} to 10^{12} cm^{-2}, the actual density depending on the oxidation process variables. Its value does not normally change with surface potential variations, thus the name _fixed_ charge.

N_f has been found to be a direct function of silicon orientation and the oxidation linear rate constant, B/A. N_f has also been found to depend on the oxidizing ambient final temperature; this N_f-oxidation ambient temperature relationship is indicated by the familiar O_2-N_f triangle shown in Fig. 19 [44,46]. Such data the first to indicate a possible relationship between the origin of N_f and the silicon oxidation process. It can be observed that the value of N_f increases with decreasing temperature, providing the high temperature treatment is in oxygen. Normally, a rapid pull (<3 sec) out of the oxidation furnace is employed to obtain the dry O_2 - N_f values indicated in Fig. 19. At any temperature, the oxide may be annealed for a particular time in an inert ambient, i.e., argon, nitrogen, etc., and the N_f value will decrease to the minimum 1200°C level. A similar N_f-temperature relationship is obtained for H_2O oxidations, with the slope of the H_2O case being less than for O_2 and the O_2 and H_2O curves intersecting at 900°C. The time required to minimize the N_f value in an inert ambient increases with decreasing temperature, ranging from less than 10 minutes at 1200°C to 90 minutes or more at 700°C. It is also possible to subsequently increase the N_f value again by retreating in oxygen (or H_2O) at any temperature, the final value again being on the O_2 (or H_2O) curve of Fig. 19.

Interface traps or states have, in the past, been referred to as fast states, surface states, and surface or interface traps. These names are based on their various properties, i.e., they can exchange charge at very high frequencies, they are related to the so-called surface states due to dangling or unsatisfied bonds on solid surfaces, and they trap charges at varying surface potentials. At least three different types of interface traps have been recognized. The first is a function of the oxidation process itself and is believed to be a structural type of defect at the $Si-SiO_2$ interface. The second type of interface trap is that induced by various types of ionizing radiation, while the third is caused by the presence of metallic impurities at the $Si-SiO_2$ interface. All three types of interface traps are designated by the symbol N_{it} (or D_{it} if expressed as a function of energy).

The structural type of interface trap is undoubtedly related to fixed oxide charge in that their densities are both a function of silicon orientation [(111) > (110) > (100)] and oxidation temperature (N_f, N_{it} ↑ as T_{ox} ↓). As a result, it has been proposed

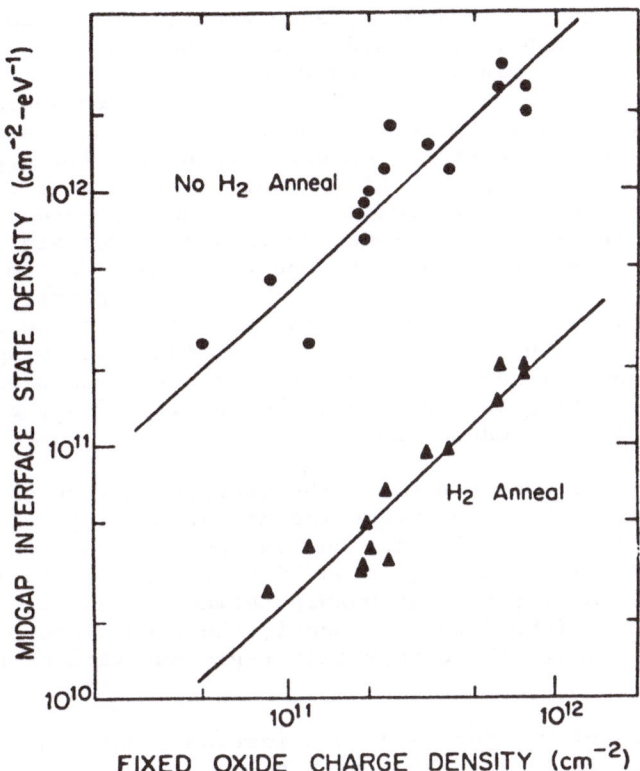

Fig. 20. Interface state density at midgap vs fixed oxide charge density for unannealed and H_2 annealed silicon oxide structures. Oxides were prepared on (111) and (100) silicon in dry O_2 at 1000°C and 1200°C. Hydrogen annealing was in 10% H_2 in N_2 for 10 min at 500°C [49].

that they have similar origins, with the major difference being that N_{it} is near enough to the Si–SiO$_2$ interface so as to be in electrical communication with the silicon. Thus, it can be charged or discharged as the silicon surface potential is varied. It has been proposed that those traps in the upper half of the band gap (near the conduction band) are acceptors and those in the lower half of the band gap are donors [47].

An important and fundamental property of the structural type of interface state is that its density can be reduced two, three, or more orders of magnitude by treatment in some hydrogen species at temperatures above 300°C [48]. This hydrogen species, possibly atomic hydrogen H, can be obtained in some equilibrium concentration in hydrogen gas H_2 or can be produced by the reaction of an active fieldplate metal such as aluminum with water present under the metal at the oxide surface. These latter reactions have been designated the "Alneal" process [44]. The active hydrogen in both cases probably reacts with the interface defect responsible for N_{it} to prevent further charge trapping. At the same time, however, it is possible for the hydrogen to be released under certain process conditions. As a result, it has been reported that certain optimum temperatures, hydrogen concentrations, and times may provide minimum effective interface trap densities [44]. The hydrogen annealing characteristic of interface traps also is refelected in the extreme sensitivity of N_{it} to cooling or pulling conditions following high temperature oxidaton. If any moisture or hydrogen-containing ambient is present during this cooling, the density of N_{it} may vary by an order of magnitude or more.

The apparent relationship between the structural type of N_{it} and N_f is indicated in Fig. 20, where the midgap density value of D_{it} is plotted against N_f [49]. Two curves are included, one for no hydrogen anneal and one for a 450°C anneal in 25% H_2 in N_2 for 10 minutes. A direct relationship between N_f and D_{it} is observed with and without the H_2 anneal, the same slope being obtained for both cases. The data points represent various process conditions.

The dependence of N_{it} (or D_{it}) on oxidation conditions is illustrated in Fig. 21 [49]. The midgap values of D_{it} are plotted against dry O_2 oxidation temperature for (111) and (100) silicon orientations. Also, two different types of cooling/pulling treatments have been employed. The slow pull in oxygen provides longer times at lower temperatures which results in higher D_{it} (and N_f) densities. The resulting relationships in Fig. 21 tend to duplicate the well-known N_f-O_2 triangle [46], again substantiating a N_f-N_{it} relationship.

Less is known about the other two types of interface states. It is well established that various types of ionizing radiation, such as electrons, x-rays, and even ion implantation, will result in additional interface states [50]. However, these states often can be annealed in inert ambients such as argon or nitrogen at temperatures as low as 350°C. Thus, no hydrogen is apparently required, as is the case for the structural type of interface state. Another difference is that, while the structural type of N_{it} often appears at specific levels of the band gap, the radiation-induced N_{it} is normally a continuous spectrum across the band gap. These characteristics would tend to indicate a less permanent more quasi-stable type of state, similar to the trapped charge in the bulk of the oxide also resulting from ionizing radiation.

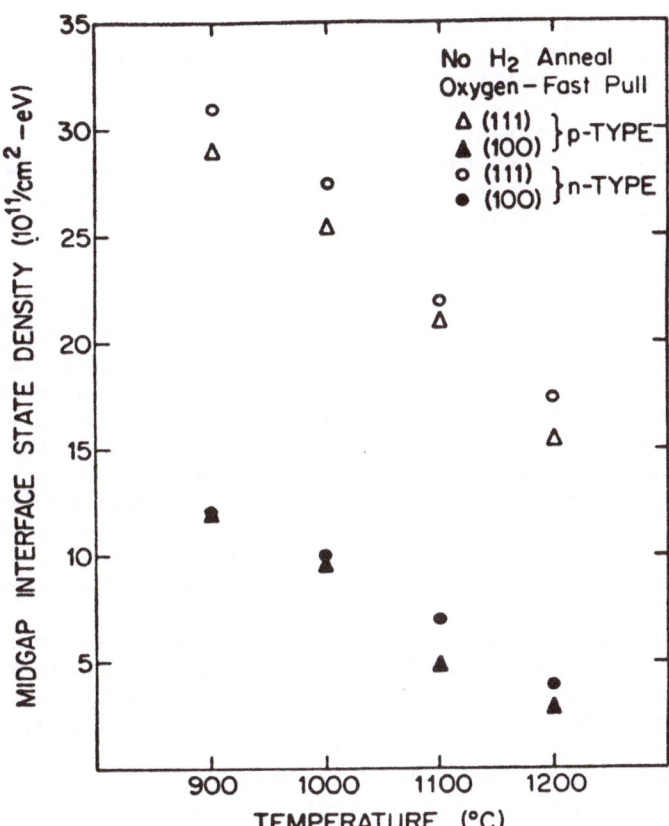

Fig. 21. Dependence of interface state density at midgap on oxidation temperature for P and N type (111) and (100) silicon, oxidized in dry O_2, and cooled in O_2 (<3 sec pull). Data are for structures with no low temprature hydrogen anneal [49].

The mobile ionic charge (N_m) is due mainly to the positively charged alkali ions Li^+, Na^+, and K^+, and possibly to the proton H^+. In the early 1960's when MOS technology was being developed, the main difficulty with MOS transistors was that they were not stable, i.e., threshold voltage would drift more than 100 volts when positive voltage was applied to the gate at 200–300°C. After considerable controversy, it was determined that the cause of the instability was alkali ions [51]. It was found that the size of the ion determined the rate of ionic migration at a given temperature and time until saturation. Thus, the relative diffusion rates are $Li^+ > Na^+ > K^+$, while Cs^+ does not drift at all up to 500°C. It was also found that if the contamination occurs before or during the high temperature oxidation process, the oxide field present during oxidation caused the positive ions to migrate towards the outer oxide surface [52]. It is thus possible to etch off 50 to 100 Å of the outer oxide and eliminate a good portion of the ionic charge. Clean processing in modern facilities has largely eliminated these mobile ionic species. Typical threshold voltage shifts due to these species today, are <0.25 volts even after applying positive gate bias at elevated temperatures.

Charge trapping in silicon oxides due to ionizing radiation has been investigated for a number of years. Much of this work has been reported at the IEEE Radiation Effects Conferences [53]. Radiation-induced carrier trapping has been a particular problem for devices which are exposed to ionizing radiation environments such as occur in space applications. However, more and more semiconductor processes themselves now produce radiation of some sort—such as electron beam heating of materials, plasma etching and deposition, sputtering processes, and, more recently, electron beam and x-ray lithography. All of these contribute to charge trapping in the dielectric films used in semiconductor structures and can lead to device degradation.

Ionizing radiation generally results in two types of charges. One is trapped holes in the oxide and is similar to oxide fixed charge. One basic difference is that these trapped holes can be annealed out in an inert ambient at 350°C. The other radiation induced charge is a type of interface state. It is a more continuous distributed charge than oxidation induced interface states and does not require hydrogen for annealing. In general, the amount of positive charge trapping increases with radiation dose and finally saturates. It is independent of dose rate, but is a function of bias across the oxide during radiation. The radiation-induced interface states also increase with dose, but are apparently independent of bias.

The origin of radiation-induced charges or traps is not clear. Some investigators believe impurity ions are responsible, and others think the trapping mechanism is related to structural effects. Discussions regarding possible origins are available in various reviews [54,55]. Whatever the origin, considerable evidence points to strong trapping dependencies on process variables. These have been demonstrated by various investigators [56,57] and involve oxidation ambient and temperature, oxide thickness, annealing conditions, hydrogen and chlorine effects, and radiation itself. Several possible schemes for improving radiation resistance of oxides have been investigated, and these include minimizing alkali ion content, incorporating metallic ions such as aluminum, and using other dielectrics [53]. It is currently believed, however, that the use of optimum process conditions in conjunction with thermal SiO_2

provides the most suitable method for minimizing radiation-induced charge trapping.

As smaller geometry circuit structures have evolved, other types of charge trapping in oxides have been demonstrated to adversely affect device performance. These have included effects due to internal photoemission, tunneling, and avalance injection, and involve both hole and electron trapping. Initially, it was believed that these charges could be annealed out at reasonably low temperatures. Now it is probably that, while the charges are neutralized, the resulting traps are quite susceptible to subsequent charging during normal device operation.

Characteristics of these types of trapped carriers are only now being investigated [47]. It is known that hole trapping efficiency at room temperature (10 to 20%) is much greater than that of electrons (as low as 10^{-6}). The trap cross section for holes is of the order of 10^{-13} cm^{-2}, and the saturated hole trap density is 10^{12} to 10^{13} cm^{-2}. These are located near the Si-SiO$_2$ interface. Trapped electrons, on the other hand, occur throughout the oxide. There are several types of electron traps with cross sections ranging from 10^{-12} to 10^{-19} cm^2. The process dependencies of both trapped holes (nonradiation) ane electrons is not known at present. Such relationships are being investigated since the ability of device manufacturers to produce submicrometer very large-scale integration (VLSI) structures will depend on the minimization of such trapping effects.

Some recent thinking concerning the origin of at least the structurally originating charges (Q_f and N_{it}), has attempted to tie these charges to the Si_I mechanism shown in Fig. 12. This would then correlate Q_f with dx/dt, just as was done for other oxidation related phenomena earlier (OED, OISF, etc). If excess Si_I are generated at the oxidizing interface, some of these (perhaps the majority [26]) should move into the SiO$_2$ layer, where eventually they would be oxidized to SiO$_2$. Q_f has historically been described as due to "excess" silicon in the SiO$_2$ near the Si/SiO$_2$ interface [44]. The Si_I could provide the source of this "excess" silicon.

If this is true, then Q_f should also be related to dx/dt, perhaps by equations of the general form of (26) an (27). Recent work tends to confirm this prediction, although the body of available data is too small to be certain. Experiments at low oxygen partial pressures [58] where dx/dt decreases, show smaller values of Q_f. (In fact the N$_2$ or Ar anneal commonly used to lower Q_f following an oxidation might be regarded as a limiting case in which $dx/dt \rightarrow 0$.)

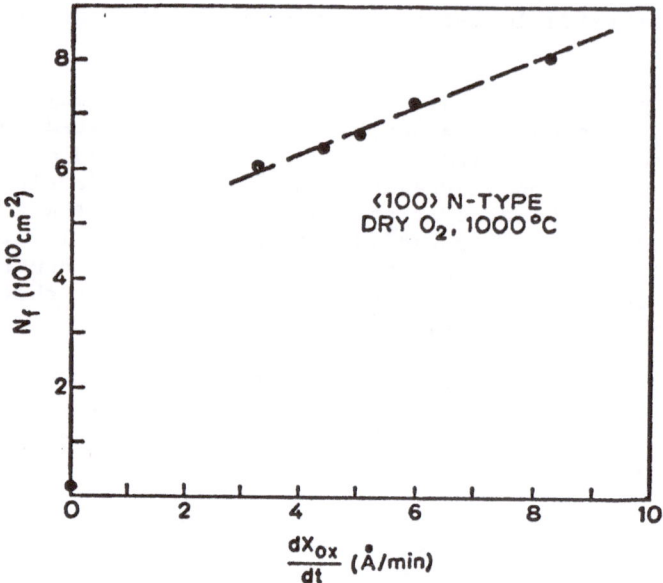

Fig. 22. Fixed oxide charge N_f vs oxide growth rate dx/dt for oxides grown in dry O_2 at 1000°C and fast pulled with no Ar anneal [21].

Experiments in the thin oxide regime (t_{ox} < 500 Å, dry O_2) in which dx/dt is known to be enhanced over that observed in thicker oxides, show enhanced values of Q_f, again in agreement with the general predictions of the model. A specific example is shown in Fig. 22 in which Q_f is plotted vs. oxide growth rate for samples grown in dry O_2 and fast pulled with no high temperature anneal. A direct correlation between the higher dx/dt for the thinner oxides and higher Q_f has been found for these samples. Similar correlations between Q_f and dx/dt have been found for thicker oxides as well although in both cases, the data at present is too sparse to deduce analytic relationships. Accurate measurement of these charge densities is also hindered at present by apparent variations of parameters like the metal-semiconductor work function difference ϕ_{MS}, with process parameters [59]. ϕ_{MS} must be known to an accuracy of better than 50 mV if Q_f is to be accurately determined for thin oxides. Reports in the literature have suggested that ϕ_{MS} can vary over a much larger range than this due to process variations [59].

2D OXIDATION EFFECTS IN SMALL STRUCTURES

We shall conclude this chapter with a brief discussion of oxidation phenomena in small structures. Essentially everything up to this point has been concerned with one dimensional phenomena. It has become apparent in recent years that 2D effects are important in small devices, with perhaps the best example taken fom the work of Lin et. al. [60].

Fig. 23 [60] illustrates a structure in which boron was implanted at a dose of $3 \times 10^{14}/cm^2$ and energy of 70 KeV into a high resistivity N type (100) wafer. Following an anneal (30' @ 900°C) to activate the implanted layer, a 500 Å thick Si_3N_4 layer was deposited on top of a 750 Å SiO_2 layer (through which the boron was implanted). The Si_3N_4 was then patterned using conventional techniques into stripes varying in width from 2 to 200 μm. Following this process, local oxidation was performed at various temperatures.

These experiments showed the expected enhancement of the boron diffusion coefficient under the oxidizing regions as revealed by junction staining. Even more interesting, however, was the behavior as the width of the Si_3N_4 masked region was shrunk. It is apparent from the figure that the enhancement effect spreads laterally as well as vertically. In fact, an effective lateral diffusion length of $\simeq 2$μm for the species involved in this phenomenon was derived from the experiment. This is in contrast to a vertical diffusion length of >25 μm which has been observed in other OED experiments [41]. If these effects are due to generation of Si_I at the oxidizing interface as has been hypothesized, then apparently lateral spread of these species is inhibited, perhaps by sinks under the Si_3N_4 masked areas (surface recombination) or by strained regions at the edges of the Si_3N_4 areas.

A recent attempt to model the 2D distribution of Si_I resulting from a local oxidation process has been made [61], with reasonable agreement with the above experimental results. The model assumed a surface regrowth process at surface kink sites under the Si_3N_4 regions to account for the reduced lateral spread of the Si_I.

In the more general context of process simulation for small (VLSI) devices, it is now clear that simulation routines will have to calculate and keep track of local concentrations of Si_I and Si_V. Evidence appears to be growing that dopants in silicon diffuse via a dual mechanism process (involving both Si_I and Si_V), even under intrinsic diffusion conditions [62]. Non equilibrium conditions like oxidation perturb the Si_I, Si_V balance by generating Si_I and consuming Si_V. As a result, diffusion coefficients in the bulk will be changed. Interstitials and vacancies may reasonably be expected to interact by annihilation [38],

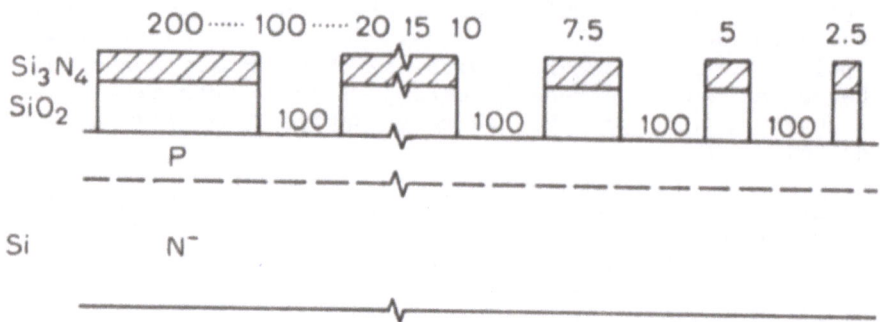

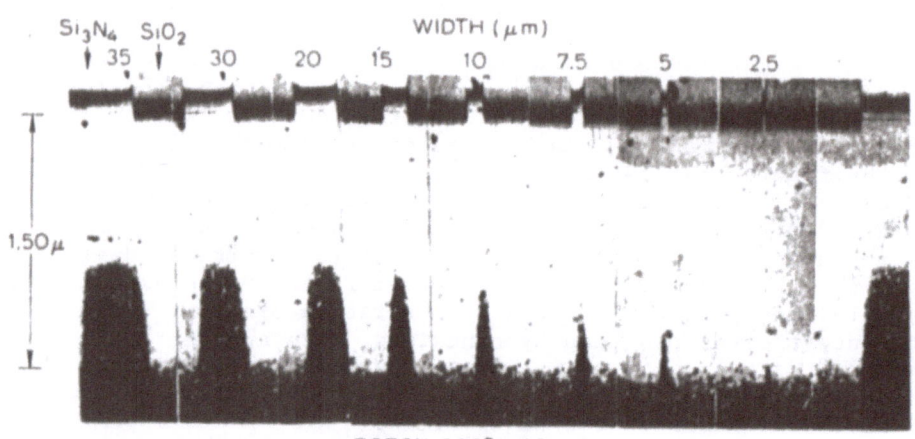

Fig. 23. Locally oxidized test structure which illustrates 2D oxidation/diffusion effects. The white regions are the boron profile after oxidation. For the narrow Si_3N_4 stripes the lateral OED effect causes enhanced boron diffusion under the Si_3N_4 stripes. [60]

implying a thermal equilibrium mass-action relation between vacancy and interstitial concentrations. However, the activation energy barrier to this reaction will determine the time-dependence of achieving the mass-action equilibrium and thus its influence in processes such as OED and ORD in which the point defect populations are disturbed by external forces.

The major conclusion to be drawn from this discussion is that process modeling programs which aim to simulate small geometry device structures, will require very robust physical models and will require as well, geometry dependent process parameters. This will imply a good deal of interaction between process and device simulation programs. There is a significant amount of basic materials related work remaining to be done in this area.

CONCLUSIONS

The thermal oxidation of silicon will continue to be an important part of device fabrication. As device geometries shrink, first order models of the process will have to be modified if device structures are to be accurately predicted. This is true both for one-dimensional effects such as crystal orientation, multiple species oxidation, high substrate doping levels, high pressure oxidation, and other situations; perhaps, more importantly, it is also true for twodimensional effects. Point defect models of thermal oxidation imply that Si_I and Si_V mechanisms are not local phenomena. Generation or consumption of these defects at localized regions during IC fabrication should be expected to impact process variables in nearby regions. Much experimental evidence indicates this to be true. Oxidation is known to affect diffusion coefficients and OISF growth in regions which may be tens of microns removed from the oxidizing interface. Such effects also likely extend to oxide charge densities and other process parameters. Only through the development of unified models physically explaining such effects will the fabrication of reliable and reproducible VLSI circuits be possible.

ACKNOWLEDGEMENT

The authors gratefully acknowledge the many contributions of their colleagues at Stanford and Fairchild, particularly C. P. Ho, who is largely responsible for the work on heavily doped oxidation; H. Z. Massoud, who carried out the studies on thin oxide kinetics in dry O_2; and R. R. Razouk and L. N. Lie, who carried out the high pressure oxidation studies. Useful input was also provided by A. M. Lin, R. W. Dutton, and W. A. Tiller. Some of the work described here was supported by R. Reynolds and S. Roosild of the Defense Advanced Research Projects Agency.

REFERENCES

1. P. J. Jorgensen, "Effect of an Electric Field on Silicon Oxidation." n. Chem. Phys. 37 (1962) 874.
2. J. R. Liginza and W. G. Spitzer, "The Mechanisms for Silicon Oxidation in Steam and Oxygen." Phys. Chem. Solids 14 (1960) 131.
3. B. E. Deal and A. S. Grove, "General Relationship for the Thermal Oxidatin of Silicon." J. Appl. Phys. 36 (1965) 3770.
4. L. Pauling, The Nature of the Chemical Bond, 3rd Ed. (Cornell University Press, 1960).
5. J. F. Norton, "Permeation of Gaseous Oxygen through Vitreous Silicon." Nature 171 (1961) 701.
6. A. J. Moulson and J. P. Roberts, "Water in Silica Glass." Faraday Soc. Trans. 57 (1961) 1208.
7. B. E. Deal and M. Sklar, "Thermal Oxidaton of Heavily Doped Silicon." J. Electrochem. Soc. 112 (1965) 430.
8. B. E. Deal, unpublished data.
9. R. R. Razouk, L. N. Lie and B. E. Deal, "Kinetics of High Pressure Oxidation of Silicon in Pyrogenic Steam." J. Electrochem. Soc. 128 (1981) 2214.
10. L. N. Lie, R. R. Razouk and B. E. Deal, "High Pressure Oxidation of Silicon in Dry Oxygen." to be published, J. Electrochem. Soc. (1982).
11. Y. J. Van der Meulen, "Kinetics of Thermal Growth of Ultra-Thin Layers of SiO_2 on Silicon. Part 1. Experiment." J. Electrochem. Soc. 119 (1972) 530.
12. K. Hamano, "Breakdown Characteristics in Thin SiO_2 Films." Jap. J. Appl. Phys. 13 (1974) 1085.
13. E. Harari, "Dielectric Breakdown in Electrically Stressed Thin Films of Thermal SiO_2." J. Appl. Phys. 49 (1978) 2478.
14. E. Taft and L. Cordes, "Optical Evidence for a Silicon-Silicon Oxide Interlayer." J. Electrochem. Soc. 126 (1979) 131.
15. R. Ghez and Y. J. Van der Meulen, "Kinetics of Thermal Growth of Ultra-Thin Layers of SiO_2 on Silicon. Part 2. Theory." J. Electrochem. Soc. 119 (1972) 1100.
16. F. P. Fehlner, "Formation of Ultra-Thin Oxide Films on Silicon." J. Electrochem. Soc. 119 (1972) 1723.
17. J. A. Aboaf, "Formation of 20-25 Thermal Oxide Films on Silicon at 950-1140°C." J. Electrochem. Soc. 118 (1971) 1370.
18. J. Blanc, "A Revised Model for the Oxidation of Si by Oxygen." Appl. Phys. Lett. 33-5 (1978) 424.
19. A. E. Irene, "Silicon Oxidation Studies: Some Aspects of the Initial Oxidation Regime." J. Electrochem. Soc. 125 (1978) 1708.
20. A. S. Grove, Physics and Technology of Semiconductor Devices, (John Wiley and Sons, New York, 1967).
21. This data was obtained jointly by Stanford (H. Z. Massoud, J. D. Plummer) and IBM (E. A. Irene), using the IBM in-situ ellipsometer.
22. R. H. Doremus, "Oxidation of Silicon by Water and Oxygen and Diffusion in Fused Silica." J. Phys. Chem. 80 (1976) 1773.

23. D. O. Rayleigh, "Transport Processes in the Thermal Oxidation of Silicon." J. Electrochem. Soc. 113 (1966) 782.

24. F. C. Collins and T. Nakayama, "Transport Processes in the Thermal Growth of Metal and Semiconductor Oxide Films." J. Electrochem. Soc. 114 (1967) 1962.

25. T. G. Mills and F. A. Kroger, "Electrical Conduction at Elevated Temperatures in Thermally Grown Silicon Dioxide Films." J. Electrochem. Soc. 120 (1973) 1582.

26. W. A. Tiller, "On the Kinetics of the Thermal Oxidation of Silicon. 1. A Theoretical Perspective" J. Electrochem. Soc. 127 (1980) 619.

27. W. A. Tiller, "On the Kinetics of the Thermal Oxidation of Silicon. II. Some Theoretical Evaluations." J. Electrochem. Soc. 127 (1980) 625.

28. P. S. Dobson, "The Effect of Oxidation on Anomalous Diffusion in Silicon." Philosophical Mag. 24 (1971) 567-576.

29. P. S. Dobson, "The Mechanism of Impurity Diffusion in Silicon." Philosophical Mag. 26 (1972) 1301-1306.

30. S. M. Hu, "Formation of Stacking Faults and Enhanced Diffusion in the Oxidation of Silicon." J. Appl. Phys. 45 (1974) 1567.

31. D. A. Antoniadis, A. G. Gonzalez, and R. W. Dutton, "Boron in Near Intrinsic (100) and (111) Silicon Under Inert and Oxidizing Ambients--Diffusion and Segregation." J. Electrochem. Soc. 125 (1978) 813.

32. A. M. Lin, D. A. Antoniadis, R. W. Dutton, and W. A. Tiller, "The Rate Control Model of Oxidation-Stacking-Faults Growth in Silicon." ECS Meeting, Oct. 1979, Los Angeles, Abstract No. 539.

33. B. E. Deal, "The Current Understanding of Charges in the Thermally Oxidized Silicon Structure." J. Electrochem. Soc. 121 (1974) 198C.

34. C. P. Ho and J. D. Plummer, "Si/SiO$_2$ Interface Oxidation Kinetics: A Physical Model for the Influence of High Substrate Doping Levels. I. Theory" J. Electrochem. Soc. 126 (1979) 1516-1522.

35. C. P. Ho and J. D. Plummer, "Si/SiO$_2$ Interface Oxidation Kinetics: A Physical Model for the Influence of High Substrate Doping Levels. II. Comparison with Experiment and Discussion." J. Electrochem. Soc. 126 (1979) 1523-1530.

36. G. D. Watkins, Lattice Defects in Semiconductors, 1974, Institute of Physics, London, 1, (1975).

37. J. A. Van Vechten, Lattice Defects in Semiconductors, 1974, Institute of Physics, London, 212, (1975).

38. W. Shockley and J. C. Moll, "Solubility of Flaws in Heavily Doped Semiconductors." Phys. Rev. 119 (1960) 1480.

39. J. A. Van Vechten and C. D. Thurmond, "Comparison of Theory with Quenching Experiments for the Entropy and Enthalpy of Vacancy Formation in Si and Ge." Phys. Rev. B 14 (1976) 3551.

40. R. B. Fair, "Recent Advances in Implantation and Diffusion Modeling for the Design and Process Control of Bipolar ICs," ECS Silicon Symposium (1977) 968.

41. A. M. Lin, D. A. Antoniadis and R. W. Dutton, "The Oxidation Rate Dependence of Oxidation-Enhanced Diffusion of Boron and Phosphorus in Silicon." J. Electrochem. Soc. 128 (1981) 1131.

42. S. P. Murarka, "Role of Point Defects in the Growth of Oxidation-Induced Stacking Faults in Silicon." Phys. Rev. B 16 (1977) 2849.

43. A. M. Lin, R. W. Dutton, D. A. Antoniadis and W. A. Tiller, "The Growth of Oxidation Stacking Faults and the Point Defect Generation at the Si-Si$_2$ Interface During Thermal Oxidation of Silicon." J. Electrochem. Soc. 128 (1981) 1121.

44. B. E. Deal, "The Current Understanding of Charges in the Thermally Oxidized Silicon Structure." J. Electrochem. Soc. 121 (1974) 198C.

45. Y. C. Cheng, "Electronic States at the Silicon-Silicon Dioxide Interface." Prog. Surf. Sci. 8 (1977) 181.

46. B. E. Deal, M. Sklar, A. S. Grove and E. H. Snow, "Characteristics of The Surface-State Charge (Q_{ss}) of Thermally Oxidized Silicon." J. Electrochem. Soc. 114 (1967) 266.

47. W. A. Pliskin and R. A. Gdula, in Handbook of Semiconductors, Vol. 3, S. Keller, ed., (Elsevier-North Holland Publishing Co., New York, to be published).

48. P. Balk, Paper No. 109 presented at the San Francisco Meeting of the Electrochemical Society, May 9-13 1965.

49. R. R. Razouk and B. E. Deal, "Dependence of Interface State Density on Silicon Thermal Oxidation Process Variables" J. Electrochem. Soc. 126 (1979) 1467.

50. See, for example, B. L. Gregory and C. W. Gwyn, "Radiation Effects on Semiconductor Devices" Proc. IEEE 62 (1974) 1264.

51. E. H. Snow, A. S. Grove, B. E. Deal, and C. T. Sah, "Ion Transport Phenomena in Insulating Films." J. Appl. Phys. 36 (1965) 1664.

52. B. E. Deal, U. S. Patent No. 3,426,422, Feb 11, 1969.

53. Special issues on device radiation effects, IEEE Trans. Nucl. Sci. Dec. issues Vol. NS-21-25 (1974-1978).

54. C. T. Sah, "Origin of Interface States and Oxide Charges Generated by Ionizing Radiation" IEEE Trans. Nucl. Sci. NS-23 (1976) 1563.

55. W. C. Johnson, "Mechanisms of Charge Buildup in MOS Structures." IEEE Trans. Nucl. Sci.. NS-22 (1975) 2144.

56. K. G. Aubuchon, "Radiation Hardening of P-MOS Devices by Optimization of the Thermal SiO$_2$ Gate Insulator." IEEE Trans. Nucl. Sci.NS-18 (1971) 117.

57. W. R. Dawes, Jr., G. F. Derbenwick, and B. L. Gregory, "Process Technology for Radiation-Hardened CMOS Integrated Circuits" IEEE J. Solid-State Circuits SC-11 (1976) 459.

58. S. P. Murarka, "Oxygen Partial Pressure Dependence of the Fixed Surface-State Charge Q_{ss} Due to Thermal Oxidation of N - <100> Silicon." Appl. Phys. Lett. 34 (1979) 587.

59. R. R. Razouk and B. E. Deal, "Hydrogen Anneal Effects on Metal-Semiconductor Work Function Difference." J. Electrochem. Soc. 129 (1982) 806.

60. A. M. Lin, R. W. Dutton and D. A. Antoniadis, "The Lateral Effect of Oxidation on Boron Diffusion in <100> Silicon." Appl. Phys. Lett. 35 (1979) 799.

61. M. Hamasaki, "On An Analytical Solution for Two-Dimensional Diffusion of Silicon Self-Interstitials During Oxidation of Silicon." Solid State Elec. 25 (1982) 1.

62. D. A. Antoniadis, "Oxidation-Induced Point Defects in Silicon." J. Electrochem. Soc. 129 (1982) 1093.

63. B. E. Deal, "Thermal Oxidation Kinetics of Silicon in Pyrogenic H_2O and 5% HCl/H_2O Mixtures." J. Electrochem. Soc. 125 (1978) 576.

THE USE OF CHLORINATED OXIDES AND INTRINSIC GETTERING TECHNIQUES FOR VLSI PROCESSING

C.L. Claeys

K.U. Leuven, ESAT Lab
Kardinaal Mercierlaan 94
B-3030 Heverlee, Belgium

1 INTRODUCTION

The world-wide research activities towards integrated circuits with larger chip sizes and smaller device geometries require a better understanding of the different processing steps and the development of very accurate models to describe the basic processes and their interactions. Especially the models for oxidation and diffusion, many of which were developed ten to fifteen years ago, have to be reviewed. In this paper some special features of oxides grown in a chlorine containing oxidizing ambient are addressed. To achieve the stringent demands placed on the device performance, it is essential to have large Si active area's which are completely free of both process-induced and material related defects. Therefore the use of several gettering techniques has been extensively studied during the last years. Good results have been achieved with internal or intrinsic gettering techniques.

Nowadays, the growth of SiO_2 layers in the presence of small amounts of a chlorine compound in the oxidizing ambient has become widespread in the semiconductor industry. This is mainly due to the improved electrical and passivation properties of the oxide layer, and to the strong impact on the silicon bulk properties. Some general features of these chlorinated oxides, including their applications for VLSI processing, are briefly reviewed in section 2.

The different aspects of oxidation-induced stacking faults (OSF's) have been reported in detail in the literature. Dependent on their electrical activity, these OSF's form the basic mechanism for yield degradation in both bipolar and MOS integrated circuits. Some characteristics of the nucleation and the growth kinetics of these defects in relation to the different oxidation processes are discussed in section 3. The defect behaviour in both the near surface region and the bulk of the wafers will be discussed.

In order to eliminate the process-induced defects, the use of a large variety of gettering techniques has been proposed. Whereas some of these gettering techniques can be carried out simultaneously with other processing steps, other techniques such as e.g. ion implantation on the back-side of the wafer or mechanically induced strain fields require additional processing steps. The implementation of intrinsic gettering techniques, which eliminate the defect formation only in the active region of the devices, has become very important for VLSI processing. This topic is addressed in section 4.

The presence of oxygen and carbon in the silicon wafers plays an important role in the nucleation and the growth of the defects. Dependent on their concentration and the way at which they are incorporated in the silicon lattice, the substrate resistivity and the wafer warpage are also influenced. The influence of oxygen on both processing and device characteristics is discussed in the last section.

2 CHLORINATED OXIDES

During the last decade many investigations have been concentrated on the use of a chlorine containing ambient in order to improve the quality of the SiO_2 layers. The work in this field was instigated by Kriegler's observation that oxide films grown in a furnace contaminated with NaCl contain less mobile ions than those grown in the presence of Na_2CO_3. This result led to the speculation that in the presence of chlorine, at least some of the sodium incorporated in the oxide had become electrically inactive. Although Kriegler observed this effect in 1970, he did not published on the neutralization of mobile ions by the addition of HCl or Cl_2 to the oxidizing atmosphere

until in 1972 [1]. The increase in minority carrier
lifetime in the silcon substrate under the oxide layer
when the oxide is grown in HCl/O_2 mixtures, however,
is first reported by Robinson and Heiman [2] in 1971.
To obtain a chlorine containing oxidizing atmosphere,
the use of either HCl gas, trichloroethylene (TCE) or
trichloroethane (C33) is extensively reported in the
literature. Some important aspects of these chlorinated
oxides are further discussed.

2.1 Oxidation Kinetics

The addition of a chlorine compound to the gas
atmosphere increases the overall oxidation rate in
relation to dry oxygen. The increased oxidation rate
in the presence of respectively HCl and Cl_2 is shown
in fig. 1. The solid lines in this figure are drawn
parallel to that for the standard dry oxide.

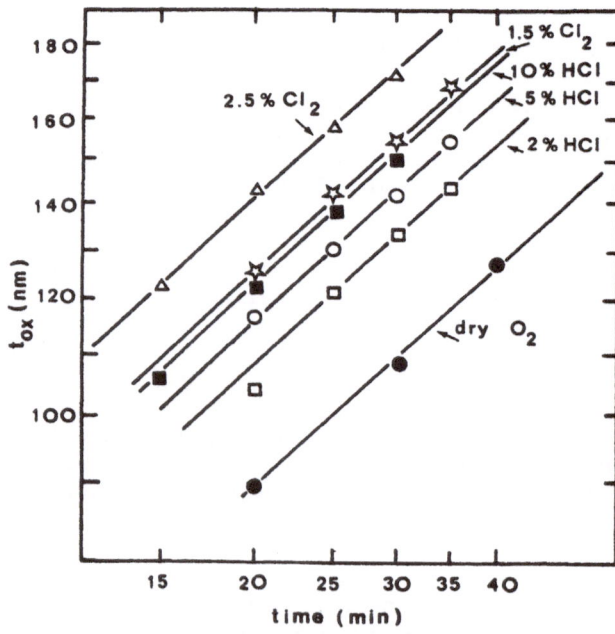

Fig. 1 : Oxidation rate for (100) Si at 1150°C in
the presence of HCl and Cl_2 [1].

As for the investigated thickness range the oxide growth is in the parabolic regime of the Deal-Grove oxidation rate equation [3], the rather good fit shows that the presence of HCl or Cl_2 increases the parabolic rate constant. Calculations also demonstrate that the formation of H_2O vapour in the oxidizing mixture is insufficient to account for the observed degree of enhancement of the parabolic rate constant, so that it is expected that chlorine is mainly responsible for the higher oxidation rate [1]. One should be careful, however, not to diminish the role of H_2O as Irene and Ghez [4] pointed out that only trace amounts of H2O have already an important influence on the parabolic rate constant.

The kinetics of HCl/O_2 oxides have been studied in detail by Hirabayashi and Iwamura [5]. Their experimental work leads to the conclusion that the considerably increase in the oxidation rate in the presence of HCl is caused by : i) an enhanced diffusion of O_2 and H_2O molecules in the oxide, due to the higher amount of non-bridging bonds, ii) the enhanced reaction at the $Si-SiO_2$ interface resulting from the chlorination reaction, and iii) the contribution of H_2O vapour formed during the oxidation reaction. These conclusions are in agreement with the experiments of Hess and Deal [6] who studied the oxidation kinetics in HCl/O_2 mixtures as a function of HCl concentration, oxidation temperature and substrate orientation. As is shown in fig. 2, the parabolic rate constant exhibits a linear increase with HCl concentration, while the linear rate constant saturates. This saturation effect has also been reported by Singh and Balk [7], who studied in detail the oxidation kinetics of TCE/O_2 and CCl_4/O_2 systems. The behaviour of the linear rate constant is governed by the catalytic action of the chlorine at the interface.

During an oxidation in a HCl/O_2 atmosphere, both H_2O and Cl_2 are formed as reaction products (see 2.2). Deal et al. [8] have studied the possible effects due to the presence of these reaction products by analyzing the oxidation kinetics of silicon in different H_2O/O_2, HCl/O_2 and Cl_2/O_2 mixtures. It is interesting to remark that they obtain almost similar results for 2% HCl + 2% Cl_2 in oxygen as with a 96 % O_2 + 4% HCl ambient, confirming the assumption that oxygen reacts with HCl to form Cl_2 and H_2O. At temperatures above 1000°C, the overall oxidation rates for the H_2O and HCl/O_2 systems

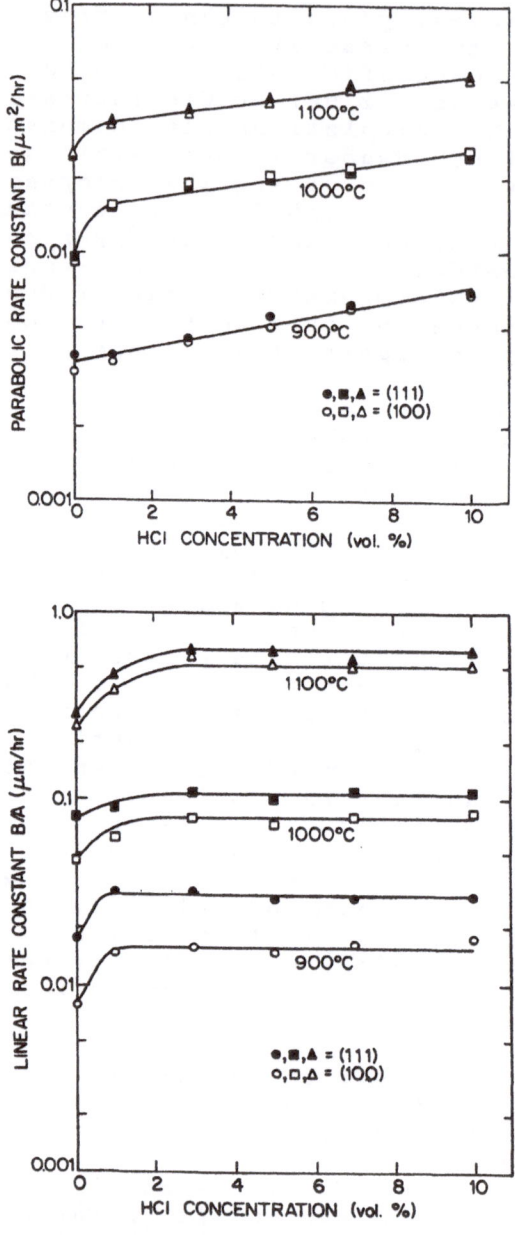

<u>Fig. 2</u> : The oxidation rate constants versus HCl
concentration for (111) and (100) n-Si at
respectively 900°, 1000° and 1100°C [6].

are the same. In this temperature range, the oxidation
rate strongly depends on the parabolic rate constant.
To obtain a clear insight into the mechanism of HCl/O_2
oxidations, however, a determination of the chlorine
distribution in the oxide layer and at the interface
for various oxidation conditions is required.

The above discussed behaviour of the rate constants
is completely different when a chlorine compound is
added to a wet O_2 oxidation system. Hattori [9] who
studied the oxidation kinetics for TCE/wet O_2 mixtures
noticed that i) the parabolic rate constant decreases
with increasing TCE concentrations, and ii) the linear
rate constant gradually increases with increasing TCE
concentrations, after a substantional addition of TCE.
As this behaviour of the rate constants is observed
also by Ehara et al. [10] for the addition of H_2 to
HCl/O_2 mixtures, it seems that the effect is not caused
by the H_2O itself, but by the H_2 gas. There exist,
however, no unambigious oxidation model which can
explain all these phenomena.

2.2 Chemical Analysis

The thermodynamic equilibrium between the oxygen
and the HCl gas added to the oxidizing atmosphere can
be written as

$$4 \ HCl + O_2 \ \rightleftharpoons \ 2 \ H_2O + 2 \ Cl_2 \qquad (1)$$

For other chlorine compounds like C_2HCl_3 (TCE) or
$C_2H_3Cl_3$ (C33) an overall combustion reaction must be
considered. For TCE this would result in

$$4 \ C_2HCl_3 + 9 \ O_2 \ \rightleftharpoons \ 2 \ H_2O + 6 \ Cl_2 + 8 \ CO_2 \qquad (2a)$$

$$2 \ H_2O + 2 \ Cl_2 \ \rightleftharpoons \ 4 \ HCl + O_2 \qquad (2b)$$

Although these reactions allow to calculate the gas
phase composition in the oxidizing ambient, no information is obtained about the concentration of the
reaction products in the oxide layer itself. The latter
would require accurate data on both the maximum solubility of the gases in the SiO_2 and their diffusivity
through the oxide layer.

A thermodynamic analysis of the gas composition in
the oxidizing atmosphere has first been reported by Van
der Meulen and Cahill [11]. By using a computer program
Tressler et al. [12] made a more extensive calculation

of the equilibrium gas composition within the furnace
under different oxidation conditions for a Cl-H-O
system. Their analysis made it possible to correlate
some of the electrical parameters of the oxide layer
with the partial pressure of Cl_2 and ClO in the gas
phase. This topic is further discussed in the next
section.
Janssens and Declerck [13] reported detailed results on
the concentration of the reaction products of HCl, H_2O
and Cl_2 as a function of the reaction temperature for
the pyrolysis of HCl, trichloroethyle (TCE), trichloroethane (C33) and dichloromethane (C22) respectively.
Their experimental results are shown in fig. 3.
It can be noticed that additives like C33 and C22 yield
the same relative concentrations of reaction products
as HCl. The reaction products of C33 with O_2 give an
oxidation atmosphere equivalent to the one obtained
when a three times higher concentration of HCl is used
(e.g. 0.1% C33 is thus equivalent to 0.3% HCl).
On the other hand, the use of TCE leads to almost
equivalent amounts of HCl and Cl_2 in the oxidizing gas
atmosphere. This can cause troubles at higher TCE
concentrations (> 1%) because of the corrosive character
of the Cl_2 gas [14]. Therefore it can be concluded that
whereas TCE has a "chlorine-like" behaviour, C33 is
equivalent to HCl. The use of C33 as an optimized
additive to improve the quality of the SiO_2 layers has
been studied in detail by Janssens [13,15].
A comparison between the electrical properties of the
different chlorinated oxides is given in the next
section.

The chlorine distribution in oxides grown in a
chlorine containing gas ambient has a strong influence
on the physical properties of the $Si-SiO_2$ system. The
incorporation of chlorine atoms in the SiO_2 layer has
been studied by means of a large variety of analytical
techniques such as nuclear backscattering of He^+ ions
[16-17], secondary ion mass spectroscopy (SIMS)
[18-20], Auger spectroscopy [14, 21], X-ray fluorescence
[17, 22-23], electron microprobe [17] and Rutherford
backscattering of alpha particles [24-25]. All the
different experimental profiles point out that most of
the chlorine piles-up in the oxide near the interface.
This is illustrated in fig. 4, showing chlorine
profiles of oxides grown in 5% HCl/O_2 at respectively
900°, 1000° and 1100°C. The oxidation times are
selected so that the oxide thickness is always in the
0.1 micron range.

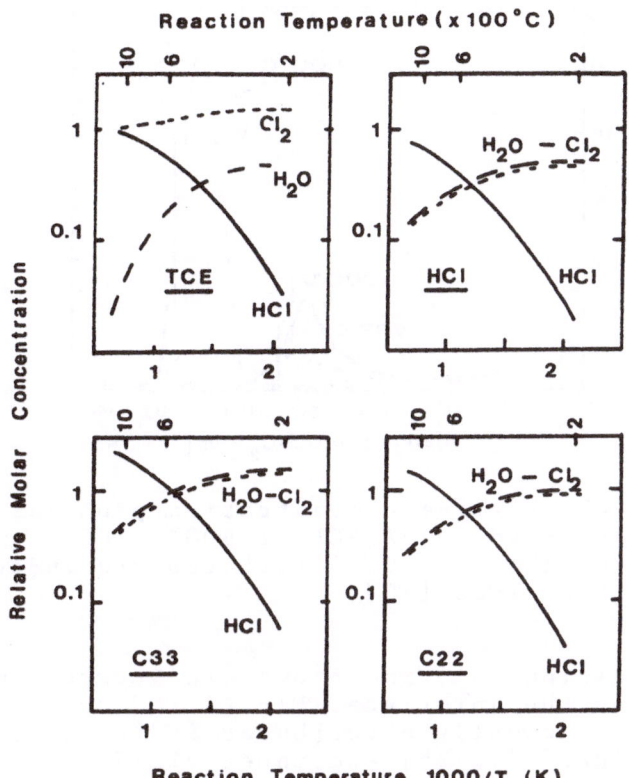

Fig. 3 : The concentration of the reaction products
HCl, H_2O and Cl_2 relative to the additive
concentration as a function of the reaction
temperature for the pyrolysis of TCE, HCl,
C33 and C22 respectively [13].

It can clearly be seen in fig. 4 that there is no
chlorine extend into the silicon, although one must
keep in mind that the depth resolution in the measurements has an uncertainty of about 5 nm. The interface
is determined from the oxide thickness and checked by
monitoring the SIMS profile of the oxygen concentration
in both the SiO_2 layer and the silicon. Frenzel and
Balk [20] have pointed out that a chlorine tail in the
silicon can be caused by an artifact of the experimental
approach.

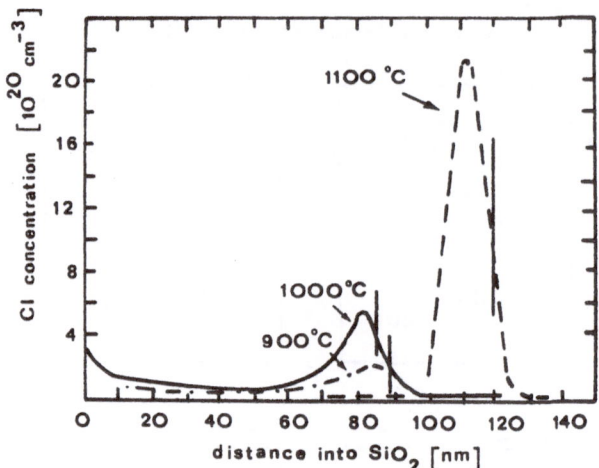

<u>Fig. 4</u> : Typical chlorine concentration profiles for
oxides grown in 5% HCl at 900°, 1000° and
1100°C. The Si-SiO$_2$ interfaces are indicated
by solid lines [19].

The majority of the chlorine atoms are located in a
thin layer near the interface. For Cl$_2$/O$_2$ oxides the
chlorine is more evently distributed in the oxide than
for HCl/O$_2$ oxides [17, 25] and under similar growth
conditions (time, temperature and additive content)
10 times as much chlorine is incorporated in Cl$_2$
oxides than into HCl oxides [17].
For oxides grown in a HCl/H$_2$O ambient, no real chlorine
incorporation in the oxide layer is noticed [19]. This
observation is in agreement with the results of Van
der Meulen et al. [17], which indicates that any Cl
build up in the oxide layer rapidly disappears by a
subsequent wet oxidation step.

Monkowski [22-23] analyzed the structure and the
composition of HCl oxides by means of TEM in conjunction with X-ray microanalysis and infrared spectroscopy.
He noticed that in the early stage of the oxidation
process a "classical" SiO$_2$ structure is formed, but
after a few minutes a second phase becomes visible.
This second phase appears as a great number of small
regions, several hundred angstroms in diameter, which
are homogeneously distributed over the entire interface
and contain most of the chlorine atoms. As the process

continues these regions, which are sandwiched between the oxide layer and the interface, grow and coalesce. For higher chlorine concentrations and/or higher temperatures an interfacial gas begins to form. Depending on the oxidation conditions, the gas may lift the oxide layer from the silicon so that bubbles are created. The growth of the second phase and the formation of the interfacial gas can be explained by the thermodynamic equilibria at the $Si-SiO_2$ interface, where two reactants, oxygen and chlorine are present. The activity gradients of these reactants are supposed to be different in the oxide and as a consequence the thermodynamic equilibrium at the interface changes during the oxidation process, allowing the formation of the chlorine-rich second phase which would not be stable under high oxygen partial pressure.

Rohatgi et al. [24] have studied the toal Cl content of various oxides as a function of oxidation time and temperature and HCl concentration. Figure 5 shows the chlorine content as a function of the chlorine concentration for various oxidation temperatures and for a fixed oxidation time. The break in each of the curves corresponds with the onset of the second phase formation. A similar set of curves is also obtained when the time parameter varies and the temperature is fixed. It should be remarked that at the very moment of complete coalescence of the second phase, complete sodium passivation is achieved.

Tsong et al. [25] used nuclear reaction and SIMS techniques to study the behaviour of both chlorine and hydrogen in oxides grown at $1100°C$ in different HCl/O_2 and Cl_2/O_2 ambients. They noticed a strong correlation between the two profiles. While the H and Cl appear to be enriched at the interface of HCl/O_2 oxides, they are higher in concentration and more evenly distributed in the oxide layer of Cl_2/O_2 oxides. This relation between the profiles seems to indicate that chlorine getters some of the hydrogen.

2.3 Electrical Properties

The use of chlorinated oxides has a direct impact on the densities of respectively fixed oxide charge, mobile ions and interface trapped charge, and under some circumstances results in radiation hardened SiO_2 layers. These oxides have a beneficial effect on the dielectric breakdown characteristics by reducing statistical variations and increasing the average

98

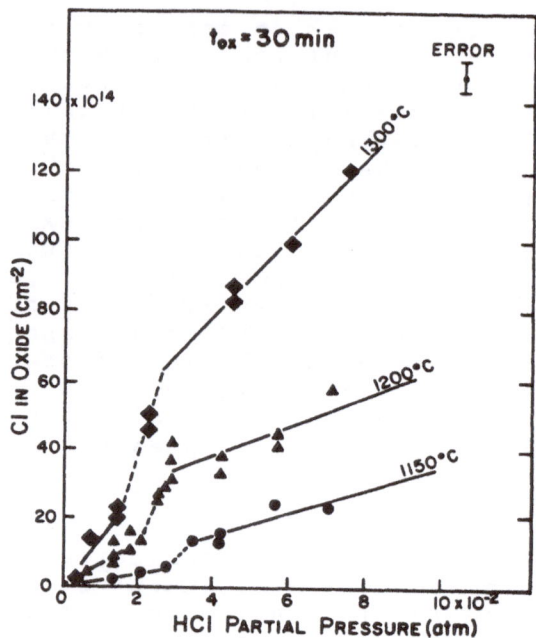

Fig. 5 : Cl content in oxide layers as a function of
         HCl partial pressure for 30 min oxidation at
         1150°, 1200° and 1300°C. The HCl concentration
         is obtained by dividing the pressure by 0.7000
         for 1150°C, 0.715 for 1200°C, and 0.749 for
         1300°C [24].

breakdown field. The minority carrier lifetime is
enhanced due to both the suppressed stacking fault
kinetics and the gettering effect of the chlorine atoms.

Kriegler et al. [1] first reported the trapping
and neutralization of mobile sodium ions by introducing
chlorine in the oxidizing ambient. They pointed out
that i) chlorine neutralizes the mobile ions, and ii)
the furnace tube can be cleaned with a prolonged
exposure to a chlorine containing atmosphere. These
two effects are generally called the "cleaning" and
"passivation" effect of a chlorine oxidation. The
cleaning effect strongly depends upon the environment
conditions such as temperature, HCl concentration and
the sodium concentration in the tube [14]. As the
effectiveness of this effect is controlled by the Cl_2
concentration in the atmosphere, the same level of
stability is obtained by using an eight times lower

TCE than HCl concentration [26-27]
The sodium neutralization effect is illustrated in
fig. 6, giving the sodium concentration as a function
of HCl concentration in the oxidizing atmosphere. It
can be seen that there exists a threshold concentration
below which almost no passiviation is observed.
Oxides grown on (111) oriented silicon require a
somewhat smaller amount of HCl to become passivated
than those grown on (100) silicon. The time and the
temperature of the oxidation have a strong influence
on the passivation effect. Very little improvement can
be observed by raising the temperature from 1150°C to
1200°C, and at 1050°C the passivation effect virtually
disappears [14]. A too high HCl concentration may lead
to a very rough substrate surface. It is believed that
the passivation is due to the trapping and neutralization of the sodium near the Si-SiO$_2$ interface. The
positive charge of the trapped ions is transferred to
the silicon. No passivation is observed, however, when
HCl is introduced in an inert atmosphere. A small
amount of oxygen is already sufficient to passivate the
structure. Kriegler [14] also pointed out that a small
amount of water vapour added to the HCl/O$_2$ mixture can
destroy the passivation effect. This is due to the fact
that no chlorine is incorporated into the oxide during
a wet chlorine oxidation.

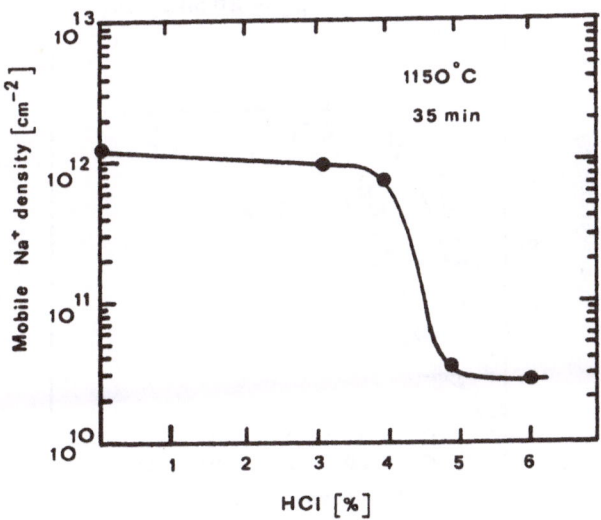

Fig. 6 : Effect of HCl on the mobile ionic charge
density for oxides grown at 1150°C [18].

100

Rohatgi et al. [28] demonstrated that the percentage of mobile sodium ions that is passivated is relatively independent of the total sodium content. The passivation factor P, which is defined as the fraction of the total mobile sodium concentration that is neutralized, can be written as [28]

$$P = \frac{N_{mo} - N_m}{N_{mo}} \qquad (3)$$

where N_{mo} is the number of ions per cm^3 transported to the $Si-SiO_2$ interface during a bias-temperature (B-T) stress and N_m is the number of ions that retain their charge after the B-T experiment. Recently, Rohatgi et al. [29] pointed out that P varies linearly with the HCl concentration in the oxide layer. This is shown in fig. 7 for oxides grown for 30 min at respectively 1150°, 1200° and 1300°C.

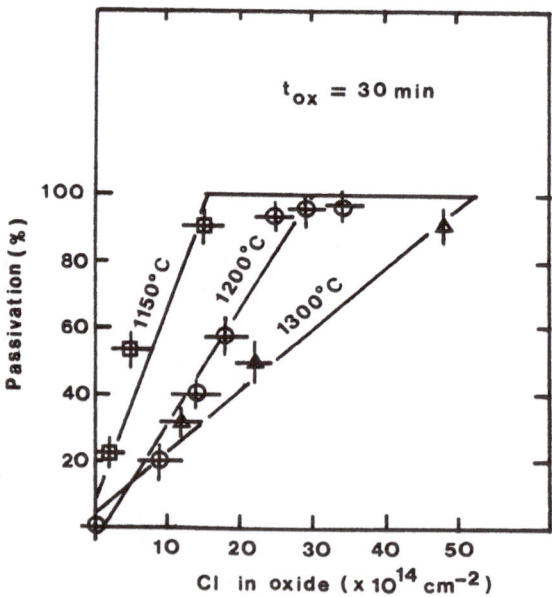

Fig. 7 : Percent passivation versus total oxide Cl concentration [29].

A similar linear relationship is observed for oxidations performed at a fixed temperature for different times. The slopes of the best-fit lines are a function of the oxidation time and temperature, but are almost not influenced by the concentration level of the mobile ions. As already discussed previously, Monkowski [22] has demonstrated that the Cl threshold concentration for sodium passivation closely correlates with the required concentration to initiate the second phase formation. This second phase coalesces and sodium passivation reaches completion at approximately the same oxidation conditions, indicating that the trapping and the neutralization of the mobile ions occurs in the Cl-rich second phase.

At very high electric fields (> 5 MV/cm) chlorine containing oxides exhibit important instabilities. For negative gate voltages, negative flat band shifts are mostly observed. These instabilities, which have been studied in detail by Van der Meulen et al. [17], are illustrated in fig. 8 for HCl and Cl_2 oxides. The magnitude of the stressing field and the chlorine concentration have a strong influence on the instability effects. For HCl and Cl_2 oxides, negative bias behaviour as a function of the oxidation conditions and the substrate parameters has been studied by Hess [30]. Similar but less pronounced instabilities than for Cl_2 oxides have also been reported for TCE [31] and for C33 oxides [32].

It can also be mentioned that the addition of a chlorine compound to the oxidizing ambient also influences the fixed oxide charge, the interface trapped charge and the oxide trapped charge. Very low interface trapped charge densities have been reported for the use of respectively HCl [34-35], TCE [36] and C33 [13, 33]. It is assumed that the reduction in interface trapped charge is due to the presence of active hydrogen or OH^- related groups, formed during the oxidation.

Osburn [37] investigated the dielectric breakdown properties of oxide films grown in mixtures of HCl/O_2 Cl_2/O_2, CCl_4/O_2, C_2HCl_3/O_2 and HBr/O_2. All these additives increase the average breakdown voltage and reduces the statistical data spread. He attributes the low-field breakdown to defects in the oxide layer. This means that the breakdown voltage can be enhanced by reducing the oxide-defect density. The defect density strongly depends on the used additive and on the halogen concentration in the gas atmosphere.

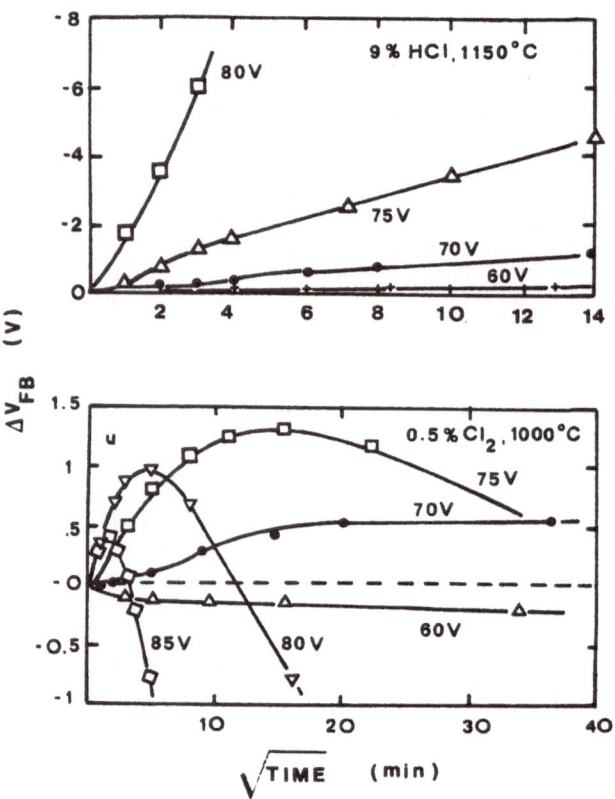

<u>Fig. 8</u> : Changes of flatband voltage shift with time
under negative bias for HCl and Cl$_2$ oxides [17].

The influence of the amount of halogen additive on the
breakdown defect density is shown in fig. 9 for various
chlorine additives. It can be seen that the different
curves are going through a minimum. It must be remarked,
however, that the lowest defect densities are obtained
at halogen concentrations of about half those required
to produce nonuniform oxide films. The defect density
increases with increasing temperature. Osburn [37]
attributes tne dielectric breakdown improvement to the
gettering or dislodging of electrically charged
impurities associated with oxide defects. Similar
curves as shown in fig. 9 are reported by Janssens and
Declerck [13] for respectively TCE and C33 oxides.

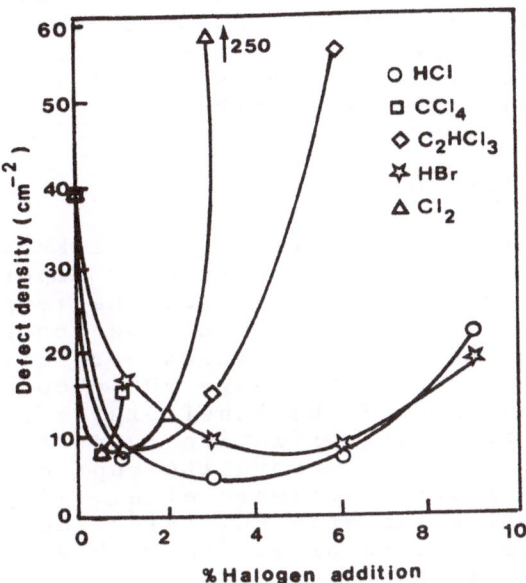

<u>Fig. 9</u> : Effect of halogen added during the oxidation
on the breakdown defect density [37]

Robinson and Heıman [2, 38] pointed out that the
addition of HCl to the gas atmosphere has a beneficial
effect on the minority carrier lifetime. They assume
that the increase in lifetime is due to the formation
of volatile metallic impurity chlorides. Young and
Osburn [39] experimentally observed that the advantageous effect of a chlorinated oxide appears to exist
even after the oxide is removed. Therefore several
investigators claim that the lifetime improvement is
due to a gettering of impurities. However, Green et
al. [40] found no marked effects on the metallic
content of deliberately contaminated wafers after an
annealing treatment in a mixture of HCl, Cl_2 and SiH_4.
As will be discussed in section 3, the addition of a
chlorine compound to the oxidizing ambient has a strong
impact on the behaviour of the oxidation-induced
stacking faults. By using the appropriate oxidation
conditions (time, temperature, chlorine concentration)
a complete suppression of the OSF's can be achieved.
This means that the lifetime improvement is due to a
dual mechanism consisting of the redistribution and
gettering of the metallic impurities, and the elimination

of oxidation-induced defects. At lower temperatures
(< 1050°C) the effect on the OSF's is very small and
the gettering of the impurities dominates, while at
higher temperatures the elimination of the OSF's is the
driving force for the redistribution and/or subsequent
removal of the impurities.

2.4 Impact on VLSI oxides

For VLSI circuits, gate oxide thicknesses in the
range 25-50 nm are becoming extremely important. To
grow these oxides one has to lower the temperature as
otherwise the oxidation time would be too short, making
the control of the oxide thickness very difficult.
Mostly temperatures in the range 600°-900°C are used.
However, to make use of the beneficial effect of a
chlorine oxidation, the oxidation temperature must be
higher than 1050°C. To improve the reproducibility of
reliable VLSI oxides, Hashimoto et al. [41] therefore
suggested to use a two step HCl oxidation technique.
First a low temperature oxidation with HCl is done,
so that the oxide thickness can be well controlled.
The oxide is then annealed at a higher temperature in
a mixture of nitrogen, HCl and oxygen in order to
activate the passivation effect of the chlorine. The
effectiveness of this treatment strongly depends on
respectively the temperature, the time and the HCl
concentration in the ambient.

To achieve the electrical performance of the VLSI
circuits, a tight control of the impurity behaviour
during the oxidation process is required. One has to
take into account that under oxidizing ambients,
i) impurities redistribute, ii) the diffusivity of
boron, phosphorus and arsenic is considerably enhanced,
and iii) the diffusivity of antimony is reduced. The
OED and ORD effects have recently been analyzed in
detail by Antoniadis and Moskowtz [42].
The diffusion enhancement, however, is reduced by the
addition of a chlorine compound to the oxidizing
atmosphere [43-45]. The reduction in OED depends on
both the oxidation temperature and the chlorine
concentration. The influence of C33 on the diffusivity
of boron is shown in fig. 10 for different temperatures.
The chlorination reaction at the interface reduces the
supersaturation of silicon interstitials by the
formation of Si-Cl bonds. It is important to remark
that the reduction of the diffusion enhancement is
closely correlated with the OSF-shrinkage observed
during chlorine oxidations.

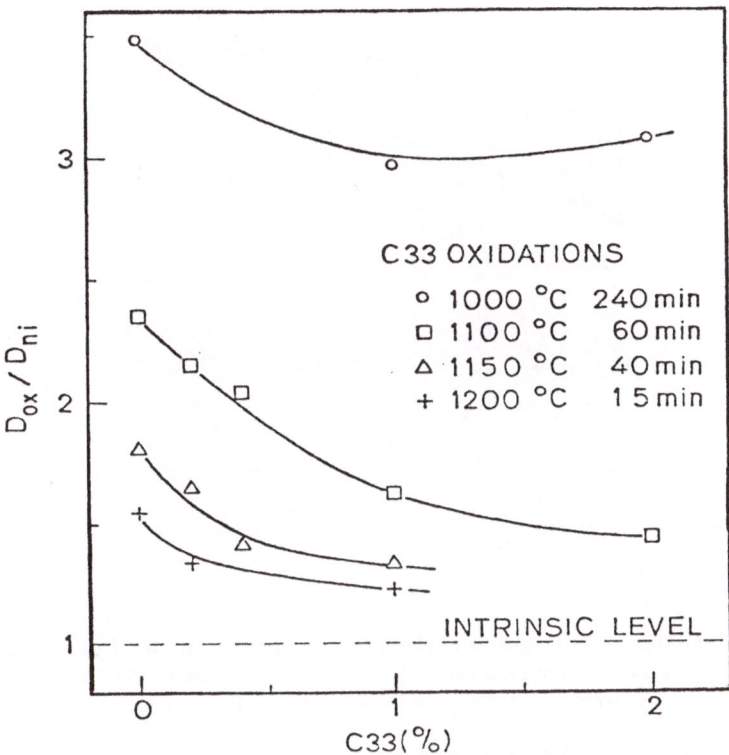

Fig. 10 : Oxidation enhanced diffusion of boron versus
          C33 concentration [45].

3 OXIDATION-INDUCED STACKING FAULTS

Thermal oxidation at high temperatures may result
in the generation of crystalline defects in the Si
substrate. Transmission electron microscopy pointed
out that most of these defects are extrinsic stacking
faults bounded by 1/3 (111) Frank partial dislocations
and lying in (111) planes. Because of the sessile
character of the dislocations, the growth of the OSF's
must occur by a climb mechanism, involving either the
emission of vacancies or the absorption of interstitials. The disturbance of the equilibrium concentration
of point defects is the driving force for the defect
behaviour.

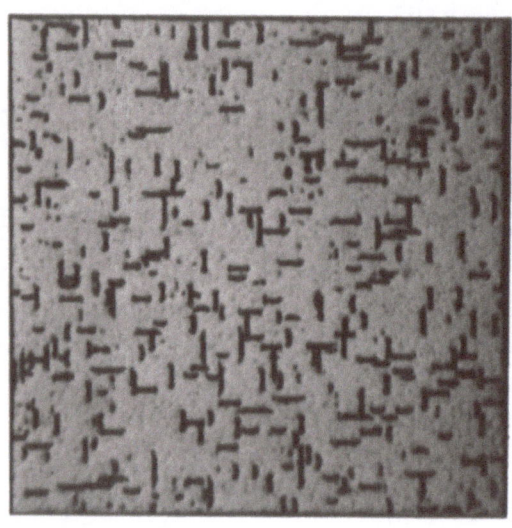

<u>Fig. 11</u> : Optical photomicrograph of a Wright etched
(100) oriented Si wafer after an oxidation
in wet oxygen at 1150°C for 150 min.

To reveal the presence of OSF's, one has to use
a preferential etching technique, such as Secco [46]
or Wright [47] etching, after stripping the oxide
layer. The OSF-length is defined as the distance
measured on the surface of the wafer between the two
ends of the OSF etch figure. The appearance of OSF's
on n-type (100) oriented silicon, after a 150 min
oxidation in wet oxygen at 1150°C, is illustrated in
fig. 11.

In recent years a strong relationship between
oxidation-induced stacking faults and yield hazards in
both bipolar and MOS integrated circuits has been
reported [48-58]. The complexity of the process induced
defects depends on the amount and the sequence of the
high temperature steps. Their harmful effect enhances
with increasing complexity and decreasing feature
sizes of the circuits. In bipolar devices they may
cause diffusion pipes between emitter and collector
regions and increased junction leakage [50-52]. For
MOS devices, their electrical activity is strongly
linked to the degree of decoration by impurity

precipitates. Reduced relaxation times of MOS capaci-
tors, dark current non-uniformities of Charge-Coupled
Devices and refresh failures of dynamic MOS memories
have been reported.

Dependent on the source of the OSF-nuclei and
the location of the OSF, one can differentiate between
surface (OSF_S), bulk (OSF_B) and process-induced (OSF_I)
defects (e.g. a pre-oxidation ion implantation step).
It is very important to make a clear distinction
between respectively the nucleation and the growth of
the stacking faults. The OSF-nucleation can be due to
residual surface damage, surface contamination, surface
reactions with hydrofluoric acid, and grown-in defects
such as swirl-defects and oxygen precipitates [59-66].
The density and the length of the stacking faults
strongly depend on both the characteristics of the
starting material (e.g. doping level, orientation,
finishing degree of the surface, crystal pulling
technique, oxygen and carbon content, thermal history
of the wafer ...) and the oxidation conditions. The
important role played by oxygen and carbon is further
discussed in paragraph 5. Without going into details,
some general features of the influence of different
oxidation processes on the OSF-kinetics are discussed
briefly.

3.1 Dry and Wet Oxidations

During a dry or a wet oxidation the OSF's grow
as a function of oxidation time, excepted at very
high temperatures where a shrinkage of the OSF's is
observed. The existence of a "growth" and "retrogrowth"
region in the OSF behaviour was first reported by Hu
[67] in 1975. The influence of both the oxidation
temperature and time on the OSF-length is illustrated
in fig. 12 for a dry and a wet oxidation of n-type,
6-8 ohm-cm (100) CZ silicon. The growth of the OSF's
is characterized by an activation energy of 2.2 eV
and has a time law exponent of 0.84. These values are
in good agreement with most of these reported in the
literature. The onset of the retrogrowth depends on
the oxidation time, temperature and gas ambient. By
reducing the partial pressure of the oxygen, both the
OSF-length and the temperature where retrogrowth
begins decrease [68]. The different models which have
been proposed in order to explain this OSF-behaviour
are discussed in section 3.3.

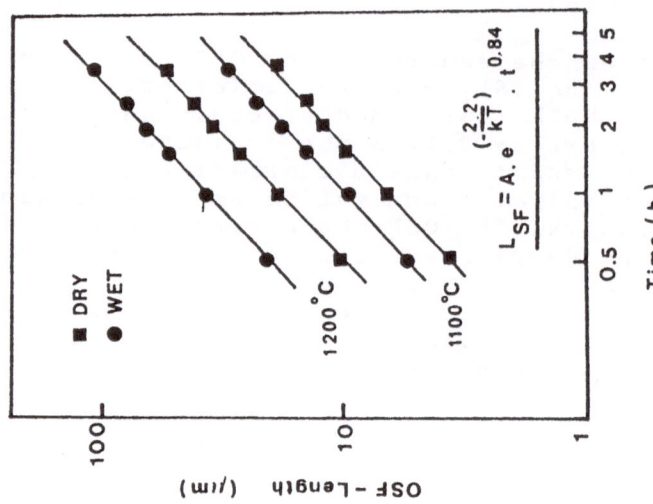

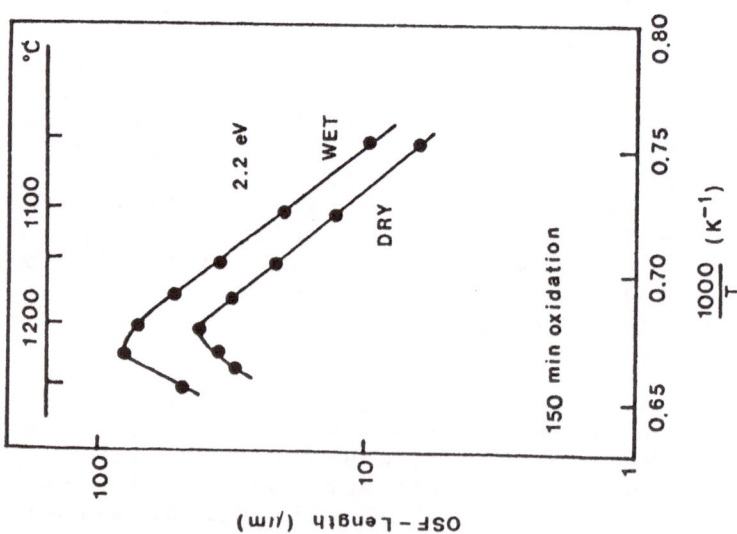

Fig. 12 : Temperature and time dependence of the OSF-growth during an oxidation in dry and wet oxygen of 6-8 ohm-cm n-type (100) oriented Czochralski silicon [57].

3.2 Chlorinated Oxides

The addition of small amounts of chlorine to the oxidizing atmosphere has a strong influence on the OSF behaviour. For oxidations in trichloroethane (C33) the influence of respectively the oxidation temperature and oxidation time is shown in fig. 13 for a 0.2% C33 oxidation of n-type (100) oriented Czochralski Si. It can be seen that the initial growth rate of the OSF's increases with increasing temperature. This effect is also observed for other C33 concentrations [57]. For temperatures higher than 1100°C the curves exhibit a growth and a shrinkage region. The maximum OSF-length and the time required to reach this value strongly depend on the C33 concentration. With increasing concentrations both the maximum length and the oxidation time become shorter. It can be noticed in fig. 13 that dependent on the oxidation conditions, it is possible to eliminate the OSF's completely. As will be discussed

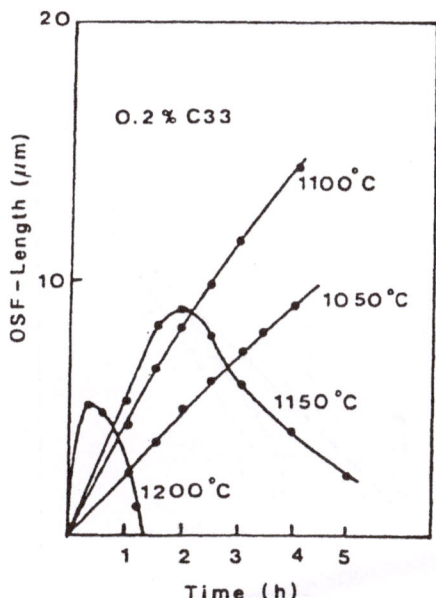

Fig. 13 : The OSF-length versus time for a 0.2% C33 oxidation at respectively 1050°, 1100°, 1150° and 1200°C [57].

110

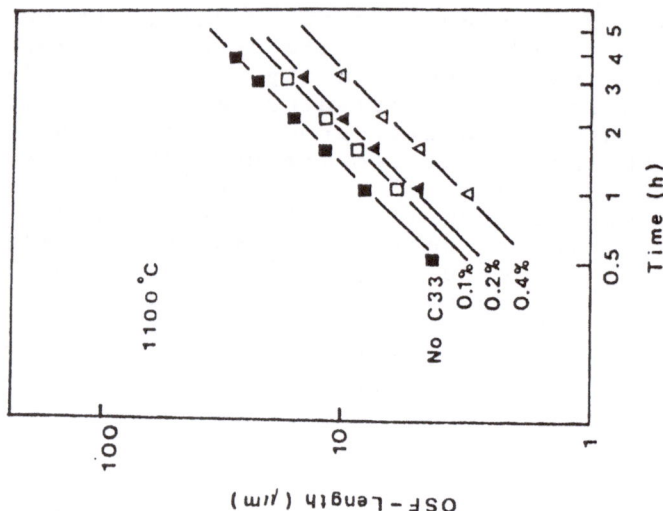

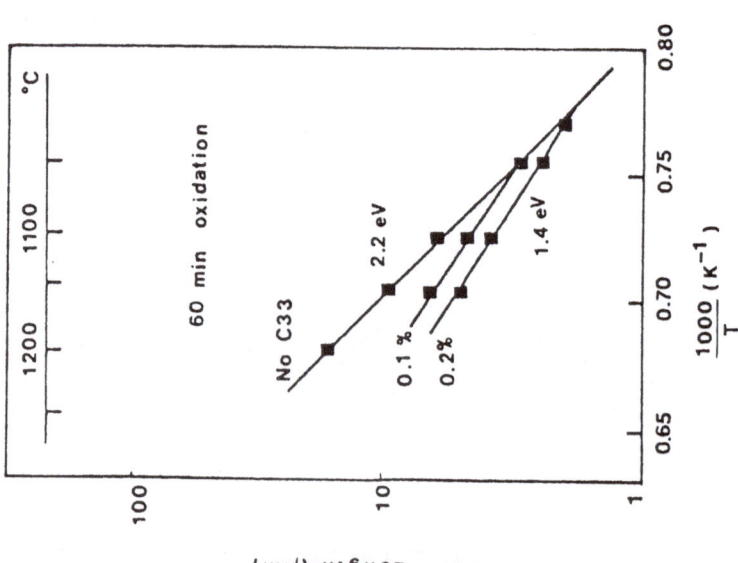

Fig. 14 : Temperature and time dependence of the OSF-length during a C33 oxidation of 6-8 ohm-cm n-type (100) CZ silicon [57].

later, the OSF behaviour in the near-surface region
of the wafer may be completely different from that
deeper into the bulk.

To obtain a better insight into the OSF kinetics
during a chlorine oxidation, it is useful to analyse
the influence of the temperature and the time parameter explicit. This is done in fig. 14, taking into
account the chlorine concentration as an additional
variable. As can be seen, for low oxidation temperatures the activation energy of 2.2 eV is the same as
for dry and wet oxidations. At higher temperatures the
activation energy decreases to 1.4 eV. This lowering
of the activation energy by the addition of C33 to the
oxidizing ambient is related to the chlorination
reaction at the Si-SiO$_2$ interface.
The time dependence of the OSF-length in the growth
region follows the same time law as for dry or wet
oxidations (see fig. 12). At higher temperatures and
or higher chlorine concentrations, the shrinkage effect
of the OSF's becomes dominant and the time law no
longer holds. Although the shrinkage rate of the OSF's
strongly depends on the gas atmosphere, its activation
energy is constant and equal to 4.8 eV [69]. This
value is close to the activation energy of the silicon
self-diffusion process.

The above reported observations for C33 oxides
are in very good agreement with those reported by
Shiraki [70-71] and Hattori [72]for respectively HCl
and TCE oxides. It should not be surprising that the
OSF behaviour during HCl and C33 oxidations is similar
as 0.3% HCl in the gas atmosphere is equivalent to
0.1% C33 in the atmosphere. This topic has been
discussed in section 2.2.

3.3 Stacking Fault Models

TEM analysis of Si wafers oxidized for short
times point out that the OSF-growth can be described
by a Bardeen Herring mechanism [73-74]. The driving
force for this mechanism is the nonequilibrium concentration of silicon interstitials. In general, the
concentration of interstitials is controlled by :
    i) the incompleteness of the oxidation process
       resulting in an excess of interstitials [67],
   ii) the chlorination reaction at the interface
       where the formation of Si-Cl bonds leads to
       an undersaturation of interstitials, and
 iii) the interaction with impurity atoms when the

diffusion process occurs via a dual vacancy
interstitial mechanism [75].
For bulk stacking faults the generation of interstitials
due to the precipitation of interstitial oxygen must
also be taken into account. The relative importance
of these different components will determine the
growth and/or shrinkage behaviour of the defects.

In the literature several models have been
proposed to describe the role of point defects on the
OSF behaviour quantitatively,   king into account a
large variety of parameters such as the dependence of
respectively the time, temperature, oxygen partial
pressure, gas ambient (e.g. chlorine or water related
species), substrate orientation, pressure (e.g. high
pressure oxidations) and retrogrowth phenomena.[76-80].
Although most of these models fit the experimental
data rather good, they are based on empirical relations
and not on a microscopical analysis of the defect
behaviour. The latter becomes very complex due to the
interaction between the different parameters and the
strong influence of the not always well defined
material parameters. As already mentioned before, there
exists a striking correspondence between the behaviour
of the OSF's and the OED and ORD phenomena. This
relationship, which has recently been reviewed by
Gösele and Frank [81], indicates that the same point
defects are controlling both the defects and diffusion
mechanism. In the literature it is not unambigiously
determined which are the predominant intrinsic point
defects in silicon, vacancies or interstitials (see
e.g. ref. 75 and 81).

It should also be remarked that most of the
reports available in the literature are dealing with
the defect behaviour only in the near-surface regions
of the wafer. The defect behaviour deeper into the
bulk, however, can be completely different [74, 82].
The depth distribution of the OSF-length after a 0.2%
C33 oxidation at 1150°C is shown in fig. 15 for
different oxidation times. It can be noticed that
dependent on the oxidation conditions, the OSF's shrink
near the surface while they continue to grow deeper
into the bulk of the wafer. For sufficient long times,
the complete distribution curve shifts downwards.
A similar behaviour of the distribution curve is seen
when the C33 concentration is taken as a parameter.
This is illustrated in fig. 16 for a 120 min oxidation
at 1200°C. It can be noticed that for a dry oxidation
without chlorine, OSF's appear throughout the whole

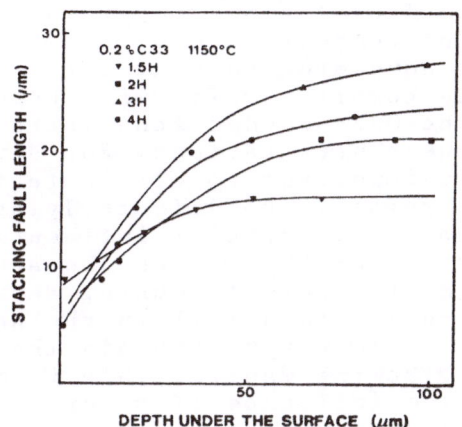

Fig. 15 : OSF-length versus depth under the surface
after a 0.2% C33 oxidation at 1150°C for
different times [82]. These curves are
obtained with p-type, (111) silicon.

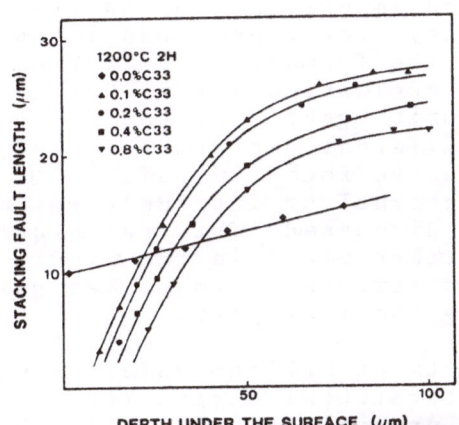

Fig. 16 : Influence of the C33 concentration on the
depth distribution of the OSF-length after
a 120 min oxidation at 1200°C [82]. These
curves are for p-type, (111) silicon.

wafer thickness. The addition of chlorine to the
oxidizing ambient results in an OSF shrinkage near the
surface while the faults grow deeper in the material.
For higher chlorine concentrations the whole distribution curve shifts downwards so that for very high
concentrations the complete wafer will become OSF free.
This means that the OSF-denuded zone increases with
increasing chlorine concentrations. For device applications it is very important to eliminate the defects
within the active region. The defects lying outside
this region can have a beneficial influence on the
electrical properties of the devices because of the
gettering effect of the defects during device processing.
This topic is discussed in detail in the next section.
No models are available yet to explain the bulk
behaviour of the stacking faults. This is mainly due
to the fact that the influence of material parameters
(e.g. interstitial oxygen, carbon, swirl-defects ...)
on the stacking fault nucleation is not well understood.
There also exists some discrepancy between the results
reported in the literature.

4 INTRINSIC GETTERING

In general, gettering is aimed at either the
removal of unwanted impurities to inactive parts of
the wafer where they are trapped and immobilized or
the prevention of the formation of lattice defects by
eliminating their nucleation sites. One can differentiate between impurity gettering and defect gettering.
A review of the different gettering techniques for
silicon devices can be found in ref. [83]. In this
report only the internal or intrinsic gettering
technique will be discussed. This technique, where the
gettering action takes place in the middle of the wafer
and no in the near surface regions, has gained a lot
of interest during the last years.

Heat treatments of silicon wafers containing a
high amount of interstitial oxygen result in the
formation of SiO_x precipitates [84-85]. These platelike
precipitates which are square shaped with <110> sides
and (100) habit planes, may be associated with prismatic dislocations generated at the $Si-SiO_x$ interface
by a prismatic punching mechanism [86]. Tice and Tan
[87] pointed out that the SiO_x-precipitate dislocation
complexes (SiO_x-PDC) are very effective sinks for
metallic impurities. The intrinsic gettering technique
is based on the reduction of these (SiO_x-PDC) in the

surface region of the wafer by an appropriate anneal step [88]. In that case the gettering will take place in the middle of the wafer, leaving the active device regions defect-free. Sometimes the anneal step is already a part of the standard processing as e.g. in the case of a subcollector diffusion used to fabricate bipolar devices. For most VLSI processes, where one wants to keep the temperature as low as possible, the intrinsic gettering anneal is performed in the beginning of the process.

Oxygen precipitates and SiO_x-PDC have also been correlated with nucleation sites for OSF's. As already pointed out in paragraph 3, avoiding oxygen precipitation results in a reduction of the OSF nucleation sites. Kishino et al. [89] use a high temperature (1000°-1100°C) nitrogen preanneal to suppress the OSF nucleation. However, as a severe N_2 anneal of bare silicon may result in a nitridation reaction at the interface, Hattori and Suziki [90] suggest to perform the pre-oxidation anneal in a $N_2/HC1/O_2$ mixture.

During the temperature treatment of the wafers, there exists a competition between the rate at which the precipitates are forming and becoming stable near the surface, and the rate at which the interstitial oxygen diffuses out of the wafer. This means that in order to form a denuded surface region the oxygen precipitation process should be suppressed in the beginning of the device processing. Craven [90], who recently reviewed internal gettering in silicon, pointed out that the denuded zone formation requires temperatures above 950°C and that the width of the denuded region increases with anneal time. A second temperature treatment in the range 650°-900°C results in an enhanced oxygen precipitation outside the denuded region. The optimum conditons for both the first and second anneal step strongly depends on the thermal history and the oxygen content of the starting material [91]. As sometimes the thermal history of the material is unknown, Tsuya et al. [92] suggest to perform a multiplestep annealing starting with an anneal at a temperature equal to or higher than 1200°C. This high temperature step resolves the precipitates which are already present in the wafer.

5 OXYGEN IN SILICON

The advanced wafer processing technology has reached a state where material inhomogenities such as doping striations and nonuniform distribution of impurities, point defect complexes and microdefects are the major yield limiting defects. Among the impurities in silicon, oxygen and carbon are playing an very important role as their incorporation into the silicon lattice is inherently related to the crystal growth technique. A review of oxygen and carbon in silicon is given in ref. [93]. This report only discusses briefly the impact of oxygen on the processing and device characteristics.

5.1 Resistivity Variations

More than two decades ago, Fuller et al. [94] already reported that a thermal anneal in the temperature range 300°-500°C of oxygen containing Si crystals results in the formation of thermal donors (TD). The generation rate of these donors strongly depends on respectively the oxygen concentration in the starting material, the anneal time and the anneal temperature. The TD generation is highest for an anneal around 450°C and can be completely annihilated by a temperature treatment above 500°C [95]. A discussion of the different models which have been proposed in order to explain the TD formation is out of the scope of this report.

Annealing at 650°C is very often used to annihilate the thermal donors which are generated duirng the cooling of the crystals. It has been reported, however, that a prolonged anneal at 650°C again leads to an increase in the donor concentration [96-97]. To avoid confusion with the donors generated at 450°C, these donors are called "new donors" (ND). Their generation rate depends on respecteively the material characteristics (oxygen and carbon content, doping type), the anneal conditions and the thermal history of the wafers [97]. Although carbon is not a part of the ND's, it seems to act as a catalyzer in the reaction for the ND formation.

Due to the generation of thermal and/or new donors the resistivity of the wafer changes during the processing. In general, the silicon wafer manufacturer implements some "soak treatments" during the crystal growth process in order to stabilize the donor formation. However, high temperature treatments and rapid cooling of the wafers may reactivate some of the donor properties. These high temperature steps are used for e.g. gettering purposes or subcollector and p-well drive-in the bipolar and CMOS process respectively. The control of the resistivity variations is extremely important for power devices (current effects) and VLSI circuits.

5.2 Dislocations and Crystal Defects

During the processing of the devices, dislocations can be generated due to respectively thermal stresses, high concentration diffusions, stresses at the interface (e.g. SiO_2 or Si_3N_4 layers) and local damage (e.g. scratches due to wafer handling). Hu [98] pointed out that the mobility of the dislocations strongly depends on the oxygen concentration. By using the method of indentation dislocation rosette (IDR) to generate dislocations in the structure, he demonstrated the dislocation pinning effect of the oxygen atoms. This effect disappears for oxygen concentrations lower than about $1E17$ cm^{-3} which means that oxygen-free silicon is not well suited for complicated IC processing as thermal stresses are almost unavoidable. It should be remarked that the dislocation pinning effect is only active when the oxygen is dissolved in the lattice. In the case that oxygen precipitates in the lattice, dislocation multiplication may occur which enhances the thermal warpage of the wafer.

As already mentioned before, the amount of oxygen in the wafer also determines the nucleation of the oxidation-induced stacking faults. These OSF's can be avoided by optimizing the processing outline and/or by implementing some gettering steps. The use of intrinsic gettering techniques, based on the oxygen precipitation in the middle of the wafer, was discussed in section 4.

5.3 Wafer Warpage

The effect of oxygen on the thermal warpage of the wafers strongly depends on the state in which the oxygen atoms are present in the silicon lattice. When oxygen is dissolved in the lattice, the warpage is smaller for wafers with a high oxygen concentration due to the hardening effect of the oxygen against plastic deformation [99]. The wafer warpage versus the dissolved oxygen content of 75 mm wafers which are heated cyclically at 1150°C for 30 min in dry N_2

118

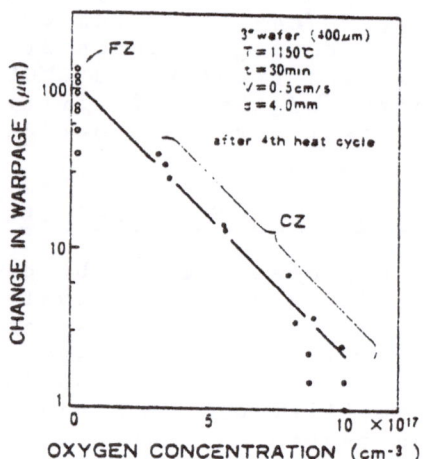

Fig. 17 : Wafer warpage versus dissolved oxygen
concentration of 75 mm wafers heated
cyclically at 1150°C for 30 min [100].

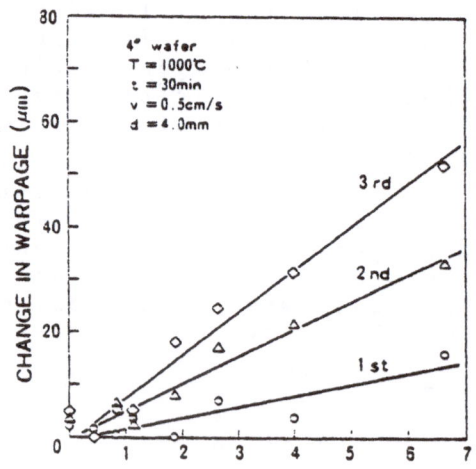

Fig. 18 : Influence of precipitated oxygen on the
wafer warpage of 100 mm wafers heated
cyclically at 1000°C for 30 min [100].

is shown in fig. 17, after Yoshihiro et al. [100].

When oxygen precipitates are formed, however, the warpage increases with increasing oxygen concentration. This is illustrated in fig. 18. A well controlled amount of oxygen-precipitates is generated by a pre-annealing at 1050°C before applying the heat cycles (30 min at 1000°C in dry nitrogen). The warpage behaviour is caused by the oxygen precipitates which act as nuclei for the dislocation multiplication due to the thermal stresses occurring during the heat treatment. The influence of several parameters such as the form and the amount of precipitated oxygen, the temperature, the temperature gradient, the direction of the initial bow of the wafers, and the thickness over diameter ratio has been investigated by Leroy and Plougonven [101]. They noticed that the higher the concentration of precipitates, the lower the critical temperature for wafer warpage.

Wafer warpage is very important for VLSI circuits as it has a direct influence on the yield of the photolithographic processes (e.g. open or shorts between lines, holes in the masking layer ...) The smaller the linewidth one wants to print, the more important the wafer flatness becomes.

6 CONCLUSION

The required gate thickness for VLSI processes limits the use of chlorinated oxides. This is mainly caused by the fact that one wants to lower the temperature in order to increase the reproducibility. To be able to make use of the beneficial effect of chlorinated oxides, some special techniques must be used.

Intrinsic gettering is becoming more and more important. This technique however, requires silicon material with a rather high oxygen content so that one also has to take into account the influence of the oxygen on respectively the substrate resistivity (thermal donors) and the wafer warpage. In general, one has to consider the tradeoff between gettering efficiency, resistivity and warpage. For each processing outline, the optimum oxygen range must be determined. This requires a good understanding of both the impact of oxygen on the process and device characteristics, and the oxygen behaviour during the heat treatments.

REFERENCES

1. Kriegler R.J., Cheng Y.C. and Colton D.R., J. Electrochem. Soc., 119 (1972) 388
2. Robinson P.H. and Heiman F.P., J. Electrochem. Soc., 118 (1971) 141
3. Deal B.E. and Grove A.S., J. Appl. Phys., 36 (1965) 3770
4. Irene E.A. and Ghez R., J. Electrochem. Soc., 124 (1977) 1757
5. Hirabayashi K. and Iwamura J., J. Electrochem. Soc., 120 (1973) 1595
6. Hess D.W. and Deal B.E., J. Electrochem. Soc., 124 (1977) 735
7. Singh B.R. and Balk P., J. Electrochem. Soc., 126 (1979) 1288
8. Deal B.E., Hess D.W., Plummer J.D. and HO C.P., J. Electrochem. Soc., 125 (1978) 339
9. Hattori T., J. Electrochem. Soc., 126 (1979) 1789
10. Ehara K., Sakuma K. and Ohwada K., J. Electrochem. Soc., 126 (1979) 2249
11. Van der Meulen Y.J. and Cahill J.G., J. Electron. Mat., 3 (1974) 371
12. Tressler R.E., Stach J. and Metz D.M., J. Electrochem. Soc., 124 (1977) 607
13. Janssens E.J. and Declerck G.J., J. Electrochem. Soc., 125 (1978) 1696
14. Kriegler R.J., in Semiconductor Silicon 1973, Huff H.R. and Burgess R.R. (Eds), The Electrochem. Soc. Inc., Princeton, 1973, 363
15. Janssens E.J., PH.D dissertation, Kath. Univer. Leuven, 1978
16. Meek R.L., J. Electrochem. Soc., 120 (1973) 308
17. Van der Meulen Y.J., Osburn C.M. and Ziegler J.F., J. Electrochem. Soc., 122 (1975) 284
18. Kriegler R.J., Aitken A. and Morris J.D., Jpn. J. Appl. Phys., 43 (1974) 341
19. Deal B.E., Hurrle A. and Schulz M.J., J. Electrochem. Soc., 125 (1978) 2024
20. Frenzel H. and Balk P., J. Vac. Sci. Techn., 16 (1979) 1454
21. Chou N.J., Osburn C.M., Van der Meulen Y.J. and Hammer R., Appl. Phys. Lett., 22 (1973) 280
22. Monkowski J.R., Ph.D dissertation, Pennsylvania State Univ., 1978
23. Monkowski J.R., Tressler R.E. and Stach J., J. Electrochem. Soc., 125 (1978) 1867
24. Rohatgi A., Butler S.R., Feigl F.J., Kraner H.N. and Jones K.W., J. Electrochem. Soc., 126 (1979) 143

25. Tsong I.S.T., Monkowski M.D., Monkowski J.R., Winterberg A.L., Miller P.D. and Moak C.D., to be published
26. Janssens E.J. and Declerck G.J., ESSDERC, Grenoble, Sept. 8-12, 1975
27. Singh B.R. and Balk P., J. Electrochem. Soc., 125 (1978) 453
28. Rohatgi A., Butler S.R., Feigl F.J., Kraner H.N. and Jones K.w., Appl. Phys. Lett., 30 (1977) 104
29. Rohatgi A., Butler S.R. and Feigl F.J., J. Electrochem. Soc., 126 (1979) 149
30. Hess D.W., J. Electrochem. Soc., 124 (1977) 740
31. Heald D.L., Das R.M. and Khosla R.P., J. Electrochem. Soc., 123 (1976) 302
32. Chen M.C. and Hile J.W., J. Electrochem. Soc., 119 (1972) 233
33. Linssen A.J. and Peek H.L., Philips J. Res., 33 (1978) 281
34. Severi M. and Soncini G., Electron. Lett., 8 (1972) 402
35. Baccarani G., Severi M. and Soncini G., J. Electrochem. Soc., 120 (1973) 1436
36. Declerck G.J., Hattori T., May G.A., Beaudouin J. and Meindl J.D., J. Electrochem. Soc., 122 (1975) 436
37. Osburn C.M., J. Electrochem. Soc., 121 (1974) 809
38. Ronen R.S. and Robinson P.H., J. Electrochem. Soc., 119 (1972) 747
39. Young D.R. and Osburn C.M., J. Electrochem. Soc., 120 (1973) 1578
40. Green J.M., Osburn C.M. and Sedgwick T.O., J. Electron. Mat., 3 (1974) 579
41. Hashimoto C., Muramoto S., Shiono N. and Nakajima O., J. Electrochem. Soc., 127 (1980) 129
42. Antoniadis D.A. and Moskowitz I., to be published
43. Hill C., in Semiconductor Silicon 1981, Huff H.R., Kriegler R.J. and Takeishi Y. (Eds), The Electrochem. Soc. Inc., Pennington, 1981, 988
44. Nabeta Y., Uno T., Kubo S. and Tsukamoto H., J. Electrochem. Soc., 123 (1976) 1416
45. Lin A.M. and Claeys C.L., unpublished
46. Secco d'Aragona F., J. Electrochem. Soc., 118 (1972) 948
47. Jenkins-Wright M., J. Electrochem. Soc., 124 (1977) 757
48. Ravı K.V., Varker C.J. and Volk C.E., J. Electrochem. Soc., 120 (1973) 533
49. Katz L.E., J. Electrochem. Soc., 121 (1974) 969
50. Kato T., Koyama H., Matsukawa T. and Fujikawa K., Solid-State Electron., 19 (1976) 955

122

51. Ashburn P., Bull C., Nicholas K.N. and Booker G.R., Solid-State Electron., 20 (1977) 735
52. Tanikawa K., Ito Y. and Sei H., Appl. Phys. Lett., 18 (1976) 285
53. Ogden R. and Wilkinson J.M., J. Appl. Phys., 48 (1977) 412
54. Jastrzebski L., Levine P.A., Fisher W.A., Cope A.D., Savoye D. and Henry W.N., J. Electrochem. Soc., 128 (1981) 885
55. Kolbesen B.O. and Strunk H., Inst. Phys. Conf. Ser., 57 (1980) 21
56. Sutherland R.R., Solid-State Electron., 25 (1982) 15
57. Claeys C.L., Laes E.E., Declerck G.J. and Van Overstraeten R.J., in Semiconductor Silicon 1977, Huff H.R. and Sirtl E., The Electrochem. Soc. Inc., Princeton, 1977, 773
58. Murarka S.P., Seidel T.E., Dalton J.V., Dishman J.M. and Mead M.H., J. Electrochem. Soc., 127 (1980) 716
59. Fisher A.W. and Amick J.A., J. Electrochem. Soc., 113 (1966) 1054
60. Pomerantz, J. Electrochem. Soc., 119 (1972) 255
61. Drum C.M. and Van Gelder W., J. Appl. Phys., 43 (1972) 4465
62. Ravi K.V. and Varker C.J., J. Appl. Phys., 45 (1974) 263
63. Takaoka H., Oosaka J. and Inoue N., Jpn. J. Appl. Phys., 18 (1979) 179
64. Chang Y.N. and Demer L.J., in Semiconductor Characterization Techniques, Barnes P.A. and Rozgonyi G.A. (Eds), The Electrochem. Soc. Inc., Princeton, 1978, 324
65. Coppus G.M., Shevlin C.M. and Demer L.J., in ref. 64, 1978, 423
66. Hu S.M., J. Vac. Sci. Techn., 14 (1977) 17
67. Hu S.M., J. Appl. Phys., 45 (1974) 1567
68. Murarka S.P., J. Appl. Phys., 49 (1978) 2513
69. Claeys C.L., Declerck G.J. and Van Overstraeten R.J., Appl. Phys. Lett., 35 (1979) 797
70. Shiraki H., Jpn. J. Appl. Phys., 15 (1976) 1
71. Shiraki H., in Semiconductor Silicon 1977, Huff H.R. and Sirtl E. (Eds), The Electrochem. Soc. Inc. Princeton, 1977, 546
72. Hattori, Denki Kagaku, 2 (1978) 122
73. Wada K., Takaoka H., Inoue N. and Kohra K., Jpn. J. Appl. Phys., 18 (1979) 1629
74. Bender H., Van Landuyt J., Amelinckx S., Claeys C., Declerck G. and Van Overstraeten R., Inst. Phys. 60 (1981) 313

75. Antoniadis D.A., in Semiconductor Silicon 1981, Huff H.R., Kriegler R.J. and Takeishi Y. (Eds), The Electrochem. Soc. Inc., Pennington, 1981, 947

76. Lin A.M., Dutton R.W., Antoniadis D.A. and Tiller W.A., J. Electrochem. Soc., 128 (1981) 1121

77. Murarka S.P., Phys. Rev. B, 16 (1977) 2849

78. Leroy B., J. Appl. Phys., 50 (1979) 7996

79. Murarka S.P., Phys. Rev. B, 21 (1980) 692

80. Hu S.M., in Narayan J. and Tan T.Y. (Eds), Defects in Semiconductors, North Holland, 1981, 333

81. Gösele U. and Frank W., ibid, 1981, 55

82. Claeys C., Declerck G., Van Overstraeten R., Bender H., Van Landuyt J. and Amelickx S., in Semiconductor Silicon 1981, Huff H.R., Kriegler R.J. and Takeishi Y. (Eds), The Electrochem. Soc. Inc., Pennington, 1981, 730

83. Claeys C., in Proc. Second Brazilian Workshop on Microelectronics, Giarola A.J. and Mammana A.P. (Eds), Campinas, 1981, 83

84. Freeland P.E., Jackson K.A., Lowe C.W. and Patel J.R., Appl. Phys. Lett., 30 (1977) 31

85. Osaka J., Inoue N. and Wada K., Appl. Phys. Lett., 36 (1980) 288

86. Tan T.Y. and Tice W.K., Phil. Mag., 34 (1976) 615

87. Tice W.K. and Tan T.Y., Appl. Phys. Lett., 28 (1976) 564

88. Tan T.Y., Gardner E.E. and Tice W.K., Appl. Phys. Lett., 30 (1977) 175

89. Kishino S., Isomae S., Tamura M. and Maki M., Appl. Phys. Lett., 32 (1978) 1

90. Hattori T. and Suzuki T., Appl. Phys. Lett., 31 (1977) 343

91. Craven R.A., in Semiconductors Silicon 1981, Huff H.R., Kriegler J.R. and Takeishi Y. (Eds), The Electrochem. Soc., Pennington, 1981, 254

92. Tsuya H., Ogawa K. and Shimura F., Jpn. J. Appl. Phys., 20 (1981) L31

93. Claeys C., in Proc. Third Brazilian Workshop on Microelectronics, Mammana A.P. and Anderson R.L. (Eds), Campinas, 1981

94. Fuller C.S., Ditzenberger J.A., Hannay N.B. and Buehler E., Phys. Rev., 96 (1954) 838

95. Kaiser W., Phys. Rev., 105 (1957) 1751

96. Liaw H.M. and Varker C.J., in Semiconductor Silicon 1977, Huff H.R. and Sirtk E. (Eds), The Electrochem. Soc. Inc., Princeton, 1977, 116

97. Kanamori A. and Kanamori M., J. Appl. Phys., 50 (1979) 8095

124

98. Hu S.M., Appl. Phys. Lett., 31 (1977) 53
99. Kondo Y., in Semiconductor Silicon 1981, Huff H.R.,
    Kriegler R.J. and Takeishi Y. (Eds), The
    Electrochem. Soc. Inc., Pennington, 1981, 220
100. Yoshihiri N., Otsuka H., Oku T and Takasu S.,
    Ext. Abstr. no. 545 presented at the Los Angeles
    ECS-Meeting, 1979.
101. Leroy B. and Plougonven C., J. Electrochem. Soc.,
    127 (1980) 961.

ION IMPLANTATION

Heiner Ryssel and Klaus Hoffmann

Fraunhofer-Institut für Festkörpertechnologie
Paul-Gerhardt-Allee 42
8000 Munich 60, Germany

1. INTRODUCTION

In the last few years, ion implantation has become the most important doping technique for integrated circuits. Especially very-large-scale integrated circuits (VLSI) are unthinkable without implantation. The reasons for this are threefold:

- ion implantation produces extremely homogeneous and reproducible doping concentrations through on-line measuring of the implanted ion current.

- ion implantation fits well into silicon planar technology. The oxide layers used for masking against diffusion can be used to mask against the ion beam. Furthermore, ion implantation can be performed through thin passivating layers (e.g. SiO_2, Si_3N_4), or using photoresist masks.

- ion implantation is a low-temperature process (although an annealing step is usually required to recrystallize the damaged lattice).

Moreover, ion implantation offers some additional possibilities, such as the tapering of windows by damage implantation [1.1], the exceeding of the chemical solubility in connection with laser annealing [1.2], the formation of chemical compounds [1.3], local doping by using focused beams [1.4] or oblique incidence [1.5], ion-beam lithography [1.6], etc., which are important supplements to standard techniques and promise to push forward the development of semiconductor technology.

Ion implantation was invented very early after the development of the bipolar transistor. The basic patent of Shockley [1.7] already describes virtually all aspects of ion implantation. Nevertheless, it took a long time (until the end of the 60's) before this technique was introduced into device manufacturing. In 1962, the first real devices, nuclear detectors, were fabricated by phosphorus implantation [1.8], and solar cells were also fabricated as early as 1963 [1.9]. At the same time, the theoretical background for ion implantation, the so-called LSS theory, was developed by Lindhard, Scharff and Schiøtt [1.10]. From about 1970 on, the implantation technique was applied more and more in semiconductor technology. At first, threshold-adjust and self-aligned source and drain structures were fabricated, and later, bipolar transistors as well [1.11-1.13]. Now, in 1982, ion implantation is, as already stated, the most important doping technology.

The purpose of this treatise is the description of the most important range theories and their application for calculation of range distributions (profiles) in semiconductors, including consideration of two-layer structures, the lateral spread of implanted ions, sputtering during implantation, and diffusion effects during implantation as well as during the necessary thermal annealing. In the Appendix, computer programs are reproduced which allow calculation of range parameters and range distributions including sputtering profiles.

2. RANGES OF IONS IN AMORPHOUS TARGETS

If fast charged particles strike a solid, different processes take place which slow down the particles during penetration, and which simultaneously damage the solid:

- inelastic impacts with bound electrons leading to ionization
- inelastic impacts with nuclei leading to bremsstrahlung, nuclear reactions or excitations
- inelastic impacts with bound electrons
- elastic impacts with atoms leading to a partial transfer of kinetic energy
- emission of Cerenkov radiation

At high energies, electronic stopping (inelastic impacts leading to ionization) is of main importance; at the low energies (up to several 100 keV) relevant for ion implantation, electronic stopping as well as nuclear stopping (elastic impacts with atoms) have to be taken into account.

In Fig. 2.1, the energy dependence of electronic and nuclear stopping is given schematically. Values for the mass dependent characteristic energies E_1, E_2, E_3 of the main doping elements in silicon are given in Table 2.1.

In the following, the classical theory of Lindhard, Scharff, and Schiøtt (LSS theory) [2.1], as well as the more recent theory of Biersack [2.2], will be treated briefly. Both theories assume that electronic and nuclear stopping are separable.

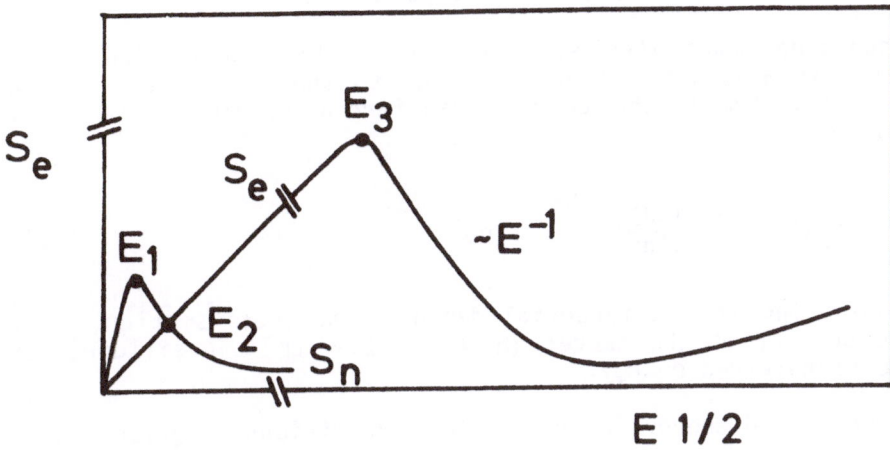

Fig. 2.1 Schematic description of energy dependence for electronic and nuclear stopping

Table 2.1 Characteristic energies E_1, E_2, E_3 for different ions in silicon [2.5]

Ion	E_1 (keV)	E_2 (keV)	E_3 (keV)
B	3	17	3×10^3
P	17	140	3×10^4
As	73	800	2×10^5
Sb	180	2000	6×10^5
Bi	530	6000	2×10^6

2.1 LSS Theory

The stopping of implanted ions thru nuclear collisions can be described as a sum of independent two-body elastic collisions.

The maximum energy which can be transferred in a head-on collision between two atoms is:

$$T_m = 4 \frac{M_1 M_2}{(M_1 + M_2)^2} E \tag{2.1}$$

where M_1 and M_2 are the masses of ion and target, respectively, and E is the energy of the striking particle.

The stopping power (the specific energy loss of an ion having energy E in a target) is proportional to the atomic density, and also to the sum of the energy transferred in all single collisions:

$$S_n(E) = - \frac{1}{N}(\frac{dE}{dx})_n = \int_0^{T_m} T_n d\sigma \tag{2.2}$$

with $d\sigma$ being the differential interaction cross section, N the atomic density of the target ($N = 5 \times 10^{22}$ cm^{-3} in silicon), and T_n the transferred energy.

The transferred energy for hard-sphere collisions is given by:

$$T_n = T_m \sin^2 \phi/2 \tag{2.3}$$

ϕ is the scattering angle in the center-of-mass system, and can be calculated if the interatomic potential is known. The easiest way would be to use an analytical expression such as the Coulomb, Bohr, or Born-Mayer potential. However, a successful description of ion collisions over a wide range of ion distances is only possible by numerical means.

Lindhard, Scharff, and Schiøtt [2.1] use in their theory an approximate analytical form of the Thomas-Fermi potential.

The electronic stopping according to their theory is calculated assuming a free electron gas. In this case, the stopping cross section is proportional to the velocity of the ions, and is therefore proportional to the square root of the energy:

$$S_e(E) = - \frac{1}{N} (\frac{dE}{dx})_e = kE^{1/2} \tag{2.4}$$

The constant k depends on the atomic weights and numbers of ion and target, and is given by (index 1: ion, index 2: target atoms):

$$k = \xi_e \frac{a^{3/2} q Z_1 Z_2 N^2}{(Z_1^{2/3} + Z_2^{2/3})^{3/4} A_1^{3/2}} \tag{2.5}$$

where a ist the screening parameter (of the order of the Bohr radius) and ξ_e is a dimensionless constant of the order of $Z_1^{1/6}$.

Equation (2.5) is valid for ion velocities below the following:

$$v < Z_1^{2/3} q^2 / h \tag{2.6}$$

For higher energies, S_e has a broad maximum and decreases subsequently according to the classical Bethe-Bloch equation.

Calculations for this theory have been performed by several authors, e.g. Gibbons et al. [2.3] and Smith [2.4], and are available in the form of tables.

2.2 Biersack Theory

In the Biersack theory [2.2], the process of slowing down implanted ions is described by a diffusion model. The mean directional cosine of the ion path during the slowing-down process is calculated. Since the ions change direction with each collision, the ion will deviate, in average, more and more from its original direction, while the nuclear energy loss (due to momentum transfer to the target atom) is directly related to the deflection angle ϕ of the ion, see Eq. (2.3). One can represent the directions of ion motion by polar and azimuthal angles ψ and ∂ and depict them on a unit sphere. Since the direction of the motion changes at random with each collision, the stochastic motion on the unit sphere is governed by a diffusion process such as a Brownian motion.

If one now considers the probability distribution function w, of either the polar angle ψ or the directional cosine $\eta = \cos \phi$, it is seen that: (i) it is originally a delta-function centered at $\psi = 0$ or $\eta = 1$; (ii) the probability distribution then spreads out in a diffusion-like manner, while the ion slows down; until (iii) finally the ion has completely "forgotten" its initial direction, i.e., at the end of the trajectory, all directions of motion are equally probable.

These arguments lead to the diffusion equation:

$$\frac{dw}{d\tau} = \frac{\partial}{\partial \eta} \left[(1-\eta^2) \frac{dw}{d\eta} \right] \tag{2.7}$$

where w is the distribution function and τ is equivalent to Dt in ordinary diffusion.

For calculating ranges, it is not necessary to determine w explicitly, but rather to calculate the average value of the directional cosine $\bar{\eta}$ directly from Eq.(2.7), which is found to be:

$$\bar{\eta} = e^{-2\tau} \tag{2.8}$$

The relation between τ and the energy loss is obtained using the two-dimensional Einstein equation, together with Eq. (2.3), as follows:

$$\tau = - \frac{M_2}{4M_1} \int_{E_0}^{E} \frac{S_n}{S_n+S_e} \frac{dE}{E} \tag{2.9}$$

leading to equations for the range R and the projected range R_p:

$$R = \int_{E_0}^{0} ds = \int_{0}^{E_0} \frac{dE}{S_e+S_n} \tag{2.10}$$

$$R_p = \int_{E_0}^{0} \bar{\eta}ds = \int_{0}^{E_0} e^{-2\tau} \frac{dE}{S_e+S_n} \tag{2.11}$$

Each path-length element of the ion trajectory is projected on the x-axis by multiplying with the directional cosine. The projected range can be obtained by first obtaining τ through Eq. (2.9), and inserting into Eq. (2.11). An algorithm, however, was developed by Biersack to give R_p, ΔR_p, and the lateral spread $\Delta R_{p,L}$.

These practical equations are

$$R_p(E+\Delta E) = R_p(E) \left(1- \frac{M_2}{2M_1} \frac{S_n}{S_n+S_e} \frac{\Delta E}{E}\right) + \frac{\Delta E}{S_n+S_e} \tag{2.12a}$$

$$\xi(E + \Delta E) = \xi(E) + \frac{2R_p}{S_e+S_n} \Delta E \tag{2.12b}$$

$$\Delta R_{p,L}^2 (E+\Delta E) = \Delta R_{p,L}^2 (E) + \left[\xi(E)-2\Delta R_{p,L}^2 (E) \frac{M_2}{M_1} \frac{S_n}{S_n+S_e}\right] \frac{\Delta E}{E} \tag{2.12c}$$

with $\xi = R_p^2 + \Delta R_p^2 + \Delta R_{p,L}^2$ (2.12d)

The Eqs. (2.12) can be programmed easily on a pocket calculator for fast iterative calculations of R_p, ΔR_p and $\Delta R_{p,L}$. For S, the LSS stopping power Eq. (2.4) is used wereas for S_n an analytic approximation e.g.

$$S_n = F \frac{0.5 \ln (1-\varepsilon)}{\varepsilon + 0.107 \varepsilon^{0.3754}}$$ (2.13)

with $F = 181 Z_1 Z_2 Na/(1+M_2/M_1)$; $a = 0.568 (Z_1^{0.23} + Z_2^{0.23})$;

$\varepsilon = f(E(keV)$; $f = 69.4 a M_2/Z_1 Z_2 (M_1 + M_2)$

can be used.

In the Appendix, a Fortran program to calculate range parameters according to this theory is reproduced.

2.3. Monte Carlo Calculations

An alternative method to the solution of transport equations (cf. sects: 2.1 and 2.2) is the application of stochastic methods, i.e. Monte Carlo calculations, for the determination of implantation profiles. With this method, not only the particle distribution, but also the spatial distribution of energy deposited in nuclear and electronic events is calculated.

The Monte Carlo simulation models the physical events during the slowing down of individual particles. The ions are, as described in sects 2.1 and 2.2, stopped by nuclear collisions and electronic friction. The nuclear collisions can be described by the Lindhard stopping formula, Eq. (2.2). The location of the scattering target atoms, and thus the impact parameter, is randomly selected. A simulation with a sufficiently high number of particle results in random distribution. For one-dimensional distributions, 1000 paricles usually lead to a satisfactory description of an implantation profile; whereas in the two-dimensioned case, considerably more particles, or else a refined interpretation of the simulation data, is required to obtain interpretable results. Since the required computer time is proportional to the number of particles, it is highly desirable to optimize the computation speed for a single particle. This is done in two ways:

- Selection of a realistic mean free path between collisions reduces the number of scattering events to be simulated without sacrificing accuracy;

- Use of an efficient algorithm to compute the scattering angle, which makes the application of realistic potentials for the nuclear interaction possible.

A very effective program for these calculations is TRIM (Transport of Ions in Matter) [2.6]. This program uses an energy-dependent free-flight path. At high energies, large deflection angles are very seldom; therefore, a large free-flight path can be chosen. The fast evaluation of the scattering angle is obtained by using an algorithm, which starts from an approximation of the scattering angle calculated by geometrical considerations. This approximation is refined by a few parameters.

An example of such a calculation, in comparsion to experimental results and computations according to the LSS and Biersack theories, is given in Fig. 2.6.

The advantage of the Monte Carlo methods is its inherent agreement with the physical events. The great disadvantage, which limits its applieabilitiy, is the excessive amount of computer time needed to obtain good statistics.

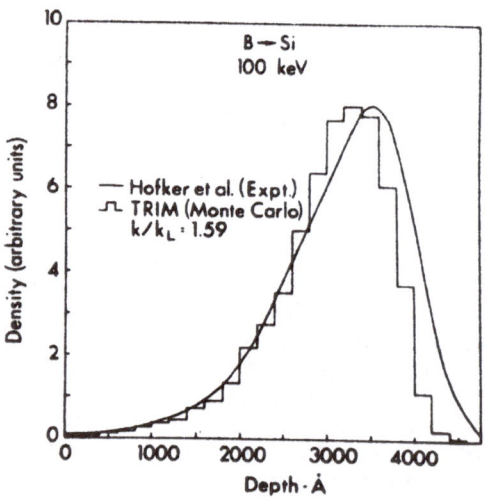

Fig. 2.2 100 keV boron depth distributions in silicon: Histograms represent Monte Carlo results (TRIM), and the full line represents the experimental data [2.7]. The agreement of the shapes of the distributions is evident, whilst the absolute values for mean projected range and standard deviation could be improved by choosing a smaller electronic stopping

3. RANGES OF IONS IN CRYSTALLINE TARGETS

All range theories [3.1, 3.2] assume an amorphous target. Semiconductors such as silicon GaAs are single crystals. Because of this
crystalline structure, ions can penetrate much deeper into the
crystal if they are implanted along a major crystal axis or
plane, since they seldom come close enough to a target atom to
lose significant energy thru nuclear collisions, see Fig. 3.1.
Therefore, electronic stopping mainly occurs, instead of a combination of electronic and nuclear stopping in a random direction.
The range is therefore proportional to the velocity of the ion,
and shows a square-root dependence on the energy. According to a
theory of Firsov [3.3], the range can be approximately calculated
taking into account the dependence of the electronic stopping on
the size of channels. However, no theory exists which is capable
of describing the dependence of channeling-profile shape on crystal orientation and angle of ion incidence.

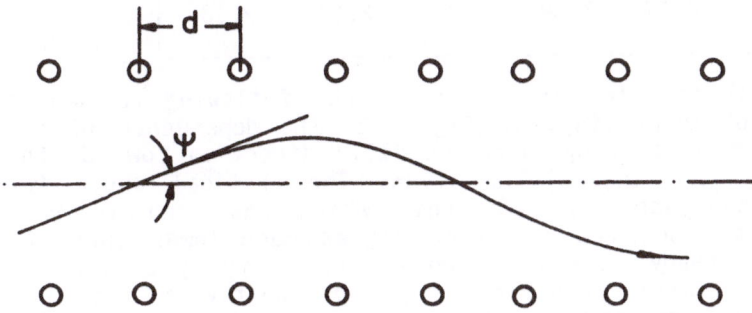

Fig. 3.1 Schematic description of the channeling effect

Lindhard calculated the angle under which an ion is accepted into
a channel [3.4]. For the energy range of ion implantation, the
following applies:

$$\psi_{c2} = (\frac{a}{d} \ \psi_{c1})^{1/2} \tag{3.1}$$

with

$$\psi_{c1} = (\frac{2Z_1 Z_2 q^2}{2\pi \ \epsilon_0 E \ d})^{1/2} \tag{3.2}$$

where d is the distance between the atoms and a is a screening
parameter which is of the same order of magnitude as the Bohr

radius ($a = 0.8853\, a_0\, (Z_1^{2/3} + Z_2^{2/3})^{-1/2}$; $a_0 = 0.529$ Å). In the energy range of ion implantation, the critical angles usually amount to several degrees. In Table 3.1, calculated values according to Eqs. (3.1) and (3.2) are given. The experimentally-determined half-angles for channeling agree well ($\psi_{c2} \approx \psi_{1/2}$) with the theoretical data.

Table 3.1 Critical angles for channeling in silicon for arsenic, antimony, boron and phosphorus

Energy (keV)	20	40	80	160	320
Arsenic	4.82	4.05	3.41	2.86	2.41
Antimony	5.12	4.30	3.62	3.04	2.56
Boron	3.50	2.95	2.48	2.08	1.75
Phosphorus	4.27	3.59	3.02	2.54	2.13

Some experimental profiles are shown in the following to depict the importance of channeling. In Fig. 3.2, the dependence of the channeling effect on tilting from the <111> direction towards the <110> direction (see insert) is given. The critical angle for channeling of phosphorus at 450 keV, which was used in this experiment, is 4°. One can see that angles much lower than the critical angle already have a marked effect on the profile. The channeling range (most probable range) is nearly a factor 3 larger than the random range R_p. For smaller energies (more nuclear stopping), the difference in the two ranges may be much greater. For 40 keV phosphorus, e.g., it is approx. a factor of 10. Since there also exist planar channels (channels between lattice planes), the rotation of the wafer is important too. This is shown in Fig. 3.3. In this case, again, phosphorus at 450 keV has been implanted with the wafers tilted by 7°. The tilting axes are indicated by the arrows.

It has been proposed to use the channeling effect to obtain deep impurity distributions, in order to compensate for the limited depth range of implanted ions, which would be worthwhile for some purposes. For this, a parallel ion beam (obtainable either by mechanical scanning of the wafers, or by an electrostatic double-deflection scanner) is required. The crystal orientation, however, has to be exact within ± 0.1°, which is very difficult to obtain.

Moreover, the channeling not only depends on the orientation, but also on the implantation dose and temperature. Ion implantation produces, as an unwanted byproduct, damage to the crystal latti-

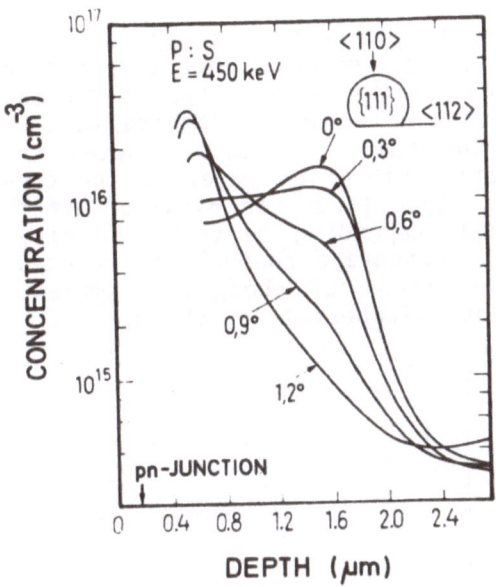

Fig. 3.2 Dependence of the channeling effect on tilting (E = 450 keV, $N_\square = 5 \times 10^{12}$ cm^{-2}) [3.5]

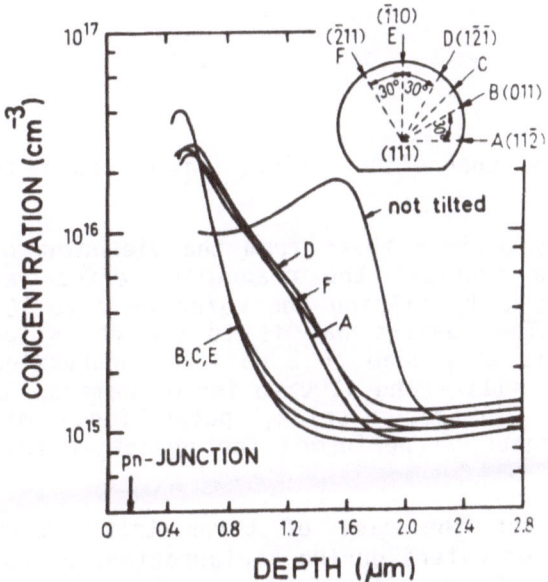

Fig. 3.3 Dependence of the channeling effect on rotation for 7° tilting for a phosphorus implantation (E = 450 keV, $N_\square = 5 \times 10^{12}$ cm^{-2}) [3.5]

ce. This damage reduces the channeling effect, and suppresses it completely if an amorphous layer is present. High-temperature implantation (due to heating the sample externally, or to the dissipated beam power) reduces the formation of lattice defects by _in situ_ annealing, but at the same time also reduces the channeling thru an increased lattice-vibration rate. If the crystal is covered with an amorphous layer (e.g., an oxide or nitride layer), the ions are scattered, depending on their velocity. Thus, the number of ions which can penetrate channels is reduced. In Fig. 3.4, the carrier concentrations of phosphorus implants through oxide layers with different thicknesses are shown.

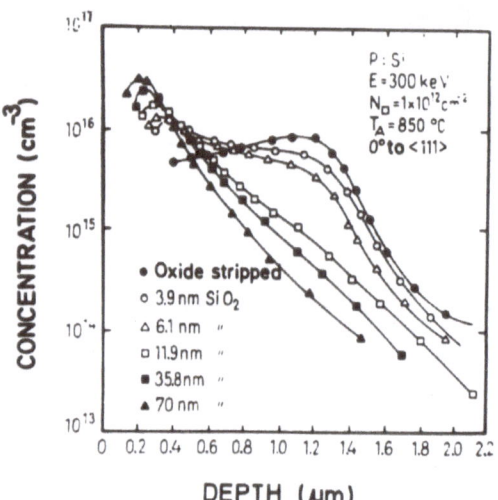

Fig. 3.4 Dependence of the channeling effect on coverage by oxide layers [3.6]

For the above reasons, it is obvious that, from the viewpoint of reproducibility, one has to suppress the channeling effect as much as possible. This is done by tilting the wafer by 7 to 10° and optimally rotating it. The easiest way is to use the <110> flat as a rotation axis. This will lead to a maximum suppression of the channeling effect for <111>- and <100>-oriented wafers. An additional scattering layer (oxide, nitride, polysilicon) may help if optimal rotation cannot be performed for practical reasons.

Figures 3.5 and 3.6 show for the case of boron that, using optimal tilting and rotation of wafers during implantation, reproducible profiles independent of crystal orientation and dose can be obtained. These boron profiles show tails which are probably due to boron ions scattered into channels. A similar good reproducibility is expected for other implanted ion species, assuming optimum control of implantation conditions.

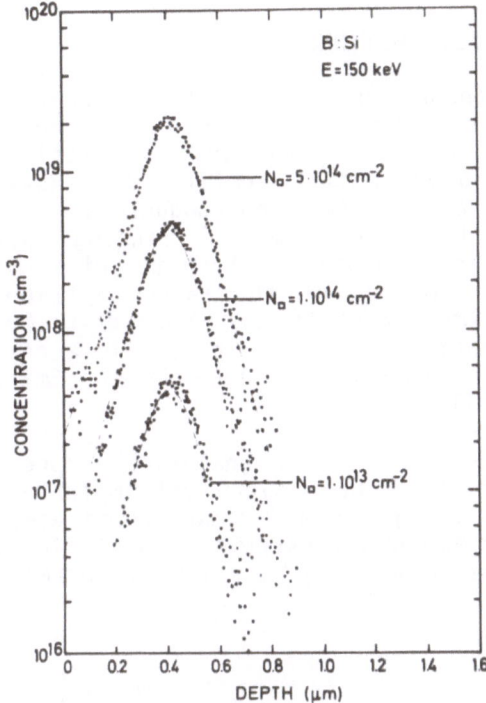

Fig. 3.5 Experimental boron profiles for doses of 10^{13}, 10^{14} and 5×10^{14} cm^{-2} and an implantation energy of 150 keV in comparison to Pearson fits; orientation <111> [3.7]

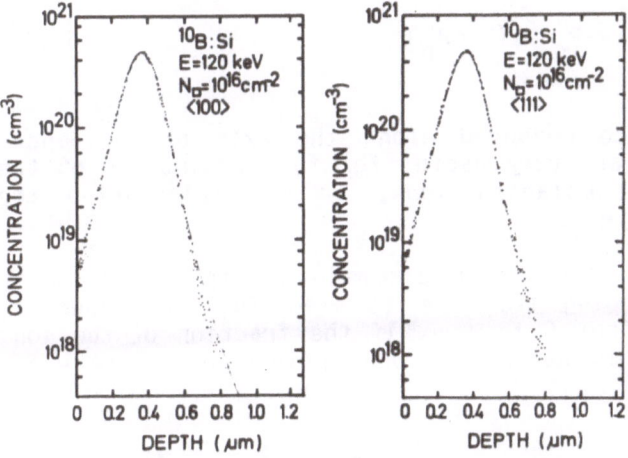

Fig. 3.6 Boron profiles for doses of 10^{16} cm^{-2} at 120 keV in <111> and <100> orientation in comparison to Pearson fits [3.7]

4. RANGE DISTRIBUTIONS

According to the classical LSS theory as well as the alternative
Biersack approach [4.1, 4.2], the range profiles of implanted
impurity distributions are given by symmetrical Gaussian distributions. Experiments have shown, however, that profiles are skewed
and possess tails, not only in crystalline semiconductors where
channeling of ions may occur. These experimental findings are
confirmed by numerical Monte Carlo simulations. Both theories can
be extended to calculate higher moments of the range distributions. Winterbon [4.3] calculated 4 moments of range distributions for many ion-target combinations; however, no easily acces-
sible tables or formulas (such as those by Gibbons [4.4], Smith
[4.5], and Biersack [4.5]) are available.

In the following section, the most widely-used analytical expressions for range profiles will be described. Since all implantations for practical applications are performed into tilted samples, the profile description for amorphous targets is sufficient.
Residual channeling tails can be modeled by a slight adjustment
of the parameters.

4.1 Gaussian Profiles

A Gaussian distribution is described by two moments, the projected range R_p and the projected standard deviation or straggling
ΔR_p. Together with the implanted dose $N_\square$, they describe the
implanted profile by:

$$C(x) = \frac{N_\square}{\sqrt{2\pi}\ \Delta R_p}\ \exp\left[-(x-R_p)^2/(2\ \Delta R_p^2)\right] \qquad (4.1)$$

where x is the distance measured along the axis of incidence.
Gaussian distributions are very useful for fast estimation of the
range distribution of implanted ions, or for calculating the
thickness of masking layers.

In deriving Eq. (4.1), it has been assumed that the ions come to
rest in a volume extending from $-\infty$ to $+\infty$. The semiconductor,
however, extends only from 0 to $+\infty$. If the fraction of the ions
which is backscattered is neglected, one arrives at a more exact
expression:

$$C(x) = \frac{2\ N_\square}{\sqrt{2\pi}\ \Delta R_p\left[(1+\mathrm{erf}\ R_p/\sqrt{2}\ \Delta R_p)\right]}\ \exp\left[-(x-R_p^2)\ /2\Delta R_p^2\right] \qquad (4.2)$$

The error in using Eq. (4.1) is small, and is usually below 1%. Only for very low-energy implants can it be appreciable. Moreover, backscattered particles have been completely neglected in Eqs. (4.1) and (4.2). Calculations and measurements by Bøttiger et al. [4.29] indicate that the backscattered particles are those lying in the part of the profile outside the semiconductor ($-\infty$ to 0). Therefore the use of Eq. (4.1) is recommended.

If the profile is only slightly asymmetrical, the third moment is sufficient to obtain a good profile description [4.6]. In this case, the profile is given by two joint half-Gaussian profiles with range R_M and range straggling ΔR_{p1} and ΔR_{p2}:

$$C(x) = \frac{2N_\square}{(\Delta R_{p1} + \Delta R_{p2})\sqrt{2\pi}} \exp\left[- \frac{(x-R_M)^2}{2\Delta R_{p1}^2} \right] \qquad x > R_M$$

$$ (4.3) $$

$$C(x) = \frac{2N_\square}{(\Delta R_{p1} + \Delta R_{p2})\sqrt{2\pi}} \exp\left[- \frac{(x-R_M)^2}{2\Delta R_{p2}^2} \right] \qquad 0 < x < R_M$$

R_M, ΔR_{p1}, and ΔR_{p2} can be calculated, e.g. from the tabulated data of Gibbons [4.4], according to:

$$R_p = R_M + 0.8(\Delta R_{p2} - \Delta R_{p1})$$

$$\Delta R_p^2 = -0.64 \ (\Delta R_{p2} - \Delta R_{p1})^2 + (\Delta R_{p1}^2 - \Delta R_{p1}\Delta R_{p2} + \Delta R_{p2}^2) \qquad (4.4)$$

$$\gamma = \Delta R_p^{-3}(\Delta R_{p2} - \Delta R_{p1})(0.218\ \Delta R_{p1}^2 + 0.362\ \Delta R_{p1}\Delta R_{p2} + 0.218\Delta R_{p2}^2)$$

γ is the skewness of the distribution (for its definition, see next section). Approximate values for ΔR_{p1} and ΔR_{p2} are obtained using Table 4.1 [4.4]. For a negative skewness, ΔR_{p1} and ΔR_{p2} have to be exchanged.

Table 4.1   Standard deviations of the joint-half Gaussian distribution

γ	0	0.1	0.2	0.3	0.4	0.5	0.6	0.7	0.8	0.9	1.0
R_{p1}/R_p	1	1.062	1.123	1.182	1.241	1.301	1.360	1.422	1.486	1.554	1.633
R_{p2}/R_p	1	0.936	0.871	0.802	0.729	0.653	0.570	0.478	0.374	0.248	0.081

A comparison of profile descriptions using the two models is given in Fig. 4.1.

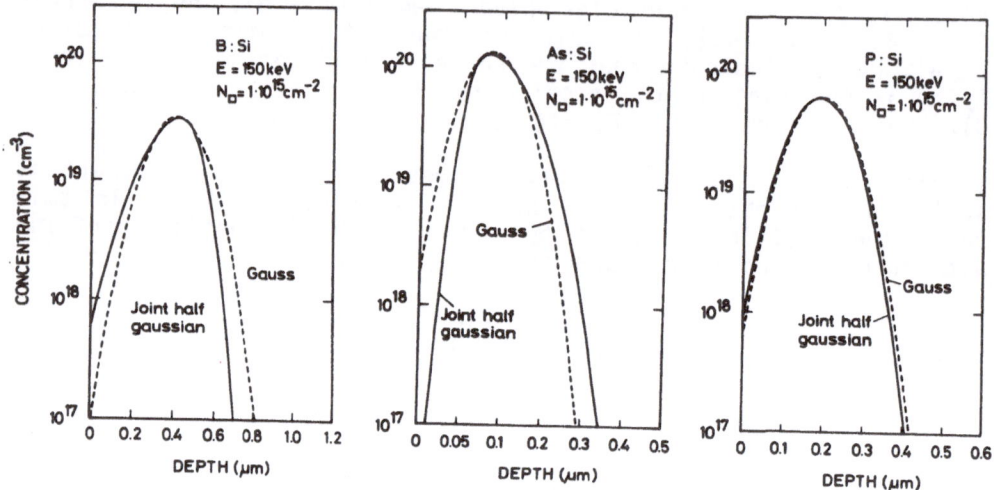

Fig. 4.1 Comparison of Gaussian and joint-Gaussian distributions for 150 keV boron, arsenic and phosphorus in silicon. The moments are according to Gibbons et al. [4.4]

4.2 Pearson Distributions

Many experimental investigations have shown that the simple description of implanted profiles given in the last section is not adequate for most ions in silicon and other semiconductors. It has been argued that this might be due to channeling because of the crystalline structure of the usual semiconductors. It has been found, however, that the profiles of many ions are asymmetrical, in amorphous targets as well, and higher moments have to be used to construct range distributions.

A particularly useful distribution is the Pearson type-IV function with four moments [4.7]. The Pearson distribution of type IV centered around the projected range R_p is given by:

$$f(x) = K\left[b_2(x-R_p)^2 + b_1(x-R_p)+b_0\right] \exp\left[\frac{b_1/b_2+2a}{\sqrt{4b_2b_0-b_1^2}} \right. \cdot$$

$$\left. \arctan \frac{2b_2\,(x-R_p)+b_1}{\sqrt{4b_2\,b_0-b_1^2}} \right. \tag{4.5}$$

where K is obtained by the constraint:

$$\int_{-\infty}^{\infty} f(x)dx = 1 \qquad (4.6)$$

The four constants a, b_0, b_1, and b_2 are given by:

$$a = -\frac{\Delta R_p \gamma (\beta+3)}{A}$$

$$b_0 = -\frac{\Delta R_p^2 (4\beta-3\gamma^2)}{A} \qquad (4.7)$$

$$b_1 = a$$

$$b_2 = -\frac{(2\beta-3\gamma-6)}{A}$$

where $A = 10\beta - 12\gamma^2 - 18$.

The parameters R_p, ΔR_p, β, γ in these equations are directly related to the four moments μ_1, μ_2, μ_3, and μ_4 of the distribution f(x).

The first moment μ_1 is well known as the average projected range:

$$\mu_1 = R_p = \int_{-\infty}^{\infty} x f(x) \, dx \qquad (4.8a)$$

The three higher moments μ_i are given by:

$$\mu_i = \int_{-\infty}^{\infty} (x-R_p)^i f(x) \, dx, \qquad i = 2,3,4 \qquad (4.8b)$$

It is customary to use the standard deviation ΔR_p, which is defined by the square root of the second moment μ_2, and dimensionless expressions for the higher moments:

standard deviation $\qquad\qquad \Delta R_p = \sqrt{\mu_2}$ $\qquad\qquad$ (4.9a)

skewness
$$\gamma = \frac{\mu_3}{\Delta R_p^{\,3}}$$
(4.9b)

kurtosis
$$\beta = \frac{\mu_4}{\Delta R_p^{\,4}}$$
(4.9c)

The skewness γ indicates the tilting of the profile, and the kurtosis β indicates flatness at the top of the profile.

The relation between the third and fourth moments has to be chosen to satisfy:

$$\beta > \beta_{min} = \frac{48 + 39\,\gamma^2 + 6(\gamma^2 + 4)^{3/2}}{32 - \gamma^2}$$
(4.10)

in order to give a Pearson distribution of type IV.

The maximum of the Pearson distribution occurs at $x = R_p + a$ unless $\gamma = 0$. For a negative skewness, the peak is deeper than R and the distribution falls off more rapidly for $x > a$ than for $x < a$. For a positive skewness, the opposite is true.

The kurtosis is 3 for a Gaussian distribution. The limit is represented by Eq. (4.10). An universal expression for the kurtosis is given by Gibbons [4.8]:

$$\beta = 2.8 + 2.4\,\gamma^2$$
(4.11)

To obtain the concentration profile $C(x)$, the Pearson distribution is mulitiplied by the dose:

$$C(x) = N_\square\, f(x)$$
(4.12)

(note that $f(x)$ as well as K have the dimension $(length)^{-1}$). In the discussion of the Pearson distribution, it has again been assumed that the profile extends from $-\infty$ to $+\infty$. To describe a more realistic, semi-infinite target, a correction of the normalizing constant K is required; K is usually computed by numerical integration.

A comparison of profiles obtained by Gaussian and Pearson distributions is given in Fig. 4.2.

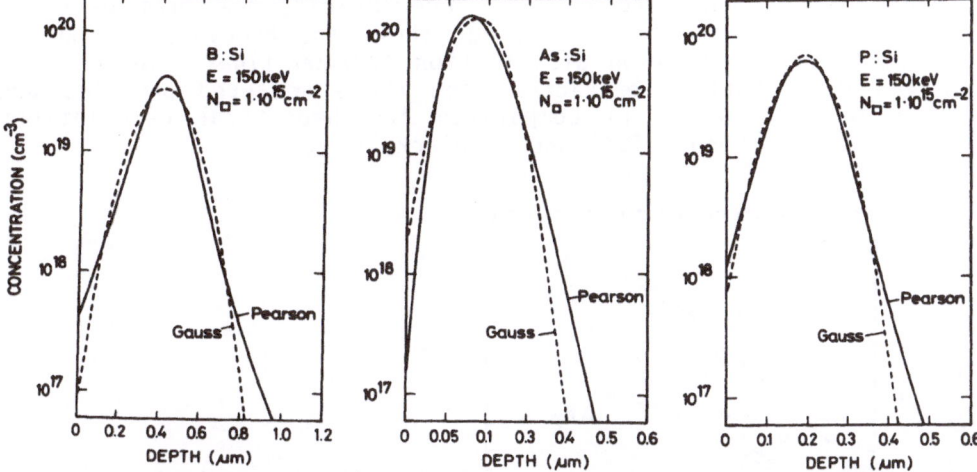

Fig. 4.2 Comparison of Gaussian and Pearson distributions for 150 keV boron, arsenic, and phosphorus in silicon. The moments are according to Gibbons et al. [4.4]

Examples for measured range distributions of boron, in silicon and SiO_2, are given in the next figures.

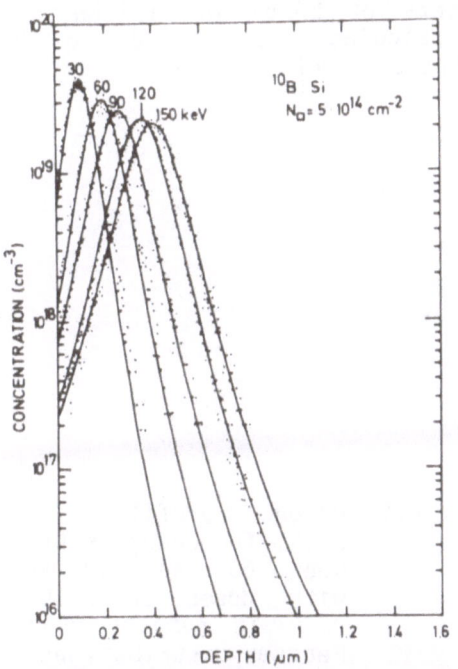

Fig. 4.3 Experimental boron profiles in silicon for energies between 30 and 150 keV with doses of 5×10^{15} cm^{-2} in comparison to Pearson distributions; orientation <111>.

144

Figure 4.3 shows boron profiles in silicon for energies of 30 to 150 keV. The samples were tilted by 7° and optimally rotated to suppress the channeling effect. Nevertheless, the profile tails are probably caused by channeling; they can, however, be modeled perfectly by the Pearson distributions (drawn lines). The moments of the distributions extracted from these measured profiles are given in Fig. 4.4 , in comparison to theoretical calculations according to Gibbons [4.4] and Biersack [4.2].

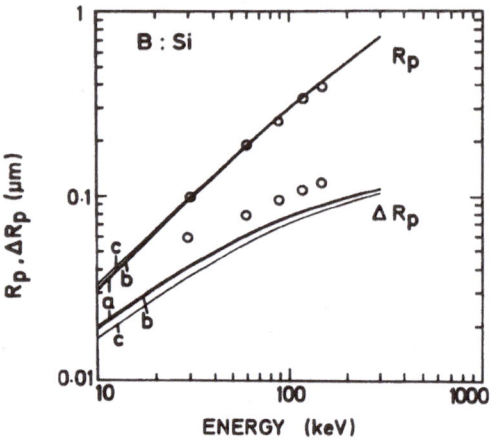

Fig. 4.4 Experimental range parameters of boron in silicon, in comparison to theoretical calculations. (a) ^{10}B, (b) ^{11}B Biersack [4.2], (c) ^{11}B Gibbons [4.4]

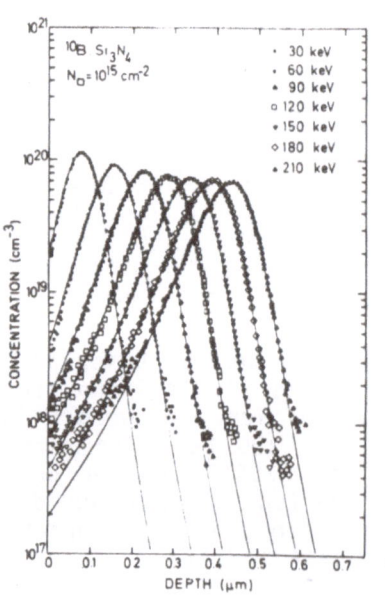

Fig. 4.5 Boron profiles in Si_3N_4 for energies between 60 and 210 keV with doses of 5x10 cm^{-2} in comparison to Pearson distributions

The R_p values agree very well within 2% with the theoretical ones, for ΔR_p , however, up to 40 % larger values are measured. The experimental skewness can be approximated by $\gamma = -E/300$ keV and the kurtosis by $\beta = 8+25(\beta_{min} - 3)$.

In amorphous targets, no channeling can take place. In Fig. 4.5, range profiles of boron in Si_3N_4 for energies between 30 and 210 keV are given. The profiles have a shallow slope towards the

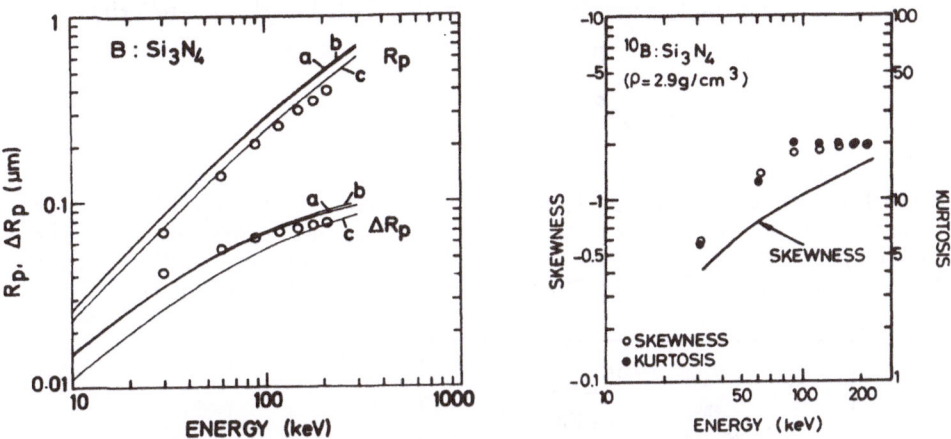

Fig. 4.6 Experimental range parameters, of boron in Si_3N_4, in comparison to theoretical calculations (a) ^{10}B, (b) ^{11}B Biersack [4.2], (c) ^{11}B , Gibbons [4.4]

surface and a steep slope towards the bulk, and show no profile tails. The corresponding moments are given in Fig. 4.6, together with theoretical calculations.

4.3 Other Distributions

Various other distribution functions have been proposed for the description of implantation profiles. The Edgeworth function [4.4] results from an expansion of the Gaussian function into Tshebyshev-Hermitian polynomials, and also uses four moments. The implantation profile is given by:

$$C(x) = \frac{N_\square}{\sqrt{2\pi}\ \Delta R_p}\ \exp\ \left[\ -\frac{(x-R_p)^2}{2\ \Delta R_p^2}\ \right]\ G(x) \tag{4.13}$$

$$G(x) = 1 + \frac{\gamma}{6}\ (Z^3 -3Z) + \frac{\beta-3}{24}\ (Z^4 -6Z^2 +3) + \frac{\gamma^2}{72}(Z^6 -15Z^4 + 45Z^2 -15) + ..$$

with $Z = (x-R_p)/\Delta R_p$.

The kurtosis is calculated according to Gibbons [4.4] by:

$$\beta = 3 + \frac{5}{3} \gamma^2 \tag{4.14}$$

In this case, for the profile, the following applies:

$$C(x) = \frac{N_\square}{\sqrt{2}\ \Delta R_p} \exp[\frac{(x-R_p)^2}{2\Delta R_p^2}] \cdot [1 + \frac{\gamma}{6}(Z^3-3Z) + \frac{5\gamma^2}{72}(Z^4-6Z^2+3) +$$

$$+ \frac{\gamma^2}{72}(Z^6 -15Z^4 + 45Z^2 - 15)] \tag{4.15}$$

The disadvantage of this function is that it is applicable only for small deviations from the Gaussian shape; moreover, it shows oscillations, with negative values of the function, and can therefore be used only close to the maximum of the distribution.

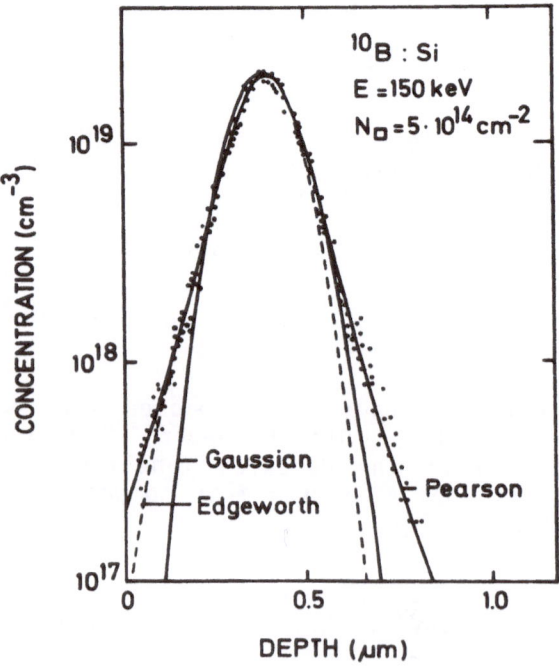

Fig. 4.7 Comparison of Gaussian, Pearson and Edgeworth fits to an experimental boron profile (E=150 keV, $N_\square$ = 5×10^{14} cm^{-2})

A comparison between optimal Gaussian, Pearson, and Edgeworth fits to an experimental boron profile is given in Fig. 4.7, and shows the superior properties of the Pearson distribution.

Other distributions such as [4.9]:

$$C(x) = cN_{\square}x^m \exp (- x/b)^n \tag{4.16}$$

are mathematically simple, but the determination of the moments requires lengthy calculations.

The Sonine expansion with three moments [4.10] gives a very good fit to Monte Carlo simulations, and probably also to experimental profiles, but is too complicated for practical applications.

4.4 Two-Layer Targets

Frequently, implantations are performed in thin dielectric layers such as SiO_2 or Si_3N_4, in order to avoid a contamination of the silicon, to provide for a scattering layer, or to adjust the depth distribution of the ions. To calculate the depth distribution of such two-layer structures in an exact way, Monte Carlo simulations or the solution of Boltzmann transport equations [4.11] are required. This is very complicated and requires much computing time.

For practical cases in silicon technology, a simple model derived by Ishiwara and Furukawa [4.12] can be applied for materials whose average atomic numbers and masses are nearly equal. This is true for the most common combinations used with silicon, e.g. SiO_2, Si_3N_4, and Al_2O_3. Moreover, for such implants, only thin layers are used, thus further reducing the possible error. If thick layers are used, they are usually required for masking and have to stop all ions completely.

Assuming Gaussian profiles, the two parts of the profile in layer 1 and substrate 2 are given by [4.12]:

$$C_1(x) = \frac{N_{\square}}{\sqrt{2\pi}\Delta R_{p1}} \exp \left[- \frac{(R_{p1}-x)^2}{2\Delta R_{p1}^2} \right] \qquad \Delta < x < d \tag{4.17a}$$

$$C_2(x) = \frac{N_{\square}}{\sqrt{2\pi}\ \Delta R_{p2}} \exp \left[- \frac{\left[d+(R_{p1}-d)\Delta R_{p2}/\Delta R_{p1}-x \right]^2}{2\ \Delta R_{p2}^2} \right] \qquad x > d \tag{4.17b}$$

where d is the thickness of the layer. The error in using Eqs. (4.17 a,b) can be calculated by:

$$\eta = \frac{R_{p2} - R'_{p2}}{R_{p2}} \quad \text{with} \quad R'_{p2} = R_{p1} \frac{\Delta R_{p2}}{\Delta R_{p1}} \qquad (4.18)$$

The profile (4.17) is derived from the profile in a thick oxide layer by a density transformation with the scaling factor $\Delta R_{p2}/\Delta R_{p1}$. This leads to a good description of the actual profile if a thick oxide is used ($d > R_{p1}$); in the case of a thin oxide ($d << R_{p1}$), however, the profile in the substrate has the range $R'_{p2} \neq R_{p2}$.

We use a different density transformation, because the profile shape in the silicon substrate is usually more critical than the profile in the oxide layer. Our model also accomodates non-Gaussian profiles, e.g. Pearson distributions, to give the two-layer profile:

$$C_1(x) = \frac{\Delta R_{p2}}{\Delta R_{p1}} \quad C \left(\frac{\Delta R_{p2}}{\Delta R_{p1}} x \right) \qquad x < d$$

$$ \qquad (4.17c)$$

$$C_2(x) = C \left(x - d \left(1 - \frac{\Delta R_{p2}}{\Delta R_{p1}} \right) \right) \qquad x > d$$

where C(x) is the profile in bare silicon without second layer.

This model gives good results if the distribution is concentrated in the second layer. A further refinement can be made to give the exact profile in both limiting cases (thin or thick mask layers). For this purpose, the profile C_1 in material 1 is calculated, and the total number of atoms in this layer ($N_{\square,1}$) is obtained by integration. Then, the profile C_1 is calculated in the target material material 2 (assuming no masking layer), and the thickness d' is determined which contains $N_{\square 1}$ atoms. The final profile is composed of profile C_1 in material 1 up to d and of profile C_2 starting from d', thus resulting in a profile containing $N_{\square}$ ions.

In Fig. 4.8, the profiles resulting from boron and arsenic implantations into SiO_2-Si structures are shown using this model, in comparison to profiles assuming silicon for the complete structure, or performing only a density tranformation. One can clearly see the necessity of the exact profile description.

Better results, but again requiring lengthy calculations, are obtained using energy-distribution functions and approximation of range distributions by joint half-Gaussian distributions [4.6].

Another model published by Satya and Palanki [4.13] has the disadvantage of predicting arbitrary deep profiles in silicon at a given energy, if a thin mask layer with low stopping power is used.

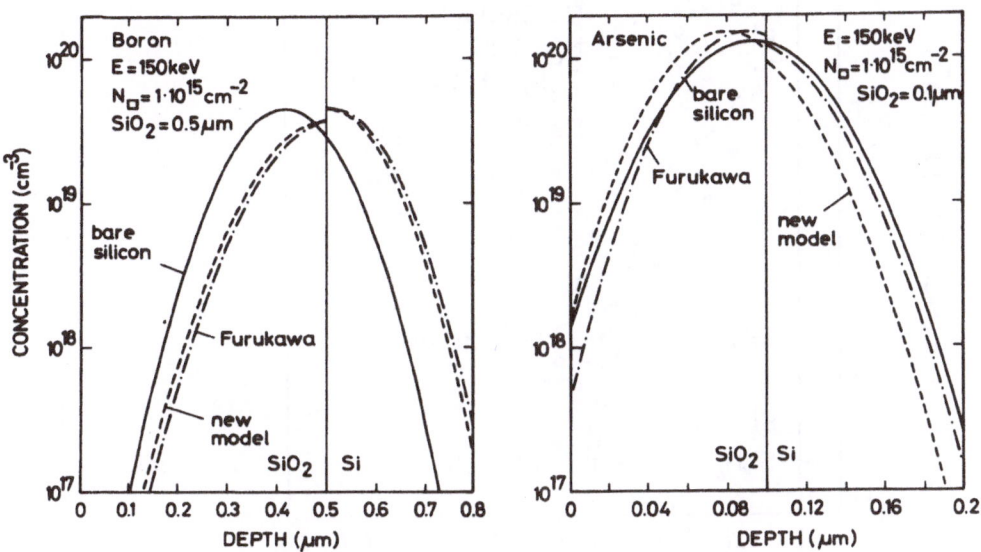

a)   b)

Fig.4.8 Boron and arsenic profiles after implantation into a SiO_2 - Si structure

The implantation of ions thru masking layers also results in recoil or knock-on implantation of ions. If the mass of the implanted ions is not too different from the mass of the atoms of the masking layer, a large fraction of the energy of the primary particles can be transferred to the atoms of the masking layer. These particles are then implanted into the substrate. The energy which is transferred in an impact is given by Eq. (2.3). Theoretical calculations of distributions of recoil implantations have been performed by Moline et al. [4.14], Fischer et al. [4.15], and recently by Sigmund [4.16] as well as by Hirao et al. [4.17]. The formalisms is too complicated to deal with it here; no simple analytical profile description exists, and also no tabulated data.

However, all theories and experiments show the same phenomena:
The recoiled atoms show an extremely shallow distribution, with
the maximum at the surface. The primary ions come to rest much
deeper in the crystal than the recoiled ones. An example of such
a profile is given in Fig.4.9.

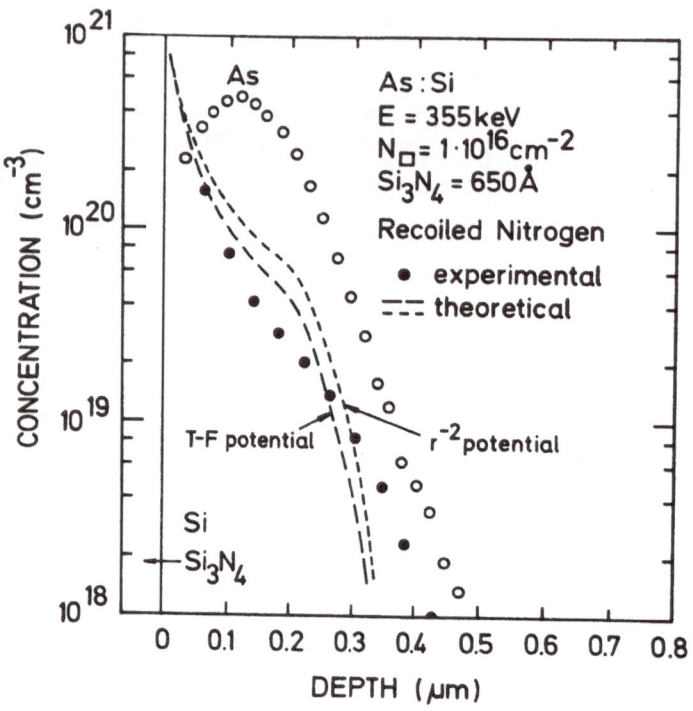

Fig.4.9 Comparison of experimental and theoretical profiles for
arsenic implantation at 335 keV with a dose of 1×10^{16}
cm^{-2}, through Si_3N_4 having 650 Å thickness [4.17]

Up to now, no detrimental effect has been found on implantation
profiles, or on the diffusion of implanted species. The damage
produced by the knocking ions, however, has to be considered. It
may extend into the area of the p-n junction, and may degrade its
properties. If, however, epitaxial layers have to be grown on
such a layer, it is necessary to get rid of the damaged layer.
This can be done by oxidation or etching.

Very recently, a new type of effect has been found to occur at
interfaces: atomic mixing [4.18 - 4.22]. At the same time, knock-
on takes place, atoms from the substrate layer can also be back-
sputtered into the masking layer, with the result that the interface layer is no longer well defined. This effect can become

remarkable in very high-dose implantations. The mixing is stronger than expected from the classical sputtering theory, and is explained thru a high atomic mobility for a short time-interval following the impact. At the same time, reactions between the atoms can also take place. These effects have been investigated especially in connection with silicide formation, and are of no concern in the present context.

4.5 Implantation and Sputtering

An effect usually neglected in ion implantation is sputtering. For shallow devices, however, low energies and high doses have to be used to obtain the desired junction depth and sheet resistivity. In this case, sputtering has to be considered. The most important parameter for dealing with this problem is the sputtering yield S.

According to Sigmund [4.23, 4.24], the sputtering yield for normal incidence is given by:

$$S = \frac{3}{4} \frac{S_n(E) \cdot \alpha(M_2/M_1)}{\pi^2 C_o U_o} \tag{4.19}$$

where C_o is a constant ($C_o = 1/2 \lambda_o a^2$; $\lambda_o = 24$; $a = 0.0219$ nm), $\alpha(M_2/M_1)$ is a numerically calculable function, U_o is the surface binding energy ($U_o = 7.81$ eV for silicon [4.24], and $S_n(E)$ is the nuclear energy deposition. $\alpha(M_2/M_1)$ is given inFig. 4.10 for two power potentials. Sputtering yields for severalions, depending on the energy, are given in Fig.4.11. Depending on the ion mass, the yield shows a maximum in the energy range of ion implantation. For a fixed energy of 45 keV, experimental sputtering yields according to Andersen et al. [4.25] are given in Fig.4.12, in comparison to theoretical values according to Sigmund. From these two figures, it is seen that the sputtering yield is usually between 1 and 5.

The sputtered silicon layer is calculated from the sputtering yield and the ion dose by:

$$d = \frac{S}{N} N_\square \tag{4.20}$$

where N is the atomic density ($N = 5 \times 10^{22}$ cm^{-3} for silicon). For high impurity concentrations, however, the sputtering yield may change. In Table 4.2, the sputtered-layer thickness, after implanting different ions at the energy for maximum sputtering, is given. One can see that the effect is small for all doses below 10^{16} cm^{-2}.

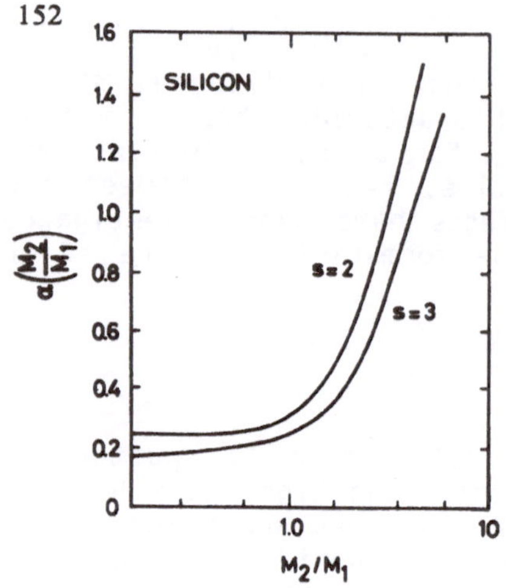

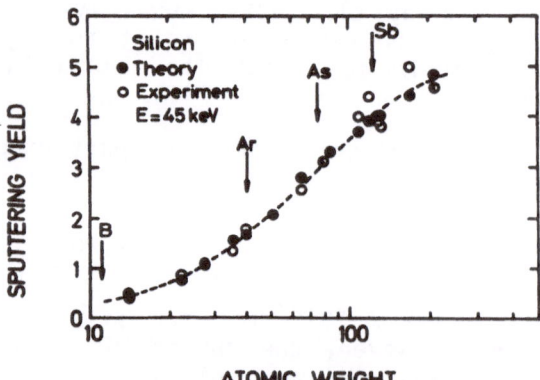

Fig. 4.10 Function $\alpha(M_2/M_1)$ for power potentials with s = 2 and s = 3; according to [4.23]

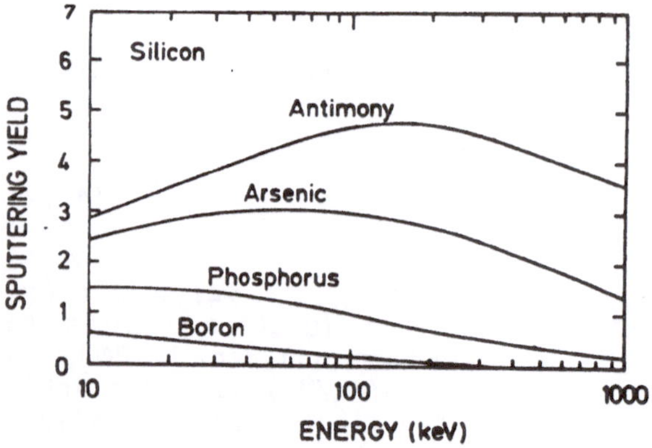

Fig.4.11 Sputtering yields of antimony, arsenic, boron, and phosphorus, as a function of energy.

Fig.4.12 Sputtering yields of ions of different mass at 45 keV, according to Andersen [4.25].

Table 4.2 Thickness of sputtered silicon layers for different ions at maximum sputtering yield. The lowest energy used for the calculations was 10 keV.

| Ion | Energy (keV) | Dose (cm^{-2}) | | |
		10^{15}	10^{16}	10^{17}
B	10	0.1 nm	1 nm	10 nm
P	10	0.3 nm	3 nm	30 nm
As	50	0.6 nm	6 nm	60 nm
Sb	150	1 nm	10 nm	100 nm

The modification of implantation profiles due to sputtering can be calculated using some simplifying assumptions:

a) the sputtering yield is constant,
b) no knock-on takes place,
c) the volume change due to the damage may be neglected, and
d) the profile is Gaussian.

Under these assumptions, the following is valid for the impurity profile:

$$C(x) = \frac{N}{2S} \left(erf \frac{x - R_p + N_\square \frac{S}{N}}{\sqrt{2}\, \Delta R_p} - erf \frac{x - R_p}{\sqrt{2}\Delta R_p} \right) \tag{4.21}$$

The saturation profile $(N_\square \to \infty)$ is given by:

$$C(x) = \frac{N}{2S}\, erfc \frac{x - R_p}{\sqrt{2}\Delta R_p} \tag{4.22}$$

with the maximum concentration at the surface:

$$C_{max} = \frac{N}{2S}\, erfc \frac{-R_p}{\sqrt{2}\Delta R_p} \approx \frac{N}{S} \text{ for } R_p > 3\Delta R_p \tag{4.23}$$

Equations (4.21) to (4.23) were derived under the assumption that the implantation profile extends from $-\infty$ to $+\infty$ (see Eq. (4.1)). Assuming that all implanted ions come to rest from 0 to $+\infty$ (see Eq. (4.2)),

154

$$C_{max} = N/S \qquad (4.24)$$

without any restrictions. This maximum concentration is independent of the implanted dose, and depends only on the relation of atomic density to sputtering yield. In silicon, e.g. the maximum concentration for arsenic (S = 3 at 60 keV) is at 1.5×10^{22} cm^{-3}, and for antimony (S = 5 at 150 keV) at 10^{22} cm^{-3}. For semiconductors with higher sputtering yields, (e.g. GaAs), this effect can be much more pronounced. In Fig.4.13, a theoretical example is given for antimony in silicon. Profile changes starting from doses of 10^{16} cm^{-2} are noticeable.

Sputtering limits in theory the maximum obtainable sheet carrier concentration. Due to solubility limits, however, this is more of theoretical interest. In Fig.4.14, examples are given for antimony (theoretical) and bismuth (experimental).

In the case of non-Gaussian profiles, sputtering-modified profiles have to be calculated numerically, solving:

$$C(x,t) = \int_{o}^{t} G(x+vt') \, dt \qquad (4.25)$$

with G being the implantation rate and v the velocity of sputtering erosion.

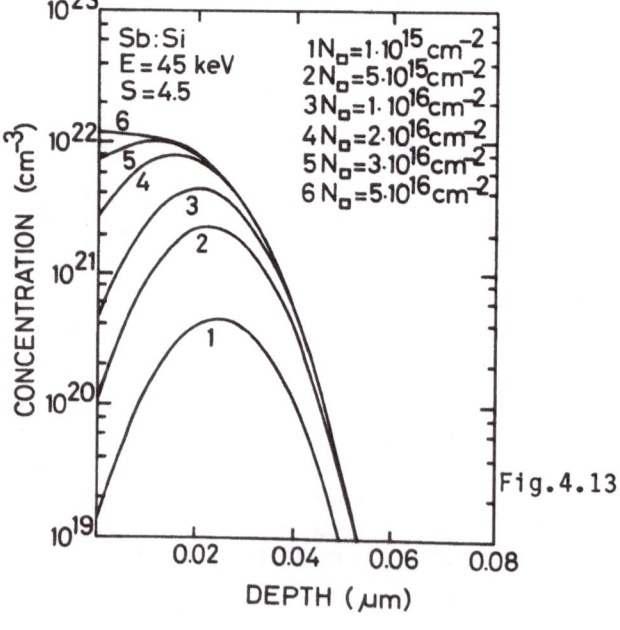

Fig.4.13 Doping profiles of implanted antimony, depending on dose and sputtering yield (E = 45 keV, S = 4.5)

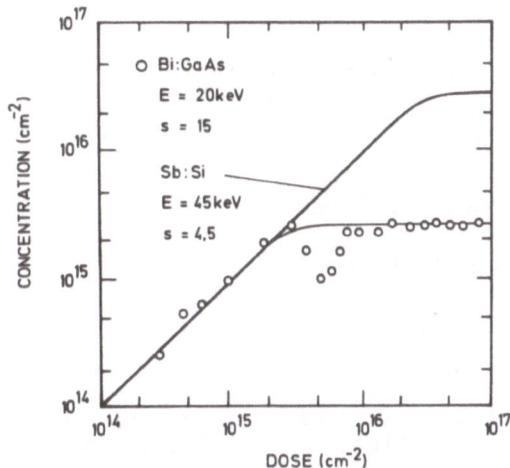

Fig.4.14 Integral doping concentration as a function of dose and sputtering yield: a) antimony in silicon (E = 45 keV, S = 4.5); b) bismuth in GaAs (E = 20 keV, S = 15); experimental data from Tinsley [4.26]

4.6 Lateral Spread

The lateral straggling of implanted ions have usually been neglected, since it is much lower than the range of the implanted ions. For VLSI circuits with dimensions in the micron or submicron range, however, it is becoming more and more important. The lateral straggling is, like the range straggling, a result of the scattering of the ions; therefore, both are of the same order of magnitude.

According to the calculations of Matsumura and Furukawa [4.27], the two-dimensional profile in case of a Gaussian implantation profile is given by

$$C(x,y) = \frac{N_\square}{(2\pi)^{1/2}\Delta R_p} \exp\left(-\frac{(x-R_p)^2}{2\Delta R_p^2}\right) \frac{1}{2}\left[\text{erfc}\,\frac{y-a}{\sqrt{2}\,\Delta R_{p,L}} - \right.$$

$$\left. -\,\text{erfc}\,\frac{y+a}{\sqrt{2}\,\Delta R_{p,L}}\right]$$

(4.26)

if the implantation was performed thru a slit of width 2a with infinite length. In Eq. (4.26) $R_{p,L}$ is the lateral spread of the ions. For real masking layers with an arbitrary shape, Runge [4.28] found for Gaussian profiles:

$$C(x,y) = \frac{N_\square}{2\pi \Delta R_{p,L} \Delta R_p} \int_{-\infty}^{\infty} \left[\exp\left(- \frac{(x-\xi)^2}{2\Delta R_{p,L}^2} - \frac{(x - d_{ox}(\xi)-R_p)^2}{2\Delta R_p^2}\right) \right] d\xi \qquad (4.27a)$$

with $d_{ox}(x)$ the thickness of the masking layer.

For non-Gaussian profiles, a multiplication with

$$\frac{1}{2} \left[\text{erfc} \frac{y-a}{\sqrt{2}\ \Delta R_{p,L}} - \text{erfc} \frac{y+a}{\sqrt{2}\ \Delta R_{p,L}} \right]$$

for an implantation thru an ideal slit of width 2a results in the desired profile. For arbitrary profiles, a convolution of the one-dimensional profile with a Gaussian lateral distribution having a standard deviation of $\Delta R_{p,L}$ is required:

$$C(x,y) = \frac{1}{2\pi R_{p,L}} \int_{-\infty}^{\infty} C(x-d(\xi)) \exp\left[- \frac{(y-\xi)^2}{2\Delta R_{p,L}^2} \right] d\xi \qquad (4.27b)$$

Examples of two-dimensional profiles are given in the following figures. In Fig. 4.15 a profile for an implantation thru an infinite steep mask is shown as an equi-concentration plot. The

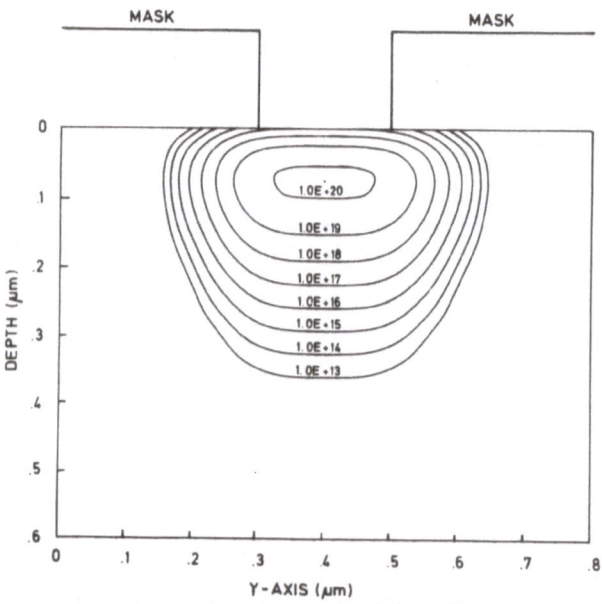

Fig. 4.15 Two-dimensional implantation profile thru an infinite steep mask of width 0.2 µm for 150 keV arsenic with 10^{15} cm^{-2}

same profile is given in Fig. 4.15 in a quasi-three-dimensional plot. The same implantation thru a tapered mask (45°) is shown in Figs.4.16 and 4.17. In both cases, Pearson-IV-distributions were assumed in vertical direction.

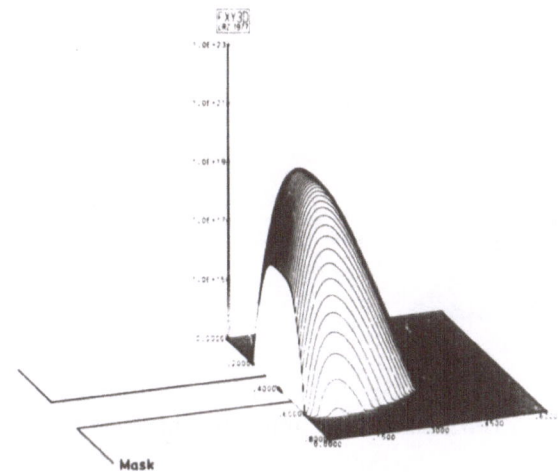

Fig. 4.16 Three-dimensional view of the profile shown in Fig. 4.15

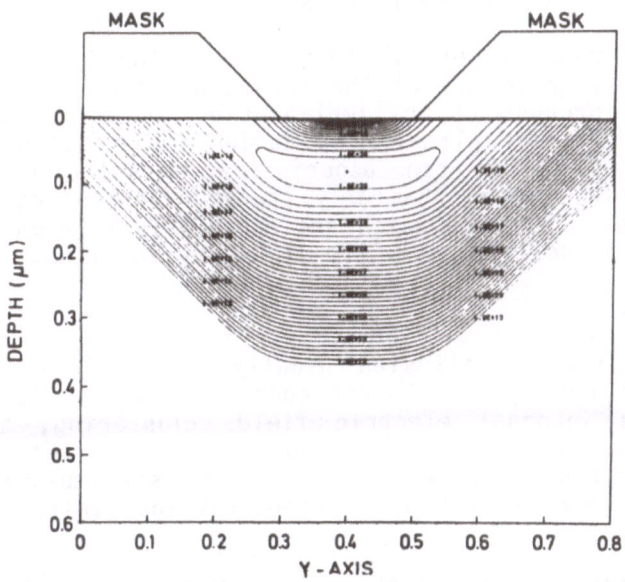

Fig. 4.17 Two-dimensional implantation profile thru a tapered mask (45°) for 150 keV boron with $10^{15} cm^{-2}$

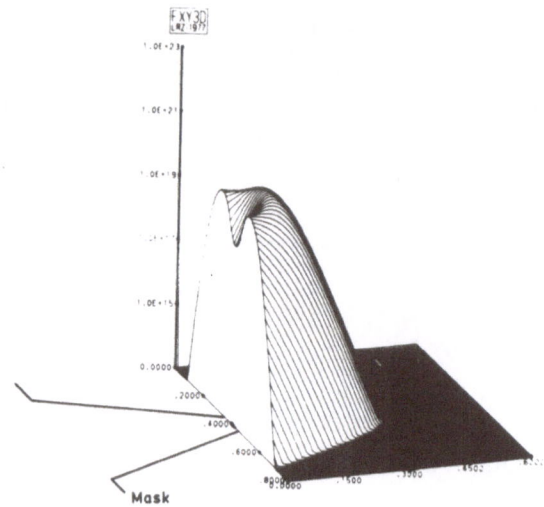

Fig. 4.18 Three-dimensional view of the profile shown in Fig. 4.17

5. DIFFUSION EFFECTS

Following ion implantation, a thermal treatment of the implanted silicon wafer is necessary to anneal the radiation damage produced as an unwanted by-product of the implantation. During annealing, diffusion usually takes place. Also during implantation at higher temperatures, diffusion can occur; and in addition, a damage-enhanced diffusion during the early stages of annealing is possible. Those diffusion effects which are especially related to ion implantation will be dealt with in the following sections.

5.1 Thermal Diffusion

During annealing, a thermal diffusion usually takes place. In general, the diffusion coefficient is not constant and depends on the Fermi level, the internal electric field, clustering, and damage enhanced diffusion [5.1,5.2]; therefore, the profiles after annealing have to be calculated using a process-simulation program. Such programs are described by Antoniadis and Tielert in this volume [5.3, 5.4].

For the purpose of estimating the influence of diffusion upon the profile, analytical solutions of the diffusion equation (Fick's second law) are possible for Gaussian implantation profiles.

Assuming an extension of the semiconductor from $-\infty$ to $+\infty$, the analytical solution is particularly simple:

$$C(x,t) = \frac{N_\square}{\sqrt{2\pi}\ \Delta R_p\ \sqrt{1+2\ Dt/\Delta R_p^2}}\ \exp\left[-\frac{(R_p-x)^2}{2\Delta R_p^2 + 4Dt}\right] \quad (5.1)$$

From Eq. (5.1), it can be seen that the profile retains the Gaussian shape, but the width of the distribution becomes larger due to the diffusion, resulting in an "effective standard deviation" of:

$$\Delta R_{p,eff} = \sqrt{\Delta R_p^2 + 2\ Dt} \quad (5.2)$$

Depending on the broadening of the profile, the maximum concentration is reduced.

An exact treatment of the problem leads to more complicated expressions. The two limiting cases:
- no out-diffusion $(\partial C/\partial x\,|_{x=0} = 0)$, and
- complete out-diffusion $(C(0,t) = 0)$
result in:

$$C(x) = \frac{N_\square}{2\sqrt{2\pi}\ \Delta R_p\ \sqrt{1+\dfrac{2\ Dt}{\Delta R_p^2}}}\Bigg\{\exp\left[-\frac{(x-R_p)^2}{2\Delta R_p^2 + 4Dt}\right]\cdot$$

$$\left[1+\mathrm{erf}\ \frac{\dfrac{R_p\sqrt{4Dt}}{\sqrt{2}\Delta R_p}+\dfrac{x\sqrt{2}\Delta R_p}{\sqrt{4Dt}}}{\sqrt{2\Delta R_p^2+4Dt}}\right] \pm\ \exp\left[-\frac{(x+R_p)^2}{2\Delta R_p^2+4Dt}\right]\cdot \quad (5.3)$$

$$\left[1+\mathrm{erf}\ \frac{\dfrac{R_p\sqrt{4Dt}}{\sqrt{2}\Delta R_p}-\dfrac{x\sqrt{2}\Delta R_p}{\sqrt{4Dt}}}{\sqrt{2\Delta R_p^2+4Dt}}\right]\Bigg\}$$

where the positive sign applies for $\partial C/\partial x\,|_{x=0} = 0$, and the negative sign for $C(0,t) = 0$. Examples of diffusion profiles, assuming no out-diffusion or complete out-diffusion, are given in Fig. 5.1 for a boron implantation at 40 keV into silicon.

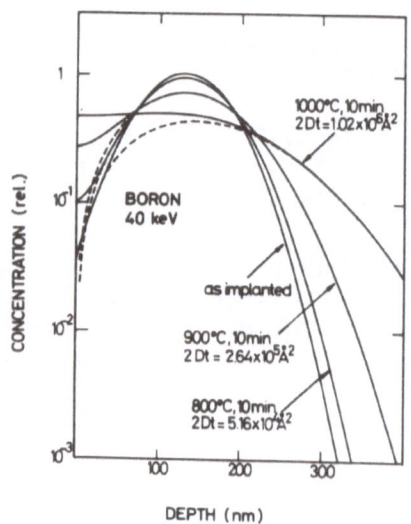

DEPTH (nm)

Fig. 5.1 Diffusion profiles of boron implanted at 40 keV with a dose of 5×10^{14} cm^{-2} for different values of Dt, assuming no out-diffusion or complete out-diffusion

For non-Gaussian profiles, the solution to the diffusion equation is given by:

$$C(x,t) = \frac{1}{\sqrt{4\pi Dt}} \int_0^\infty C(x') \left\{ \exp\left[-\frac{(x'-x)^2}{4\,Dt}\right] \pm \exp\left[-\frac{(x'+x)^2}{4\,Dt}\right] \right\} dx' \qquad (5.4)$$

were again the positive sign applies for no outdiffusion and the negative for outdiffusion where G(x) is the implantation profile, and has to be evaluated numerically.

5.2 Diffusion During Implantation

Two effects can cause diffusion during ion implantation. High-current implantation can heat silicon wafers to considerable temperatures, and wafers are sometimes implanted at elevated temperatures in order to anneal them already during the implantation. Certain elements, such as indium or gallium, already diffuse at these temperatures of about 400 to 600°C.

The second effect may play an equally important part in diffusion during implantation: Due to the stopping of the implanted ions, many point defects are created. The diffusion of impurity atoms is correlated to the point-defect concentration regardless of whether they diffuse via an interstitial or a substitutional mechanism (see, e.g., the contribution of Antoniadis in this volume). Therefore, the diffusion coefficient may be higher than its purely thermal value. Since point defects have a higher diffusion coefficient than impurity atoms, their increased diffusion coefficient can be assumed to be constant. Fick's second law, is written, in the case of diffusion during implantation, as:

$$\frac{\partial C}{\partial t} = - D \frac{\partial^2 C}{\partial x^2} + G(x) \qquad (5.5)$$

where $G(x)$ is the generation function of the implanted ions.

For the two cases:
- no out-diffusion $(\partial C/\partial x \,|_{x=0} = 0)$, and
- complete out-diffusion $(C(0,t) = 0)$,
the solution for an arbitrary generation rate is given by:

$$C(x,t) = \int_0^t \int_0^\infty \frac{G(x')}{\sqrt{4\pi D(t-\tau)}} \{ \exp[- \frac{(x'-x)^2}{4D(t-\tau)}] \qquad (5.6)$$

$$\pm \exp[- \frac{(x'+x)^2}{4D(t-\tau)}]\} \, dx' d\tau$$

where the positive sign holds for the first and the negative for the second case.

In the case of a Gaussian generation rate of the ions:

$$G(x) = \frac{j}{q\sqrt{2\pi}\, \Delta R_p} \exp[- \frac{(R_p-x)^2}{2\,\Delta R_p^2}] \qquad (5.7)$$

and assuming an infinitely extended semiconductor, an analytical solution to Eq. (5.5) is possible, yielding:

$$C(x,t) = \frac{j/q}{2\sqrt{\pi}\,D} \{ \sqrt{2\Delta R_p^2 + 4Dt} \, \exp[- \frac{(R_p - x)^2}{2\,\Delta R_p^2 + 4Dt}] -$$

$$(5.8)$$

$$- \sqrt{2}\Delta R_p \exp[- \frac{(R_p-x)^2}{2\,\Delta R_p^2}] - \sqrt{\pi}(R_p-x) \, \mathrm{erf} \frac{R_p-x}{\sqrt{2}\,\Delta R_p} +$$

$$+ \sqrt{\pi}\,(R_p -x) \, \mathrm{erf} \frac{R_p - x}{\sqrt{2\Delta R_p^2 +4Dt}} \}$$

162

An example of resulting impurity profiles, assuming no out-diffusion for the same diffusion parameters as in Fig. 5.1, is given in Fig. 5.2.

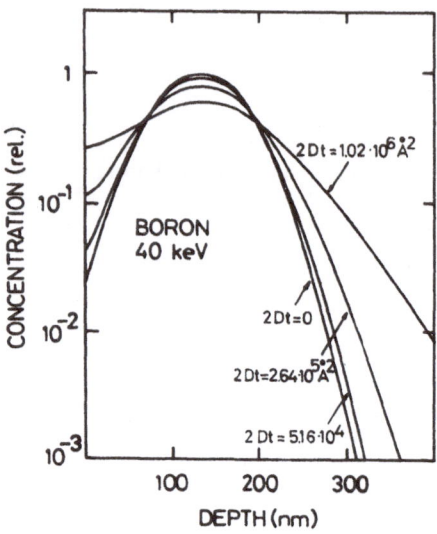

Fig. 5.2 Diffusion during implantation, assuming the same diffusion parameters as in Fig. 5.1

The diffusion effect is less than in the case of thermal diffusion during annealing. This is caused by the fact that, in the case of post-implantation diffusion, all atoms diffuse in the same time; whereas in the case of diffusion during implantation, most atoms are brought into the crystal during diffusion, and their diffusion time is shorter. The effect has been found to apply in the cases of indium, antimony and gallium [5.7].

5.3 Damage-Enhanced Diffusion

During annealing of implanted layers, point defects are released from the damage, thus enhancing the diffusion coefficient. After the crystal lattice has recrystallized, the emission of point defects stops. This effect has been found to be extremely pronounced in the cases of aluminum and boron [5.5],[5.6]. Especially at low temperatures, damage-enhanced diffusion can be more important than thermal diffusion. In Fig. 5.3, enhancement of boron diffusion is plotted, assuming a constant enhancement for a period of 10 min [5.6]. In the case of boron, this enhancement is independent of the implantation dose. This is probably due to the fact that, up to the highest doses, no amorphous layers are formed during boron implantation; resulting in a damage concentration proportional to the dose, which in turn causes the same enhancement for all doses.

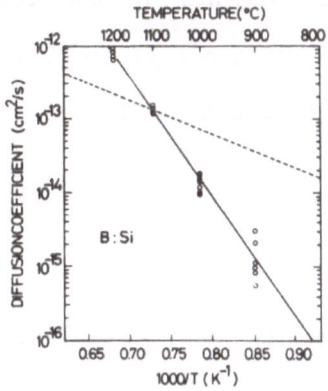

Fig. 5.3 Enhancement of boron diffusion caused by point-defect release during annealing

For other elements such as arsenic and phosphorus, a dose dependence and a dose-rate dependence (because of self-annealing during implantation) of the damage-enhanced diffusion have to be expected.

At temperatures above approx. 950°C, thermal diffusion prevails.

6. APPENDIX

6.1 Range Program

Biersack's algorithm described in 2.2 was implemented in the computer program DIMUS, which calculates ion ranges and straggling in arbitrary compounds. To compute the stopping power of the compound we use Bragg's rule; this is motivated by the fact that the interesting energy range is well above the chemical binding energies. In the program, the correct stopping powers are used particularily beyond the electronic stopping maximum but without relativistic corrections (i.e. up to ≈ 10 MeV/mm). Also higher order energy loss moments in both nuclear and electronic stopping are used.

To start the program one has to log on the computer and type:

    run dimus
    The computer asks now for the required information:

```
TARGET
<arbitrary target description>
RHO, NTA, (ZT(I), MT(I), N(I))
```

<target density>, <number of different elements in the compound>
<atomic number >, <atomic mass>, <number of atoms of the first element>
<...............>. <............>, <...of second element>
ION

<arbitrary ion specification>

ZI, MI, COR

<atomic number>, <atomic mass>, <correction factor for the LSS-stopping power, defaults to 1>

A sample session is shown for Krypton in PMMA ($C_5H_8O_2$, $\rho = 1.19$ g cm^{-3})

```
run dimus

TARGET

polymethylmethacrylat

RHO, NTA, (ZT(I), MT(I), N(I))
1.19, 3
6, 12.010, 5
1, 1.008, 8
8, 15.999, 2

ION

kr

ZI, MI, COR

36, 83.912, 1

ION
```

<either a new ion specification for the same target or ctrl Z>

The program is terminated by typing ctrl Z. To obtain a print-out of the range listing, one types "print for002.dat".

```
0001          PROGRAM DIMUS
0002          DIMENSION X(100),ITARG(20)
0003          COMMON /WPAR/NTA,D(10),CA(10),F(10),RK(10),EPSK(10),SRK(10)
              1 ,OT(10),RMT(10),ZT(10)
0004          REAL MY,MI,KL,IO,N,K,
              1 M111,M112,M021,M022,M221,M222
0005          DOUBLEPRECISION ION,SCR
0006          DATA EE/1001./,COR/1./
       * ,X
       * / 10.,   20.,   30.,   40.,   50.,   60.,   70.,   80.,   90.,  100.,
       * 110.,  120.,  130.,  140.,  150.,  160.,  170.,  180.,  190.,  200.,
       * 220.,  240.,  260.,  280.,  300.,  320.,  340.,  360.,  380.,  400.,
       * 420.,  440.,  460.,  480.,  500.,  550.,  600.,  650.,  700.,  750.,
       * 800.,  850.,  900.,  950., 1000., 1500., 2000., 2500., 3000., 3500.,
       * 4000.,4500.,5000.,47*0./
       * ,ION/'BOR'/, RHO/2.32/
       * ,NTA,ZT(1),RMT(1),OT(1)   /1,14.,28.,1./
       * ,ZI,MI,COR/5.,11.,1.59/
0007          CALL ASNLUN(6,'TI',0)
0008   500    WRITE(6,*)'TARGET'
0009          READ(5,1005,END=9999)ITARG
0010   1005   FORMAT(20A2)
0011          WRITE(6,*)'RHO,NTA,(ZT(I),MT(I),N(I))'
0012          READ(5,*,END=9999)RHO,NTA,(ZT(I),RMT(I),OT(I),I=1,NTA)
0013   600    WRITE(6,*)'ION'
0014          READ(5,1010,END=9998)II,ION
0015          IF(II.EQ.0)GOTO 500
0016   1010   FORMAT(Q,A8)
0017          WRITE(6,*)'ZI,MI,COR'
0018          READ(5,*,END=9998)ZI,MI,COR
0019          WRITE(2,2001)
0020   2001   FORMAT('1')
0021          WRITE(2,2000)
0022   2000   FORMAT(5X,'I',51(1H ),'I',51(1H ),'I')
0023          WRITE(2,2010)
0024   2010   FORMAT(5X,'I',51X,'I',51X,'I')
0025          WRITE(2,2020)ITARG,ION
0026   2020   FORMAT(5X,'I',3X,'TARGET :',20A2,'I',2X,
              1 'ION :',A8,36X,'I')
0027          WRITE(2,2030)INT(.5+ZT(1)),RMT(1),RHO,OT(1),
              1 INT(.5+ZI),MI,COR
0028   2030   FORMAT(5X,'I',3X,'ZT :',I3,4X,'MT :',F7.3,4X,'RHO :',F7.3,
              1 2X,F7.3,1X,'I',3X,'ZI :',I3,5X,'MI :',F7.3,10X,'COR :'
              1 ,F7.3,3X,'I')
0029          IF(NTA.LE.1)GOTO 550
0030          DO 540 I=2,NTA
0031   540    WRITE(2,2035)INT(.5+ZT(I)),RMT(I),OT(I)
0032   2035   FORMAT(5X,'I',3X,'ZT :',I3,4X,'MT :',F7.3,5X,'N :',F7.3,
              1 10X,'I',51X,'I')
0033   550    WRITE(2,2010)
0034          WRITE(2,2000)
0035          WRITE(2,2040)
0036   2040   FORMAT(5X,'I',6(1H ),'I',3(14(1H ),'I'),
              1 4(12(1H-),'I'))
0037          WRITE(2,2050)
0038          WRITE(2,2055)
```

```
0039   2050   FORMAT(5X,'I',2X,'E ',2X,'I',6X,'SE',6X,'I',
              1 6X,'SN',6X,'I',6X,'W2',6X,'I',5X,' R',5X,'I',
              2 5X,'DR',5X,'I',5X,'RP',5X,'I',5X,'DRP',4X,'I')
0040   2055   FORMAT(5X,'I',[KEV],1X,'I',3X,'[KEV/MY]',3X,'I'
              1 3X,'[KEV/MY]',3X,'I',1X,'[KEV**2/MY]',2X,'I',4X,'[MY]',4X,'I'
              2 4X,'[MY]',4X,'I',4X,'[MY]',4X,'I',4X,'[MY]',4X,'I')
0041          WRITE(2,2040)
       C
```

```
        C****    RECHENPROGRAMM - ANFANG
        C
0042            ELZ=0.
0043            QT=0.
0044            DO 10 I=1,NTA
0045            QT=QT+RMT(I)*OT(I)
0046     10     ELZ=ELZ+OT(I)
0047            QT=QT/ELZ
0048            N=.6022*RHO/QT
0049            DO 20 I=1,NTA
0050            CALL COEF(COR,N,MI,ZI,QMT(I),ZT(I),CA(I),B(I),F(I)
0051     20     OT(I)=OT(I)/ELZ
0052            MY=QT/MI
0053            B=.1
0054            E1=R
0055            SN1=WI(0,1,E1)
0056            W21=WI(0,2,E1)
0057            W31=WI(0,3,E1)
0058            W41=WI(0,4,E1)
0059            W51=WI(0,5,E1)
0060            W61=WI(0,6,E1)
0061            SE1=WI(1,0,E1)
0062            S1=SE1+SN1
        C
0063            P11=.5*MY*SN1/E1-.25*(.5-MY)*W21/E1**2-.125*(1.-1.5*MY)*W31
               1 /E1**3-5./32.*(.75-MY)*W41/E1**4-7./64.*(1.-1.25*MY)*W51
               2 /E1**5-105./1024.*(1.-1.2*MY)*W61/E1**6
        C
0064            P21=1.5*MY*SN1/E1-.375*(1.-MY)**2*(W21/E1**2+W31/E1**3
               1 +W41/E1**4+W51/E1**5+W61/E1**6)
        C
0065            R1=.5*B/S1
0066            DR1=SQRT(.5*R*W21/S1**3)
0067            EX11=EXP(-.5*B*P11/S1)
0068            EX21=EXP(-.5*B*P21/S1)
0069            M111=R1
0070            M021=R*M111/S1
0071            M221=B*M111/S1
        C
0072            DO 100 I=1,100
0073     3      C=10.-E1
        C
0074            Q=(10.-B)/16.
        C
0075            DE=5.
0076            IF(I.LE.45)DE=1.
0077            IF(I.LE.35)DE=.5
```

```
0078            IF(I.EQ.1)DE=AMIN1(B,Q,C)
0079            E2=E1+DE
        C...
0080            IF(E2.GT.EE)GOTO 600
        C...
0081            SN2=WI(0,1,E2)
        C
0082            W22=WI(0,2,E2)
0083            W32=WI(0,3,E2)
0084            W42=WI(0,4,E2)
0085            W52=WI(0,5,E2)
0086            W62=WI(0,6,E2)
        C
0087            SE2=WI(1,0,E2)
        C
0088            S2=SE2+SN2
        C
0089            P12=.5*MY*SN2/E2-.25*(.5-MY)*W22/E2**2-.125*(1.-1.5*MY)*W32
               1 /E2**3-5./32.*(.75-MY)*W42/E2**4-7./64.*(1.-1.25*MY)*W52
               2 /E2**5-105./1024.*(1.-1.2*MY)*W62/E2**6
```

```
          C
0090              P22=1.5*MY*SN2/E2-.375*(1.-MY)**2*(W22/E2**2+W32/E2**3
         1    +W42/E2**4+W52/E2**5+W62/E2**6)
          C
0091              R2=.5*DE*(1./S2+1./S1)+R1
0092              DR2=SQRT(.5*DE*(W22/S2**3+W21/S1**3)+DR1**2)
0093              EX12=EXP(-.5*DE*(P11/S1+P12/S2))
0094              EX22=EXP(-.5*DE*(P21/S1+P22/S2))
          C
0095              M112=M111*EX12+.5*DE*(1./S2+EX12/S1)
0096              GM11=.5*DR2**2*((P12**2+S2*(P12-P11)/DE)*M112-P12)
0097              RP=M112+GM11
0098              M022=M021+DE*(M112/S2+M111/S1)
0099              GM02=DR2**2*(1.-P12*M112)
0100              RM02=M022+GM02
          C
0101              M222=M221*EX22+DE*(M112/S2+M111*EX22/S1)
          C
0102              GM22=.5*DR2**2*((P22**2+S2*(P22-P21)/DE)*M222-2.*(P12+P22)
         1    *M112+2.)
          C
0103              RM22=M222+GM22
0104              DRP=SQRT(1./3.*RM02+2./3.*RM22-RP**2)
0105              E1=E2
0106              SE1=SE2
0107              SN1=SN2
0108              W21=W22
0109              W31=W32
0110              W41=W42
0111              W51=W52
0112              W61=W62
          C
0113              S1=S2
0114              P11=P12
0115              P21=P22
0116              R1=R2
```

FORTRAN IV-PLUS V3.0   11:46:00   5-Jul-82   Page 4
DIMUS.FTN;63   /TR:BLOCKS

```
0117              DR1=DR2
0118              M111=M112
0119              M021=M022
0120              M221=M222
          C
0121              IF(E1.NE.X(I))GOTO 3
          C
0122              WRITE(2,2060)INT(.5+E1),SE1,SN1,W21,R1,DR1,RP,DRP
0123      2060    FORMAT(5X,'I',1X,I4,1X,'I',3(1X,1PE12.3,1X,'I'),
         1    0P,4(1X,F10.4,1X,'I'))
          C
0124              IF(I.EQ.20.OR.I.EQ.35.OR.I.EQ.45.OR.I.EQ.53)
         1    WRITE(2,2040)
          C
0125      100     CONTINUE
0126              GOTO 600
0127      9999    STOP 'EOF AFTER TARGET'
0128      9998    STOP 'EOF AFTER ION'
0129              END
```

PROGRAM SECTIONS

Name	Size		Attributes
$CODE1	005576	1471	RW,I,CON,LCL
$PDATA	001340	368	RW,D,CON,LCL
$IDATA	000042	17	RW,D,CON,LCL
$VARS	001270	348	RW,D,CON,LCL
$TEMPS	000026	11	RW,D,CON,LCL
WPAR	000552	181	RW,D,OVR,GBL

Total Space Allocated = 011270   2396

```
FORTRAN IV-PLUS V3.0          11:46:35     5 Jul 82        Page 5
DIMUS.FTN;63              /TR:BLOCKS

0001              FUNCTION WI(IE,I,EN)
0002              COMMON /WPAR/NTA,D(10),GA(10),F(10),QK(10),EPSK(10),SBK(10)
                 1 ,OT(10),RMT(10),ZT(10)
      C
      C****       POTENTIALE
      C
0003              DIMENSION WPA(4,6)
0004              DATA
                 1 WPA /4*0.,.18696,-1.6828,6.9825,-.94342,
                 2       11.340, .9569 ,.1033 ,-1.230 ,
                 3       15.8   , .9512 ,.29   ,-1.71  ,
                 4       20.495,-.9538 ,.2697 ,-1.752 ,
                 5       24.95 , .957 ,.2989 ,-1.76    /
0005              WI=0.
0006              IF(IE.NE.0)GOTO 10
0007              IF(I.LE.2)GOTO 20
0008              DO 1 J=1,NTA
0009     1        WI=WI+OT(J)*D(J)*GA(J)*(GA(J)*EN)**(I-2)/(F(J)**2*
                 1 (4*I-4+WPA(1,I)*(F(J)*EN)**WPA(2,I)+
                 2      WPA(3,I)*(F(J)*EN)**WPA(4,I)))
0010              RETURN
0011     20       IF(I.EQ.1)GOTO 30
0012              DO 2 J=1,NTA
0013     2        WI=WI+OT(J)*D(J)*GA(J)/(F(J)**2*
                 1 (4+WPA(1,2)*(F(J)*EN)**WPA(2,2)+
                 2      WPA(3,2)*(F(J)*EN)**WPA(4,2)))
0014              RETURN
0015     10       DO 3 J=1,NTA
0016     3        WI=WI+OT(J)/(1./(QK(J)*SQRT(EN))+EPSK(J)*EN/(SBK(J)*
                 1 ALOG(1.+EPSK(J)*EN+(5./(EPSK(J)*EN)))))
0017              RETURN
0018     30       DO 4 J=1,NTA
0019     4        WI=WI+OT(J)*D(J)*.5*ALOG(1.+F(J)*EN)/(F(J)*(F(J)*EN+.051953*
                 1 (F(J)*EN)**.32011))
      C
      C****       ENDE DER POTENTIALE
      C
0020              RETURN
0021              END
```

PROGRAM SECTIONS

Name	Size		Attributes
$CODE1	001256	343	RW,I,CON,LCL
$PDATA	000010	4	RW,D,CON,LCL
$IDATA	000004	2	RW,D,CON,LCL
$VARS	000142	49	RW,D,CON,LCL
$TEMPS	000016	7	RW,D,CON,LCL
WPAR	000552	181	RW,D,OVR,GBL

Total Space Allocated = 002224   506

```
FORTRAN IV-PLUS V3.0          11:46:44     5 Jul 82        Page 6
DIMUS.FTN;63              /TR:BLOCKS

0001              SUBROUTINE COEF(COR,N,MI,ZI,MT,ZT,GA,D,F,K,EPSK,SBK)
0002              REAL MY,MI,MT,KL,N,K,IO
0003              A=.0053*.529/(SQRT(ZT)+SQRT(ZI))**(2./3.)
0004              GA=4.*MI*MT/(MI+MT)**2
0005              D=3.1415927*A**2*N*GA*1E4
0006              F=A*MT/(ZT*ZI*14.41*(MI+MT))*1E3
0007              KL=38.33*ZI**(7./6.)*ZT/(ZI**(2./3.)+ZT**(2./3.))**1.5/SQRT(MI)
0008              K=COR*KL*N/1000.*1E4
0009              IO=12.17./ZI
```

```
0010            IF(ZI.GF.13.)10=9.76+50.5*ZI**(-1.19)
0011            EPSK=1./(.454*ZI*MI*I0)
0012            SBK=8.*3.1415927*N*(ZT*14.41)**2/(1000.*I0)*JE4
0013            RETURN
0014            END
```

PROGRAM SECTIONS

Name	Size		Attributes
$CODE1	000632	205	RW,I,CON,LCL
$PDATA	000044	18	RW,D,CON,LCL
$VARS	000020	8	RW,D,CON,LCL
$TEMPS	000004	2	RW,D,CON,LCL

Total Space Allocated = 000722   233

6.2 Pearson Program

The pearson program is the implantation processor of the process-modeling program ICECREM. The program is started by typing:
    run pearson

The computer responds

ELEM	ENERGY	DOSE	RANG	STDV	GAMMA	BETA2	SPUT
<e.1>	<z.2>	<z.3>					[<z.8>]

or <e.1>   <z.3>   <z.4> <z.5   [<z.6> <z.7>] [<z.8>]

<e.1> indicates which element is to be implanted (boron, phosphorus, Sb, As), in the first form, the profile is constructed with 4 moments from the stored tables. <z.2> indicates the energy in keV, <z.3> is the dose per square centimeter.

In the second form, a Pearson distribution is produced,

    RANG = <range>
    STDV = <standard deviation>
    GAMMA = <third moment>/STDV**3
    BETA2 = <fourth moment>/STDV**4

If <z.6> <z.7> are not specified, a Gaussian distribution is produced (GAMMA = 0, BETA2 = 3).

SPUT is the sputtering ratio, i.e.
SPUT = <number of sputtered atoms> <number of primary ions>

The program results in a simple profile output on the line printer. For practical applications, a plotting routine for the plotter used has to be added.

170

```
0001            PROGRAM MAINIT
0002            REAL*8 ITEML(8)
0003            DIMENSION ITYPL(8),Y(130)
0004            INTEGER*4 ISEML(2,3)
0005            COMMON /IMPL/ISEM,DATAF(16)
0006            COMMON /PEAR/B0,B1,B2,AMX,R,S
0007         DATA
          2      ITEML/
          4      'ELEM    ',  'ENERGY  ',  'DOSE    ',  'RANG    ',   ! IMPLNT
          5      'STDV    ',  'GAMMA   ',  'BETA2   ',  'SPUT    '/
0008            DATA ITYPL/4,7*1/
          6      ISEML/ '177,7, '177,'35, '177,'175/
0009            CALL SYNLIN(DATAF,ISEM,ITEML,ITYPL,8,IERR)
0010            IF(IERR.NE.0)STOP
0011            DO 10 I=1,3
0012      10    IF((ISEM.AND.ISEML(1,I)).EQ.ISEML(2,I))GOTO 20
0013            STOP'SEMANTICAL ERROR'
0014      20    CALL IMPLIT
0015            DPMX=R+3*S
0016            WRITE(5,*)'PLOTDEPTH= ',DPMX,' ENTER REQUIRED VALUE OR /'
0017            READ(5,*)DPMX
0018            DX=DPMX/130
0019            H=0
0020            SUM=0
0021            VMAX=0
0022            DO 30 I=1,130
0023            Y(I)=PEARS(H)
0024            VMAX=MAX(Y(I),VMAX)
0025            SUM=SUM+Y(I)
0026      30    H=H+DX
          C
          C      SPUTTERING?
          C
0027            SPUTDP=0
0028            IF(ISHFT(ISEM,-7).NE.0)SPUTDP=DATAF(3)*DATAF(8)*(1E4/5E22)
          C    5E22 ATOME/CM**3,1E4 CM/MY
0029            NSPUT=SPUTDP/DX
0030            SUM=SUM*DX*1E-4
0031            VMAX=VMAX*DATAF(3)/(SUM*(NSPUT+1))
0032            IF(NSPUT.EQ.0)GOTO 55
0033            DO 50 I=1,130-1
0034            DO 50 J=I+1,MIN(I+NSPUT,130)
0035      50    Y(I)=Y(I)+Y(J)
0036      55    CONTINUE
0037            DO 40 I=1,130
0038      40    Y(I)=LOG(MAX(Y(I)*DATAF(3)/(SUM*(NSPUT+1)),VMAX*1E-4))/LOG(10.)
0039            CALL DRUPLO(Y,DPMX/130,7)
0040            CALL PLOTNDD(1)
0041            WRITE(1,100)DPMX
0042     100    FORMAT(' DEPTH TO ',1PG20.4,T130,'I')
0043            END
```

```
0001         SUBROUTINE IMPLIT
          C
          C    IMPLNT MODELS THE IMPLANTATION STEP
          C
          C    THE CONCENTRATION ARRAY C IS SET UP WITH A PEARSON DISTRIBUTION
          C    OR A GAUSSIAN DISTRIBUTION ACCORDING TO INPUT SPECIFICATION
          C
          C    MATERIAL ASSIGNMENT:
          C    MAT=1:BOR, MAT=2:PHOSPHOR, MAT=3:ANTIMON, MAT=4:AS, MAT=5: GA
          C
0002            COMMON /PEAR/B0,B1,B2,AMX,RR,SS,ICTL
0003            COMMON /IMPL/ISEM,DATAF(16)
0004            PARAMETER PI=3.1415926535
```

```
0005          REAL HSP(400),SGMO(46,5),
             ,DRPBO(46),DRPASO(46),DRPGAO(46),
             ,DRPSBO(46),DRPPO(46),RPGA(46),DRPGA(46),GMGA(46),
             ,RNG(46,5),SGM(46,5),GAM(46,5),RPB(46),S2(46),S3(46),S4(46),
             ,DRPB(46),S6(46),S7(46),S8(46),GMB(46),S10(46),S11(46),S12(46)
       C
       C
       C
0006          EQUIVALENCE(RNG(1,1),RPB(1)),(RNG(1,2),S2(1)),(RNG(1,3),S3(1)),
             ,(RNG(1,4),S4(1)),(SGM(1,1),DRPB(1)),(SGM(1,2),S6(1)),(SGM(1,3),
             ,S7(1)),(SGM(1,4),S8(1)),(GAM(1,1),GMB(1)),(GAM(1,2),S10(1)),
             ,(GAM(1,3),S11(1)),(GAM(1,4),S12(1)),(R,DATAF(4)),(S,DATAF(5)),
             ,(G,DATAF(6)),(B,DATAF(7)),(ENERG,DATAF(2)),(DOSE,DATAF(3)),
             ,(SGMO(1,1),DRPBO(1)),(SGMO(1,2),DRPASO(1)),(SGMO(1,3),DRPSBO(1)),
             ,(SGMO(1,4),DRPASO(1)),(SGMO(1,5),DRPGAO(1))
       C
0007          DIMENSION FM(5),FM1(5),FMO(5),FMO1(5)
0008          DIMENSION NEN(11),IFNE(11)
0009          DATA NEN/0,10,20,25,30,35,37,39,41,43,45/,
             1    IENE/0,50,100,150,200,300,400,500,750,1000,1250/
0010          DATA FM /25,2*1.5,0.,1.5/,FM1/8.,2*4.5,0.,4.5/
0011          DATA S2/ 0.,
             1        0.0139, 0.0253, 0.0368, 0.0486, 0.0607, 0.0730, 0.0855,
             2        0.0981, 0.1109, 0.1230, 0.1367, 0.1497, 0.1627, 0.1757,
             3        0.1888, 0.2019, 0.2149, 0.2279, 0.2409, 0.2539, 0.2798,
             4        0.3054, 0.3309, 0.3562, 0.3812, 0.4060, 0.4306, 0.4549,
             5        0.4790, 0.5029, 0.5265, 0.5499, 0.5730, 0.5959, 0.6186,
             6        0.6744, 0.7288, 0.7819, 0.8338, 0.8846, 0.9343, 0.9829,
             7        1.0306, 1.0773, 1.1231/
0012          DATA S3/ 0.,
             1        0.0088, 0.0141, 0.0187, 0.0230, 0.0271, 0.0310, 0.0347,
             2        0.0385, 0.0421, 0.0457, 0.0493, 0.0529, 0.0564, 0.0599,
             3        0.0634, 0.0669, 0.0704, 0.0739, 0.0773, 0.0808, 0.0870,
             4        0.0947, 0.1017, 0.1086, 0.1156, 0.1227, 0.1297, 0.1368,
             5        0.1439, 0.1510, 0.1581, 0.1653, 0.1725, 0.1797, 0.1870,
             6        0.2052, 0.2236, 0.2421, 0.2608, 0.2796, 0.2985, 0.3175,
             7        0.3366, 0.3559, 0.3752/
0013          DATA S4/ 0.,
             1        0.0097, 0.0159, 0.0215, 0.0269, 0.0322, 0.0374, 0.0426,
             2        0.0478, 0.0530, 0.0582, 0.0634, 0.0686, 0.0739, 0.0791,
             3        0.0845, 0.0898, 0.0952, 0.1005, 0.1060, 0.1114, 0.1223,
             4        0.1334, 0.1445, 0.1558, 0.1671, 0.1785, 0.1900, 0.2015,
             5        0.2131, 0.2247, 0.2364, 0.2482, 0.2600, 0.2718, 0.2837,
```

```
FORTRAN IV-PLUS V3.0              16:39:21    5 Jul-82          Pase 3
SCR.#1                     /TR:BLOCKS

             6        0.3135, 0.3435, 0.3736, 0.4038, 0.4340, 0.4643, 0.4946,
             7        0.5245, 0.5552, 0.5854/
0014          DATA S6/ 0.,
             1        0.0069, 0.0119, 0.0166, 0.0212, 0.0256, 0.0298, 0.0340, 0.0380,
             2        0.0418, 0.0456, 0.0492, 0.0528, 0.0562, 0.0595, 0.0628, 0.0659,
             3        0.0689, 0.0719, 0.0747, 0.0775, 0.0829, 0.0880, 0.9228, 0.9974,
             4        0.1017, 0.1059, 0.1098, 0.1136, 0.1172, 0.1206, 0.1239, 0.1271,
             5        0.1301, 0.1330, 0.1358, 0.1424, 0.1484, 0.1539, 0.1590, 0.1636,
             6        0.1680, 0.1721, 0.1758, 0.1794, 0.1827/
0015          DATA S7/ 0.,
             1        0.0026, 0.0043, 0.0058, 0.0071, 0.0084, 0.0096, 0.0107, 0.0118,
             2        0.0130, 0.0140, 0.0151, 0.0162, 0.0172, 0.0183, 0.0193, 0.0203,
             3        0.0213, 0.0224, 0.0234, 0.0244, 0.0264, 0.0283, 0.0303, 0.0322,
             4        0.0342, 0.0361, 0.0380, 0.0400, 0.0419, 0.0438, 0.0457, 0.0475,
             5        0.0494, 0.0513, 0.0531, 0.0578, 0.0623, 0.0669, 0.0713, 0.0758,
             6        0.0802, 0.0845, 0.0888, 0.0930, 0.0972/
0016          DATA S8/ 0.,
             1        0.0036, 0.0059, 0.0080, 0.0099, 0.0118, 0.0136, 0.0154, 0.0172,
             2        0.0189, 0.0207, 0.0224, 0.0241, 0.0258, 0.0275, 0.0292, 0.0308,
             3        0.0325, 0.0341, 0.0358, 0.0374, 0.0407, 0.0439, 0.0470, 0.0502,
             4        0.0533, 0.0564, 0.0594, 0.0624, 0.0654, 0.0684, 0.0713, 0.0741,
             5        0.0770, 0.0798, 0.0826, 0.0894, 0.0960, 0.1024, 0.1087, 0.1147,
             6        0.1206, 0.1263, 0.1319, 0.1373, 0.1425/
```

172

```
0017          DATA S10/ 0.,
      1   0.641,   0.549,   0.459,   0.384,   0.319,   0.261,   0.209,   0.162,
      2   0.119,   0.080,   0.044,   0.011,  -0.021,  -0.049,  -0.075,  -0.100,
      3  -0.122,  -0.141,  -0.158,  -0.181,  -0.224,  -0.263,  -0.298,  -0.332,
      4  -0.365,  -0.396,  -0.424,  -0.451,  -0.475,  -0.501,  -0.523,  -0.544,
      5  -0.566,  -0.588,  -0.608,  -0.662,  -0.711,  -0.757,  -0.799,  -0.839,
      6  -0.876,  -0.911,  -0.945,  -0.976,  -1.007/
0018          DATA S11/ 0.,
      1   0.638,   0.584,   0.574,   0.570,   0.567,   0.565,   0.562,   0.560,
      2   0.558,   0.556,   0.554,   0.552,   0.550,   0.547,   0.545,   0.543,
      3   0.541,   0.538,   0.536,   0.534,   0.525,   0.517,   0.510,   0.503,
      4   0.496,   0.490,   0.483,   0.476,   0.469,   0.464,   0.459,   0.453,
      5   0.445,   0.442,   0.438,   0.421,   0.405,   0.390,   0.376,   0.362,
      6   0.350,   0.340,   0.328,   0.317,   0.307/
0019          DATA S12/ 0.,
      1   0.675,   0.647,   0.635,   0.627,   0.619,   0.611,   0.604,   0.597,
      2   0.590,   0.583,   0.576,   0.569,   0.563,   0.557,   0.551,   0.544,
      3   0.537,   0.529,   0.526,   0.522,   0.499,   0.478,   0.459,   0.443,
      4   0.427,   0.412,   0.393,   0.377,   0.361,   0.348,   0.344,   0.336,
      5   0.328,   0.320,   0.317,   0.287,   0.260,   0.236,   0.215,   0.197,
      6   0.183,   0.174,   0.163,   0.155,   0.147/
0020          DATA DRPBO/ 0.,
      * 0.0187,0.0320,0.0441,0.0534,0.0611,0.0677,0.0735,0.0784
      *,0.0829,0.0868,0.0904,0.0936,0.0965,0.0992,0.1017,0.1040
      *,0.1061,0.1081,0.1099,0.1117,0.1140,0.1176,0.1202,0.1225
      *,0.1246,0.1265,0.1283,0.1299,0.1315,0.1329,0.1342,0.1354
      *,0.1366,0.1377,0.1388,0.1412,0.1433,0.1451,0.1468,0.1484
      *,0.1498,0.1511,0.1523,0.1534,0.1544/
0021          DATA DRPASO/ 0.,
      * 0.0042,0.0068,0.0092,0.0115,0.0138,0.0160,0.0183,0.0205
      *,0.0226,0.0247,0.0268,0.0290,0.0310,0.0332,0.0352,0.0373
      *,0.0393,0.0413,0.0434,0.0495,0.0534,0.0572,0.0611
      *,0.0649,0.0687,0.0724,0.0760,0.0797,0.0832,0.0867,0.0901
```

```
      *,0.0936,0.0969,0.1003,0.1084,0.1163,0.1237,0.1310,0.1381
      *,0.1448,0.1515,0.1579,0.1640,0.1699/
0022          DATA RPB/ 0.,
      * 0.0309,0.0632,0.0952,0.1263,0.1563,0.1852,0.2131,0.2401,
      * 0.2661,0.2914,0.3159,0.3397,0.3629,0.3854,0.4075,0.4290,
      * 0.4500,0.4706,0.4908,0.5106,0.5490,0.5861,0.6220,0.6548,
      * 0.6905,0.7234,0.7554,0.7867,0.8172,0.8471,0.8763,0.9049,
      * 0.9330,0.9606,0.9877,1.0534,1.1167,1.1777,1.2367,1.2940,
      * 1.3496,1.4038,1.4567,1.5083,1.5588/
0023          DATA DRPB/ 0.,
      * 0.0259,0.0428,0.0555,0.0657,0.0742,0.0814,0.0876,0.0929,
      * 0.0977,0.1019,0.1057,0.1091,0.1123,0.1152,0.1179,0.1203,
      * 0.1224,0.1246,0.1266,0.1285,0.1319,0.1349,0.1376,0.1401,
      * 0.1423,0.1445,0.1463,0.1481,0.1497,0.1512,0.1527,0.1540,
      * 0.1553,0.1565,0.1577,0.1602,0.1625,0.1646,0.1663,0.1681,
      * 0.1696,0.1710,0.1723,0.1735,0.1747/
0024          DATA GMR   / 0.,
      *-0.0333,-0.0667,-0.1000,-0.1333,-0.1667,-0.2000,-0.2333,-0.2667,
      *-0.3000,-0.3333,-0.3667,-0.4000,-0.4333,-0.4667,-0.5000,-0.5333,
      *-0.5667,-0.6000,-0.6333,-0.6667,-0.7333,-0.8000,-0.8667,-0.9333,
      *-1.0000,-1.0667,-1.1333,-1.2000,-1.2667,-1.3333,-1.4000,-1.4667,
      *-1.5333,-1.6000,-1.6667,-1.8333,-2.0000,-2.1667,-2.3333,-2.5000,
      *-2.6667,-2.8333,-3.0000,-3.1667,-3.3333/
0025          DATA DRPGAO/ 0.,
      * 0.0034,  0.0056,  0.0076,  0.0096,  0.0115,  0.0134,  0.0152,  0.0170,
      * 0.0189,  0.0207,  0.0224,  0.0242,  0.0260,  0.0277,  0.0295,  0.0312,
      * 0.0329,  0.0346,  0.0363,  0.0379,  0.0413,  0.0445,  0.0478,  0.0509,
      * 0.0541,  0.0571,  0.0602,  0.0632,  0.0661,  0.0690,  0.0719,  0.0747,
      * 0.0775,  0.0802,  0.0829,  0.0894,  0.0957,  0.1017,  0.1075,  0.1130,
      * 0.1184,  0.1236,  0.1285,  0.1333,  0.1379/
0026          DATA DRPGRO/ 0.,
      * 0.0027,  0.0041,  0.0054,  0.0065,  0.0077,  0.0088,  0.0098,  0.0109,
      * 0.0119,  0.0130,  0.0140,  0.0150,  0.0160,  0.0170,  0.0180,  0.0190,
      * 0.0200,  0.0210,  0.0219,  0.0229,  0.0248,  0.0267,  0.0286,  0.0305,
      * 0.0324,  0.0343,  0.0361,  0.0380,  0.0398,  0.0417,  0.0435,  0.0453,
      * 0.0471,  0.0489,  0.0507,  0.0551,  0.0595,  0.0639,  0.0681,  0.0724,
      * 0.0765,  0.0806,  0.0847,  0.0887,  0.0927/
```

```
0027            DATA DRPPO/ 0.,
       * 0.0063, 0.0111, 0.0157, 0.0201, 0.0244, 0.0286, 0.0326, 0.0365,
       * 0.0402, 0.0439, 0.0474, 0.0508, 0.0541, 0.0572, 0.0603, 0.0633,
       * 0.0662, 0.0689, 0.0716, 0.0742, 0.0792, 0.0839, 0.0883, 0.0925,
       * 0.0964, 0.1002, 0.1037, 0.1071, 0.1103, 0.1133, 0.1162, 0.1190,
       * 0.1216, 0.1242, 0.1266, 0.1322, 0.1374, 0.1420, 0.1463, 0.1502,
       * 0.1538, 0.1572, 0.1603, 0.1632, 0.1660/
0028            DATA RPGA/ 0.,
       * 0.0102, 0.0169, 0.0233, 0.0295, 0.0357, 0.0418, 0.0479, 0.0540,
       * 0.0601, 0.0662, 0.0724, 0.0786, 0.0848, 0.0910, 0.0972, 0.1035,
       * 0.1098, 0.1161, 0.1224, 0.1287, 0.1415, 0.1543, 0.1672, 0.1802,
       * 0.1932, 0.2063, 0.2194, 0.2326, 0.2458, 0.2590, 0.2723, 0.2856,
       * 0.2989, 0.3123, 0.3256, 0.3591, 0.3926, 0.4260, 0.4595, 0.4929,
       * 0.5263, 0.5595, 0.5927, 0.6257, 0.6586/
0029            DATA DRPGA/ 0.,
       * 0.0039, 0.0064, 0.0086, 0.0108, 0.0129, 0.0150, 0.0170, 0.0191,
       * 0.0211, 0.0230, 0.0250, 0.0270, 0.0289, 0.0308, 0.0327, 0.0346,
       * 0.0365, 0.0384, 0.0402, 0.0421, 0.0457, 0.0493, 0.0529, 0.0564,
       * 0.0598, 0.0632, 0.0666, 0.0699, 0.0732, 0.0764, 0.0795, 0.0827,
```

```
       * 0.0857, 0.0888, 0.0917, 0.0990, 0.1060, 0.1127, 0.1192, 0.1254,
       * 0.1315, 0.1373, 0.1429, 0.1483, 0.1535/
0030            KEV(I)=NEN(1+I/100)+(NEN(2+I/100)-NEN(1+I/100))*
       *MOD(I,100)/100
0031            ENV(I)=IENE(1+I/5)+(IENE(2+I/5)-IENE(1+I/5))*MOD(I,5)/5
        C**** START
0032            IEL=DATAP(1)
0033            IF(IAND(ISHFT(ISEM,-5),1).NE.0) GOTO 106
0034            B=0.
0035            G=0.
0036  106       IF(IAND(ISEM,2).EQ.0)GOTO 200
        C
        C       PEARSON DISTRIBUTION
        C
0037            ENERG=MAX(0.,MIN(1000.,ENERG))
0038            IKEV=KEV(INT(ENERG))
0039            H=(ENERG-ENV(IKEV))/(ENV(IKEV+1)-ENV(IKEV))
0040            R=RNG(IKEV+1,IEL)+H*(RNG(IKEV+2,IEL)-RNG(IKEV+1,IEL))
0041            S=SGM(IKEV+1,IEL)+H*(SGM(IKEV+2,IEL)-SGM(IKEV+1,IEL))
0042            G=GAM(IKEV+1,IEL)+H*(GAM(IKEV+2,IEL)-GAM(IKEV+1,IEL))
0043  200       BMIN=6*((G**2+4.)**1.5-8.+7*G**2)/(32.-G**2)
0044            RR=R
0045            SS=S
0046            IF(IAND(ISEM,2).NE.0) B=FM(IEL)*BMIN+FM1(IEL)
0047            B=AMAX1(B,BMIN+3)
0048            B1=-G*S*(B+3)/(10*B-12*G*G-18)
0049            BO=-S*S*(4*B-3*G*G)/(10*B-12*G*G-18)
0050            B2=-(2*B-3*G*G-6)/(10*B-12*G*G-18)
        D      TYPE*,'R,S,G,B',R,S,G,B,'SO',SO
        D      TYPE*,'BO,B1,B2',BO,B1,B2
0051            IF(B.EQ.BMIN+3) GOTO 240
0052            AMX=ALOG(ABS(BO))/
       */(2*B2)-(B1/B2+2*B1)/SQRT(4*B2*BO-B1**2)*ATAN((B1)/
       */SQRT(4*B2*BO-B1**2))
0053            ICTL=1
0054            RETURN
        C
        C       B1**2-4*BO*B2 .EQ. 0
        C
0055  240       ICTL=2
0056            AMX=ALOG(ABS(BO))/
       */(2*B2)+1/B2+2
        C
        C       GAUSSIAN DISTRIBUTION
        C
0057            IF(G.EQ.0)ICTL=3
0058            RETURN
0059            END
```

174

PROGRAM SECTIONS

Name	Size		Attributes
$CODE1	001464	410	RW,I,CON,LCL
$PDATA	000016	7	RW,D,CON,LCL

FORTRAN IV-PLUS V3.0   16:39:21   5-Jul-82   Page 6
SCR.#1   /TR:BLOCKS

$IDATA	000004	2	RW,D,CON,LCL
$VARS	013520	2984	RW,D,CON,LCL
$TEMPS	000004	2	RW,D,CON,LCL
PEAR	000032	13	RW,D,OVR,GBL
IMPL	000102	33	RW,D,OVR,GBL

Total Space Allocated = 015366   3451

FORTRAN IV-PLUS V3.0   16:40:03   5-Jul-82   Page 7
SCR.#1   /TR:BLOCKS

```
0001            FUNCTION PEARS(X)
0002            COMMON /PEAR/B0,B1,B2,AMX,R,S,ICTL
0003       F(X)=ALOG(ABS(B0+B1*X+B2*X*X))/
           /(2*B2)-(B1/B2+2*B1)/SQRT(4*B2*B0-B1**2)*ATAN((2*B2*X+B1)/
           /SQRT(4*B2*B0-B1**2))
0004       F0(X)=ALOG(ABS(B0+B1*X+B2*X*X))/
           /(2*B2)+(B1/B2+2*B1)/(2*B2*X+B1)
0005       FE(X)=-(X/S)**2*.5
0006            PEARS=0
0007            GOTO(10,20,30)ICTL
0008       10   H=F(X-RP)-AMX
0009            GOTO 100
0010       20   IF(2*B2*(X-R)+B1.GE.0)RETURN
0011            H=F0(X-RP)-AMX
0012            GOTO 100
0013       30   H=FE(X-RP)-AMX
0014       100  IF(H.LT.-37)RETURN
0015            PEARS=EXP(H)
0016            RETURN
0017            END
```

PROGRAM SECTIONS

Name	Size		Attributes
$CODE1	000714	230	RW,I,CON,LCL
$PDATA	000010	4	RW,D,CON,LCL
$IDATA	000010	4	RW,D,CON,LCL
$VARS	000010	4	RW,D,CON,LCL
$TEMPS	000004	2	RW,D,CON,LCL
PEAR	000032	13	RW,D,OVR,GBL

Total Space Allocated = 001002   257

FORTRAN IV-PLUS V3.0   16:40:09   5-Jul-82   Page 8
SCR.#1   /TR:BLOCKS

```
0001            SUBROUTINE SYNLIN(DATAF,ISEM,ITEML,ITYPL,LENG,IERR)
0002            BYTE INBUF(64),INBUF1(8)
0003            INTEGER*4 ISEM
0004            REAL DATAF(8)
0005            REAL*8 ITEML(8)
0006            DIMENSION ITYPL(8)
0007            IERR=0
0008            WRITE(5,13) (ITEML(I),I=1,LENG)
0009            READ(5,7) INBUF
0010            IPOI=1
         C
```

```
      C      INTERPRETE DATA LIST
      C
0011         DO 104 I=1,LENG
0012         IANF=IANF+1
0013         IA=IA+1
0014         ITP=ITYPL(IANF)
0015         DO 107 J=1,8
0016  107    IHBUF1(J)='  '
0017         IPOI1=1
0018         DO 105 J=1,8
0019         IPOI=IPOI+1
0020         IF(INBUF(IPOI-1).EQ.9) GOTO 205
0021         IF(INBUF(IPOI-1).EQ.' '.OR.INBUF(IPOI-1).EQ.0) GOTO 105
0022         IHBUF1(IPOI1)=INBUF(IPOI-1)
0023         IPOI1=IPOI1+1
0024  105    CONTINUE
0025  205    NCHA=IPOI1-1
0026         IF(NCHA.EQ.0) GOTO 104
0027         ISEM=ISEM.OR.ISHFT(1,IA-1)
0028         GOTO (211,212,213,214,215)ITP
      C
      C      NUMERICAL DATA
      C
0029  211    IF((NCHA.EQ.1).AND.(IHBUF1(1).EQ.2H* )) GOTO 235
0030         DECODE(NCHA,2,IHBUF1,ERR=216) DATAF(IA)
0031         GOTO 104
0032  235    WRITE(5,22) I
0033         READ(5,*,ERR=235) DATAF(IA)
0034         GOTO 104
      C
      C      HOLLERITH DATA
      C
0035  212    DECODE(8,23,IHBUF1) DATAF(IA),DATAF(16)
0036         GOTO 104
      C
      C      LOGICAL DATA
      C
0037  213    IF(IHBUF1(1).EQ.'Y') GOTO 236
0038         IF(IHBUF1(1).NE.'N') GOTO 216
0039         DATAF(IA)=0.
0040         GOTO 104
0041  236    DATAF(IA)=1.
0042         GOTO 104
      C
      C      ELEMENT DESCRIPTORS
```

```
      C
0043  214    CALL ELMTCH(IHBUF1,J,JD)
0044         IF(JD.EQ.0.OR.J.LT.4)GOTO 216
0045  217    DATAF(IA)=J-3
0046         GOTO 104
      C
      C      MODEL SPECIFICATIONS
      C
0047  215    MDHP=0
0048         J=1
0049  110    CALL ELMTCH(IHBUF1(J),L,JD)
0050         IF(JD.EQ.0.AND.IHBUF1(J).NE.' ')GOTO 216
0051         IF(JD.EQ.0)MDHP=MDHP.OR.ISHFT(1,L)
0052         J=J+MAX(1,JD)
0053         MDHP=MDHP.OR.ISHFT(1,L)
0054         IF(J.LE.8)GOTO 110
0055         DATAF(IA)=MDHP
0056  104    CONTINUE
0057         RETURN
0058  216    WRITE(5,10)IA
0059         IERR=1
0060         RETURN
0061  2      FORMAT(E8.0)
```

```
0062    3     FORMAT(I1)
0063    7     FORMAT(44A1)
0064    8     FORMAT('(',I1,'X,',I1,'A2)')
0065    9     FORMAT(A4)
0066   10     FORMAT(X,'ERROR IN DATA FIELD NO. ',I1)
0067   13     FORMAT(X,8A8)
0068   15     FORMAT(X,'ILLEGAL COMBINATION OF ITEMS')
0069   16     FORMAT(/X,4A2,3X,6(8A2,2X))
0070   17     FORMAT(G15.7,X)
0071   18     FORMAT(A4,12X)
0072   19     FORMAT(12X,6(8A2,2X)/)
0073   20     FORMAT(12X,6(8A2,2X))
0074   21     FORMAT(/X,4A2,3X,36A2)
0075   22     FORMAT(X,'ENTER HIGH PRECISION DATA FOR FIELD ',I1,$)
0076   23     FORMAT(2A4)
0077   24     FORMAT(8A1)
0078          END
```

PROGRAM SECTIONS

Name	Size		Attributes
$CODE1	001632	461	RW,I,CON,LCL
$PDATA	000172	61	RW,D,CON,LCL
$IDATA	000064	26	RW,D,CON,LCL
$VARS	000136	47	RW,D,CON,LCL
$TEMPS	000004	2	RW,D,CON,LCL

Total Space Allocated = 002252   597

FORTRAN IV-PLUS V3.0   16:40:21   5-Jul-82   Page 10
SCR.#1   /TR:BLOCKS

```
0001          SUBROUTINE DRUPLO(Y,DX,ICTL)
0002          PARAMETER NX=130,NY=68
0003          DIMENSION Y(NX)
0004          BYTE DRU(0:NX,0:NY)
0005   110    FORMAT(I3)
0006          IF((ICTL.AND.2).EQ.0)GOTO 11
0007          DO 10 I=0,NX
0008          DO 10 J=0,NY
0009   10     DRU(I,J)=' '
0010   11     IF((ICTL.AND.1).EQ.0)GOTO 12
0011          YMIN=Y(1)
0012          YMAX=Y(1)
0013          DO 20 I=2,NX
0014          YMIN=MIN(Y(I),YMIN)
0015   20     YMAX=MAX(Y(I),YMAX)
0016          DY=NY/(YMAX-YMIN)
0017          DO 30 I=INT(YMIN+.9999),INT(YMAX)
0018   30     ENCODE(3,110,DRU(0,(I-YMIN)*DY+.5))I
0019   12     IF((ICTL.AND.4).EQ.0)RETURN
0020          DO 40 I=0,NX
0021   40     DRU(I,MAX(0,MIN(INT((Y(I)-YMIN)*DY),NY)))='*'
0022          RETURN
0023          ENTRY PLOTNOD(IUNIT)
0024          DO 50 I=NY,0,-1
0025   50     WRITE(IUNIT,120)(DRU(J,I),J=0,NX)
0026   120    FORMAT(1X,140A1)
0027          END
```
PROGRAM SECTIONS

Name	Size		Attributes
$CODE1	000714	230	RW,I,CON,LCL
$PDATA	000016	7	RW,D,CON,LCL
$IDATA	000012	5	RW,D,CON,LCL
$VARS	021540	4520	RW,D,CON,LCL
$TEMPS	000002	1	RW,D,CON,LCL

Total Space Allocated = 022506   4771

```
FORTRAN IV-PLUS V3.0          16:40:20    5-Jul-82      Page 11
SCR.11                   /TR:BLOCKS

0001          SUBROUTINE ELMTCH(IBUF,IEL,NCH)
         C
         C    IBUF : 2 CHARS TO BE COMPARED
         C    NCH=0: KEIN ELEMENT GEFUNDEN
         C    NCH=1: oder
         C    NCH=2: MATCH MIT 1 ODER 2 CHARS
         C    IEL=1...8 NUMMER DES ELEMENTS/MODELS
         C
0002          DIMENSION MDLST(8)
0003          BYTE IBUF(2),IMDLST(2,8)
0004          EQUIVALENCE(IMDLST(1,1),MDLST(1))
0005          DATA MDLST/'D','W','X','R','P','SB','AS','GA'/
0006          NCH=0
0007          DO 10 I=1,8
0008          IF(IBUF(1).NE.IMDLST(1,I))GOTO 10
0009          NCH=1
0010          IEL=I
0011          IF(IBUF(2).EQ.IMDLST(2,I))GOTO 20
0012    10    CONTINUE
0013          RETURN
0014    20    NCH=2
0015          RETURN
0016          END


PROGRAM SECTIONS

   Name      Size              Attributes

 $CODE1    000170     60       RW,I,CON,LCL
 $IDATA    000012      5       RW,D,CON,LCL
 $VARS     000022      9       RW,D,CON,LCL


Total Space Allocated = 000224     74

No FPP Instructions Generated
```

References:

1.1 J.C. North, T.E. McGahan, D.W. Rice, and C. Adams, IEEE Trans. Electron Devices ED-25, 809 (1978)

1.2 A. Lietoila, J.F. Gibbons, T.J. Magee, J. Peng, and J.D. Hong, Appl. Phys. Letters 35, 532 (1979)

1.3 T. Tsujide, M. Nojiri, and H. Kitagawa, J. Appl. Phys. 51, 1605 (1980)
P. Bayerl, H. Ryssel, and M. Ramin, Rad. Effects 47, 217 (1980)
E. Sano, P. Kasai, K. Ohwada, and H. Ariyoshi, IEEE Trans. Electron Devices ED-27, 2043 (1980)

1.4 R.L. Kubena, C.L. Anderson, R.L. Seliger, R.A. Jullens, and E.H. Stevens, IEEE Electron Device Letters EDL-2, 152 (1981)

1.5 H. Shibata, H. Iwasaki, T. Oku, and Y. Tarui, IEDM Technical Digest, p. 395 (1977)

1.6 H. Ryssel and K. Haberger, Microcircuit Engineering 1981, Lausanne, Switzerland (Ed. A. Oosenbrug), p.299
R.L. Seliger and P.A. Sullivan, Electronics, March 27, p. 142 (1981)

178

1.7 W. Shockley, U.S. Patent No. 2787, 564 (1957)

1.8 T. Alväger and N.J. Hansen, Rev. Sci. Inst. 33, 367 (1962)

1.9 W.J. King, J.T. Burrel, S. Harrison, F. Martin, and C.M. Kellett, Nucl. Inst. & Methods 38, 178 (1965)

1.10 J. Lindhard, M. Scharff, and H.E. Schiøtt, Kgl. Danske Videnskab. Selskab., Mat.-Fys. Medd. 33, No. 14 (1963)

1.11 R.W. Bower and H.G. Dill, IEEE Int. Electron Devices Meeting, Washington (1966)

1.12 K.G. Aubuchon, Int. Conf. on Prop. and Uses of MIS Structures, Grenoble (1969)

1.13 R.S. Payne, R.J. Scavuzzo, K.H. Olsen, J.M. Nacci, and R.A. Moline, IEEE Trans. Electron Devices ED-21, 273 (1974)

2.1 J. Lindhard, M. Scharff, and H.E. Schiøtt, Kgl. Danske Videnskab., Selskab., Mat. Phys. Medd. 33, No. 14 (1963)

2.2 J.P. Biersack, Nucl. Inst. & Methods 182/183, 199 (1981)

2.3 J.F. Gibbons, W.S. Johnson, and S.W. Mylroie, Projected Range Statistics in Semiconductors, Dowden Hutchinson and Ross, Academic Press, Stroudsburg (1975)

2.4 B. Smith, Ion Implantation Range Data for Silicon and Germanium Device Technologies, Learned Information (Europe) Ltd., Oxford (1977)

2.5 J.W. Mayer, L. Eriksson, and J.A. Davies, Ion Implantation in Semiconductors, Acad. Press, New York (1970)

2.6 J.P. Biersack and L.G. Haggmark, Nucl. Inst. & Methods 174, 257 (1980)

2.7 W.K. Hofker, D.P. Oosterhoek, N.J. Joeman, and H.A.M. De Grefte, Rad. Eff. 24, 223 (1975)

3.1 J. Lindhard, M. Scharff, and H.E. Schiøtt, Kgl. Danske Videnskab., Selskab., Mat. Phys. Medd. 33, No. 14 (1963)

3.2 J. P. Biersack, Nucl. Inst. & Methods 182/183, 199 (1981)

3.3 O.B. Firsov, Sov. Phys. JETP 36, 1076 (1959)

3.4 J. Lindhard, Kgl.Danske Videnskab.Selskab.,Mat.-Fys. Medd. 34, No. 14 (1965)

3.5 V.G.K. Reddi and A.Y.C. Yu, Solid State Technology 15, 35 (1972)

3.6 R.A. Moline and G.W. Reutlinger, in Ion Implantation in Semiconductors (Eds. I. Ruge, J. Graul), Berlin, Heidelberg, New York (1971), p. 58

3.7 H. Ryssel, K. Müller, K. Haberger, R. Henkelmann, and F. Jahnel, Appl. Phys. 22, 35 (1980)

4.1 J. Lindhard, M. Scharff, and H.E. Schiøtt, Kgl. Danske Videnskab., Selskab., Mat. Phys. Medd. 33, No. 14 (1963)

4.2 J.P. Biersack, Nucl. Inst. & Methods 182/183, 199 (1981)

4.3 K.B. Winterbon, Ion Implantation Range and Energy Deposition Distributions, Vol.2: Low Incident Ion Energies, IFI/Plenum Press (1973), New York

4.4 J.F. Gibbons, W.S. Johnson, and S.W. Mylroie, Projected Range Statistics in Semiconductors, (Eds.: Dowden, Hutchinson, and Ross), Academic Press, Stroudsburg (1975)

179

4.5 B. Smith, <u>Ion Implantation Range Data for Silicon and Germ-anium Device Technologies</u>, Learned Information (Europe) Ltd., Oxford (1977)

4.6 J.F. Gibbons and S. Mylroie, Appl. Phys. Lett. 22, 568 (1973)

4.7 W.K. Hofker, Philips Res. Repts., Suppl. No.8 (1975)

4.8 J.F. Gibbons, in: <u>Handbook on Semiconductors</u>, Vol.3 (Ed. T.S. Moss), North Holland Publ.Comp., Amsterdam (1980)

4.9 E.F. Krimmel and H. Pfleiderer, Rad. Effects 19, 83 (1973)

4.10 Y. Yamamura and H. Inuma, Rad. Effects 38, 2511 (1978)

4.11 L.A. Christel, J.F. Gibbons, and S. Mylroie, J. Appl.Phys. 51, 6176 (1980)

4.12 H. Ishiwara, S. Furukawa, J. Yamada, and M. Kawamura in: <u>Ion Implantation in Semiconductors</u> (Ed. S. Namba) p.423, Plenum Press, New York (1975)

4.13 A.V.S. Satya and H.R. Palanki in: <u>Ion Implantation in Semi-conductors</u> (Ed. S. Namba) p.405, Plenum Press, New York (1975)

4.14 R.A. Moline and A.G. Cullis, Appl. Phys. Lett. 26, 551 (1975)

4.125 G. Fischer, G. Carter, and R. Webb, Rad. Effects 38, 41- (1978)

4.16 P. Sigmund, J. Appl.Phys. 50, 726 (1979)

4.17 T. Hirao, K. Inoue and Takayanagi, J. Appl.Phys. 50, 193 (1979)

4.18 W.K. Chu, M.J. Sullivan, S.M. Ku and M. Shatzkes, First Int. Conf. on Ion Beam Modification of Materials, Budapest (1978)

4.19 T. Ito, S. Hijya, H. Nishi, M. jShinoda, and T. Furuya, Jpn. J.Appl. Phys. 17, 201 (1978)

4.20 J.M. Poate and T.C. Tisone, Appl. Phys. Lett. 24, 391 (1974)

BEAM ANNEALING OF ION IMPLANTED SILICON

James F. Gibbons

Solid-State Electronics Laboratory
Stanford University

INTRODUCTION

Extensive research has been performed over the past several years on the use of lasers, electron beams and arc sources for annealing damage created in silicon by ion implantation. This field, originally identified as "beam annealing" because its focus was on the removal of defects introduced by ion implantation, has rapidly broadened to encompass much more diverse situations, many of which do not involve ion implantation and may not even involve defect removal. In particular, lasers and electron beams have been used to recrystallize thin films of vapor deposited poly-silicon with substantial improvements in its electronic properties [1]; to facilitate the formation of metal silicides [2]; and to perform a number of other processing functions that are of increasing importance in the fabrication of fine geometry integrated circuits and high speed devices. In this paper we will concern ourselves entirely with the annealing process; specifically, on the mechanisms by which laser, electron beam and arc source annealing proceeds and the basic metallurgical and electronic properties of beam annealed material.

BASIC BEAM ANNEALING MECHANISMS

Two distinctly different beam annealing mechanisms have been identified, depending on the duration of the beam exposure. For Q-switched lasers or pulsed electron beams, exposure times are typically in the range of 5 ns - 500 ns and the annealing process then involves the formation of a thin molten layer of silicon that recrystallizes on the underlying substrate when the radiation is removed. If the irradiated sample is an ion implanted single crystal and the depth of the molten layer is sufficient to envelop the implantation damaged region, the molten layer regrows by a very high speed liquid phase epitaxial process on the crystalline substrate, producing material with a very high degree of structural perfection and very superior electronic properties.

For cw systems, on the other hand, the silicon surface is typically exposed to the beam for 0.1–10 ms and in some cases for durations of several seconds. The annealing of ion implanted material can then proceed by a process similar to that which occurs in conventional furnace annealing; i.e., a solid phase epitaxial regrowth process at temperatures that are well below the melting point. As in the pulsed beam case, a very high degree of crystalline perfection and very superior electronic properties can be obtained under appropriate annealing conditions. However, the absence of melting proves to be of interest since no redistribution of the implanted impurity profile then occurs during the annealing process, whereas significant impurity redistribution occurs when annealing is effected by a pulsed laser or electron beam. These and other differences in the two annealing processes make it convenient to discuss them separately. In what follows we first consider the pulsed beam annealing process and then take up the cw alternative. Conventional furnace annealing will be seen to be a special case of cw beam annealing.

CENTRAL FEATURES OF PULSED BEAM ANNEALING OF ION IMPLANTED SILICON

We will begin with a brief review of experimental data that characterize the annealing of ion implanted silicon using a Q-switched laser pulse.

Pulsed laser annealing has been demonstrated using a variety of sources and pulse durations. In most of the experiments reported so far, implanted samples have been annealed directly in the laboratory ambient; i.e., no special precautions have been taken to immerse the wafer in an inert environment during annealing. The area irradiated by the beam is in most cases large compared to both the diffusion length for heat in the solid and the thickness of the wafer, so the process can be treated as being basically one dimensional.

Experimental results from various laboratories are now reasonably consistent and are summarized in Table 1. For convenience, the data are selected to illustrate the annealing of As^+-implanted silicon with the As^+ implantation conditions chosen to provide amorphous regions of differing thicknesses at the sample surface. In the cases reported in Table 1, the beam intensities employed were in the range of ~40–60 MW/cm^2.

TABLE 1. Pulse Energies for Annealing As[+]-implanted Silicon [3]

Laser	As$_+$ Impl. Par.	Amorphous Layer Thickness, X_d	Melt Initiated	Fully Annealed
Ruby, 50 ns	5x10^{15}, 400 keV	4300 Å	0.6 J/cm^2	2 J/cm^2
Ruby, 150 ns	1.4x10^{16}, 100	1700 Å	0.64	1.4
Nd:YAG, 110 ns	8x10^{15}, 100 keV	~1500 Å		6.0
40 ns	10^{15}, 130 keV	468	3.5	4.5
Nd:YAG, Doubled 40 ns	10^{15}, 30 kev	468	0.2	0.85

The principal features of the data are as follows. For pulses of 50-100 ns duration (typical of a Q-switched laser):

1. Surface melting occurs at a threshold energy of 0.2 - 3.5 J/cm^2, depending on the laser wavelength and the thickness of the amorphous layers.

2. Full annealing of an ion implanted layer (as judged by crystal recovery and electrical activity) requires additional energy in an amount that also depends on both the laser wavelength and the thickness of the amorphous layer.

Qualitative Analysis of the Melt Threshold

The initial interaction of a light beam with a material is always with its electrons. Energy absorbed by the electrons is ultimately shared with the atoms of the material as the electronic excitation is transformed to heat. Under most conditions the transfer of electronic excitation to heat is accomplished with relaxation times on the order of 1 ps and may therefore be considered to be instantaneous. The basic process of pulsed beam annealing then consists of the absorption of sufficient energy to melt a layer of silicon having a thickness at least equal to the thickness of the implantation-damaged region, followed by liquid phase epitaxial regrowth of the melted layer. The laser energy required to bring the surface of the sample to its melting temperature is therefore a critical parameter in the process.

To estimate the threshold energy required to produce surface melting, we consider the sample to consist of an implantation-damaged layer of thickness X_d and optical absorption coefficient is α_d resting on a crystalline solid with optical absorption coefficient α_c. A laser pulse of intensity I_0 and duration τ_p illuminates the implanted surface of the sample. The reflection coefficient at this surface is R. The laser intensity is assumed to be sufficiently low that ordinary one-photon optical absorption processes are dominant.

The relevant thermal properties of the material are its specific heat C_v, its mass density ρ and its thermal conductivity K. The heat diffusivity is then given by $D = K/\rho C_v$ and determines a characteristic length $(D\tau_p)^{1/2}$. This length gives an estimate of the distance over which the temperature profile is spread by heat diffusion during the laser pulse.

To estimate the surface temperature rise produced in a limit-ing case of practical interest, we suppose that the laser energy is absorbed in a distance small compared to the diffusion length for heat $[\alpha_d^{-1} < (D\tau P)^{1/2}]$. Under these conditions the absorbed laser energy can be considered to heat a slab of thickness $(D\tau_p)^{1/2}$. The corresponding temperature rise ΔT is then obtained from

$$(I_0\tau_p)(1-R) = \rho C_v \, \Delta T(D\tau_p)^{1/2} \tag{1}$$

This formula suggests that the surface temperature is proportional to the energy in the laser pulse $(I_0\tau_p)$ and predicts a melt threshold energy of

$$E_m^{\circ} = \frac{C_v \, \rho \, (D\tau_p)^{1/2}}{1-R} \quad (T_m - T_o) \tag{2}$$

where T_m is the melting temperature and T_0 is the ambient tempera-ture.

The extreme temperature change envisaged ($\sim 1400°C$) of course makes it necessary to account for the temperature dependence of the thermal and optical parameters appearing in Eq. 2 in order to arrive at a meaningful estimate of E_m° for practical use. The parameters that are customarily used for process analysis are given in Table 2.

The thermal conductivity is an especially sensitive function of temperature and its temperature dependence must be taken into account to obtain accurate theoretical predictions [4]. For num-erical estimates we use the average value of K calculated from

TABLE 2.   Thermal   and   Optical   Parameters   for   Solid   Silicon

Parameter	Value at T = 300°K	Value at T = 1685°K
C_v, J/gm-°K	0.65	0.95
ρ, gm/cm^3	2.33	2.33
K, W/cm-°K	1.45	0.25
R	0.35	0.7
*α_d (λ = 0.69 μm)	3×10^4	–
*α_d (λ = 1.06 μm)	$3-6 \times 10^3$	–
α_c (λ = 0.69 μm)	3×10^3	–
α_c (λ = 1.06 μm)	50	–

*The optical absorption coefficients for implantation-amorphized silicon have not been measured; values quoted for α_d are for sputtered silicon.

$$\bar{K} = \frac{1}{T - T_o} \int_{T_o}^{T} K(T) \, dT \tag{3}$$

with K(T) for silicon approximated by

$$K(T) = \frac{299}{T - 99} \quad \left(W/cm°K \right) \tag{4}$$

Substitution of numerical values from Table 2 leads to values of E_m that are in satisfactory agreement with experiment for short wavelength irradiation such as that from ruby or frequency-doubled Nd:YAG lasers, where the basic assumption of heat transport made in the analysis is valid [i.e., $\alpha_d^{-1} < (D\tau_p)^{1/2}$].

For Q-switched Nd:YAG lasers operated at the fundamental frequency, however, the optical absorption depth (α_c^{-1}) in crystalline silicon is substantially greater than the diffusion length for heat. An order of magnitude estimate of the melt threshold energy E_m^* can be made for this case by assuming that the thickness of the slab that is heated by the radiation is (α_c^{-1}). Replacing $(D\tau_p)^{1/2}$ with α_c^{-1} in Eq. 2, we obtain

$$E^*_m = \frac{\rho C_v (T_m - T_o)}{\alpha_c (1 - R)} \tag{5}$$

In practical circumstances, however, the melt threshold predicted by Eq. 5 proves to be too large for a variety of reasons. For the particular case of ion implanted silicon with damage depths of 0.1 - 0.3 μm, Lietoila and Gibbons [5] have shown that free carrier absorption in the underlying crystalline material and bandgap narrowing during the heating cycle have a profound effect on the surface temperature.

Impurity Profiles in Pulse Annealed Silicon

Clear insight into the basic device implications of pulsed laser (or electron beam) processing is provided by the early work of White and co-workers [6].

Figure 1 shows B profiles in as-implanted and laser annealed samples taken from their early work. The measurements were made by secondary ion mass spectroscopy (SIMS) and therefore yield the total concentration of boron without regard for lattice location. The samples were implanted at 35 keV to a dose of $1 \times 10^{16}/cm^2$. The as-implanted profile is approximately Gaussian as expected, and is only marginally changed by conventional furnace thermal annealing at 900°C for 30 minutes. In contrast, the profile measured after laser annealing (1.6 J/cm^2, 60 ns pulse duration) exhibits a substantial redistribution of boron. In particular, after the pulsed laser annealing cycle, the impurity profile becomes almost uniform from the surface down to a depth of approximately 0.2μm in the crystal, and significant quantities of boron are observed at depths of approximately 0.5μm.

The redistribution of the implanted boron is found to be both pulse energy density and pulse number dependent. Figure 2 shows experimental results measured after laser annealing with different pulse energy densities in the range 0.64 J/cm^2 to 3.1 J/cm^2. At a laser energy density of approximately 0.6 J/cm^2, only surface melting is initiated and the boron profile is indistinguishable from that of the as-implanted sample. Hall effect measurements and transmission electron microscopy show about 30% of the expected electrical activity and significant damage remaining in the form of dislocation loops under this pulse energy. At 1.1 J/cm^2 and greater, the impurity profiles are almost flat-topped in the surface region, and the boron spreads deeper in the sample as the energy density is increased.

The substantial redistribution of boron induced by pulsed laser annealing cannot be explained by thermal diffusion in the solid phase because the time duration is too short. However, theoretical calculations using a one-dimensional heat conduction

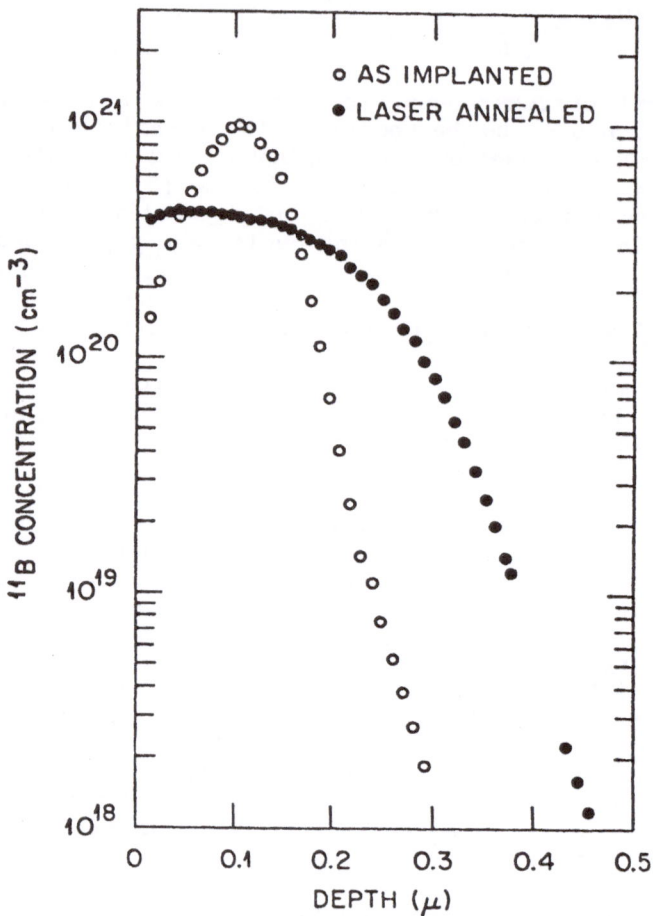

FIG. 1   Boron redistribution produced by pulsed laser anneal
[after White, et al. (6)].

equation [7] show that a region several thousand angstroms deep
can be melted for pulse energies greater than about 1 J/cm2.
Calculations of the melt front position as a function of time
for silicon irradiated with a 60 nanosecond, 1.5 J/cm2 pulse from
a ruby laser are shown in Fig. 3, where the melt front is seen to
reach a maximum penetration of approximately 0.95μm in a time
only slightly greater than the pulse duration. The melt front
then sweeps back toward the surface, recrystallizing the material
epitaxially as it proceeds. While the crystal is molten, the dopant atoms have a very high diffusion coefficient and the implanted
profile can change markedly. For boron in silicon, the experimental data can be fit by using a liquid diffusion coefficient

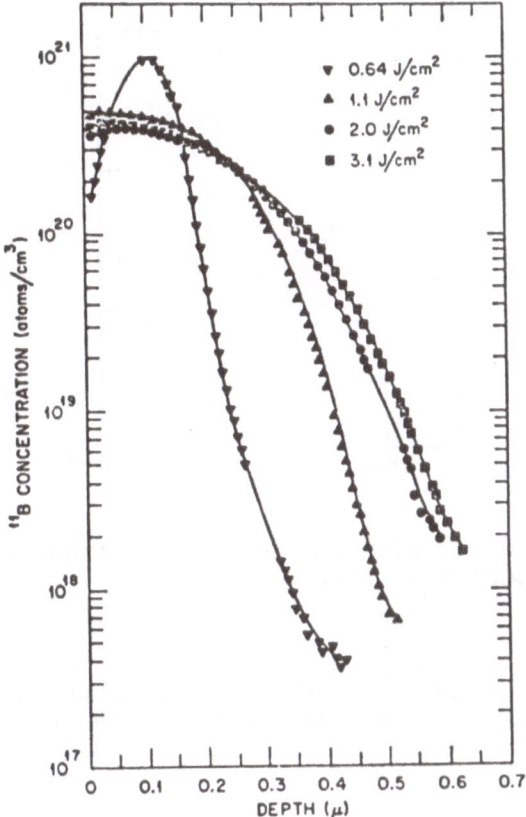

FIG. 2 Boron profiles after pulsed laser annealing with increasing pulse energy [after White, et al. (6)].

of $2.4 \pm 0.7 \times 10^4$ cm^2/sec [8] and a diffusion time of 180 ns. These parameters provide the calculated fit shown in Fig. 4 and adequately justify the conclusion that normal diffusion processes in the molten state account for the spreading of the implanted profile during pulsed laser annealing.

Hall effect and transmission electron microscopy measurements made on implanted material annealed in this way show 100% electrical activity and no defects to a resolution of at least 50 Å, provided that the amorphous layer is sufficiently thin. However, it should be mentioned that, since the reflection co-efficient R at the surface of an irradiated sample changes from energy 0.35 to 0.7 when the surface becomes molten, only 30% of the pulse over and above the melting threshold is absorbed by the silicon. The result is that it is difficult to melt layers that are in excess of approximately 1μm without heating the

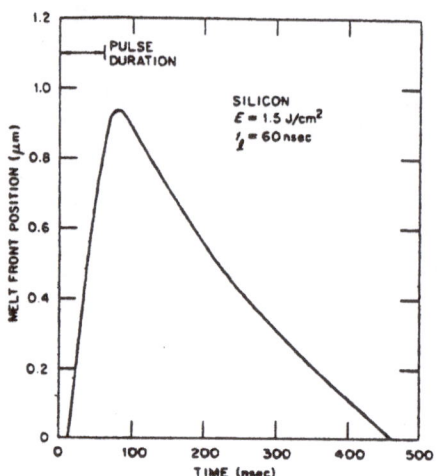

FIG. 3   Calculated position of the melt front as a function of time [after Wang et al. (7)].

surface to the boiling point, which leads to very poor surface morphology. As a result, pulsed laser annealing is generally useful only for the annealing of relatively shallow layers.

Summary

To summarize, the mechanism of annealing for Q-switched laser pulses involves melting of a surface layer followed by a liquid phase epitaxial regrowth on the underlying substrate. The critical parameter for this process is the pulse energy density. The principal electrical characteristics of the annealed layer are:

1. 100% substitutionality of the implanted dopant, even for concentrations that exceed the solid solubility.

2. No residual defects in TEM to 50 Å resolution.

3. Redistribution of the implanted dopant via diffusion in a liquid layer during the recrystallization process.

4. Residual point defects below the recrystallized layer that require a subsequent low temperature (800°C, 30 min) anneal for their removal. These defects may be thought of as arising from the fact that silicon vacancies and other crystalline defects diffuse rapidly from the high temperature portion of the crystal toward the cooler interior where they form point defects. These

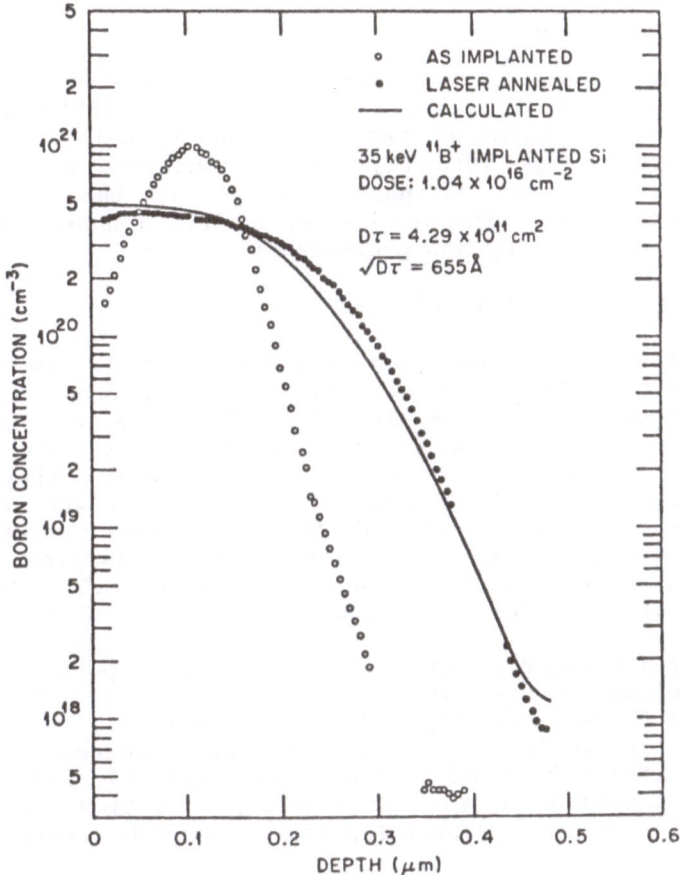

FIG. 4   Comparison of experimental and calculated B profiles in pulse laser annealed Si assuming liquid phase diffusion of B during recrystallization [after Wang et al. (7)].

defects have been shown to reduce the lifetime of minority carriers in the bulk of the silicon substantially, though this reduction can be largely removed by an appropriate thermal anneal.

CW BEAM PROCESSING

In contrast to the pulse annealing process, a scanning laser or electron beam provides an extremely convenient and highly controllable means for heating the _surface_ of a semiconductor to a given temperature while holding the body of the material at a convenient low temperature (typically 350-500°C). Recognition of this fact leads naturally to a comparison of the annealing mechanism with the annealing process for furnace annealing.

Furnace Annealing Process

Conventional furnace annealing of implantation-amorphized silicon proceeds by a solid phase epitaxial (SPE) recrystallization process. Atoms at the amorphous–crystalline interface re-arrange themselves into crystalline locations by a process that appears to involve vacancy diffusion in the crystalline and amorphous layers. The process is described quantitatively by an activation energy E_a and proceeds at a rate

$$r(t) = r_o \exp(-E_a/kT) \tag{6}$$

where $r(t)$ is normally measured in $Å$/second. Measurements of $r(t)$ made by Cspregi et al. [9] in the temperature range 400°C–600°C on undoped <100> silicon (amorphized by the implantation of Si^+) show an activation energy of E_a = 2.35 eV and a pre-exponential multiplier of 3.22 x 10^{14} $Å$/sec. Using these values at a temperature of 800°C gives a furnace regrowth rate of about 0.3 µm/sec for <100> undoped Si. In other words, a 0.5 µm thick amorphized layer can recrystallize by the solid phase epitaxial process on the underlying substrate in less than two seconds if the layer is held at a temperature of 800°C.

The implanted dopants are normally incorporated into substitional sites during this recrystallization process, leading to high electrical activity. However, point defects and trapping centers form during the recrystallization process (related to the vacancy motion by which the process proceeds), and these defects must be annealed out to obtain good pn junction characteristics; i.e., high carrier mobilities and reasonable carrier lifetimes.

The motion and agglomeration of defects during the annealing is especially troublesome if the implant dose and energy are not sufficient to produce an amorphous layer that envelops the implanted dopant. Such a condition is most often obtained when B is implanted into Si, a case which has been thoroughly studied. A 30 minute anneal in the temperature range 800°C–1000°C then produces a combination of dislocation loops, rods and precipitates that are nearly immune to further furnace annealing [10,11].

Partly as a result of these problems it is customary to use a so-called "two-stage" annealing process [12]. The first stage is carried out at ~600°C for 1/2 hour and is intended only to recrystallize the implanted region. A number of point and line defects remain after this stage, which lead to low electrical activity and poor carrier mobility and lifetime. The low temperature anneal is then typically followed by a 1000°C, 10 minute anneal to remove point defects, increase carrier mobility and lifetime, and provide for some impurity diffusion so that pn junctions formed by the process are located outside the region of residual damage. Such a process is especially important in the annealing of implanted <111> Si; i.e., for most bipolar applications. In contrast a single stage anneal can be sufficient for less demanding applications in <100> Si.

Basic CW Laser Annealing Systems

The basic cw laser annealing process is identical to the furnace annealing process just described, except that the sample is maintained at the annealing temperature for such a short time that only the fastest thermal processes, including solid phase epitaxy, can be carried to completion. Dopant precipitation and the formation of dislocation loops and rods are generally not observed since there is insufficient time for them to form.

The basic system used for scanned cw laser annealing is shown in Fig. 5 and consists of an Ar cw laser that is passed through a lens and deflected by X and Y mirrors onto a sample that is mounted in the focal plane of the lens [13]. The X mirror is mounted on a galvonometer that is driven with a triangular waveform and the Y mirror is mounted on a galvonometer that is driven by a staircase waveform. This arrangement permits the beam to be scanned across the target in the X direction, stepped by a controlled Y increment and then scanned back across the target in reverse X direction. Individual scan lines can be overlapped or not by appropriate adjustments of the Y step. As an alternative the sample can be mounted on an XY table and mechanically scanned underneath a stationary laser beam.

The samples are mounted on a sample holder that can be heated to about 500°. Control of the annealing ambient can be obtained by placing a cylindrical quartz jacket around the sample holder

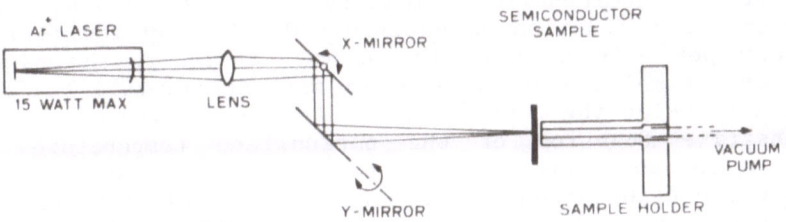

FIG. 5 A general schematic of the annealing apparatus, including Ar+ laser, lens perpendicular X and Y mirrors, and a vacuum sample holder.

and pumping appropriate gases into this jacket. A variety of lenses, laser powers, scan rates, and sample temperatures have been found to produce essentially perfect annealing of ion implanted semiconductors. If the samples are held at room temperature, a typical set of annealing conditions consists of a laser output (Ar, multi-line mode) of 7 W focused through a 79 mm lens into a 38 micron spot on the target. The spot is typically scanned across the target at a rate of approximately 2.5 cm per second. Adjacent scan lines are overlapped by approximately 30% to produce full annealing in the overlapped areas. The laser power for full annealing can be reduced (and the width of the annealed line increased) by increasing the sample temperature.

Surface Temperature Profiles

The beam geometry and scanning conditions listed above lead to a spot dwell time on the order of 1 ms. This is to be compared to a thermal time constant (for a 40 μm cube of Si) of approximately 10 μs, from which it follows that the semiconductor surface has adequate time to come to thermal equilibrium with the scanning heat source. It is therefore possible to calculate exactly the surface temperature that will be produced at the center of the spot by simply solving the steady state equation for heat flow. Calculations of this type have been carried out for both circular and elliptical scanning beams [14]; the results for a cylindrical beam irradiating a silicon substrate are shown in Fig. 6. The calculations presented there account for the temperature dependence of the thermal conductivity and specific heat, and have proven to give very accurate estimates of the surface temperatures achieved with the laser beam.

Two features of these curves are worth particular attention. First, the horizontal axis is in units of power per unit radius of the beam, this being due to the fact that the heat flow problem has essentially hemispherical symmetry. As a result the surface temperature is a function of <u>power divided by spot radius</u> rather than power per unit area of the beam. Secondly, the sizeable decease of thermal conductivity of Si with temperature leads to a situation in which the temperature at the irradiated surface is a very sensitive function of the backsurface temperature. For example, a surface temperature at the center of a 40 micron spot of 1000°C can be obtained with a 5 watt laser output if the backsurface temperature (or thermal bias) is 350°C. If the backsurface temperature is reduced to 150°C, the temperature in the center of the irradiated area drops by nearly 500°C. Hence it is very important to control the backsurface temperature accurately, and it is also possible to use relatively low laser powers if substantial thermal bias can be employed.

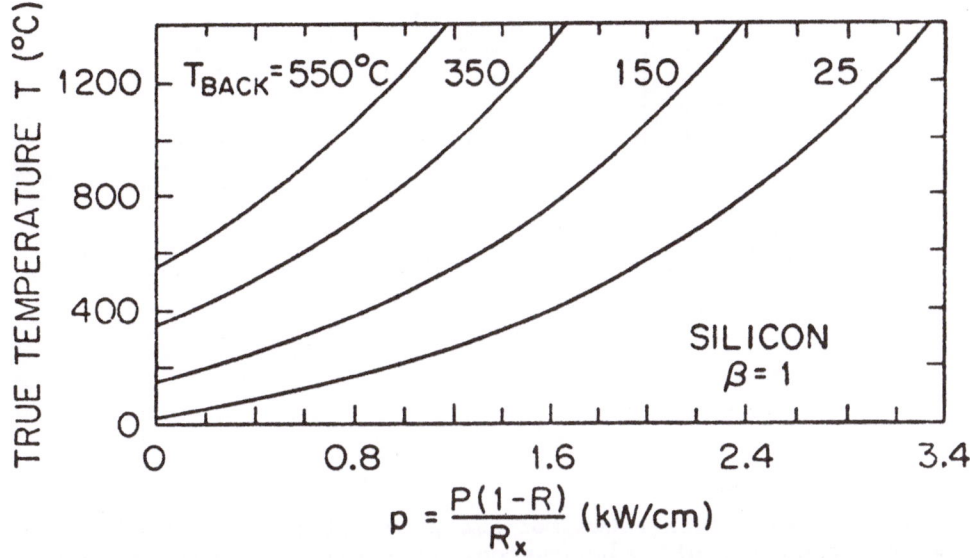

FIG. 6   The true   maximum   temperature ($X = Y = Z = V = 0$) in Si
is plotted versus the normalized power p ($p=P(1-R)/R_x$) for different substrate backsurface temperatures.

Basic Mechanism of CW Beam Annealing

As mentioned earlier, the basic mechanism of cw beam annealing is solid phase epitaxy. The most convincing illustration of
this fact is provided by the work of Roth et al. [15] and illustrated in Fig. 7. Here the regrowth rate of both implantation-
amorphized and UHV deposited Si layers on a <100> Si substrate
is plotted as a function of the surface temperature produced by
the laser. The data show exactly the form expected for a process
described by Eq. 6, and in fact the growth rate at 800°C is found
to be 0.3 µm/sec, in excellent agreement with the furnace annealing result.   Hence the regrowth process is quantitatively
identified as solid phase recrystallization.

Impurity Profile

Figure 8 shows the impurity profiles obtained by SIMS under
as-implanted, laser annealed and thermally annealed conditions
for As[+] implanted into <100> Si under typical conditions. The
most striking feature of this figure is that the laser annealed

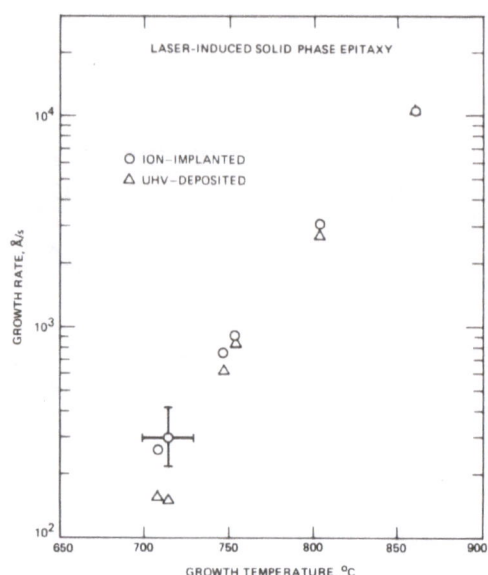

FIG. 7   SPE growth rates for amorphous Si layers formed by ion-implantation and UHV evaporation, determined from time-resolved reflectivity measurements.

profile is identical to the as-implanted profile. In other words, there has been no diffusion of the implanted species during the laser anneal. Furthermore, the as-implanted impurity distribution is matched exactly by the Pearson type IV distribution function using LSS moments [16], so the experimental and theoretical profiles are in excellent agreement. The thermal anneal shows the well-catalogued impurity redistribution given by the open triangles in Fig. 8.

Majority carrier profiles and carrier mobilities were obtained by sheet resistance and Hall effect measurements and are shown in Fig. 9. As can be seen the carrier concentration profile also fits the Pearson type IV distribution quite well. Under very high dose conditions (> $10^{16}/cm^2$), As precipitation can occur near the peak of the profile, leading to a residual, nonsubstitutional As content of approximately 5% [17].

Transmission Electron Microscopy

Typical results of TEM performed on thermally annealed and laser annealed samples are shown in Fig. 10. The thermally annealed sample (Fig. 10a) shows a single crystal diffraction pattern with a bright field micrograph containing the usual variety

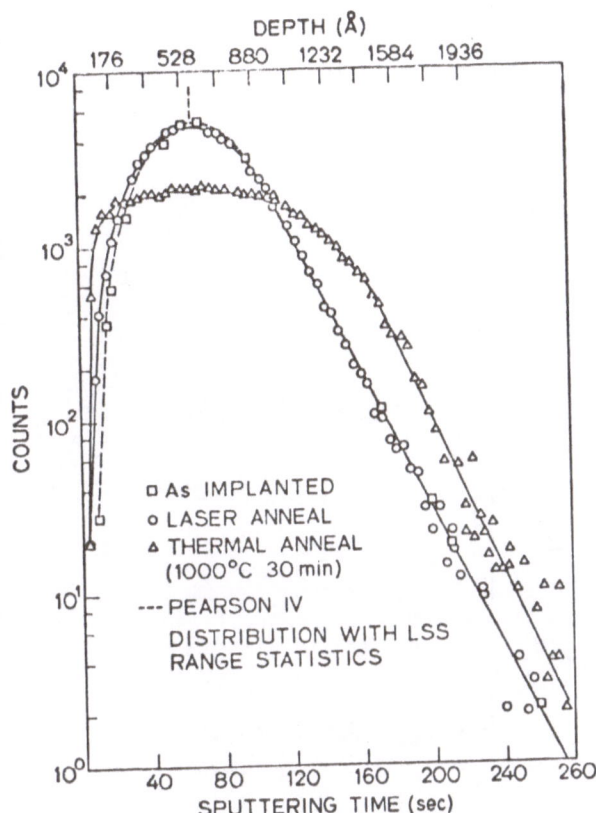

FIG. 8   Carrier concentration and mobility profiles obtained on As-implanted samples annealed with a scanning Ar cw laser beam.

of ~200Å diameter defect clusters and dislocation loops.   The The laser laser annealed sample shown in Fig. 10b is essentially free of any defect observable in TEM except near the boundary between crystallized and amoprhous regions.   It should be emphasized also that no defects are observed in TEM in the region where adjacent scan lines overlap.

B-implanted Single Crystal Si

A similar set of experiments has been performed in B-implanted Si.   The central results, reported in Ref. 18, are identical to those described above for As-implanted Si.   In particular, 100% electrical activity can be obtained with no diffusion of the implanted species from its as-implanted profile.   Recrystallization is also perfect as judged by TEM to a resolution of ~ 20 Å.

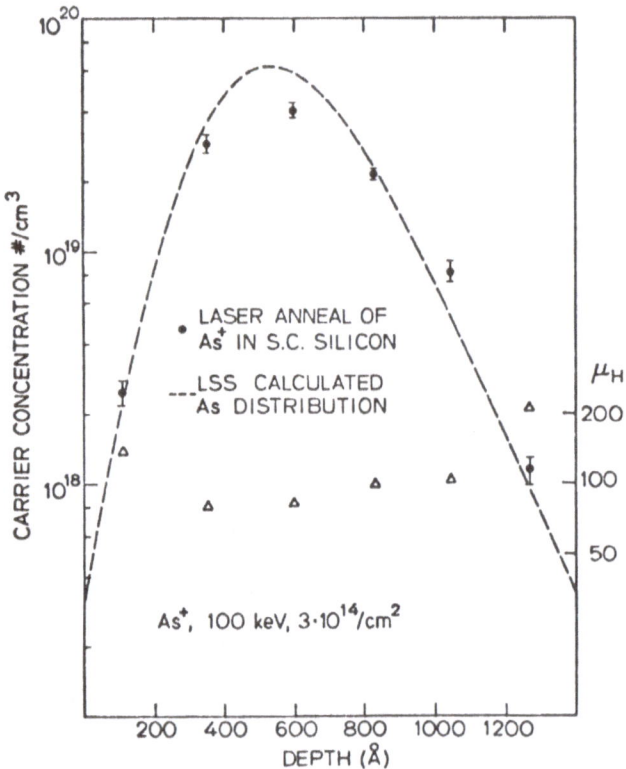

FIG. 9   As   concentration profile in   As–implanted silicon after laser anneal and thermal anneal.

These results are independent of whether the B is implanted into pre-amorphized Si or directly into single crystal material.

Summary

To summarize the foregoing, laser cw annealing proceeds by solid phase epitaxial recrystallization. For "small" irradiated areas (40 μm spot), the critical parameter is the beam power per unit radius. The principal results of the annealing are:

1. 100% substitutionality of the implanted dopant even for concentrations that exceed the solid solubility.

2. No residual defects in TEM to 50 Å resolution.

3. No dopant redistribution during annealing.

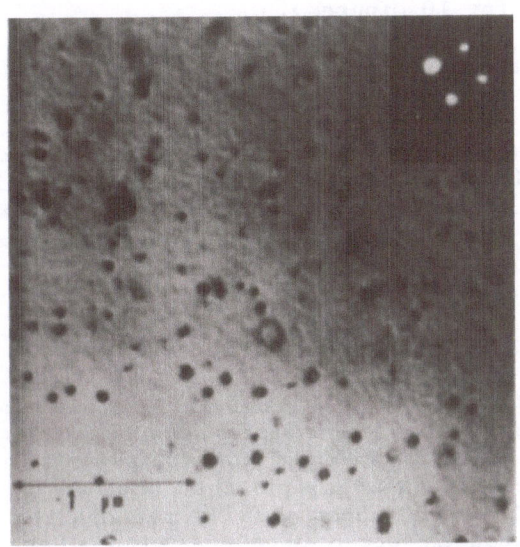

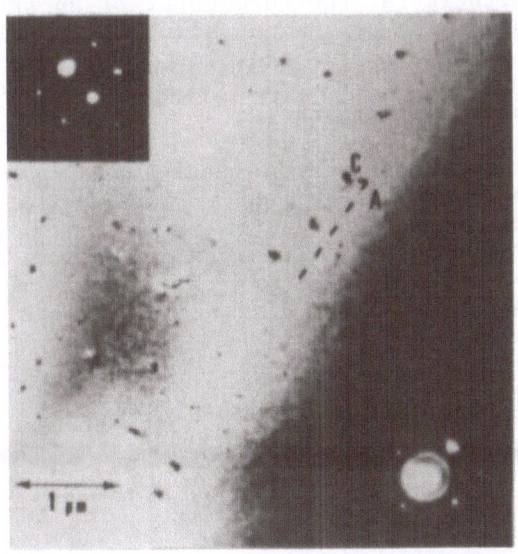

FIG. 10   Electron micrographs of As-implanted silicon subjected
to thermal anneal of   1000°C 30 min   (a) and laser   annealing;
(b) inserts show diffraction patterns which are typical to their
regions.

4. Residual point defects remaining below the recrystallized layer that require a subsequent low temperature anneal for their removal (800°C for 10 minutes).

For VLSI applications, the absence of dopant redistribution may be a significant feature of this annealing process. In any case it is an ideal companion for the implantation process because the dopants are annealed into their as-implanted sites. Since the implanted impurity profile can be calculated with considerable accuracy, this process provides a means of assuring that computer simulations of impurity profiles based on the implantation process can be realized with precision.

ANNEALING WITH LARGE DIAMETER CW SOURCES

A very attractive alternative for cw laser annealing is the use of cw arc sources. Arc sources can be configured to produce both large diameter annealing spots and ribbon-shaped beams. These beams are scanned at speeds that produce dwell times on the order of seconds, which substantially reduces the severity of the thermal shock that is often obtained with a cw laser. Such an annealing time is, however, still short enough to prevent excessive dopant redistribution during annealing of ion implanted semiconductors. Furthermore, the throughput of an arc lamp or ribbon electron beam annealing system can also be very high (hundreds of 4" wafers per hour) and, in addition, both of these annealing techniques are insensitive to anti-reflective dielectric coatings, unlike their laser counterpart.

Both arc lamps and electron beams can be used for annealing in two possible modes, an isothermal mode and a heat sunk mode. In the isothermal mode the entire wafer reaches the annealing temperature. The dominant heat loss mechanism in this case is radiation; the time required to reach a given temperature is typically on the order of 5-10 seconds. This method has the advantage of requiring only low beam power. However, in certain cases it is desirable to reach annealing temperature more rapidly or to confine the heating to the sample surface. The heat sink mode must then be employed [19], in which the backsurface of the sample is kept at a specified temperature T_0. The principal heat loss mechanism under these conditions is conduction through the wafer.

Stationary Arc Source Irradiating a Thermally Isolated Wafer

A convenient geometry for this form of annealing is obtained by imagining either an arc source or simply a quartz halogen lamp to be located near the pole of a spherical cavity having walls of reflectivity $R_s \approx 1$ with the sample mounted on thermal insulators in the center of this volume. Under these conditions, the wafer will receive radiation on both surfaces from all angles and will be cooled by radiation only. The total power Q (Watts) of the source (as well as radiation from the cavity walls) is coupled to the wafer with an efficiency η which is given by

$$\eta = \frac{2A_w(1-R_w)}{2A_w(1-R_w)+A_s(1-R_s)} \tag{7}$$

where A is area (cm^2) and R the reflection coefficient. The subscripts w and s refer to the wafer and sphere respectively. Since the radiation losses from the wafer are similarly recoupled back to the wafer, these losses are reduced by the factor $1-\eta$ as compared with the losses which would occur in vacuum. Assuming that the walls of the sphere remain at temperature T_0 (by external cooling or high thermal mass), the equation describing the temperature rise of the wafer is given by

$$\rho C_p L A_w \frac{dT}{dt} = \eta(Q + \varepsilon_s \sigma A_s T_0)^4 - (1-\eta)\sigma \varepsilon_w A_w T^4 \tag{8}$$

where σ is the Stefan-Boltzmann constant ($5.67 \cdot 10^{-12} Wcm^{-2} {}^\circ K^{-4}$) and the emissivity of silicon is taken to be $\varepsilon_w = 0.5$. (Note we assume $\varepsilon = 1-R$ for both wafer and sphere.) The steady state temperature T_f is

$$T_f = \left[T_0^4 + \frac{Q}{(1-R_s)A_s\sigma} \right]^{1/4} \tag{9}$$

By substituting for Q, Eq. (9) can be written

$$\frac{dT}{dt} = \frac{(1-\eta)2\sigma\varepsilon_w}{L\rho C_p} (T_f^4 - T^4) \tag{10}$$

As an experimental guide it is useful to note that, because of the T^4 dependence, (dT/dt) is within 90% of its initial value until $T = 0.6T_f$. Hence a first order estimate of the time required for T to reach its final value is obtained by extrapolating the initial slope to the final temperature. This gives

$$\Delta t = \frac{(T_f-T_0)L\rho C_p}{2\varepsilon_w\sigma(1-\eta)T_f^4} \tag{11}$$

For a cavity of radius 15cm. and reflectivity of 99% containing a 300 μm thick 4" diameter wafer we obtain a coupling efficiency η of 0.741. For $T_f = 1000°C$, such a wafer reaches steady state in a time (from Eq. 11) of $\Delta t = 16.3$ sec.

Although Eq. 11 predicts that the time to reach a given steady state temperature increases as η increases, it should be noted that for constant <u>power</u> Q, Δt will decrease rapidly as η approaches unity.

This mode of annealing was first studied by Nishiyama et al. [20] and more recently by Powell et al. [21], both of whom obtain high electrical activation without dopant redistribution for ion implanted silicon samples. The residual defect densities, as judged by pn junction diode leakage currents, remain higher than for a furnace annealed sample, however.

Arc Source Irradiation in the Heat Sink Mode

This mode, where the backsurface of the sample is kept at a constant temperature, was first described by Gibbons [22]. The power densities required are substantially higher than in the iso-thermal mode, since the high thermal conductivity of silicon causes a large heat flow through the sample. On the other hand, the time constant describing the approach to equilibrium is sub-stantially shorter (because of the high intensity used), and very rapid anneals are possible. Also, this method is preferred for applications where melting of the sample surface is desired; the melt depth is easier to control because there is a temperature gradient across the wafer.

The temperature calculations in this mode are complicated by the broad wavelength spectrum of arc radiation sources, which makes makes a single absorption coefficient inapplicable. This problem can be solved by utilizing the superposition theorem as follows. First, the wavelength spectrum of the radiation source is divided into small intervals, each of which is assumed to have a constant absorption coefficient. The linear temperatures produced by these intervals are calculated and summed. The inverse Kirchoff transform is then applied to the sum to obtain the real temperature.

Lietoila et al., [23] have applied this procedure to calcu-late the temperatures achieved in silicon irradiated by a xenon arc lamp. It was shown that <u>steady state</u> solutions to the tem-perature can be used, if the dwell-time of the annealing spot is about 30 ms or more. The results are given in Fig. 11 for two different backsurface temperatures and wafer thicknesses. Note that the independent variable is now the absorbed power <u>density</u> in W/cm^2, rather than the power per unit radius.

We can see that use of a thicker sample and higher backsur-face temperature results in a saving of beam power. However, even in the best case shown in Fig. 11, the absorbed intensity required to reach useful annealing temperatures is about 5 kW/cm^2,

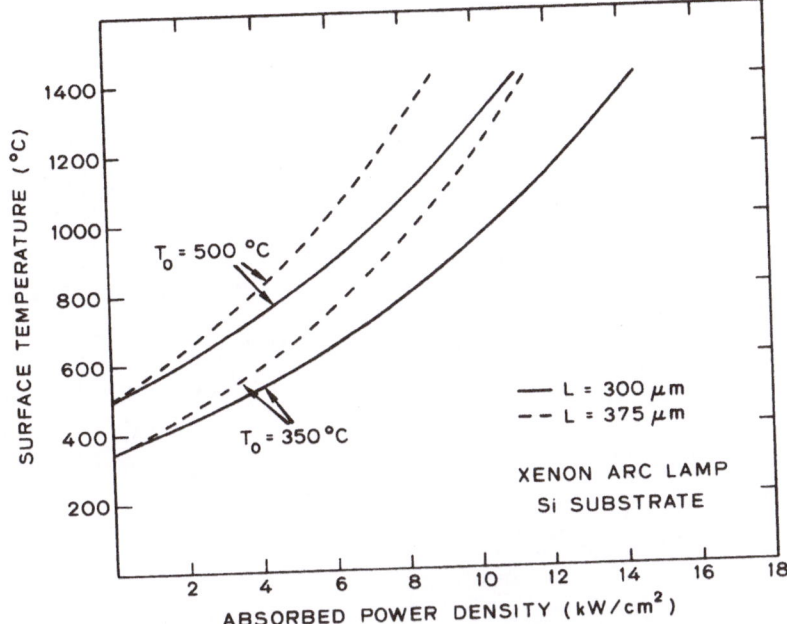

FIG. 11   Surface temperature as a function of absorbed power density for a heat sunk wafer.

which may be difficult to achieve with practical arc sources. An easy way to achieve the temperature needed for annealing, while still maintaining a well-controlled temperature gradient across the wafer, is to place a poorly conducting layer between the semiconductor sample and the heat sink. A wafer of quartz serves well as such a layer. Lietoila et al. [23] have calculated temperatures achieved for a 375 μm thick wafer and a backsurface temperature $T_0 = 500°C$ using quartz wafers of different thicknesses. The results are given in Fig. 12 and show that the required intensities are in a range much more easily achievable than if the wafer were placed directly on the heat sink.

Heat Sink Analysis

To provide a more quantitative description of the importance of a good thermal contact for arc source annealing, we show in Table 3 the effect of both a liquid tin interfacial layer and a layer of air between the sample and the heat sink. In each case the layer thickness is calculated to limit the temperature rise at the backsurface of the wafer to a prescribed value (100°C and 200°C). The air layers are sufficiently thin to show that a very thorough contact between the sample and the heat sink is required in practice to approximate the "heat sink" boundary conditions.

202

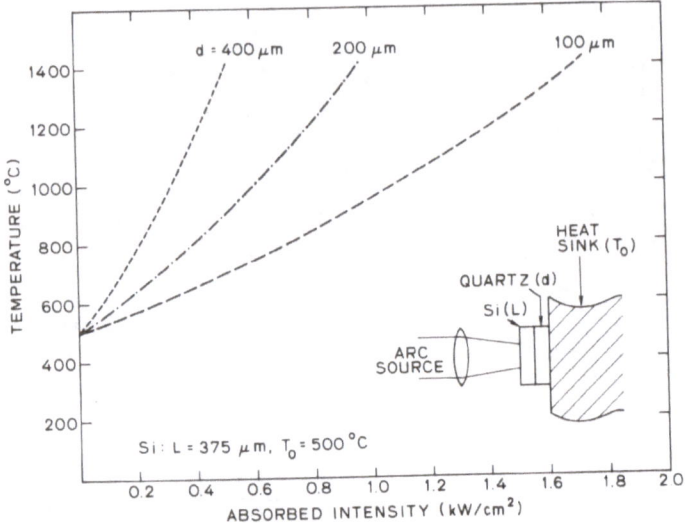

FIG. 12 Surface temperature of a 375 μm thick silicon wafer with a quartz film interposed between the heat sink and the sample.

TABLE 3. Illustrating the properties of the thermal interface between a silicon sample and a heat sink necessary to minimize backsurface temperature changes during irradiation. W = 1 mm, L = 400 μm, V = 1 cm/s.

	$Q_{melt} \frac{KW}{cm^2}$	$T_{back}(°C)$	$T_o(°C)$	$t_{max}(μm)$
LIQUID TIN	7.9	450	350	43
$K = 0.34W\ cm^{-1}\ °K^{-1}$	6.6	550	350	103
AIR	7.9	450	350	.059
$K = 0.47mW\ cm^{-1}\ °K^{-1}$	6.6	550	350	.142

Rapid Annealing of Silicon With a Scanning CW Hg Lamp

An extremely practical form for a scanned annealing system has been built using a 3" long mercury arc lamp with an elliptical reflector. A variety of arc sources can be used, though a mercury lamp is recommended because the spectral distribution of this lamp is most heavily weighted in the uv range. Consequently, the mean absorption depth for this source is much shallower than that for the other lamps. Specifically, the fraction of light absorbed versus depth is given by:

$$F = I(\lambda)(1-R(\lambda))(1 - \exp(-\alpha(\lambda)x) \tag{12}$$

where

$I(\lambda)$ is the wavelength dependent radiation intensity;
$R(\lambda)$ is the wavelength dependent reflectively;
$\alpha(\lambda)$ is the wavelength dependent absorption coefficient.

Using published absorption and reflectivity data for silicon and the spectral distributions for xenon, krypton and mercury lamps, the percent of incident light absorbed versus depth in silicon was calculated for each of the lamps. These data are given in Fig. 13. In thin film processes such as annealing and recrystallization, the region of interest is generally less than 1 μm thick. As shown in Fig. 13, the fraction of light absorbed from a mercury lamp in this region is more than twice that for the other lamps.

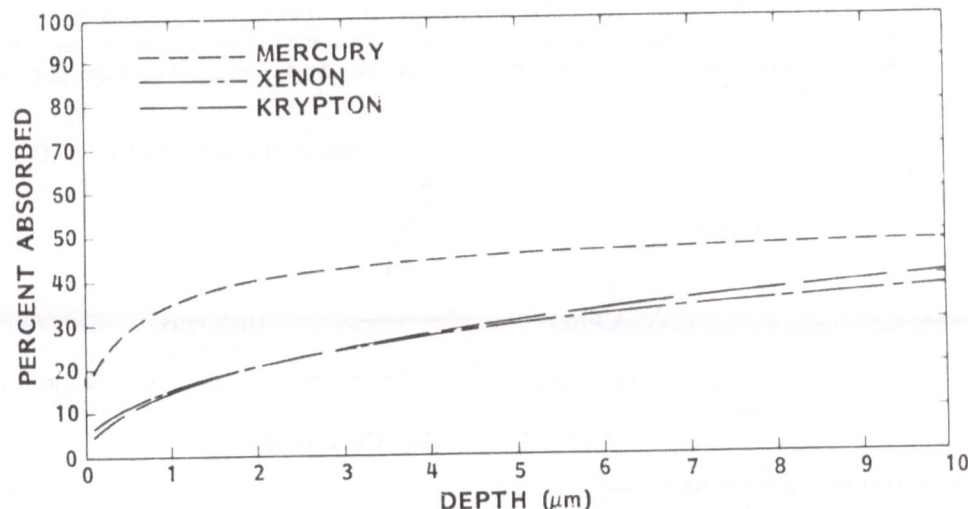

FIG. 13 Percent of incident light absorbed versus depth into silicon for three different sources.

The light from the lamp is focused into a narrow (< 5mm wide) ribbon by a 4" long reflector with an elliptical cross section. By using this shape of reflector to collect and focus the light, an intense linear heat source can be created while still maintaining a reasonable working distance between the sample and the lamp. The reflector has major and minor axes of 10 cm and 8.6 cm, respectively. This configuration gives a magnification factor of about 3.2 and a working distance of about 26 mm between the edge of the reflector and the focal plane.

The wafers are placed on a heated sample stage, which can be translated beneath the lamp at speeds up to 10cm/s. A schematic representation of the arc lamp annealing system is shown in Fig. 14.

Annealing experiments were carried out using 2 and 3 inch 1-8 Ω-cm p-type (100) silicon wafers which were implanted with $^{75}As^+$ at 100keV to 1 x $10^{15}cm^{-2}$. Several combinations of arc lamp power, scan rate and substrate temperature can be used to achieve a wafer temperature sufficient for good annealing, ~ 1000°C. In general, scan rates from 5mm/s to 1cms, substrate temperatures of 400° C to 600° C and arc lamp input power of 1Kw/inch to 1.5Kw/inch gave good results.

Because of the length of the arc lamp, an entire wafer can be annealed in a single scan. Figure 15 is a photograph of a 3" wafer in which the scan was stopped at the center of the wafer. Because of the difference in reflectivities between single crystal and amorphous silicon, the demarcation between the annealed and unannealed regions is readily seen. Transmission electron microscope (TEM) diffraction patterns made from samples taken from the unannealed and annealed regions clearly indicate the amorphous

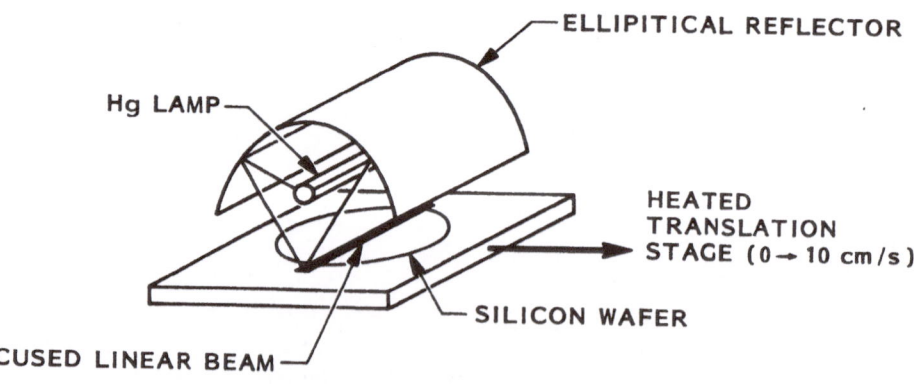

FIG. 14   Schematic representation of scanning cw arc lamp annealing system.

and completely recrystallized nature of the regions, respectively. (The picture is somewhat distorted due to the angle at which the photo was taken.)

The recrystallized layer was also analyzed by Rutherford Backscattering Spectroscopy (RBS) and directly compared to a sample which was thermally annealed at 850° C for 30 minutes followed by a 10 second 1000° C anneal. Figure 16 shows the backscattering spectra from an as-implanted, arc lamp annealed and thermally annealed sample. As shown, there is no detectable difference between the quality of the regrown material annealed by the scanned arc lamp and the thermally processed sample.

Carrier concentration and mobility of the scanning arc lamp annealed material as a function of depth were determined by means of differential sheet resistivity and Hall effect together with an anodic oxidation stripping technique. These results are shown shown in Fig. 17, along with the as-implanted profile calculated using LSS theory and published bulk silicon mobilities for the respective impurity concentrations. As shown, no measurable dopant redistribution occured during the annealing process and

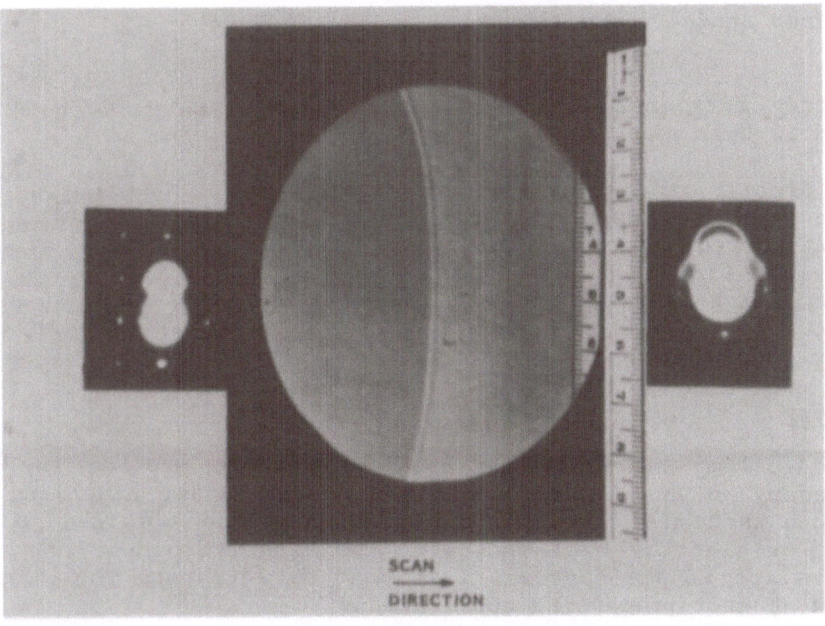

SCAN
DIRECTION

FIG. 15   Photograph of a partially annealed silicon wafer, and TEM diffraction patterns from the annealed and unannealed regions.

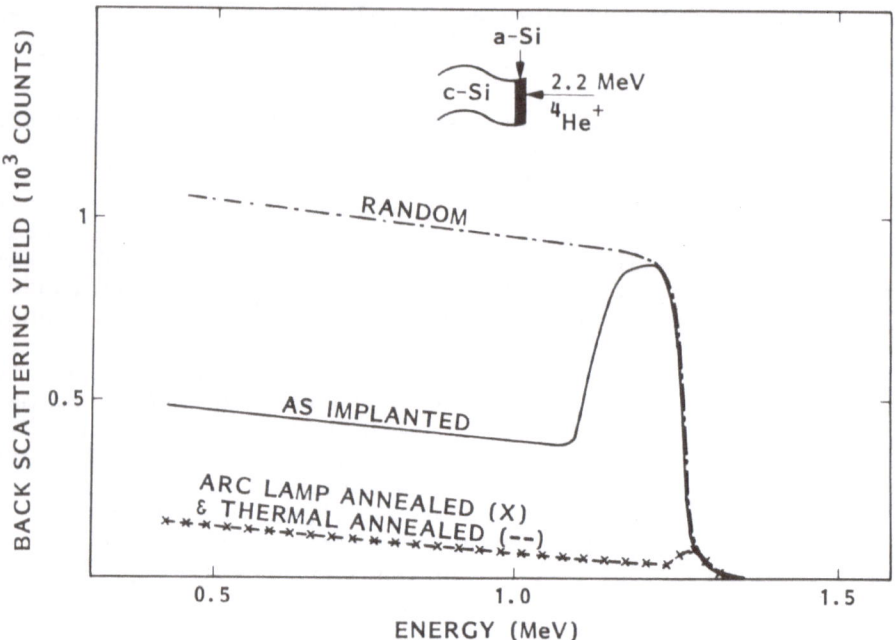

FIG. 16 Rutherford backscattering spectra from as-implanted, arc arc lamp annealed and thermally annealed samples.

the free carrier mobility in the annealed region is as good as found in bulk silicon.

Finally, to assess the uniformity of the anneal, four point probe sheet resistivity measurements were made on a wafer which was completely annealed in a single scan. The variation across the wafer from side to side and top to bottom (with respect to the scanning beam) showed no significant variation in either direction, indicating a very uniform annealing of the implanted wafer. These data are shown in Fig. 18.

<u>Summary</u>

Arc lamp annealing systems can readily provide the power necessary to anneal ion-implanted silicon under annealing conditions that are physically similar to those obtained with cw laser or electron beam sources. The high throughput capability of the arc source systems gives them an advantage for practical semiconductor processing applications.

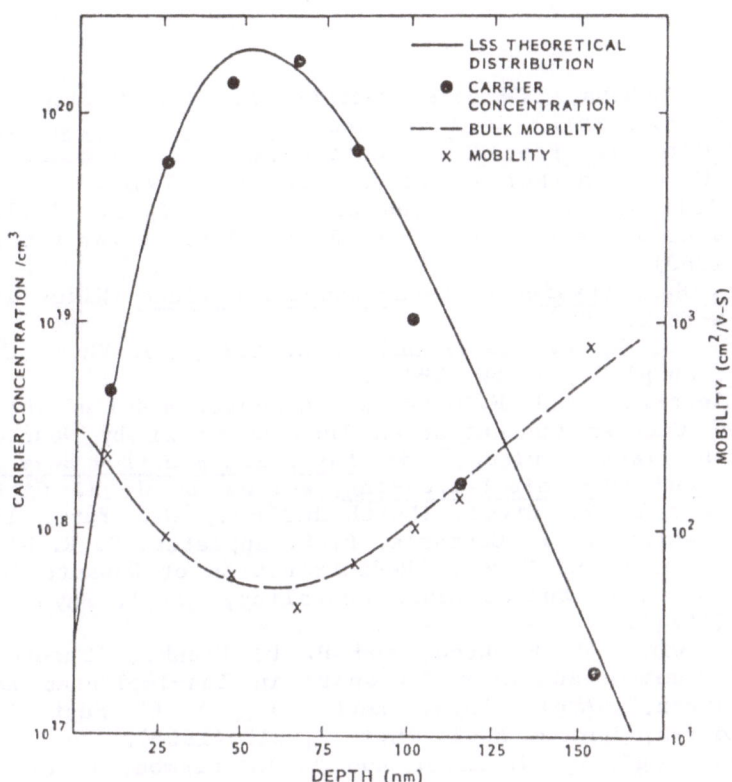

FIG. 17   Carrier concentration and mobility versus depth from an arc lamp annealed sample.

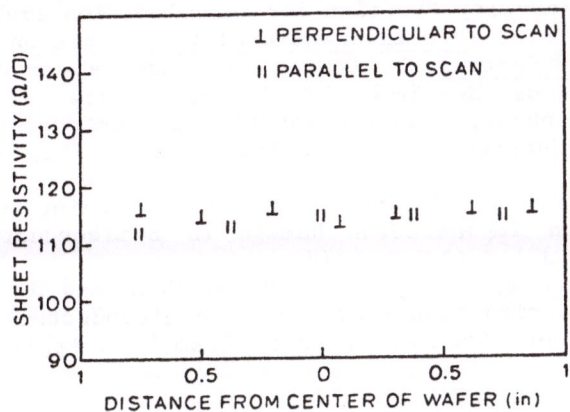

FIG. 18   Sheet resistivity profiles across an arc lamp annealed wafer.

REFERENCES

1. J. F. Gibbons, "CW Laser-Recrystallized Polysilicon as a Device-Worthy Material," in Laser and Electron Beam Solid Interactions and Materials Processing, edited by J. F. Gibbons, L. D. Hess and T. W. Sigmon (North Holland, New York, 1981).

2. T. Shibata, J. F. Gibbons and T. W. Sigmon, "Silicide Formation Using a Scanning cw Laser Beam," Appl. Phys. Lett. 36, 7 (1 April 1980).

3. S. M. Sze, Physics of Semiconductor Devices (Wiley Interscience, New York, 1969).

4. C. Y. Ho, R. W. Powell and P. E. Liley, J. Phys. Chem., Ref. Data 3, Suppl. 1, I-588 (1974).

5. A. Lietoila and J. F. Gibbons, "Computer Model of the Temperature and Carrier Concentration Induced in Si by Nanosecond and Picosecond Laser Pulses," in Laser and Electron Beam Solid Interactions and Materials Processing, edited by J. F. Gibbons, L. D. Hess and T. W. Sigmon (North Holland, New York, 1981).

6. C. W. White, W. H. Christie, B. R. Appleton, S. R. Wilson, P. P. Pronko and C. W. Magee, "Redistribution of Dopants in Ion Implanted Silicon by Pulsed Laser Annealing," Appl. Phys. Lett. 33, 7, 662 (1978).

7. J. C. Wang, R. F. Wood, and P. P. Pronko, "Theoretical Analysis of Thermal and Mass Transport in Ion-Implanted Laser-Annealed Silicon," Appl. Phys. Lett. 33, 5 (1 Sept 1978).

8. H. Kodera, Jpn. J. Appl. Phys. 2, 212 (1965).

9. L. Csepregi, J. W. Mayer and T. W. Sigmon, Phys. Lett. 54A, 157 (1975).

10. R. W. Bicknell and R. M. Allen, Proc. of the First International Conf. on Ion Implantation in Semiconductors, Thousand Oaks, Calif. (Gordon and Breach Pub., New York, 1970).

11. A. Chu and J. F. Gibbons, "A Theoretical Approach to the Calculation of Impurity Profiles for Annealed, Ion Implanted Boron and Silicon," Proc. 1976 International Conference on Ion Implantation in Semiconductors, F. Churno, J. Borders and D. K. Price, Eds., (Plenum Press, New York, 1977), pp. 711 ff.

12. J. F. Gibbons, "Ion Implantation in Semiconductors - Part II: Damage Production and Annealing," Proc. IEEE 60, 9, 1062 (1972).

13. A. Gat and J. F. Gibbons, "A Laser Scanning Apparatus for Annealing of Ion Implantation Damage in Semiconductors," Appl. Phys. Lett. 32, 3 142 (1 Feb 1978).

14. Y. I Nissim, A. Lietoila, R. B. Gold and J. F. Gibbons, "Temperature Distributions Produced in Semiconductors by a Scanning Elliptical or Circular cw Laser Beam," J. Appl. Phys. 51, 1 (1 Jan 1980).

15. J. A. Roth, G. L. Olson, S. A. Kokorowski, and L. D. Hess, "Laser-Induced Solid Phase Epitaxy of Silicon Deposited Films," Laser and Electron-Beam Solid Interactions and Materials Processing, J. F. Gibbons, L. D. Hess and T. W. Sigmon, Eds., (North-Holland, 1981), pp. 413-426.

16. J. F. Gibbons, W. S. Johnson and S. W. Mylroi, Projected Range Statistics in Semiconductors, distributed by John Wiley & Sons, (Dowden, Hutchinson and Ross, 1975).

17. W. Brown, "Laser Effects in Ion Implanted Semiconductors," E. Rimini, Ed., Inst. di Struttura della Materia, Universita di Catania, Corso, Italy (1978).

18. A. Gat, J. F. Gibbons, T. J. Magee, J. Peng, V. R. Deline, P. Williams and C. A. Evans, Jr., "Physical and Electrical Properties of Laser-Annealed Ion Implanted Silicon," Appl. Phys. Lett. 32, 2767 (1 Mar 1978).

19. H. S. Carlsaw and J. C. Jaeger, Conduction of Heat in Solids, 2nd Edition (Clarendon Press, 1959), p. 11.

20. K. Nishiyama, M. Arai and N. Watanabe, Jap. J. Appl. Phys. 19, 10 (1980).

21. R. A. Powell, T. O. Yep and R. T. Fulks, "Activation of Arsenic-Implanted Silicon Using an Incoherent Light Source," Appl. Phys. Lett. 39, 2 (15 July 1981).

22. J. F. Gibbons, "CW Laser-Recrystallized Polysilicon as a Device-Worthy Material," Laser and Electron Beam Solid Interactions and Materials Processing, J. F. Gibbons, L. D. Hess and T. W. Sigmon, Eds., (North Holland, New York, 1981) pp. 449-463.

23. A. Lietoila, R. B. Gold and J. F. Gibbons, "Temperature Rise Induced in Si Continous Xenon Arc Lamp Radiation," J. Appl. Phys. 53, 2 (Feb 1982).

MATERIALS CHARACTERIZATION

C. R. Helms

Stanford Electronics Laboratories
Electrical Engineering Department
Stanford University
Stanford, California 94305

INTRODUCTION(1,2)

Semiconductor technology requires probably more sophisticated characterization techniques than any other modern technology. For example, many applications require atomic sensitivities of 1 ppb or less, and we may be interested in analyzing a volume smaller than $10^{-16}cm^3$ (hopefully not at the same time). In addition to elemental analysis of very small volumes, determination of the chemical state (oxidation state, etc.) of the constituents is many times critical. Of equal importance, especially for the high density of devices contemplated for very large scale integration, is the detection and characterization of defects present in wafer starting materials. Many techniques have been applied to these problems. In this paper, I will describe the capabilities and limitations of some of these techniques for studies of important systems in semiconductor technology. The goal of this paper will be to provide the reader with the information necessary to choose among the techniques for a particular analysis application and to show how each can be applied in a complimentary fashion for a specific problem.

In many cases, materials characterization is performed using electrical measurements which can at best provide an indirect determination of the fundamental materials property of interest. This is in contrast to spectroscopic techniques where a fundamental materials property is measured directly. This direct measurement approach is clearly desired and therefore semiconductor characterization with techniques that can directly measure a fundamental semiconductor property (doping level, point defect density, interface morphology, etc.) will be emphasized in this paper.

Included in this discussion will be a detailed comparison of four techniques which have shown great value in semiconductor characterization problems. They are Auger electron spectroscopy (AES), Rutherford backscattering (RBS), secondary ion mass spectroscopy (SIMS), and x-ray photoelectron spectroscopy (XPS).

Reviews of the techniques to be covered here have been presented in detail elsewhere. It will not be my intention, therefore to repeat detailed material presented there. I will concentrate instead on a description of each techniques capabilities and limitations.

The factors to be considered will include sensitivity and data acquisition time, spatial resolution, depth resolution, quantitative analysis, chemical state determination, and nondestructive testing. Many of these factors are related for the various techniques and these relationships will be discussed. The discussion will center on the capabilities of equipment that is commercially available but near state of the art. Therefore, much of the equipment in the field will not have the capabilities described here and new equipment may appear shortly to extend these capabilities. I will attempt to indicate where new developments may be forthcoming. Each technique will be discussed separately and then a general comparison made. AES and XPS will be discussed together since they are both electron spectroscopies.

Examples of Characterization Problems

I will classify semiconductor characterization problems into two categories. First are those problems that do not require any significant spatial resolution parallel to a samples surface. These might include problems in process development where one dimensional profiles might be sufficient to characterize a process. Second are those problems which require a significant spatial resolution parallel to the surface. These would include any measurements on actual device structures which would be necessary for trouble-shooting or the assessment of any two or three dimensional effects in device processing.

As an illustration of what types of analysis capabilities are necessary, a typical silicon integrated circuit structure is shown in figure 1. This is a cross-section of a typical n channel MOS device. The substrate is p type with a doping level of $10^{14}-10^{16}$ cm^{-3} of boron. The n$^+$ regions are the source and drain of the device which would be doped with As to a peak concentration of $\sim 10^{21}$ cm^{-3} at the surface and a diffusion profile with a junction depth of <0.5 m. The channel region between the two n$^+$ regions would be implanted with additional boron (concentration profile determined by device size). Above this channel region is a thermally grown gate oxide with thickness determined by the lateral dimensions. Note the interface between the channel region and gate oxide must be extremely well controlled for proper device operation.

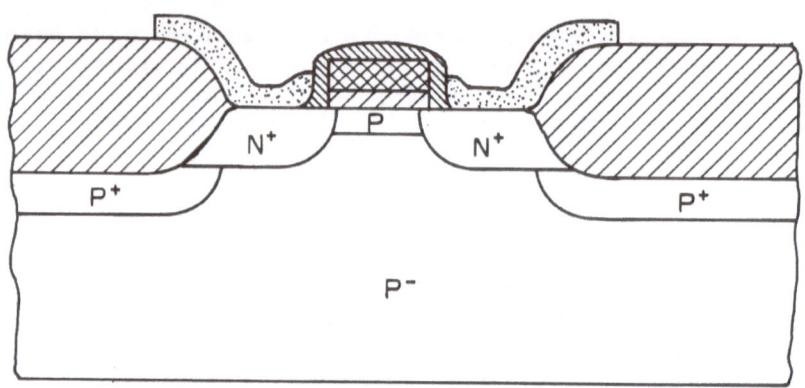

Figure 1. Cross-section of n channel MOS device, with p-substrate, n$^+$ source and drain, a p region under the gate oxide. SiO$_2$ regions shown as regions with slanted lines, polysilicon as cross hatched regions and Al as dotted regions.

Electrical connection to the gate is typically made by heavily doped polysilicon (refractory metal silicides will probably at least partially replace polysilicon in submicron structures). Electrical contact to the source and drain regions is typically made by Al as indicated. Finally the devices are passivated with a grown or partially grown and partially deposited SiO$_2$ layer. Typical values of some of the parameters as a function of the channel width are shown in Table 1.

Channel length	Gate oxide thickness	Channel implant peak conc/junction depth	field oxide thickness
4μm	500 Å	2x10cm^{-3}/3000 Å	8000 Å
1μm	125 Å	8 x 10^{16}/750 Å	2000 Å
0.25μm*	30 Å	3.2x10^{17}/200 Å	500 Å

*These values listed for comparison only; they would be nearly impossible to achieve in practice

Classical Characterization Techniques(3)

Although the major thrust of this paper is the more modern characterization techniques, a brief discussion of more classical techniques is in order. These include four point probe, Hall effect, Van der Pauw measurements, ellipsometry, interferometry, optical staining techniques, spreading resistance, and capacitance voltage measurements, among others.

The first three of these measurements, four point probe, hall effect, and Van der Pauw provide a means to determine carrier concentration and carrier mobility on bulk samples. They have limited applicability to thin film structures unless the details of the film profiles are already known accurately. In many cases of current interest actual device structures may be required for accurate measurement of carrier concentrations and mobilities.

The second two techniques mentioned, ellipsometry and interferometry, along with stylus techniques, provide information concerning film thickness. Ellipsometry is nondestructive and can be used to determine film thickness for films transparent at the wavelength of interest, if the optical properties of the substrate material are sufficiently known. The last two techniques require that a step be etched to define a sharp edge of the film whose thickness is to be measured.

Profiles of electrical conductivity can be obtained using optical staining techniques, spreading resistance measurements, and capacitance/voltage measurements. The optical staining technique does not actually give a profile but shows at what point on a sample (usually after angle lapping) the junction between n type and p type occurs. Spreading resistance is also performed using angle lapping with mechanical probes, to measure differential resistivity as a function of depth. CV techniques on pn functions and MOS structures can also be used to infer carrier profiles by measuring the voltage required to deplete the region near the junction.

Most of the techniques mentioned above are inadequate for sufficiently characterizing semiconductor device structures. A major limitation of most of these techniques is insufficient depth resolution. As discussed in a previous section, in many cases depth resolutions better than 10 Å are required. Another limitation for the techniques used to measure dopant profiles is the presence of electrically inactive or compensated impurities which go undetected in measurements of carrier concentrations. In the next few sections some techniques which overcome these problems, especially with respect to impurity profile determination, will be discussed.

Secondary Ion Mass Spectrometry (SIMS)(4-7)

Of all the techniques to be discussed, the main advantage
of SIMS is its potential sensitivity to small impurity concentrations. The technique, however, does have two major disad-
vantages. First, it is by its nature destructive requiring
sputtering to generate the secondary ions. Second, the yield
of emitted ions is extremely sensitive to the chemical composition of the surface making quantitative analysis (and sometimes
qualitative analysis) of multilayered structures extremely difficult.

SIMS experiments are performed in ultra-high vacuum systems
by bombarding the sample to be analyzed with ions with energies
from 500 eV to 20 keV. This causes sputtering to occur and
secondary ions that are emitted from the surface are collected
and mass analyzed to obtain the elemental composition of the
sample's surface. Since sputtering is used to generate the
data, depth profiling is performed automatically. Typical profiling rates range from a few to a few hundred Å/min.

Questions of sensitivity, data acquisition time, spatial
resolution, and depth resolution are not independent of each
other for this technique. In general, the signal from a particular ion will appear as a counting rate or tatal number of
counts which can be given as follows:

Counts = (sputter rate) (data acquisition time per point)
(analysis area) (volume concentration) (ion yield)
(instrument efficiency).

The first three factors are determined by the investigator within the limits of the articular instrument being employed. The
fourth is the unknown that we wish to determine and the last two
depend on the sample itself and the instrument being used. The
product of sputtering rate (Å/min) and data acquisition time
(min) gives the minimum depth resolution (other factors can also
limit depth resolution). As an example, consider a sample where
we chose 100 counts (S/N = 10 assuming no background) as the
minimum acceptable signal. We will use a sputtering rate of
100 Å/min and a data acquisition time of one min per point giving a depth resolution of 100 Å (at this rate one micron would
require 100 mins to profile). We will use an analysis area of
100 x 100 m with an element that gives an ion yield of 10% and
an instrument with an efficiency of 10%. These last two
factors can be considered the best case for the ultimate in
instrumentation and are almost never achieved in practice.
Using the above relationship within these constrains the mini-
mun concentration that could be detected would be 10^{14} cm^{-3}. A

number of factors will decrease this sensitivity, including much lower ion yields, large background signals at the mass of interest, interference from other molecular ions with the same mass, and instruments with poorer efficiencies than 10%. In addition, if a smaller analysis area was required for the same sample and experimental equipment either the sensitivity or depth resolution would be degraded. This effect is shown in Figure 2 for a typical case (10^{15} cm^{-3} sensitivity for 100 x 100 μm analysis area) where the left hand axis is a plot of sensitivity vs analysis area for constant depth resolution of 100 Å and the right hand axis is a plot the depth resolution vs analysis area for a constant sensitivity of 10^{17} cm^{-3}.

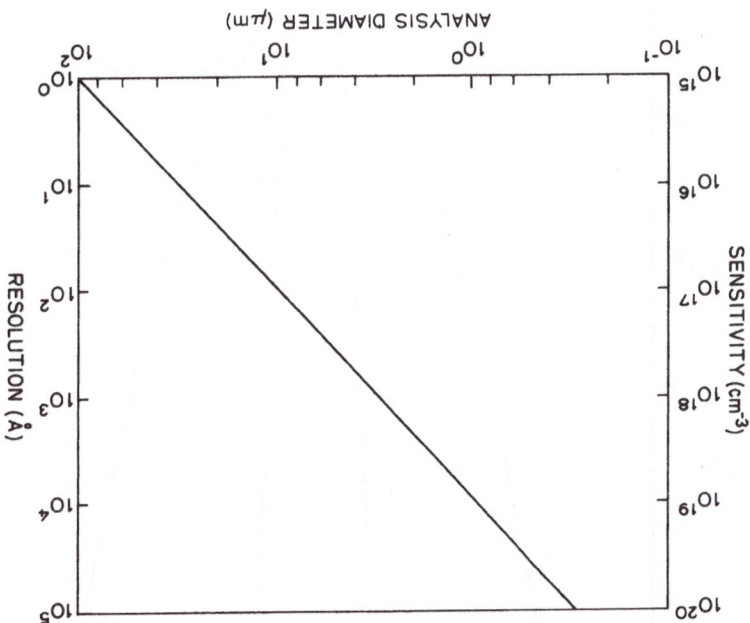

Figure 2. Relationship between SIMS sensitivity, depth resolution, and analysis area. Left hand axis is the sensitivity versus analysis area for a 100 Å depth resolution. Right hand axis is the depth resolution possible for a constant sensitivity of 10^{17} cm^{-3}.

The fundamental limit on the depth resolution of the technique is determined primarily by two factors: ion knock-on mixing or enhanced diffusion, and sputter induced nonuniformities due to nonuniform ion flux distribution or roughness induced by

216

the sputtering process.(8,9) The effect of the ion knock-on mixing and enhanced diffusion on the resolution of a 10^{15} cm^{-2} concentration delta function profile is shown in Figure 3 for Xe
ions at 500 eV and Ne ions at 10 keV(10). These curves show the
general trend in depth resolution: the best resolution is obtained for heavier bombarding ions at lower energies.

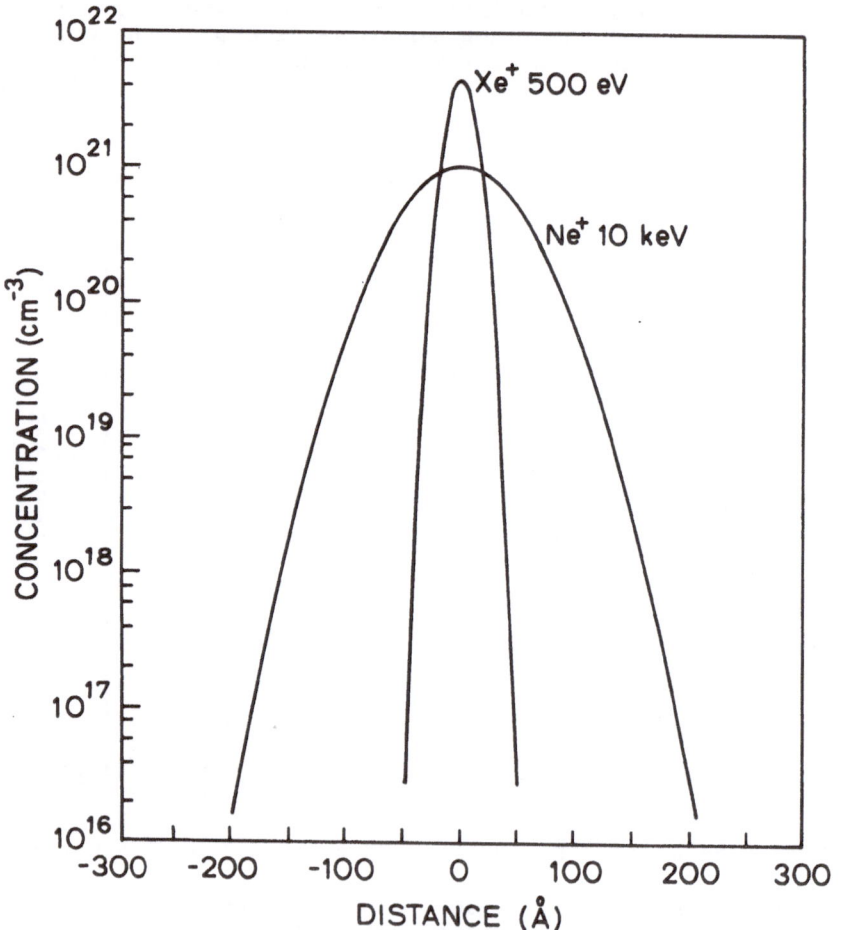

Figure 3. Measured concentration of a δ function of concentration 10^{15} cm^{-2} versus distance away from the δ function for Ne
ions at 10 keV and Xe ions at 500 eV.

Sputter induced nonuniformities can have a particularly severe effect on depth resolution especially for thicker samples. This contribution to the depth resolution is typically at least 1% of the thickness that has been removed by sputtering. For some crystalline materials, sputter induced roughness can be quite severe as in the case of Al sputtered with an inert gas where large hillocks are formed making profiling difficult.(9)

One of the more difficult problems for SIMS is accurate qualtitative analysis. The use of SIMS for quantitative analysis is predicted on a knowledge of the ion yield (no. of ions emitted/total no. of atoms emitted) alluded to earlier in this section. This parameter is very sensitive to the substrate and the gas used for sputtering and can vary by over 10^4 in going from one material to another.(4-7) To perform accurate quantitative analysis data from standards of composition similar to the sample of interest must be obtained. In addition, residual levels of reactive gasses present during profiling (especially O_2, CO, and H_2O) must be kept low or at least constant since the ion yield can be very sensitive to oxygen adsorbed on the surface.

Finally are the questions of nondestructive testing and analysis of chemical state. Obviously, since SIMS measures ions emitted during sputtering, it is by its nature destructive. It also has little chemical selectivity due to the displacements caused by the sputtering process.

Rutherford Backscattering Spectroscopy (4,11-14)

RBS is a semi-nondestructive technique for measuring elemental profiles as well as crystal perfection (using channeling). The experiments are performed by bombarding the sample with light ions (usually He) at energies of a few meV. Any ions that are backscattered are energy analyzed to determine the atom responsible for the scattering as well as the depth below the surface at which the atom is located.

For a 180° backscattering geometry the energy of the backscattered particles can be given by

$$E = \left(\frac{M-m}{M+m}\right)^2 E_o \text{ for } M > m$$

where M is the mass of the scattering atom, m is the mass of the scattered atom, E_o is the initial beam energy and E is the measured energy of the scattered particle. This assumes no other energy loss mechanisms which would be the case for an atom on a samples surface. For atoms below the surface the incident

beam will lose energy on the way into the sample, will scatter
and then lose energy on the way out of the sample. A profile is
then obtained where the high energy scattered particles come from
the surface; lower energy particles come from various depths below the surface. For RBS, profile information is therefore more
difficult to extract than for SIMS since the information concerning both elemental identification and depth must be extracted
from a single energy.

As was the case for SIMS, for RBS, questions of sensitivity,
data acquisition time, spatial resolution and depth resolution
cannot be independently dealt with. The signal strength from a
particular element would again be given by a numnber of counts
associated with the backscattered particle at a particular energy
as

counts = (ion flux) (analysis area) (volume concentration)
(cross-section) (time) (depth resolution) (geometrical factors).

Typical sensitivities in a silicon substrate are shown in Figure
4 versus atomic number for an analysis area of a few $(mm)^2$. The
discontinuity that occurs is due to the large background of the
silicon substrate at masses 28, 29, 30 that appears for the low
Z elements. In addition, due to the detectors used, phosphorus
is difficult to distinguish from silicon because of the proximity of their masses. The depth resolution for this curve is
~200 Å. Since sputtering is not being used for profiling, as
the case for SIMS, the flux, time, and depth resolution are independent. The signal can therefore be increased by counting
for longer times so that the sensitivity α $(time)^{1/2}$ (neglecting effects of the RBS ion beam). The curve of Figure 4 was
determined for a data acquisition time of 2 hours. In addition,
the data from all depths is being collected simultaneously so
that this data collection time would be comparable to the example mentioned above the a SIMS profile through 1 μm.

The depth resolution of RBS is limited by the energy resolution of the detector and the experimental geometry. For a
typical detector of 15 keV resolution, a 2 meV He ion beam,
and the sample placed 45° with respect to the beam, a depth
resolution in Si of ~200 Å is obtained. For more grazing
angle geometries, resolutions down to 10 Å are possible.

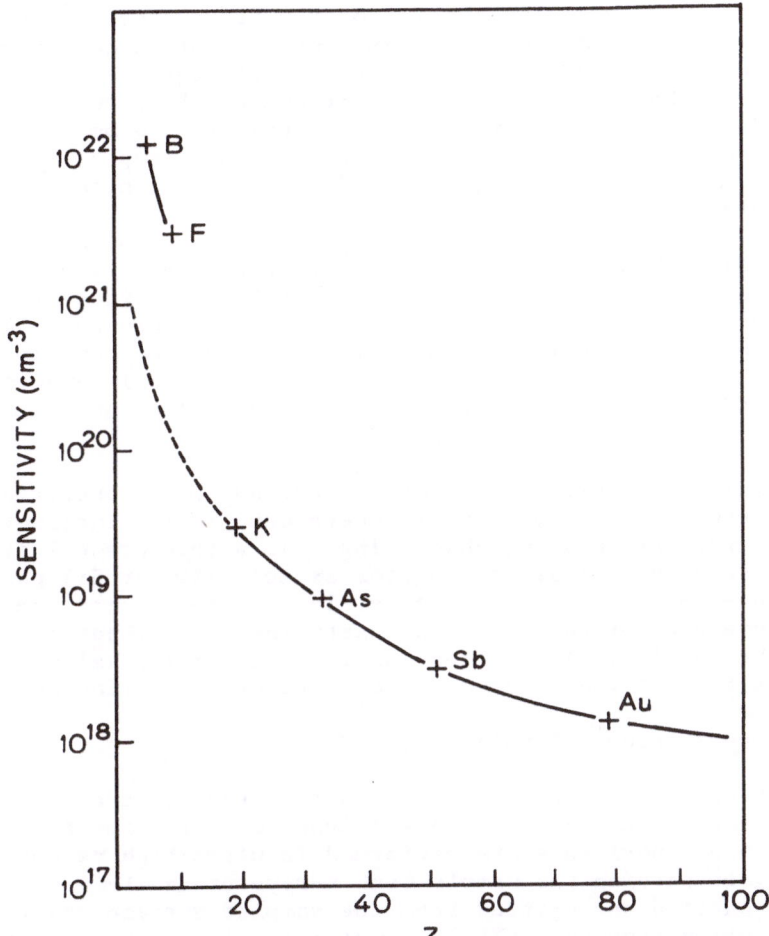

Figure 4. Typical sensitivity of RBS in a Si substraté versus atomic number. The extension of the heavier elements by the dashed curve would be possible if there was no substrate inter-ference.

As was the case for SIMS, the lateral resolution of RBS is limited by the beam size. Virtually all systems operate with beam sizes a few mm in diameter. Focused systems have, however, been developed with beam sizes down to 1 μm(2,15) but at present are not generally available.

The two major advantages of RBS over the other techniques discussed are its capabilities for quantitative analysis and nondestructive testing. The physics of the RBS process is quite well understood and scattering cross-sections can be determined to great accuracy. In addition, there are no major matrix effects as for SIMS so elemental compositions can be accurately determined. For the beam sizes normally used (a few mm) and typical doses of He ions, the damage caused to a sample is minimal. This makes this technique suited to quality control use on actual device wafers.

In addition to its poor sensitivity to elements with Z < the substrate, another major limitation of the technique is its sensitivity to chemistry since the scattering cross-sections don't depend on chemical environment. The technique can, in many cases, be used to infer the presence of chemical compounds however since accurate measurement of atomic ratios can be performed.

In addition to elemental profiles RBS equipment outfitted with accurate angular sample positioners can provide information on crystal perfection using channeling. If a thin crystal is aligned with a channel direction (for example 110 for Si) parallel to the beam the backscattering yield will be reduced considerable since many of the impinging particles will "channel" through the lattice. This technique is particularly valuable in looking at interstitial concentrations and other lattice defects.

Electron Spectroscopy (16-21)

The techniques to be discussed in this section are x-ray photoelectron spectroscopy (XPS) and Auger electron spectroscopy (AES). These experiments are performed in ultra-high vacuum systems by bombarding the sample with x-rays or an electron beam. Electrons excited and emitted from the samples surface due to single electron processes (XPS) or multiple electron Auger processes (AES) are energy analyzed. The measured electron energy distributions can then be used to identify the samples constituents. In addition, the energy distributions are sensitive to the electrical potential near a particular atom so that changes in chemistry can be measured with these techniques.

Both these techniques are surface sensitive with inherent depth resolution of from 5 to 30 Å depending on the energy of the transition used. For depth profiling, either sputtering or other stripping methods (chemical etching, etc.) must be used in conjunction with the technique. Sputtering is typically used for most applications and the subsequent discussion will be limited to sputter profiling.

Auger Electron Spectroscopy

A similar relationship for the sensitivity, data acquisition time, spatial resolution, and depth resolution can be written down as follows

signal = (electron flux) analysis area) (volume concentration) (excitation efficiency) (time) (depth resolution) (instrument efficiency)

where the volume concentration is an average over the depth resolution and the excitation efficiency is similar to a cross section. The relationship to signal-to-noise ratio here, however, is not as simple as the previous techniques discussed due to the large background level that is always present, which introduces considerable noise.

We will consider the typical case discussed previously: a one micron deep profile with a depth resolution of 100 Å, a data acquisition time of 100 minutes and a signal to noise ratio of 10. Using an excitation current of 100 μA and near state of the art electron spectrometers would give a minimum detection limit in profiling of $\sim10^{18}$ cm^{-3}. As is also the kcase for SIMS, the sensitivity decreases as the lateral resolution (or depth resolution) is increased. For electron excited AES, the beam current versus beam size is shown in Figure 5 as the solid line. For currents above 1 μamp, the beam must be defocused to the diameter shown by the dashed line to reduce sample heating to acceptable levels. The sensitivity is proportional to the square of the current, therefore if a sample area 1 μm square were analyzed the sensitivity would be reduced by a factor of 20 to 2 x 10^{19} cm^{-3}.

The depth resolution of AES used with sputter profiling is typically the same as SIMS due to the inherent limitations of the sputtering process. However, hisotrically, AES profiles have been measured using lower energy (down to 500 eV) ions than is typical for SIMS so that in general better depth resolutions have been demonstrated. The best depth resolution that has been routinely demonstrated is 20 Å for thinner (<1000 Å) layers or 1% of the thickness for thicker layers.

Nondestructive testing in a profiling mode cannot be performed by AES due to the etching necessary. If surface concentrations are required without the need for profiling and electron beam damage is not a problem, AES can be applied in a nondestructive sense.

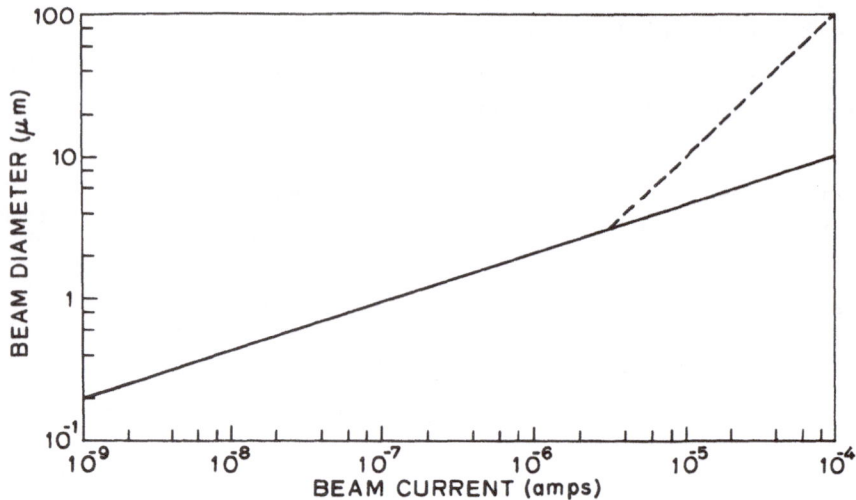

Figure 5. Typical beam diameter versus beam current with LaB_6 emitter--solid curve. Dashed segment corresponds to the approximate increase in beam size necessary to keep the local sample temperature down to reasonable levels in silicon.

Quantitative analysis can be performed to ± 10% if spectra are available for standards. This is especially important in sputter profiling due to preferential sputtering effects that can significantly alter a samples surface composition. As was mentioned above, qualitative analysis (within about a factor or two) is possible from literature comparisons due to the insensitivity of the techniques to matrix effects.

X-Ray Photoelectron Spectroscopy

The main advantage of XPS over AES is the reduction of radiation (electron beam vs X-ray beam) damage to the sample being investigated. In addition, for many elements, both the Auger and core level spectra can be obtained simultaneously, giving an additional method for chemical state determination. Its major disadvantage is our inability to easily focus the x-ray beam to analyze small areas and in many cases poor surface sensitivity due to longer mean free path of the emitted electrons at the electron kinetic energies of the core level transitions.

The assessment of XPS capabilities and limitation concerning sensitivity, speed, etc., can be performed as above for the previous techniques.

At present, XPS equipment is not available to perform a 1 μm deep profile and retain a 100 Å resolution. This is due to the broad beam used for the excitation of the XPS spectra and the inavailability of sputtering guns with ion flux uniformity over the large areas required. For purposes of comparing sensitivities if we take the previous case limited to thinner layers with an analysis time of 1 minute per data point, the maximum XPS sensitivity would be $\sim 10^{19}$ cm^{-3} using spectrometers similar to what would be used for AES. New energy analyzers using parallel data collection techniques are becoming available which increase the sensitivity of XPS by a factor of ten and AES sensitivities for small beam currents (small spot sizes) by a similar factor. In addition, typical x-ray sources are limited to the 0.5 to 1k watt range and in many cases have not been optimized for maximum x-ray flux; more powerful, efficient sources would provide better sensitivity capabilities for XPS.

As mentioned above, the depth resolution of XPS used with sputter profiling is degraded significantly due to the large spot size of the x-ray beam and the difficulty of providing a uniform ion flux for sputtering over the few mm x-ray beam size. Good success has been obtained with chemical etching, but this is a much slower process due to the sample handling in and out of the vacuum system that is necessary.

Comments concerning nondestructive testing and quantitative analysis are similar to those for Auger electron spectroscopy.

Comparison Summary

In the previous few sections the capabilities and limitations of RBS, XPS, AES and SIMS have been discussed. A summary of the author's views is shown in Table 2 for the various categories of importance. The last column is an indication of which technique would provide the best performance for that capability. Note that this table is prepared assuming the use of sputter profiling.

Acknowledgment

The author would like to thank DARPA for continuing support through contract MDA 903-79-C-0257 which has made much of this work possible.

References

1. C. R. Helms in "Optical Characterization Techniques for Semiconductor Technology", SPIE Vol. 276 (Society of Photo-Optical Instrumentation Engineering, Elllingham Washington 1981).
2. C. R. Helms, J. Vac. Sci. Tech. 20, (1982) 948
3. see for example A.B. Glaser and G. E. Subak-Sharpe "Integrated Circuit Engineering", (Addison Wesley Reading Massachusetts 1977) Chapt. 11.
4. W. Reuter and J. E. E. Baglin, J. Vac. Sci. Tech., 18, (1981) 282.
5. C. A. Evans, J. Anal. Chem., 44, (1972) 67A.
6. K. Wittmaach, Nucl. Instrum. Meth., 168, (1980) 343.
7. K. Wittmaach, Surf. Sci., 89, (1979) 668
8. C. F. Cook, C. R. Helms, and D. C. Fox, J. Vac. Sci. Tech., 17, (1980) 44.
9. T. Adachi and C. R. Helms, J. Vac. Sci. Tech., 19, (1980) 119.
10. S. A. Schwarz and C. R. Helms, J. Vac. Sci. Tech., 16, (1979) 781.
11. M. A. Nicolet, J. W. Mayer, and I. V. Mitchell, Science, 177, (1972) 841.
12. W. K. Chu, M. A. Nicolet, J. W. Mayer, and C. A. Evans, J. Anal. Chem. 46 (1974) 2136.
13. J. W. Mayer and J. M. Poate, "Thin Films Interdiffusion and Reactions" J. M. Poate, K. N. Til, and J. W. Mayer, eds. (Wiley, New York (1978).
14. W. K. Chu, J. W. Mayer, and M. A. Nicolet, "Backscattering Spectrometry", (Academic Press, New York, 1978).
15. J. A. Cookson and F. D. Polling, Thin Solid Films, 19, (1973) 381.
16. C. C. Chang, J. Vac. Sci. Tech., 18 (1981) 276.
17. K. Siegbahn, C. Nordling, and A. Fahlman, "Atomic Mlecular and Solid State Structure", studied by means of Electron Spectroscopy, ESCA, (Almquist and Wiksell Boktryckeri AB., Uppsala 1967).
18. "Electron Spectroscopy", C. R. Brundle and A. D. Baker, eds. (Academic Press, New York, 1980).
19. C. C. Chang, "Characterization of Solid Surfaces", P. F. Kane and G. B. Larrabee, eds. (Plenum, New York, 1971).
20. D. Chattarji, "The Theory of Auger Transitions", (Academic Press, New York, 1976).
21. D. T. Hawkins, "Auger Electron Spectroscopy, A Bibliography: 1925-1975", (Plenum, New York, 1977).

TABLE 2

TECHNIQUE COMPARISON

Capability	RBS	XPS	AES	SIMS	Best Technique
Sensitivity/Speed	Fair–Good	Fair–Good	Good	Excellent	SIMS
Spatial Resolution	Poor	Poor	Excellent	Good	AES
Depth Resolution	Fair*	Good**	Excellent**	Good–Excellent	AES
Nondestructive	Excellent	Poor**	Poor	Poor	RBS
Quantitative	Excellent	Good	Good	Poor	RBS
Chemical Sensitivity	Poor	Excellent	Good	Poor	XPS

*Excellent when grazing angle geometries can be employed

**when used with sputter profiling

ONE-DIMENSIONAL SIMULATION OF IC FABRICATION PROCESSES

D. A. Antoniadis

Massachusetts Institute of Technology
Cambridge, MA 02139

1. INTRODUCTION

This chapter presents important aspects of technology modeling which allow numerical simulation of migration of single or multiple dopant species as well as redistribution effects associated with moving boundaries during oxidation and epitaxy in silicon.

This chapter is organized as follows. An overall computer program structure which makes it possible to simulate a complete sequence of fabrication steps is described. Next, the features of process models for ion implantation, impurity migration, oxidation and segregation phenomena and epitaxy are discussed with emphasis on physical effect such as extrinsic diffusion impurity clustering, coupled species diffusion and enhanced oxidation rates. These effects are crucial for realistically modeling current and future technologies. Last, numerical aspects of the simulation are discussed.

2. PROCESS SIMULATION

Figure 1 illustrates the structure of the pocess simulation program SUPREM II (1). The program is designed so that steps can be simulated either individually or sequentially, just as they would occur during the actual fabrication of an IC. The output of the program, available at the end of each step, consists of the one-dimensional profiles of all the dopants present in the silicon and silicon dioxide materials. These profiles may be displayed in various formats including line-printer output, line-printer plots, and high-resolution plot. It is understood that, in sequential step simulation, the output of a processing

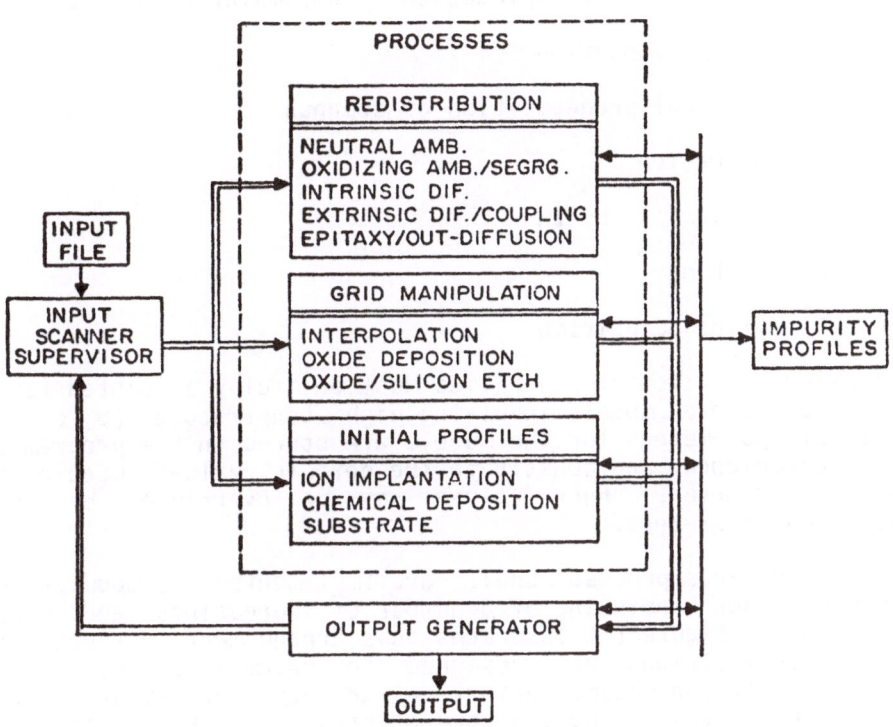

Fig. 1   General block diagram of the SUPREM process simulator.

step constitutes the initial conditions for the subsequent step. The junction depths and sheet resistance of all n or p layers formed during the process are also calculated.

The fabrication step simulation is based on several process models. Typical models implemented in a simulator include:

(a)   ion implantation

(b)   chemical predeposition of dopants

(c)   oxidation

(d)   epitaxial growth

(e)   etching

(f)   oxide deposition

Diffusive migration of impurities must be fully accounted for in all of the above models involving high temperature (b, c, d). Physical parameters for the models are stored in the program for user convenience, and constitute the default values used by the models. However, these values may be overridden by user-specified parameters.

The various process models are implemented as modular sub-programs, each consisting of a number of subroutines and special functions. Figure 1 illustrates this arrangement. Input parameter specifications are designed to resemble actual process runsheet data and documentation. A run may consist of a series of process steps. The sequence of steps and the specification of the correct model parameters are controlled by a supervisor program that evokes the appropriate step program. All communication between the various subprograms is directed through the common variable area of the computer memory.

One essential part of the program common area contains the impurity concentration arrays. Capability for handling up to three different impurity species has been found to be suitable (1). The impurity concentration is stored in terms of a discrete profile with some maximum number of points (400 is typical (1)). Each concentration value corresponds to a point in a discrete space (spatial grid) defined along a vertical axis, with its origin at the surface of the solid, i.e. silicon (Si) or silicon dioxide (SiO_2). Very often, during processing, the physical dimensions of the simulated discrete space may change, as happens for example during oxidation, etch, deposition, and epitaxy. The distance between spatial grid points may not be uniform during any of the above processes. A cubic spline

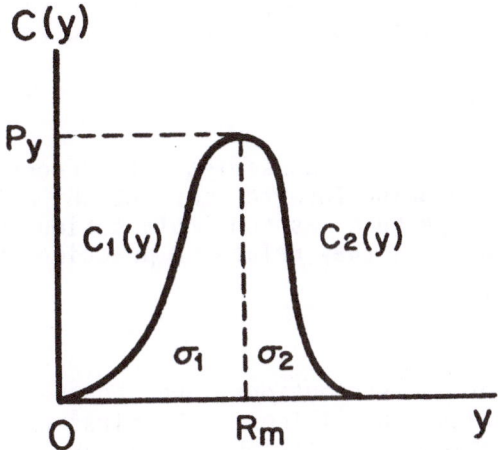

Fig. 2   The joint half Gaussian representation of as-implanted impurity profiles.

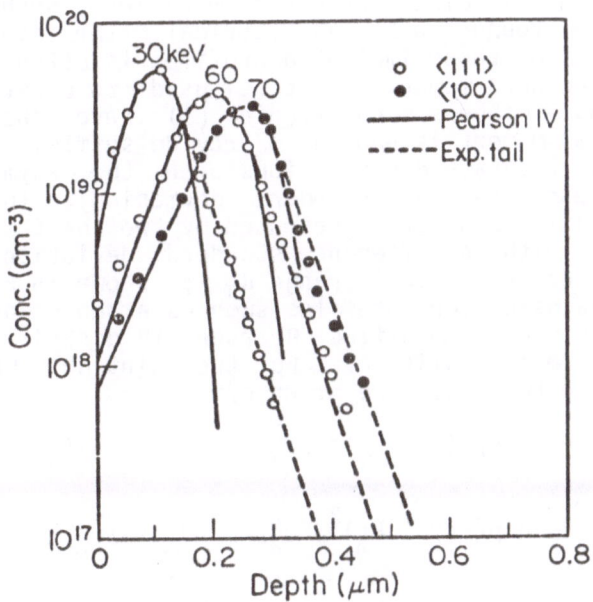

Fig. 3   Boron as-implanted profiles in <111> and <100> silicon in a random direction. Pearson IV and modified Pearson IV distributions for representation of these profiles. Dose is 10^{13} cm^{-2}.

interpolation, in the Grid Manipulation block of Fig. 1, is used at the end of each of the processing steps to restore uniformity of the spatial grid.

3. PROCESS MODELS

This section describes a series of process models. The overall thrust is to bring forward current thinking about models for technology steps such as ion implantation, diffusion, oxidation and epitaxy as they relate to practical process simulation.

3.1 Ion Implantation

Two types of distributions are considered here for implanted impurities in silicon: (a) first order double half-Gaussian approximations for arsenic and phosphorus, and (b) modified Pearson IV distribution for boron.

The simplest description of an implanted impurity profile in silicon or silicon dioxide is a symmetrical Gaussian curve with first two moments, the projected range, R_p, and the standard deviation, σ_p, calculated from the LSS theory (2). However, experimental distributions of many ions, such as boron or arsenic, are found to by asymmetrical. The simple Gaussian approximation of those implanted profiles is often inadequate so that higher-order moments must be used to construct impurity distributions. Gibbons and Mylroie (3) have shown that the third central moment is enough to provide sufficient information to construct accurate distributions when the asymmetry is not excessive (less than the standard deviation). In these cases, the distribution can be represented by two half-Gaussian profiles, each with a different standard deviation, σ_1 and σ_2, joined together at a modal range R_m as shown in Fig. 2. This method can be used for profiles such as arsenic and phosphorus. However, for boron a modified Pearson IV distribution (1) is found to be more realistic. For the joint half-Gaussian distribution, the two sides are given by

$$C_1(y) = P_y \exp[-(y - R_m)^2/2\sigma_1^2] \qquad 0 < y < R_m$$

$$C_2(y) = P_y \exp[-(y - R_m)^2/2\sigma_2^2] \qquad R_m < y < \infty$$

$$(1)$$

The parameters are determined from table look-up and interpolation (3) (4).

Hofker et al (5) have shown that the implanted boron profiles in amorphous silicon before annealing, may be described by a Pearson type IV distribution. Later Ryssel et al (6) found that this is true for other elements and targets as well. Although this distribution describes well the profile of boron from the surface to a distance somewhat beyond the peak of the distribution, Fig. 3 shows that the approximation fails to model the observed exponential tail that is due to a small residual random scattering of boron ions along channeling directions even when silicon wafer targets are properly tilted to avoid channeling. Based on experimental results (7-9) an empirically modified Pearson IV distribution can be used by adding an exponential tail with a fixed characteristic length (0.045 μm), independent of dose energy and crystalline surface orientation (1). The tail is attached to the shoulder of the standard Pearson IV distribution where the concentration drops to 50% of the peak value. Of course, after the addition of the tail, renormalization of the distribution to the implanted ion dose is necessary. Typical resulting profiles from this modification are shown in Fig. 3.

3.2 Impurity Migration During Thermal Processing

The redistribution of impurities in the space of the silicon-silicon dioxide system, during thermal processing, is governed by the general continuity equation which can be written as

$$\frac{d}{dt}\int_{V(t)} CdV = \int_{V(t)} (g-\ell)dV - \oint_{S(t)} \vec{F}\cdot\vec{n}dS \qquad (2)$$

where C = impurity concentration in atoms per unit volume, t = time, $S(t)$ = closed surface (function of time), $V(t)$ = volume enclosed by $S(t)$, $\vec{F}$ = impurity flux vector, n = outward unit normal to $S(t)$, g = impurity generation rate per unit volume, and ℓ = impurity loss rate per unit volume.

The left-hand side of Eq. (2) represents the time rate of change of the impurity content in a volume, $V(t)$, and it is equated to the net impurity generation rate in the same volume minus the net impurity outflow through the surface $S(t)$ enclosing $V(t)$. The reason for using the continuity equation in integral form is that it simplifies treatment of volume changes which occur during silicon oxidation and epitaxy processes. The generation and loss mechanism have been included to account for the exchange of impurity atoms between different states in the silicon lattice as in the case of arsenic where atoms may coexist in substitutional and clustered states.

For one-dimensional flow along the y-axis which is defined perpendicular to the silicon surface pointing inward, Eq. (2) can be written

$$\frac{d}{dt}Q(y_1,y_2) = U(y_1,y_2) - [F(y_2) - F(y_1)]$$ (3)

where

$$Q(y_1,y_2) = \int_{y_1}^{y_2} C(y)dy$$ (4)

is the impurity atom content between y_1 and y_2,

$$U(y_1,y_2) = \int_{y_1}^{y_2} (g-\ell)dy$$ (5)

is the net generation between y_1 and y_2, and the impurity flux $F(y)$ is positive in the y-direction. Physically, the impurity flux may arise from solid-state diffusion, from interface phenomena such as evaporation or segregation, and from the motion of interfaces, as in the case of silicon oxidation and epitaxy.

In the sections that follow we discuss the models that describe the physical processes that enter in Eq. (3) as well as the numerical implementation of this equation.

3.2.1 Solid state diffusion in silicon. Solid state diffusion is the physical mechanism responsible for impurity migration within the silicon body during high-temperature processing steps. At any point, y, the diffusive flux $F_D(y)$, of impurities is related to their concentration and diffusivity gradient by Fick's first law. For one-dimensional flow the relation is

$$F_D(y) = - D(y)\frac{d}{dy}C(y)$$ (6)

where $D(y)$ is the diffusion coefficient of the impurity. Under the assumption of single-state migration there is no generation or loss within the material. Under the further assumption of uniform diffusivity, the well known Fick's second law can be derived from equation (2) and (6), i.e.

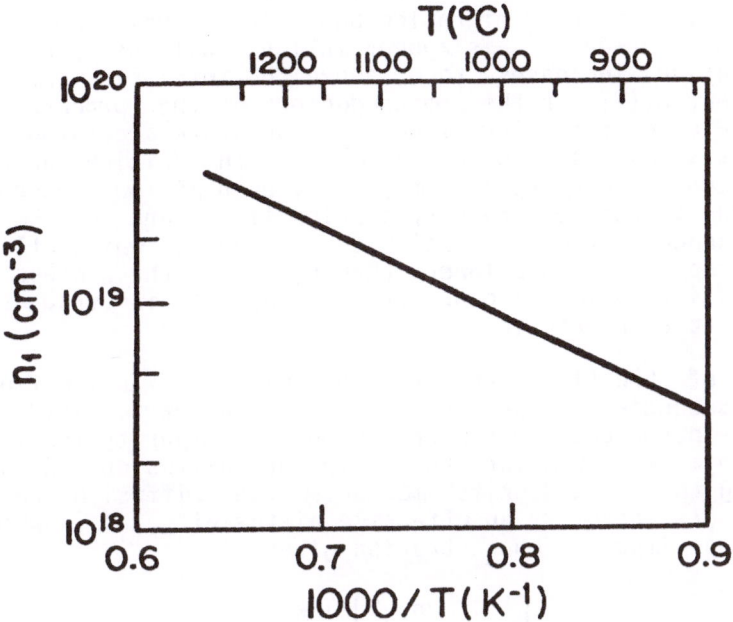

Fig. 4   Intrinsic carrier concentration in silicon vs. temperature.

$$\frac{\partial C}{\partial t} = D\frac{\partial^2 C}{\partial y^2}$$

Fick's second law is generally adequate for the calculation of diffusive impurity migration under low impurity concentration conditions. However, this approximation fails as the impurity concentration increases to or above the intrinsic carrier concentration, $n_i(T)$, in the semiconductor at the process temperature. Fig. 4 is a plot of n_i vs temperature according to Morin and Maita (10), as used in SUPREM II. In addition equation (7) for a given impurity species may fail even at low concentration conditions if another impurity species is present in the silicon at high concentration. We refer to silicon in which all dopants exist at concentrations lower than $n_i(T)$ at the process temperature, as intrinsic silicon. If the opposite is true then we refer to it as extrinsic.

One of the first attempts to explain diffusive flux under extrinsic conditions was to include the "electric field effect" of the introduced free carriers on the impurity ion migration (11), in a way similar to ambipolar diffusion in plasmas. Including this "field drift" mechanism the diffusion coefficient in eq. (7) becomes an effective diffusivity, D_e, which is a function of impurity concentration given by

$$D_e = Df_e = D\left\{1 + \left[1 + 4\left(\frac{n_i}{C}\right)^2\right]^{-1/2}\right\} \tag{8}$$

As can be seen, the maximum value of f_e is 2, for $C \gg n_i$ and this is clearly inadequate to explain diffusivity enhancements of the order of 10 to 20 often observed with most of the common impurities at high concentrations.

At present, it is generally accepted that extrinsic diffusion phenomena are the result of impurity migration by interaction with charged point defects in silicon (see chapter tiled "Diffusion in Silicon"). All common impurities diffuse in silicon by means of interaction with the lattice point defects such as silicon atom vacancies and interstitials. Thus, the diffusion coefficeint is proportional to the concentration of such point defects. Although the concentration of neutral defects at any given temperature is independent of the impurity concentration (so long as it does not approach that of silicon atoms), the concentration of defects at various charge states (which have been identified within the silicon bandgap), depends on the Fermi level position in the bandgap and therefore is a function of impurity concentration. Thus, as shown in the chapter on

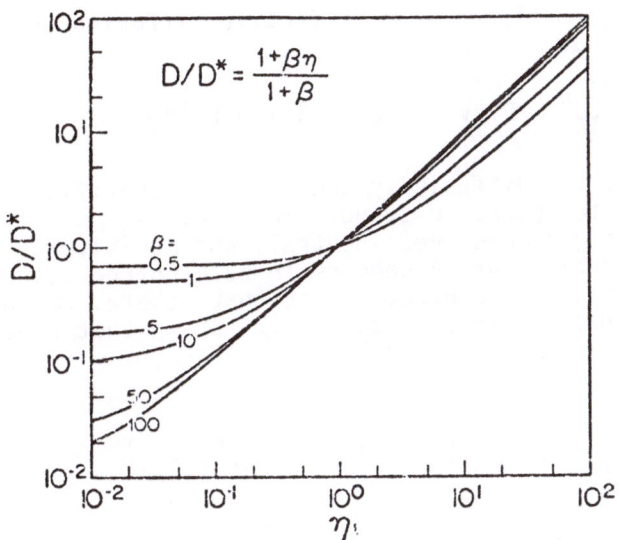

Fig. 5   Normalized diffusivity vs. normalized carrier concentration for different values of β.

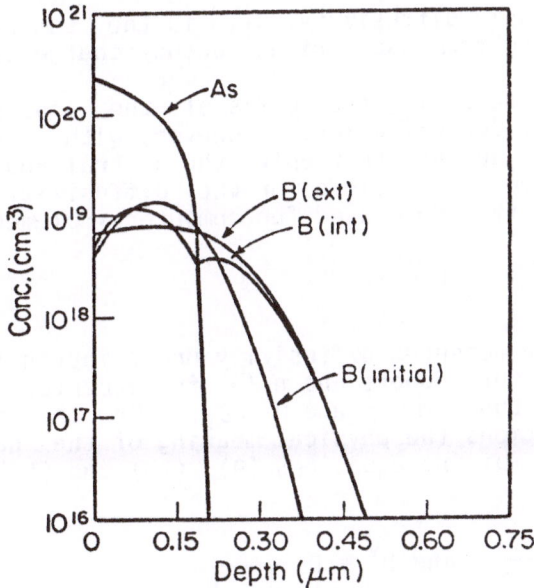

Fig. 6   Illustration of effects of heavy arsenic doping on the redistribution of boron.

"Diffusion in Silicon", the effective diffusion coefficient can be expressed as

$$D_e = f_e \left\{ D^X + D^-[V^-] + D^=[V^=] + D^+[V^+] \right\}$$

(9)

where D^C is the diffusivity due to each identified charge state, c, of vacancies (here c stands for: =, -, x, +, i.e. doubly negative, singly negative, neutral, and positive), and $[V^C]$ is the concentration of vacancies in each charge state, normalized to the intrinsic concentration at that state. Using the Boltzmann approximation it can be easily shown that these normalized concentrations may be given by

$$[V^-] = \frac{n}{n_i} \quad , \quad [V^=] = \left(\frac{n}{n_i}\right)^2 \quad \text{and} \quad [V^+] = \frac{n_i}{n}$$

(10)

where n is the free electron concentration. Thus, under intrinsic conditions ($n = n_i$), equation (9) becomes

$$D_i = D^X + D^- + D^= + D^+$$

(11)

i.e., the intrinsic diffusivity, D_i, is the sum of the diffusivities resulting from the various vacancy charge states.

The above model is the basis of the appropriate process simulation model for diffusion. However, with the exception of phosphorus, it appears that only the neutral and one charged defect state are responsible for the diffusivity of impurity atoms. The specific form used for computer implementation is:

$$D = D_i(1 + \beta\eta)/1 + \beta)$$

(12)

where D_i is the measured diffusivity under intrinsic conditions and $\eta = n/n_i$ for donors and n_i/n for acceptors. Thus, under intrinsic conditons $\eta = 1$ and $D = D_i$. On the other hand for extrinsic conditions the physical meaning of the parameter β can be derived by combining equations (9), (10) and (12), to obtain

$$D^X = D_i \frac{1}{1 + \beta} \quad \text{and} \quad D^V = D_i \frac{1}{1 + \beta}$$

where v stands for - or +. Thus, $\beta = D^V/D^X$, is an index of the effectiveness of charged vacancies relative to neutral ones in impurity diffusion. Fig. 5 is a plot of normalized diffusivity vs. η. Although it might be expected that β for any impurity element is a function of temperature, no definite characterizations exist at present. Typical values (1) are β = 3 for boron and β = 100 for arsenic while for phosphorus a different model is used as described later.

It is well known that when different impurity atoms are present in silicon there is direct interaction among them (14). Fig. 6 illustrates one such simulated case where high concentration arsenic affects the distribution of boron. In all practical circumstances it is the impurity of high concentration that affects the migration of the lower concentration impurity species while the opposite effect is negligible. For the example shown in Fig. 6 the diffusive flux of boron in the region which is heavily n-doped is reduced because $\eta(boron) = n_i/n \to 0$, while the dip is produced by the electric field that exists at the front of diffusing arsenic profile. It has been shown (15) that under these conditions the migration of the affected impurity species can be modeled by a modified form of Fick's first law as follows:

$$F_D = - \frac{d}{dy} [D(y)C(y)] \tag{13}$$

It is important to note that eq. (13) is not physically correct (although intuitively appealing) and is used in SUPREM for reasons of mathematical expedience only (see also chapter on "Diffusion in Silicon"). In addition, eq. (13) is mathematically valid only for impurities whose diffusivity can be modeled by eq. (12) with $\beta \gtrsim 3$. Thus, in SUPREM II, the diffusive flux for arsenic and boron is modeled by equation (13). However, for phosphorus the proper form is used. This form is

$$F_D(y) = - D_e(y) \frac{d}{dy} C(y) \tag{14}$$

3.2.2 Diffusion of Phosphorus. A model for the diffusive migration of phosphorus has been presented by Fair and Tsai (16). The model predicts with reasonable accuracy the phosphorus kink formation as well as the base push effect, commonly observed during heavy emitter diffusions in bipolar technology. According to this model the physical explanation of these "anomalous" effects lies in the enhancement of vacancy concentration in the silicon caused by dissociation of phosphorus-doubly ionized vacancy pairs that flow from the surface into the silicon bulk. A typical high concentration phosphorus profile is composed of three regions as shown schematically in Fig. 7. In SUPREM II, each region is identified according to the cri-

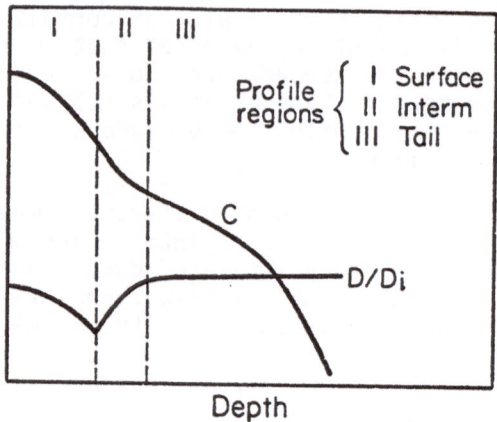

Fig. 7   A typical phosphorus doping profile with a demarkation
of the three regions considered by the Fair and Tsai
model. Also shown is the local diffusivity profile derived by the model.

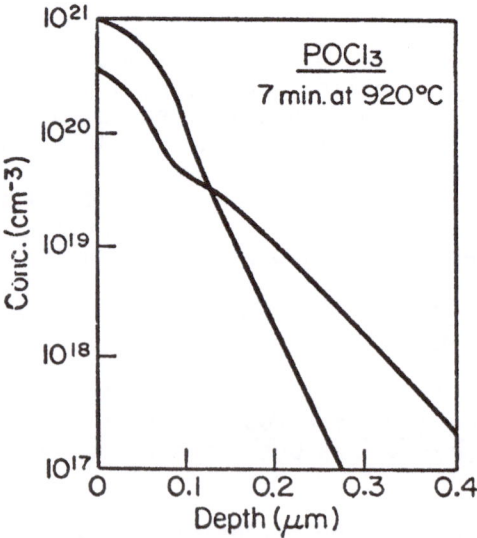

Fig. 8   Calculated phosphorus profiles for two different surface concentrations. Slower diffusion of the higher
concentration profile is due to the silicon stress-
induced bandgap narrowing model.

teria outlined below. For more detailed discussion, the reader is referred to references (16), (17), and (18). Briefly, the diffusivity in the three regions is given as:

I) Surface Region: region between surface and position where electron concentration, $n(y)$, becomes equal to n_e, given by

$$n_e = 4.65 \times 10^{21} \exp -\left(\frac{0.39 \text{ eV}}{kT}\right) \qquad (15)$$

n_e is the electron concentration at or below which it is assumed that dissociation of the phosphorus-doubly ionized vacancypair takes place. In this region the phosphorus diffusivity given by

$$D_{eI} = f_e\left\{D^X + D^=\left(\frac{n}{n_{ie}}\right)^2\right\} \qquad (16)$$

where n, the electron concentration is related to the concentration of phosphorus, C, by

$$C = n + 2.04 \times 10^{-41} n^3 \qquad (17)$$

and n_{ie} is the effective intrinsic carrier concentration accounting for the heavy doping stress induced bandgap narrowing and is given by

$$n_{ie} = n_i \exp\left[\frac{1.5 \times 10^{22}(C_{TS} - 3 \times 10^{20} \text{ cm}^{-3}) \text{ eV}}{kT}\right] \qquad (18)$$

for C_{TS}, total phosphorus surface concentration greater than 3×10^{20} cm^{-3}.

II) Intermediate Region: Region starts where $n = n_e$. Diffusivity here is given by

$$D_{eII} = f_e\left\{D^X + D^=\left(\frac{n_e}{n_{ie}}\right)^2\right\}\left(\frac{n_{ie}}{n}\right)^2 \qquad (19)$$

Region ends where $D_{eII} = D_{eIII}$, where D_{eIII} is given below.

III) Tail Region: Region begins where $D_{eII} = D_{eIII}$, where

$$D_{eIII} = f_e\left\{ D^X + D^- \frac{n_s^3}{n_e^2 n_{ie}}\left[1 + \exp\left(\frac{0.3\ eV}{kT}\right)\right]\right\} \tag{20}$$

where n_s is the surface electron concentration. This region is assumed to extend to infinity. At present, lacking complete experimental characterization, the diffusivity of all impurities in this region is multiplied by an enhancement factor f_{enh} given by

$$f_{enh} = \frac{D_{eIII}}{f_e D^X} \tag{21}$$

Fig. 8 shows a typical application of the phosphorus diffusion model using SUPREM II.   Note the formation of the kink and the fact that diffusion slows down as the surface concentration is increased because of the stress induced silicon bandgap narrowing mentioned above.

3.2.3 Diffusion of Arsenic. Almost all dopants may exist in silicon in more than one state particularly when the concentrations are high. Typically one of these states is substitutional and therefore mobile while the others, if present, may be some form of precipitate or cluster which is immobile.   Exchange between these states gives rise to the generation and loss terms in the continuity equation (2) of the mobile species. Clustering of arsenic has been discussed by many authors (e.g. 19, 20, 21, 22).   It is now discussed as an essential model for arsenic diffusion. The assumed chemical reaction in SUPREM II (version 05) is according to Tsai et al. (21)

$$3As^+ + e^- \underset{k_d}{\overset{k_c}{\rightleftharpoons}} As_3^{+2} \xrightarrow{25\ °C} As_3 \tag{22}$$

where k_c and k_d are the clustering and declustering rate coefficients. Defining the concentration of clustered atoms, C_c, as

$$C_c = C_T - C \tag{23}$$

where C_T is the total concentration and C the substitutional concentration, the conservation equation for C_c may be written as

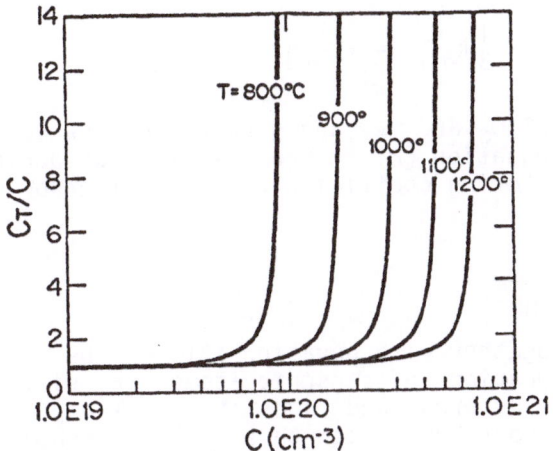

Fig. 9   Normalized total arsenic concentration vs. the substitutional arsenic concentration, resulting from clustering model, as a function of process temperature, under
thermal equilibrium.

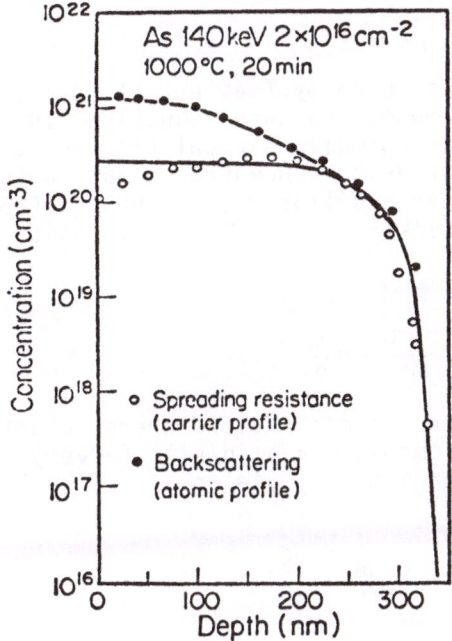

Fig. 10   Total atomic concentration (dashed curve) and carrier
concentration (solid curve) of implanted and diffused
arsenic.   After Ref. (21).

$$\frac{\partial C_c}{\partial t} = k_c nC^3 - \frac{1}{3}k_d C_c = \ell - g \tag{24}$$

where $n \sim C + 2/3 \, C_c$ and ℓ and g refer to the loss and generation terms in equation (2). From simple mass action law, the equilibrium clustering coefficient, k_e, is given by

$$k_e = \frac{k_c}{k_d} = \frac{C_c}{3nC^3} \tag{25}$$

Equation (24) together with equation (2) can be used to describe the thermal migration of arsenic atoms in silicon and their partition into clustered and substitutional populations. Actually, SUPREM II does not use this dynamic scheme for arsenic diffusion calculation. Instead it assumes that the clustered atoms are always in equilibrium with unclustered ones. This is a good approximation for temperatures above 900 °C. Then using equations (23) and (24) the following relationship is derived

$$C_T = \frac{C + k_e \, C^4}{1 - 2k_e \, C^3} \tag{26}$$

Since eq. (13) was used by Tsai et al (21) for the characterization of their model the same equation is used in SUPREM II. Fig. 9 plots the normalized total arsenic concentration vs the substitutional arsenic concentration as a function of process temperature, under equilibrium, using eq. (26) and $k_e = 1.26 \times 10^{-70} \exp(2.062 \text{ eV}/kT)$ cm^{-9} (21). Equation (13) can now be rewritten as

$$F_D = - D\frac{dC}{dC_T} \frac{dC_T}{dy} - C\frac{dD}{dy} \tag{27}$$

Assuming that for concentrations where clustering is important ($C > 5 \times 10^{19}$ cm^{-3}), $D = D_i(n/n_i)$ (a very good approximation for arsenic), equation (27) yields

$$F_D = -D\left[\frac{dC}{dC_T} + \frac{C}{n} \frac{dn}{dC_T}\right]\frac{dC_T}{dy} \tag{28}$$

Thus, the problem of diffusion of arsenic is simplified into a single population problem (total arsenic) diffusing with effective diffusivity given by

$$D_e = D_i \frac{n}{n_i} \left[\frac{dC}{dC_T} + \frac{C}{n} \frac{dn}{dC_T} \right] \qquad (29)$$

Equation (28) is then used in equation (3), with $U = 0$, to model the migration of arsenic.

Although the clustering model predicts formation and dissolution of clusters during predeposition and subsequent drive-in, an assumption must be made about the initial cluster concentration when arsenic is implanted. In this case, it is arbitrarily assumed that the initial clustered arsenic profile is simply that determined by thermal equilibrium at the annealing temperature, given the known implanted profile of total arsenic atom concentration.

An example of use of this model to simulate arsenic diffusion is shown in Fig. 10. The experimental conditions being simulated are as follows: arsenic was implanted at 140 keV to a dose of 2×10^{16} cm^{-2} through 25 nm of oxide; it was subsequently diffused for 20 min at 1000 °C. Both total and substitutional arsenic concentrations are shown in the figure in comparison with measured data (21). As can be seen the agreement between model and experiment is very good.

3.2.4 Oxidation enhanced diffusion.

It has been observed by several authors (e.g. 23, 24) that the diffusivity of boron and phosphorus is enhanced when the silicon surface is oxidized. Recently the same effect has been observed for arsenic (25). This phenomenon of oxidation enhanced diffusion (OED) is generally attributed to enhancement of silicon point defects due to the oxidation. Perhaps the most plausible model has been proposed by Hu (26) and subsequently quantified by Antoniadis et al. (24) and more recently by Antoniadis (27). This model relates OED with oxidation stacking fault (OSF) growth, by invoking a dual diffusion mechanism for impurities in silicon whereby both vacancy and interstialcy effects are responsible for diffusion, and by postulating an enhancement, due to oxidation, of the concentration of silicon self-interstitials. Thus, according to the model, during oxidation the interstitialcy component of impurity diffusivity is enhanced leading to OED while increased interstitial precipitation leads to OSF. For boron, in the present version of SUPREM II, a temperature dependent but time independent OED is assumed for any ambient causing silicon oxidation. For phosphorus a fixed enhancement of a factor 1.8 is assumed for dry O_2 and 3.3 for wet O_2. Recent work (28) has shown that for both boron and phosphorus OED is related to the rate of oxidation. An improvement to SUPREM II has been suggested by Taniguchi et al. (29), where the diffusivity of phosphorus and boron at low concentrations under oxidizing conditions is given by

$$D = D_i + E\left(\frac{dZ_{ox}}{dt}\right)^{0.3} \exp\left(-\frac{y}{25\ \mu m}\right)\exp\left(-\frac{2.08\ eV}{kT}\right) \tag{30}$$

where $E = 1.7 \times 10^{-5}$ cm^2/sec for <100> Si and 6.1×10^{-6} cm^2/sec for <111>.

For more details on the theory of OED the reader is referred to the chapter on "Diffusion in Silicon."

3.3 Thermal Oxidation

The rate of SiO$_2$ growth on silicon is described by the well known formula (30),

$$Z_{ox}^2 + AZ_{ox} = B(t + \tau) \tag{31}$$

where Z_{ox} is the oxide thickness, t is time and A and B are related to the linear and parabolic growth coefficients K_L and K_P and normalized partial pressure, P_{O_2}, of O_2 by

$$A = P_{O_2} K_P / K_L$$

$$B = P_{O_2} K_P$$

The parameter τ is related to the initial oxide thickness by

$$\tau = \frac{Z_{ox}^2(t=0) + AZ_{ox}(t=0)}{B} \tag{32}$$

Under relatively low dopant concentrations, K_P and K_L depend only on silicon crystal orientation and on the oxidizing ambient, and they are singly activated functions of temperature. It is, however, well known that under high surface concentration conditions, such as in a MOSFET source and drain or bipolar emitter region, the oxidation rate of silicon is enhanced. A detailed description of the phenomenon has been given by Ho et al. (31).

According to the model, the linear rate coefficient can be written as

$$K_L = K_L^i 1 + \gamma(C^T - 1) \tag{33}$$

where K_L^i is the intrinsic (i.e. low concentration) coefficient, γ is an experimentally determined parameter given by

$$\gamma = 2.62 \times 10^3 \exp\left[-\frac{1.10 \text{ eV}}{kT}\right] \tag{34}$$

c^T the normalized total vacancy concentration given by

$$c^T = \frac{1 + c^+\left(\frac{n_i}{n}\right) + c^-\left(\frac{n}{n_i}\right) + c^=\left(\frac{n}{n_i}\right)^2}{1 + c^+ + c^- + c^=} \tag{35}$$

with

$$c^+ = \exp[(E^+ - E_i)/kT] \quad ; \quad E^+ = 0.35 \text{ eV}$$

$$c^- = \exp[(E_i - E^-)/kT] \quad ; \quad E^- = E_g - .57 \text{ eV}$$

$$c^= = \exp[(2E_i - E^=)/kT] \quad ; \quad E^= = E_g + E^- - .11 \text{ eV}$$

The above expressions should be recognized as the normalized intrinsic concentrations of vacancies in the three charge states with their corresponding state energies in the silicon bandgap. Finally, the silicon energy bandgap, E_g, and intrinsic level, E_i, are given as functions of temperature by

$$E_g(T) = 1.17 - 4.3 \times 10^{-4}[T^2/(T + 636)] \text{ eV}$$

$$E_i(T) = E_g/2 - kT/4$$

For n-type dopants the enhancement of the parabolic oxidation rate has also been obtained by Ho and Plummer (32). K_p is given as

$$K_p = K_p^i(1 + \delta c_T^{0.22}) \tag{36}$$

where

$$\delta = 9.63 \times 10^{-16} \exp\left[-\frac{2.83 \text{ eV}}{kT}\right] \tag{37}$$

K_p^i is the intrinsic parabolic rate and C_T is the total n-type dopant atom concentration.

Since during oxidation the impurity surface concentration changes due to diffusion and segregation distribution, the calculated enhanced values of K_L and K_p may generally be time dependent. Thus for simulation purposes, using the classical oxide growth equation (31) an incremental form of the same equation is used, namely

$$\Delta Z_{ox} = \frac{1}{2}\left[-(2Z_{ox} + A) + \sqrt{(2Z_{ox} + A)^2 + 4B\Delta t}\right] \tag{38}$$

Thus, as the simulation time proceeds in small time increment, Δt, and the impurities redistribute in the silicon, the coefficients A and B are obtained from the surface impurity concentrations in each time increment, and from those values the corresponding increment of oxide thickness, ΔZ_{ox}, is calculated.

3.4 Impurity Transfer Across Interfaces

As has been suggested earlier, impurity segregation plays a major role during oxidation in altering impurity profiles in the silicon and the resulting device properties. In addition to the SiO_2 - Si interface, segregation type phenomena can occur at gas-solid interfaces including the cases of inert as well as chemical vapor deposition environments. Although the actual details of the chemistry taking place at such interfaces may not be known, generally the dopant atom flux across interfaces may be phenomenologically described by means of a first order kinetic model as

$$F_s = h(C_1 - C_2/m_{eq, 1-2}) \tag{39}$$

where F_s is the dopant flux defined positive from region 1 to 2, C_1 is the dopant concentration of the interface in region 1 and C_2 the same in region 2. The factor $m_{eq, 1-2}$ is the well known equilibrium segregation coefficient for the specific impurity species in the system of regions 1-2 and is defined as

$$m_{eq, 1-2} = \frac{C_2}{C_1} \tag{40}$$

Finally, h which has units of velocity, is the interface mass-transfer coefficient.

The form given in equation (39) can serve for simulating a broad range of process steps including the following:

3.4.1 Evaporation.
The mass-transport coefficient becomes the impurity evaporation coefficient which is a function of temperature. The same evaporation coefficient is used for SiO and for silicon (1). Also, C is assumed to be zero and $m_{eq} = 1$.

3.4.2 Chemical deposition. Chemical deposition is simply modeled by assuming (arbitrarily) that $h \to \infty$, (actually h is typically set equal to 1 µm/sec), $m_{eq} = 1$ and C_1 equal to either the dopant solid solubility or to any other specified concentration. Thus, the simulated surface concentration of silicon becomes very rapidly equal to C. It is recognized that the model may be overly simplistic but there exist too many different processes by which dopants may be deposited in silicon, making it impractical to attempt to accurately model each of them specifically. This is no longer a serious limitation since increasingly modern processes rely on ion implantation for doping.

3.4.3 SiO$_2$ - Si interfacial flux. Under non-oxidation conditions the SiO$_2$ - Si interface is stationary and equation (39) is sufficient to model the impurity flux exchanged between the two regions. Unfortunately, there exists practically no characterization of the flux in this stationary system with the exception of the phosphorus doped SiO$_2$ - Si system (used for silicon doping), which has been carefully explored by Ghoshtagore (33). In this case h was derived as a singly activated function of temperature, while m_{eq} was assumed infinite. Actually, the exact value of m_{eq} would not alter the observed result as long as it is kept large (say > 50).

In the case of a moving interface as in silicon oxidation there also exists a motion-induced interfacial flux resulting from the differnt dopant concentration, across the interface. This flux denoted by F_b, is given by

$$F_b = -v_{ox}(C_1 - \alpha C_2) \qquad (41)$$

where $v_{ox} = dZ_{ox}/dt$ is the oxide growth rate and α is the ratio of oxidized silicon to resulting oxide thickness (equal to 0.44). Generally this flux competes with the flux F_s. If $h \gg v_{ox}$, then $C_2 \to m_{eq}C_1$, while if $h \ll v_{ox}$, then $C_1 \to \alpha C_2$. In all characterizations of the moving interface system to date, the first condition has been implicitly assumed, i.e. that the equilibrium segregation condition prevails. Thus, in the absence of any meaningful values, h has been arbitrarily assumed equal to 0.1µm/min and thus the condition $h \gg v_{ox}$ is always satisfied (1).

3.4.4 Silicon epitaxy. Impurity redistribution during epitaxial growth includes both a diffusive component as well as an interfacial flux, F_s. A detailed derivation of the form of F_s has been recently given by Reif and Dutton (34). Briefly, the interfacial flux is given by

$$F_s = K_{mf}\left[P_D^{\circ} - C_I/K_P\right] - gC_i - K_A \frac{dC_I}{dt} \tag{42}$$

where C_I is the impurity concentration at the solid surface, g is the epitaxial growth rate, P_D° is the input dopant partial pressure, and K_{mf}, K_P and K_A are parameters related to the specific epitaxial reactor system and must thus be determined for each such system. A fourth parameter must be specified for proper simulation of autodoping; namely, the initial value of C_I. Procedures for the determination of these parameters are given in the above cited reference.

4. NUMERICAL SIMULATION OF IMPURITY MIGRATION

Up to this point all impurity fluxes were assumed due to solid state diffusion. However, surface and interface effects can often play a significant role in the redistribution of impurities. An example is shown in Fig. 15 which shows an as-implanted boron profile in bare silicon and the distribution resulting from that initial profile after oxidation, which produces a moving Si/SiO_2 interface. About 65% of the total boron atoms have migrated into the oxide. These profiles were generated by SUPREM II which implements the techniques that are described here.

The previous section has outlined the models that may be used to describe impurity fluxes in a $Si - SiO_2$ system. In the present section, the numerical implementation of these process models is discussed with particular emphasis on the discrete formulation of the impurity continuity equation under the moving boundary conditions encountered during silicon oxidation and epitaxy. However, in order to establish a basis for the discussion of this issue, the numerical solution of the continuity equation under stationary conditons is outlined first.

4.1 Stationary Boundaries

The space over which the impurity continuity Eq. (3) is to be solved is partitioned into discrete cells. The impurity concentration, $C(y)$, is evaluated at points (or nodes) lying in the middle of each of the discrete cells. Figure 11 illustrates this space discretization. For instance, taking cell $i = 6$ as an example, the discrete continuity equation (neglecting generation-loss terms) is written as

$$\frac{d}{dt} Q_6 = -[F_D(y_{6.5}) - F_D(y_{5.5})] \tag{43}$$

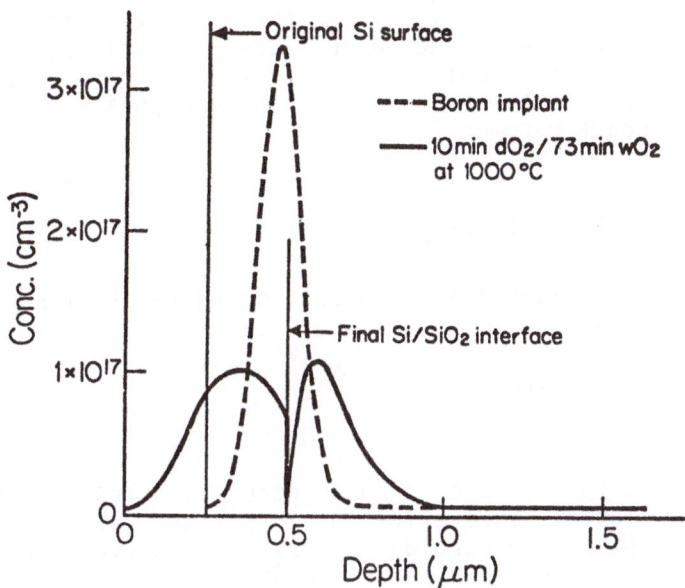

Fig. 15 Boron profiles in silicon: 1, implanted into bare silicon and 2, after diffusion in the presence of oxidation. In the second case the boron distribution in
both silicon and silicon dioxide are shown.

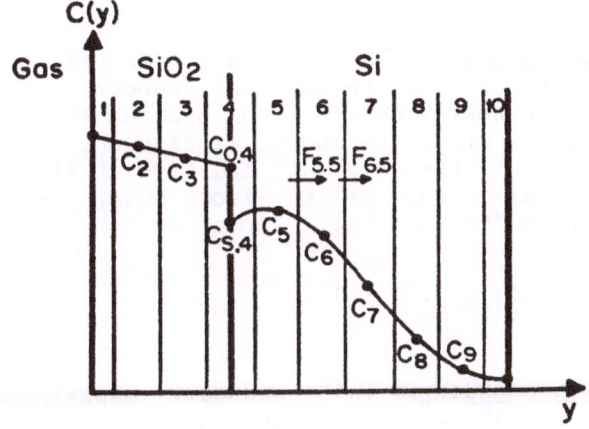

Fig. 11 Illustration of the partitioning of the simulation
space into discrete cells with impurity fluxes across
cell boundaries.

where $F_D(y_{6.5})$ and $F_D(y_{5.5})$ are the impurity diffusive fluxes at the right- and left-hand boundaries of cell 6, and Q_6 is the impurity content of cell 6. Thus, for the general ith cell not lying at any of the boundaries, the continuity equation discretized in space becomes

$$\frac{d}{dt} Q_i = -[F_D(y_{i+1/2}) - F_D(y_{i-1/2})] \tag{44}$$

Where F_D is evaluated at the cell boundary from Eq. (6) or (14) and Q_i is given by

$$Q_i = \int_{y_{i-1/2}}^{y_{i+1/2}} c_i dy \tag{45}$$

For simplicity we have ignored generation-loss terms in (44). Both the spatial derivative in Eq. (6) or (14) and the integration in (45) are carried out using numerical approximations. Specifically, the flux F_D is evaluated by replacing (6) or (14) by a difference equation while Q_i is evaluated using midpoint integration. The set of relevant discrete equations is given in the Appendix.

The cells at the two space extreme boundaries, as well as at the SiO_2 - Si boundary (when it exists), deserve attention at this point:

4.1.1 Top boundary. The first discrete cell is actually a half-cell with its node at the physical boundary. Generally, at this boundary an evaporation or incorporation flux, $F_S(0)$, governed by Eq. (39) as discussed above, is assumed. At the inner boundary of this cell there exists a diffusive flux that may be evaluated as for all other cells, from Eq. (6) or (14). The continuity equation for this cell becomes

$$\frac{d}{dt} Q_1 = -[F_D(y_{1+1/2}) - F_S(0)] \tag{46}$$

4.1.2 Deep boundary. This boundary usually lies inside the silicon substrate at the point where the simulated space terminates. The last cell in this end is also a half-cell similar to the first one. Typically it is convenient to assume a reflecting boundary (implied by setting h = 0 in Eq. (39)) at this point. Because the depth of simulation is often specified rather arbitrarily, care must be taken to ensure that the presence of a reflecting boundary at that point does not affect the simulation results. On the other hand, this reflecting boundary acquires actual physical significance in simulations of Si on sapphire or of poly-Si on SiO_2, where it may be desirable to study the effect of the reflecting deep boundary on impurity distribution. The continuity equation for this last cell becomes

$$\frac{d}{dt} Q_n = F_D(y_{n-1/2}) \qquad (47)$$

4.1.3 SiO_2 - Si Interface. Since this interface is a boundary point where impurity concentrations must be evaluated it must always lie on a node $(i = I)$. This node is shared by two half-cells, the one in SiO and the other in Si. Since the impurity concentration is generally discontinuous across the interface (because of thermodynamic segregation), the interface node contains two different concentrations, one for the oxide half-cell and one for the silicon half-cell. An interfacial flux, F_S, described by Eq. (39), is assumed to flow between these two cells, while diffusive flux calculated from Eq. (6) or (14) with the appropriate diffusivities for the two materials, is flowing across the other two boundaries. The continuity equation for the two interfacial half-cells becomes

$$\frac{d}{dt} Q_{I,ox} = -[F_S - F_D(y_{I-1/2})] \qquad (48)$$

$$\frac{d}{dt} Q_{I,Si} = -[F_D(y_{I+1/2}) - F_S] \qquad (49)$$

The established boundary cell equations (46), (47), (48), and (49) together with the set of Eq. (44) for all other cells constitute a system of equations that describes the temporal evolution of the impurity distribution. Numerically, this evolution is obtained by replacing the time derivative in this system of equations by a discrete approximation. All the above equations are of the type

$$H_i(t) = \frac{d}{dt} Q_i(t) \qquad (50)$$

where $H_i(t)$ denotes in abbreviated form the cell boundary flux difference due to either F_S or F_D's. Assuming that at time t_0 the concentration distribution is known, the distribution at a future time t_1 may be derived by solving the equation

$$\int_{t_0}^{t_1} H_i(t)\, dt = Q_i(t_1) - Q_i(t_0) \qquad (51)$$

Various methods exist for performing this integration numerically. A suitable method is the second-order implicit method,
which assumes that during the time interval $(t_1 - t_0)$, fluxes
are constant, i.e., $H_i(t) = [H_i(t_0)]/2$. Eq. (51) becomes

$$[H_i(t_0) + H_i(t_1)]/2 = [Q_i(t_1) - Q_i(t_0)]/(t_1 - t_0) \qquad (52)$$

There are as many equations of the form given above, as the
number of discrete space cells. Also, since for each cell the
flux function H_i, is evaluated at t_1 (i.e., at the future time),
it involves not only the unknown cell concentration $C_i(t_1)$ but
also the two neighboring cell concentrations $C_{i\pm1}(t_1)$. Thus,
the resulting equations are mutually coupled and form a system
with a tridiagonal matrix. This system of equations is solved
for the unknowns C, successively in small (simulated) time
increments by means of Gaussian elimination. Also, since the
system may be nonlinear due to the dependence of diffusivity on
impurity concentration, Newton-Raphson iterations of the solution are performed until convergence of the results (concentrations) is achieved.

4.2 Moving Boundary: SiO_2/Si

When silicon is oxidized, the interface between SiO_2 and Si
moves into the silicon and the oxide layer expands with a velocity $v_{ox}(t)$. Under these conditions, the moving boundary
induced impurity flux, F_b, given by Eq. (41) also flows across
the interface.

The presence of the moving boundary complicates the numerical formulation of the continuity equation in two ways. First,
because of the nonunity volumeric ratio of Si to SiO_2, $\alpha = 0.44$,
there is expansion of the discrete volume cells as they become
part of the SiO_2. Fig. 12 illustrates this effect; for the
cells around the interface, volume is a function of time and the
integration boundaries of Eq. (45) change between times t_0 and
t_1. Second, the existence of the two interfacial fluxes F_s and
F_b creates a jump discontinuity in the impurity flux that pro-
pagates through space during oxidation. In the discrete space-
time domain, as illustrated in Fig. 13, the interface moves by
small steps in short intervals of time within which the system
may be considered linear. It is therefore possible to consider
the effect of interfacial fluxes as a superposition of two distinct processes each containing only one of the two flux terms.
The first process consists of instantaneous motion of the
interface at time t_0^+ to its new node position, I. Thus, the
interfacial flux, F_s, is considered as flowing across the boundary I for the entire time interval (t_0, t_1). The same is true
for the diffusive flux $F_D(y_{I-1}/2)$, flowing across the cell
boundary between the two SiO_2 cells near the interface. On the

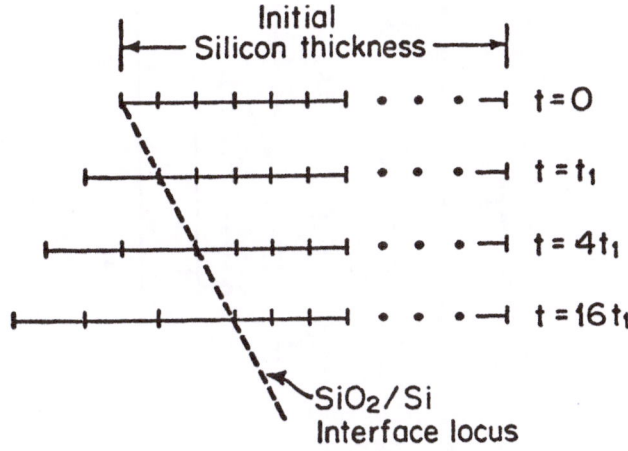

Fig. 12   Relationship of SiO$_2$ and Si volumes when silicon is
         oxidized at four time increments. Here oxidation rate
         is assumed purely parabolic. The tick marks identify
         nodes.

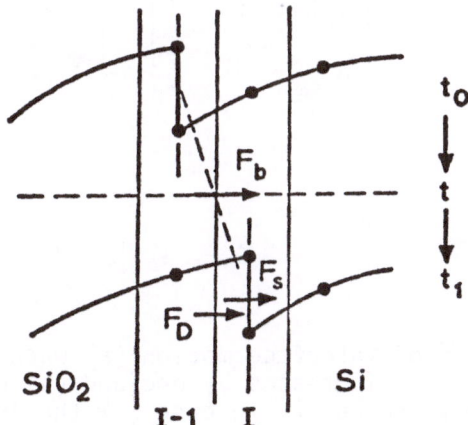

Fig. 13   Interfacial cells and moving boundary flux at two in-
         crements of time. For simplicity the volumetric ratio,
         α, is assumed unity in this figure.

254

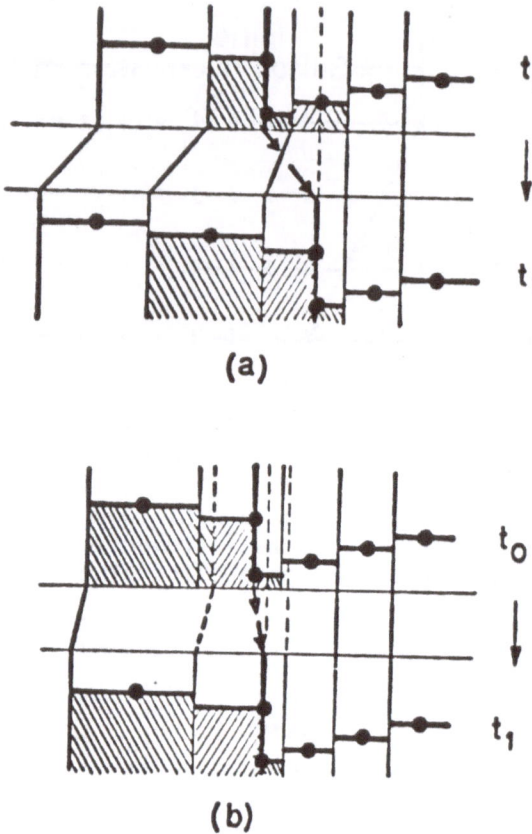

(a)

(b)

Fig. 14 Examples of interface motion (a) when the number of
oxide nodes increases by one and (b) when the number
does not increase. In case (a) the interface crosses a
cell boundary while in (b) it does not.

other hand, the second process consists of the redistribution of the impurity contents of these two SiO_2 cells, due to the moving interface flux, F_b. As shown in Fig. 13, the moving interface crosses the cell boundary at time t in the interval (t_0, t_1). Thus, in the interval (t_0, t_1), F_b is only internal to cell I - 1/2, and only in (t, t_1) does it give rise to impurity flux between the two cells. Superposition of the two processes discussed above requires that for moving boundary conditions, Eq. (51) must be modified only for the cells I and I - 1/2 by the addition of F_b to H_i for these two cells. Carrying out the integration as before and remembering that F_b is nonzero at the boundary only in (t, t_1) yields

$$[H_i(t_0) + H_i(t_1)]/2 \pm \frac{t_1-t}{t_1-t_0}F_b(t_1)=[Q_i(t_1) - Q_i(t_0)]/t_1-t_0)$$

(53)

Referring to Fig. 13, i in the above equation may be equal to I or I - 1, where I identifies the interface node. The + sign is used when the equation represents the continuity in the interfacial cell I in SiO_2 and the - sign when the continuity in cell I - 1, also in SiO_2, is represented.

To complete this discussion, it remains now to outline the interrelation of time and space discretization that arises from the consideration of the moving boundary. Given the constraints that the interface must always lie on a node and that the number of oxide nodes may increase by at most one node in any time step (while at the same time the silicon node number is reduced by one), two distinct possibilities for advancement of the interface exist, depending on the time step and on the oxidation rate which is typically a quadratic function of time. These two possibilities are shown in Fig. 14 (a and b). This Figure illustrates the discrete impurity distribution at consecutive times t_0 and t_1. The solid vertical lines indicate the cell boundaries and thus trace the evolution of the discrete system space. The arrows show the motion of the SiO_2/Si interface. In Fig. 14(a) the number of oxide nodes increases by one, while in Fig. 14(b) it remains fixed. In the first case, (a), the interface crosses the left-hand boundary of the cell that contains the interface at time t_1, while in the second case, (b), it does not cross any boundary. When case (a) arises, Eq. (53) must be used in the two cells that share the crossed boundary as already discussed, while in case (b) it should not be used because the cell boundary in SiO_2 never sees F_b, and thus Eq. (52) is valid just as in all other cells. Referring back to Fig. 14, it can be seen that due to the volumetric difference between SiO_2 and Si, the volume of three cells changes as Si gets converted to SiO_2. In both (a) and (b) the impurity content of these cells is crosshatched, and the hatching polarization is used to indi-

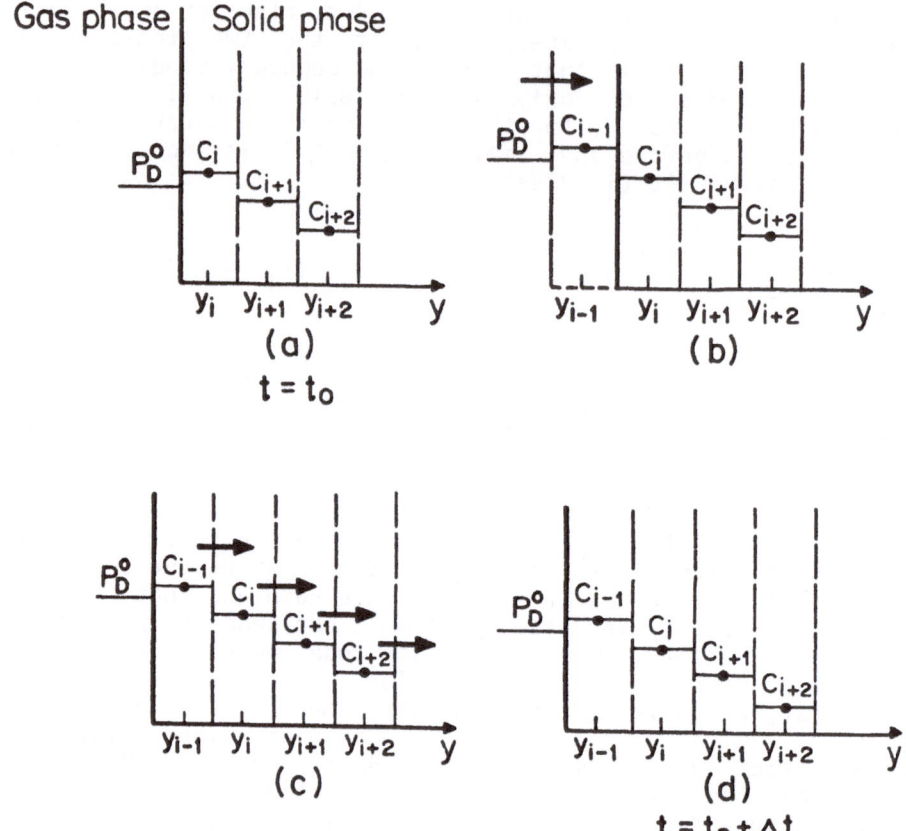

Fig. 16 Implementation of the numerical technique used to solve the continuity equation with the surface boundary condition dictated by the epitaxial process. For explanation see text.

cate the corresponding volumes at the two times. Since volume expands from t_0 to t_1, the cell boundaries and thus the limits of integration of Eq. (45) must be traced from time t_1 back to time t_0. Given that at t_1 the cell boundaries must lie half-way between nodes, in case (b) the cell boundaries around the interface do not trace back onto the boundaries at t_0. The same is true also in case (a) where the two interfacial half-cells at time t_0 become a single SiO_2 cell at t_1 while the full Si cell at t_0 becomes two half-cells at t_1. Using always midpoint integration to establish $Q(t_1)$, the hatching polarization and broken lines serve to indicate the impurity volume content at t_0, $Q(t_0)$, that is used in the calculation of the content difference in Eq. (52) and/or (53). Of course other integration rules can also be applied, but midpoint has been found very satisfactory.

4.3 Moving Boundary: Epitaxy

Fig. 16 shows the discretization of the simulation space for epitaxy as implemented in SUPREM (34). In Fig. 16 the solid line separates the gas phase and the solid phase. P_D^o is the dopant partial pressure in the gas phase. The solid silicon is shown partitioned into discrete cells with broken lines delineating the cell boundaries. The dopant concentration within each cell (i.e. C_i, C_{i+1}, C_{i+2}, etc.) is considered uniform. The coupling between the surface boundary conditin (Eq. (42)) and the main body of SUPREM is carried out in two steps, and is described below.

Step 1: At time $t = t_0$ (Fig. 16) the doping profile in the solid silicon is known. This profile is either the initial condition or the result of the simulation up to the time t_0. The cycle now starts by adding a new cell y_{i-1} (Fig. 16). In order to calculate the dopant concentration C_{i-1} of this new cell a numerical routine is used to solve Eq. (42) with the left-hand side of the equation set to zero, i.e.

$$0 = k_{mf}\left[P_D^o - \frac{C_{i-1}}{K_P}\right] - gC_{i-1} - K_A\frac{dC_{i-1}}{dt} \qquad (54)$$

This is equivalent to accounting only for dopant introduction into the newly added cell without computing the simultaneous impurity redistribution in the solid silicon. This step is illustrated in Fig. 16, with the arrow representing the net flux of dopant atoms entering the new cell.

Step 2: The thermal redistribution of impurities that occurs during the growth of cell y_{i-1} is now computed. This is done by entering the impurity profile shown in Fig. 16 into

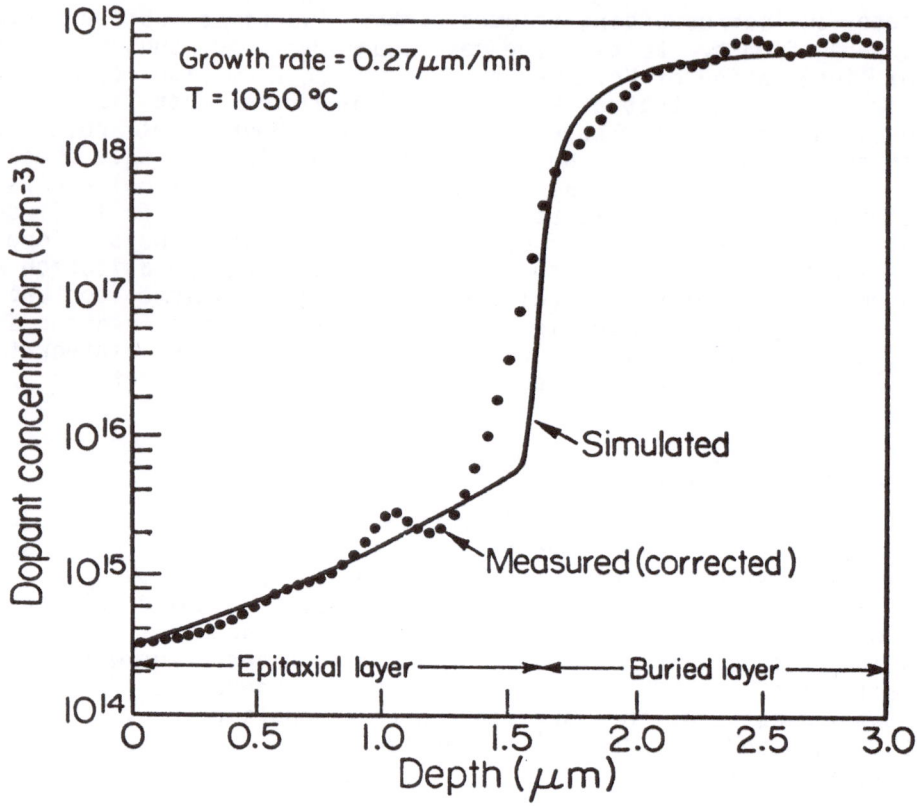

Fig. 17 Measured and simulated arsenic doping profile corresponding to a typical autodoping situation observed as a result of silicon epitaxy on a heavily doped layer.

SUPREM, which then computes the diffusive redistribution of impurities during the interval Δt under consideration. This is illustrated in Fig. 16. The arrows in the figure represent diffusive fluxes crossing cell boundaries. Notice that no flux is shown crossing the gas-solid interface. This is because the net introduction of impurities during this interval Δt was already considered in connection with Fig. 16 and Eq. (54) when C_{i-1} was first determined. After the dopant concentration in each cell is rearranged due to these diffusive fluxes, one time increment Δt has been advanced (Fig. 16), and the cycle of operations just described is repeated.

An example of application of this technique is shown in Fig. 17 which shows the measured (dotted line) and simulated (solid line) profile (34). In this experiment arsenic was implanted (3×10^{15} cm^{-2}, 100 keV) into a boron-doped, 10 Ω-cm, <100> silicon wafer, and then redistributed for 2 hours at 1250 °C. The substrate was vapor etched with HCl (0.5% by volume, 2 min., 1200 °C) and then baken in hydrogen (32 min, 1200 °C) before the epitaxial deposition step. The growth rate was approximately 0.27 μm/min and the deposition time was 6 min. The epitaxial layer deposited in this experiment was intended to be intrinsic, i.e., no arsine entered the reactor. However, as can be seen in Fig. 17 an exponential tail of the burried As layer is observed and is reproduced by the model with good agreement. This is a typical case of autodoping which is observed when epitaxy is carried out over heavily doped silicon.

APPENDIX

The second-order implicit difference equation for the diffusion flux resulting from Eq. (14) is

$$F_D(y_{i+1/2}) = -\frac{1}{2}\left[f_E\left(C^{m-1}_{i+1/2}\right)\frac{C^{m-1}_{i+1/2} - C^{m-1}_i}{y_{i+1} - y_i}\ D\left(C^{m-1}_{i+1/2}\right)\right.$$

$$\left. + f_E\left(C^m_{i+1/2}\right)\frac{C^m_{i+1} - C^m_i}{y_{i+1} - y_i}\ D\left(C^m_{i+1/2}\right)\right] \tag{A-1}$$

where $i = 1, 2, \ldots, (n-1)$, m is the discrete time step index, and $c_{i+1/2} = (C_{i+1} + C_i)/2$.

The impurity content of the general ith cell using midpoint integration is given by

$$Q_i = C_i(y_{i+1} - y_{i-1})/2 \tag{A-2}$$

where $i \neq 1, I, n$. For the various boundary cells the impurity content equations are

$$Q_1 = C_1(y_2 - y_1)/2 \tag{A-3}$$

$$Q_{I-1/2} = C_{I,ox}(y_I - y_{I-1})/2 \tag{A-4}$$

$$Q_{I+1/2} = C_{I,Si}(y_{I+1} - y_I)/2 \tag{A-5}$$

$$Q_n = C_n(y_n - y_{n-1})/2 \tag{A-6}$$

REFERENCES

(1)   D. A. Antoniadis, S. E. Hansen, and R. W. Dutton, "SUPREM
       II -- A Program for IC Process Modeling and Simulation,"
       SEL 78-020 Stanford Electronics Labs, Stanford University,
       June 1978.

(2)   J. Lindhard, M. Scharff and M. Schiot, Mat. Fys. Medd.
       Dan. Vid. Sclsk. (33), p. 1 (1963).

(3)   J. Gibbons and S. Mylroie, Appl. Phys. Letts. (22),
       p. 568, June 1973.

(4)   Mayer, Erihasen and Davies, Ion Implantation in Semicon-
       ductors, Academic Press, New York (1974).

(5)   W. K. Hofker, D. P. Oosthoek, N. J. Koelman and H. A. M.
       DeGrefte, Rad. Effects (24), p. 223 (1975).

(6)   H. Ryssel, K. Haberger, K. Hoffmann, G. Prinke, R. Dumcke
       and A. Sachs, IEEE, ED-27 (8), p. 1484 (1980).

(7)   W. K. Hofker, H. W. Werner, D. P. Oosthoek and H. A. M. De
       Grefte, Appl. Phys. (2), Springer-Verlag, p. 265 (1973).

(8)   H. Ryssel, H. Kranz, K. Muller, R. A. Henkelmann and J.
       Biersack, Appl. Phys. Lett. (30), p. 399, April 1977.

(9)   H. Ryssel, private communication.

(10)   F. J. Morin and J. P. Maita, Phys. Rev. (96), p. 28
       (1954).

(11)   K. Lehovec and A. Slobodskoy, Solid-State Electron (3),
       p. 45 (1961).

(12)   J. S. Makris and B. J. Masters, J. Appl. Phys. (42),
       p. 3750 (1971).

(13)   R. B. Fair, Proc. of Third International Symp. on Silicon
       Materials Science and Technology (77-2), The Electrochem.
       Soc., p. 968, May 1977.

(14)   A. F. W. Willoughby, J. Phys. D: Appl. Phys. (10), p. 455
       (1977).

262

(15) F. Morehead, Extended Abstracts, Electrochem. Soc. Meeting, Abstract No. 139, p. 366, Spring 1980.

(16) R. B. Fair and J. C. C. Tsai, J. Electrochem. Soc. (124), p. 1107 (1977).

(17) R. B. Fair, J. Appl. Phys. (50), p. 860 (1979).

(18) D. A. Antoniadis and R. W. Dutton, IEEE, ED-26, p. 490 (1979).

(19) R. O. Schwenker, E. S. Pan and R. F. Lever, J. Appl. Phys. (42), p. 3195 (1971).

(20) R. B. Fair and G. R. Weber, J. Appl. Phys. (44), p. 280 (1973).

(21) M. Y. Tsai, F. F. Morehead, J. E. E. Baglin and A. E. Michel, J. Appl. Phys. (51), p. 3230 (1980).

(22) R. B. Fair, Proceedings of the Fourth International Symposium on Silicon Materials Science and Technology, H. R. Huff, R. J. Kriegler and Y. Takeishi Editors, The Electrochemical Society, p. 963 (1981).

(23) G. Masetti, S. Solmi and G. Soncini, Solid St. Electron (16), p. 1419 (1973).

(24) D. A. Antoniadis, A. G. Gonzalez and R. W. Dutton, J. Electrochem. Soc. (125), p. 813 (1978).

(25) D. A. Antoniadis, A. M. Lin and R. W. Dutton, Appl. Phys. Lett. (33), p. 1030 (1978).

(26) S. M. Hu, J. Appl. Phys. (45), p. 1567 (1974).

(27) D. A. Antoniadis, J. Electrochem. Soc. (129), p. 1093 (1982).

(28) A. M. Lin, D. A. Antoniadis and R. W. Dutton, J. Electrochem. Soc. (128), p. 1131 (1981).

(29) K. Taniguchi, K. Kurosawa, and M. Kashiwagi, J. Electrochem. Soc. (127), p. 2243 (1980).

(30) B. E. Deal and A. S. Grove, J. Appl. Phys. (36), p. 3770 (1965).

(31) C. P. Ho, J. D. Plummer, J. D. Meindl and B. E. Deal, J. Electrochem. Soc. (125), p. 813 (1978).

(32) C. P. Ho, J. D. Plummer, J. Electrochem. Soc. (126),
      p. 1523 (1979).

(33) R. N. Ghoshtagore, Solid St. Electron (17), p. 1065
      (1974).

(34) R. Reif and R. W. Dutton, J. Electrochem. Soc. (128),
      p. 909 (1981).

MODELING OF POLYCRYSTALLINE SILICON STRUCTURES FOR INTEGRATED CIRCUIT FABRICATION PROCESSES

Len Mei[*], Robert W. Dutton, and Stephen E. Hansen

Integrated Circuits Laboratory

Stanford University, Stanford, Calif.

Abstract

The increasing complexity of VLSI fabrication often requires the use of multilayer structures above the silicon substrate. Electrical and metallurgical properties of multilayer structures have an important effect on the circuit performance and reliability. Polycrystalline silicon is one of the most important elements of these multilayer structures. A new process model has been developed to simulate the properties of structures involving polycrystalline silicon and its oxide. This paper discusses the physics as well as the results of simulation supported by experimental data. The model can simulate many desirable properties of multilayer structures involving polycrystalline silicon. These include grain growth, resistivity and oxidation rate for the polysilicon layer, the impurity redistribution across multilayers after high temperature processing, the segregation of impurities at both grain boundaries and at interfaces, and the interdependent phenomena of dopant-dependent oxidation/diffusion.

---

[*]Present address: Fairchild Camera and Instrument, Palo Alto, CA.

1. Introduction

The development of very-large-scale integration (VLSI) in the fabrication of integrated circuits has required the use of multilayer structures above
the silicon substrate. Multilayer structures may consist of many materials
including silicon dioxide, silicon nitride, polycrystalline silicon, silicides and
metals. Of these materials, polycrystalline silicon [1] has the greatest variety
of applications, for example: interconnects [2], gate electrodes [3], high valued
resistors [4], diffusion sources for shallow junctions [5], as well as for buried
contacts [6], while the thermal oxide of polysilicon may serve as isolation between interconnection layers [7]. For each application, different properties of
the polysilicon layers, as a function of processing conditions, must be understood in order to control the process. For example the differential oxidation
rate of doped polysilicon relative to that of the silicon substrate, the sheet
resistance of the layer and the junction depth in the bulk after diffusion from
polysilicon all depend on detailed characterization of the material properties
of polysilicon and the process conditions.

A new modeling capability has been developed for multilayer structures
which allows the simulation of many of the important characteristics of the
polysilicon layer: its resistivity, impurity segregation at both the interfaces
and grain boundaries, the instantaneous oxidation rate of polysilicon and
the penetration of impurities from the doped polysilicon layer through a
thin oxide layer during thermal processing. Many physical models, such as
grain growth, electrical conduction in the polysilicon layer, segregation of
impurities, and oxidation-enhanced diffusion through polysilicon have been
developed. Property changes in the polysilicon layer after laser annealing can
also be modeled. This paper reviews many of the physical models related to
the processing aspects and physical properties of polycrystalline silicon and
discusses the structure of the SUPREM [8] simulation program in which the
individual physical models are implemented.

To illustrate the range of models necessary to simulate modern integrated circuit structures, device structures utilizing polycrystalline silicon

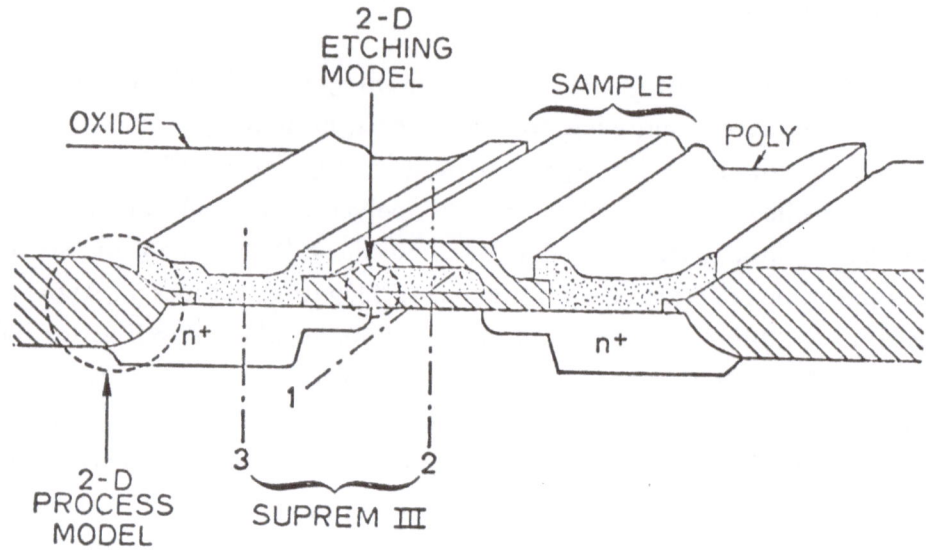

Fig. 1. Cross-section of a typical MOS device showing structures involving simulation of polycrystalline silicon.

are analyzed. Figure 1 shows a typical MOS device structure. Cross-section 1 is through the polysilicon layer, in which the grain sizes, grain boundary segregations and therefore the resistivities of the layer are of primary interest. Cross-section 2 cuts vertically across a four-layer structure of poly-oxide, polysilicon, gate oxide and bulk silicon substrate. In such a structure, one wishes to know the oxidation rate of the polysilicon under different processing conditions, and the interface segregation and redistribution of impurities after thermal processing. Cross-section 3 cuts vertically across the polysilicon layer deposited directly over the silicon substrate, providing buried contacts to the source and drain regions of the device. The property of interest in this cross-section is the impurity distribution in the silicon substrate under different processing conditions. Figure 2 shows a representative self-aligned bipolar device using polycrystalline silicon [9]. In cross-section

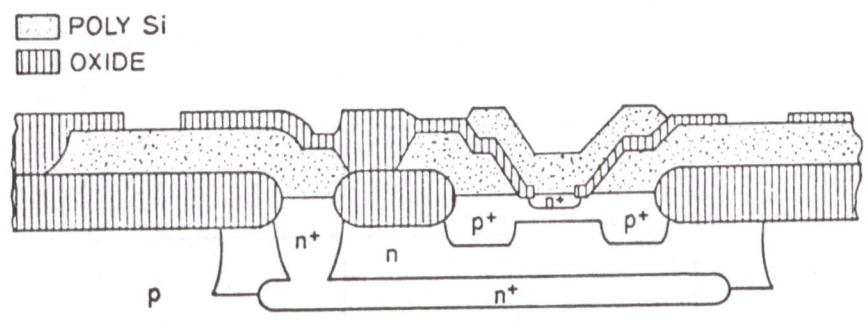

POLY Si
OXIDE

NEC's SELF-ALIGNED STRUCTURE

Fig. 2. A representative self-aligned bipolar device using polysilicon layers.

1, the polysilicon is used to form both emitter and base contacts. The self-aligned isolation between emitter and base contacts is provided by the thermal oxide grown from the first polysilicon layer. In practice, both MOS and bipolar devices tend to have similar cross-sections involving polycrystalline silicon. These cross-sections are summarized in Figure 3, where the first and second cross-sections consist of polysilicon deposited on a thermal oxide layer, while the third and fourth cross-sections have the polysilicon layer deposited directly on the silicon substrate. In the second and third cross-sections, the polysilicon has been oxidized, while in the other two cross-sections the polysilicon layer was not exposed to an oxidizing ambient during high temperature processing.

To further illustrate the modeling needs in state-of-the-art device design, a structure similar to that in cross-section 3 from Figure 3 is modeled [10]. Figure 4(a) shows an initial 3100 Å thick layer of polysilicon deposited directly over single crystal silicon by atmospheric pressure chemical vapor deposition (APCVD) at 776°C. After deposition, phosphorus was ion-implanted to a dose of $1.0 \times 10^{16}/cm^2$ at an energy of 100 keV. The phosphorus atoms were

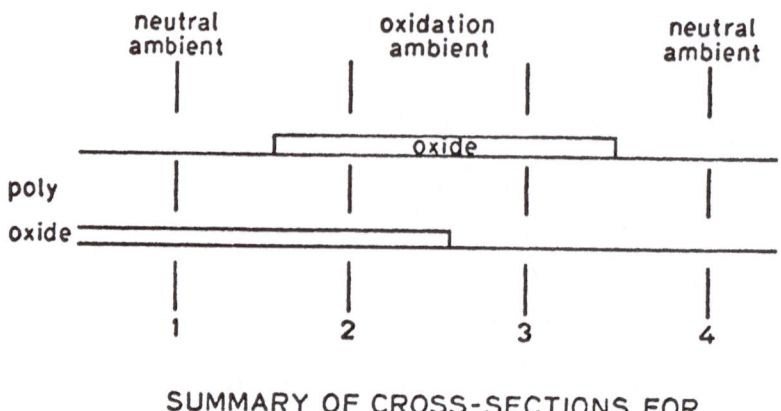

SUMMARY OF CROSS-SECTIONS FOR
MULTILAYER SIMULATION

Fig. 3. Four types of multilayer structures involving polysilicon.

primarily confined to the polysilicon layer. The wafer was then subjected
to a wet oxidation at 1000°C for 90 minutes during which an oxide layer approximately 0.5 μm thick was grown with a remaining polysilicon thickness
of almost 500 Å. The simulated impurity distribution throughout the structure is shown in Figure 4(b). Some of the impurities from the polysilicon
layer have been incorporated into the grown oxide, but much more of it
has diffused into the bulk silicon. Simulated profiles of both the total and
electrically active concentrations are shown in the bulk silicon. The total
impurity concentration is higher than the electrically active concentration
near the polysilicon/bulk substrate interface due to phosphorus precipitation
at high concentrations. As shown in the figure, the simulated profile of the
electrically active concentration is in agreement with the profile determined
by spreading resistance measurements. It may be noted that the phosphorus
profile in bulk silicon does not exhibit the kink and tail regions typical of
a $POCl_3$ predeposition directly into single crystal silicon. The formation
of the kink and tail region in predeposited phosphorus diffusions is believed

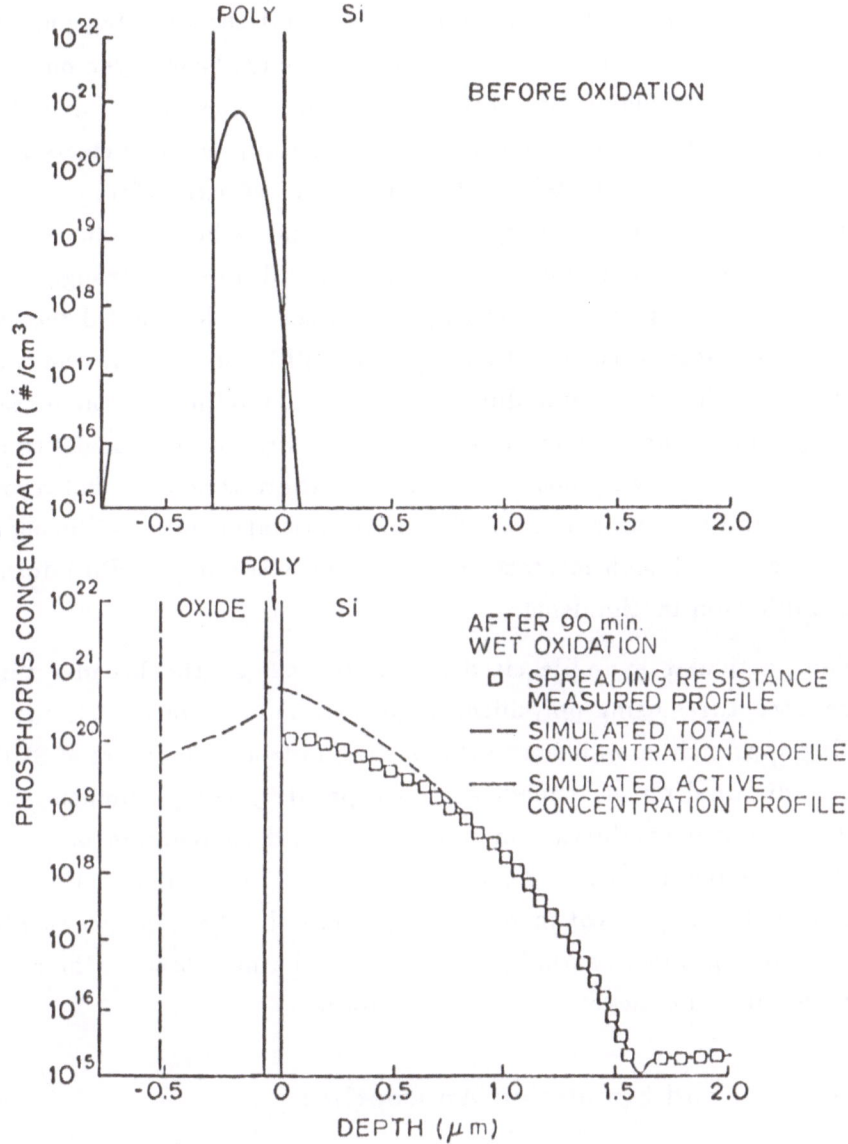

Fig. 4. Simulation of cross-section 3 in Figure 3: (a) distribution of phosphorus after ion implantation into 3100 Å of polysilicon over single crystal silicon, (b) structure and impurity distribution after wet oxidation. Experimental data were obtained by spreading resistance measurement.

to be due to the interaction of point defects and the diffusing phosphorus. It has been reported that the presence of a polysilicon layer on top of a bulk single crystal silicon layer interacts strongly with the growth of stacking faults and oxidation-enhanced diffusion [11,12]. Point defects in single crystal silicon are also factors in the growth of stacking faults and in the oxidation rate. It is therefore expected that the polysilicon can be regarded as either a sink or source of point defects, which will interact strongly with the point defects in bulk silicon, consequently affecting the point-defect-related properties. In later sections of this paper, it will be shown that the enhancement of diffusivity in the bulk due to oxidation is modified by the presence of a polycrystalline silicon layer at the substrate surface. The model to be discussed in this paper emphasizes particularly the interactions between grain boundaries and any impurities or silicon interstitials generated during oxidation. The study of such interactions can yield extremely useful information on basic diffusion mechanisms.

The particular example cited above as well as the broader range of physical structures using polysilicon requires a new set of models and a new modeling approach in addition to the basic simulation structure of SUPREM II. The individual physical models treat separately the phenomena of grain growth, electrical conduction in polysilicon, and segregation of impurities both at grain boundaries and at the interfaces. The organization and interaction of these physical models emulate closely the sequence of physical effects occurring during actual processing. In the next section the approach of the overall model development is presented.

2. Modeling and Simulation Approaches

To develop a process simulation program, the desired properties to be simulated must first be defined and then the suitable inputs must be specified. In the present prototype multilayer modeling program, the inputs are both the device structure and the processing conditions. Following the process steps, the resulting device structure will correspond to one of the four types

```
TITLE  **** EXAMPLE OF POLY SIMULATION *****

SUBS  ELEM=B, CONC=1.0E15, ORNT=100

STEP  TYPE=OXID, TEMP=1000, TIME=20, MODL=DRY0

POLY  THICK=0.35, ELEM=P, DOSE=1.0E15, GSIZE=800

STEP  TYPE=OXID, TEMP=1000, TIME=20, MODL=DRY0

PRINT TOTL=Y

MEASU TEMP=20

END
```

At present, no attempt has been made to model the gas-phase doping process (such as PH_3 or $POCl_3$), nor to model the grain growth mechanism during polysilicon deposition. Therefore the initial dose and grain size in the polysilicon layer (POLY card) must be provided as inputs. The desirable outputs from a simulation are both the electrical and metallurgical properties of the multilayer structure, including layer resistivities, thicknesses, dopant distributions, junction depths and other selected outputs.

The organization of the simulation model is illustrated in Figure 5 with emphasis on discrete time representations. At the beginning of each time step, the electrically active concentration in the polysilicon layer determines various dopant related process parameters such as the grain growth rate, the impurity diffusivity, and the oxidation rate. From the oxidation rate, the oxide thickness at a given time step is calculated, which in turn determines the total oxide thickness and the remaining polysilicon layer thickness. If the device structure is either the third or the fourth type defined in Figure 3, the oxidation enhancement to the diffusivity is determined not only from the oxidation rate but also the thickness of the polysilicon layer and grain size at the appropriate time step. The diffusion algorithm determines the impurity fluxes out of the polysilicon layer into the adjacent layers. With oxidation, the flux of impurities incorporated into the oxide is also affected by

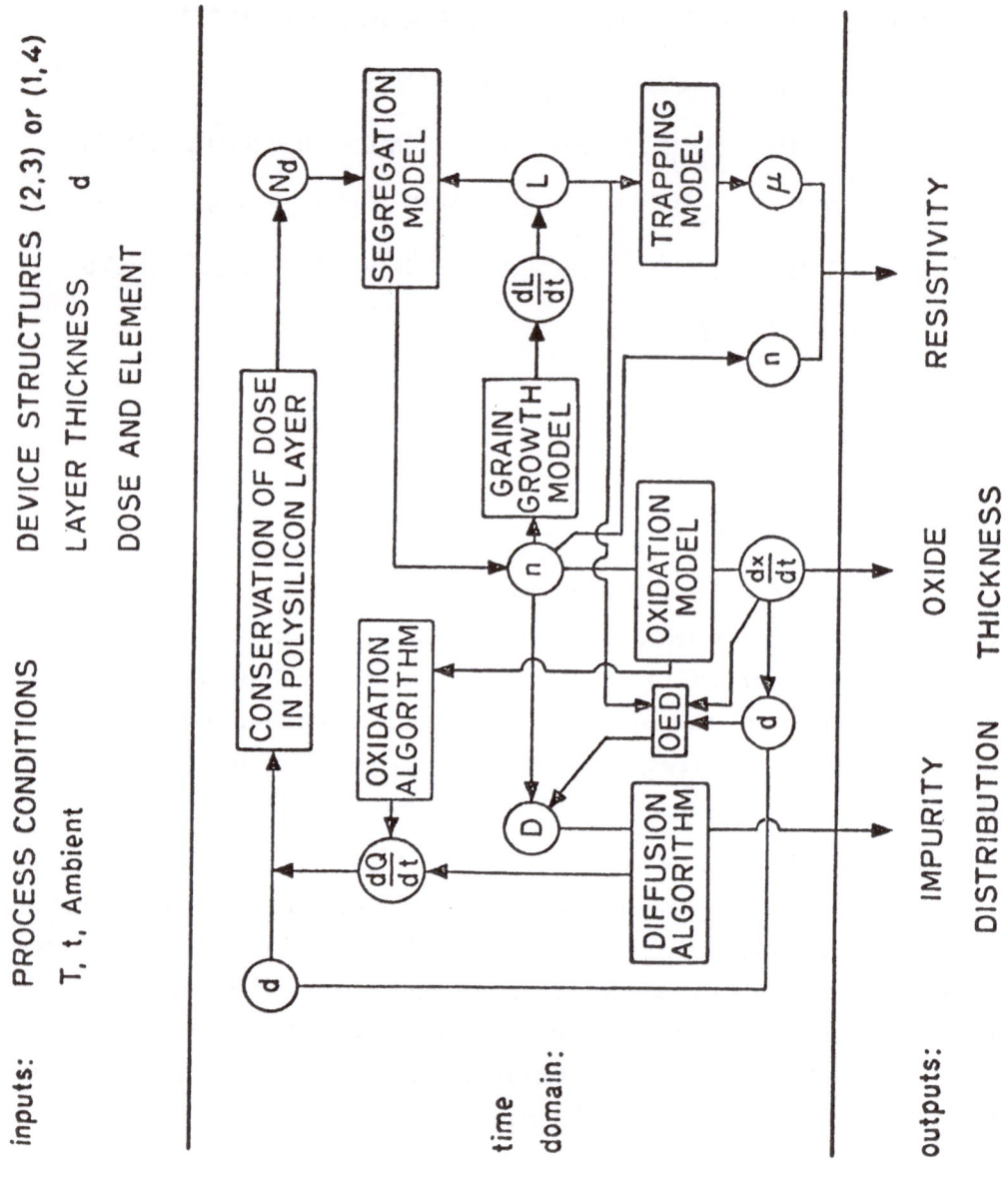

Fig. 5. Structure of the multilayer simulation model.

the oxide growth algorithm. Using the equation of conservation of dopant, the total impurity concentration in the polysilicon layer is calculated for the next time step. The active concentration in the polysilicon layer is then determined from the segregation model with the inputs of grain size, processing temperature, total doping concentration and thickness of the polysilicon layer, as determined by the most recent time step. The average grain size of the polysilicon layer is determined by the grain growth model, which calculates the increment of grain size at the present time step. The electrically active concentration and the grain size are then input into the conduction model to determine the mobility and carrier concentration, and therefore the resistivity.

It is essential that the model emulate the real events during processing. From such a model, it can be understood that the oxidation rate for the polysilicon layer is neither parabolic nor linear, because the rate depends on the doping concentration in the polysilicon layer, which changes as oxidation proceeds as a result of the impurity distribution. Other phenomena associated with the polysilicon/bulk silicon structure, such as the dependence of the diffused profiles on the oxidation rate and thickness of the polysilicon layer, can also be explained. In the following sections, the inidividual physical models are discussed in detail and the simulation results are compared with experimental results.

3. Electrical Conduction in Polysilicon Layer

Conduction in polycrystalline silicon depends on physical (structural and metallurgical) properties of the material as well as electrical effects of the crystalline defects such as trapping-induced conduction barriers [13]. Due to the presence of a high density of crystalline defects, such as grain boundaries, the electrical conductivity of polysilicon is always lower than that in single-crystal silicon for the same doping concentration. The reduction in conductivity is attributed both to the reduced carrier concentration and to mobility. Figure 6 lists the principal models which have been developed in the past decade to describe the mobility and carrier concentration effects in

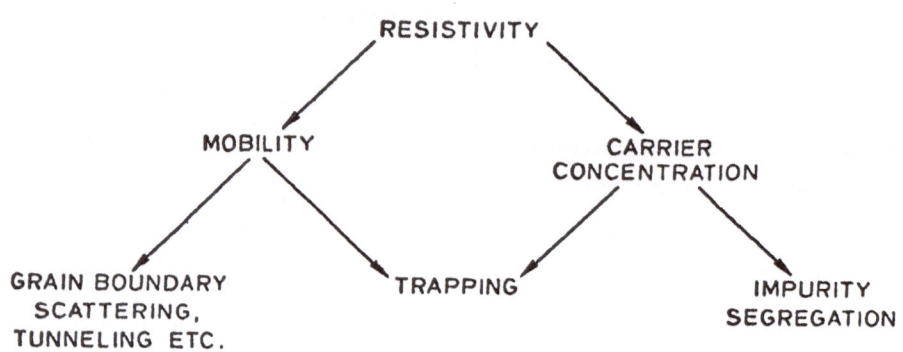

Fig. 6. Principle models for electrical conduction in polysilicon layer.

polysilicon. These include dopant segregation, carrier trapping and scattering, and tunneling. Depending on the doping concentration, average grain size and ambient temperature, different mechanisms can dominate the conductivity. The trapping model is discussed in this section. The segregation model will be discussed in the next section.

A comprehensive carrier trapping model for polysilicon was first proposed by J. Seto [13], and later improved by Baccarani et al. [14], and Segeal et al. [15]. This model proposes that the current transport through a polysilicon layer is controlled by the thermionic emission across the energy barrier at each grain boundary. This energy barrier is formed by the depletion of carriers at the grain boundaries resulting from carrier trapping at grain boundary sites. The energy barrier height depends on the carrier concentration, density of trapping states, and the grain size. The barrier energy can be determined by solving Poisson's equation with the following results:

$$E_B = \frac{q^2 W^2 N_g}{8\epsilon} \tag{1}$$

where W and N_g are the grain boundary depletion width and the ionized doping concentration in the grain, respectively. When the grains are partially depleted, the depletion layer width can be determined by solving simultaneously equation (1) and the Fermi-Dirac distribution function. When the grains are completely depleted of carriers, the depletion layer width is given by the grain size. Experimental results show that a single level trapping state with energy level near the mid-bandgap is found to be most appropriate in describing the energy band formation [13]. In the present model, the energy level of the traps is assumed to be at mid-bandgap and density of trap states to be $3.8 \times 10^{12}/cm^2$ [13]. The simulated barrier height for a grain size of 1000 Å is plotted as a function of doping concentration for two densities of trap states (Figure 7): 1.0×10^{12} and $1.0 \times 10^{13}/cm^2$. The change of barrier height as a function of doping concentration is clearly separated into two regions with a maximum barrier height at a specific intermediate concentration. In the low concentration region, the trap states are more numerous than the carriers. Thus more carriers become trapped at higher concentrations and the barrier height increases with the doping concentration. From equation (1), it can be seen that when W is fixed to be the average grain size L, the energy barrier height is proportional to the concentration N_g. On the other hand, in the high concentration region, the grains are partially depleted of carriers. Therefore, the width of the depletion layer, and thus the barrier height, decreases with increasing doping concentration.

The average density of majority carriers, n_a, can be determined by integrating over one grain:

$$n_a = n_b \left[\left(1 - \frac{Q_t}{LN_g} \right) + \frac{1}{qL} \sqrt{\frac{2\epsilon kT}{N_g}} \, \mathrm{erf} \left(\frac{qQ_t}{2} \sqrt{\frac{1}{2\epsilon kT N_g}} \right) \right] \tag{2}$$

where n_b is the carrier concentration in the single-crystal silicon with doping

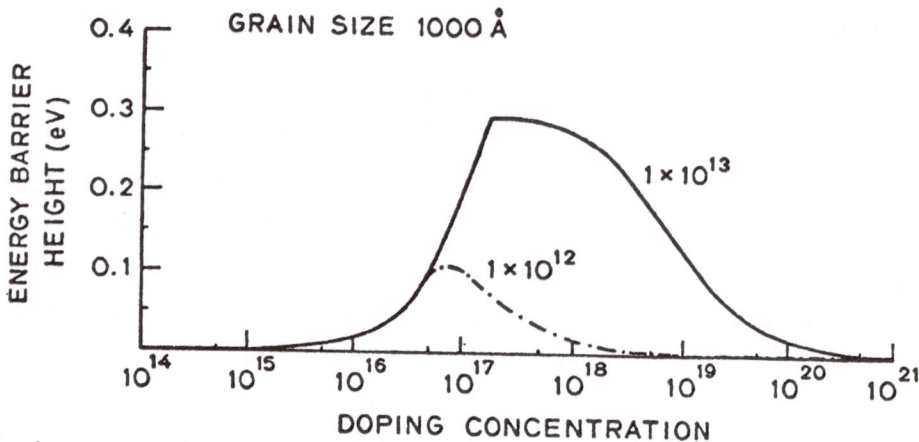

Fig. 7. Energy barrier height at grain boundary in polycrystalline silicon
as a function of doping concentration simulated by the trapping
model for two densities of trap states.

concentration N_g, and Q_t is the density of trapping states at grain boundaries. The current density flowing through the grain is then given by,

$$J = qn_a \sqrt{\frac{kT}{2\pi m^*}} \exp\left(-\frac{E_B}{kT}\right) \exp\left(\frac{qV_g}{kT}\right) \tag{3}$$

where V_g is the voltage drop across the grain, and m^* is the effective mass
of majority carriers.

For small applied voltage, the resistance becomes linear and the mobility
can be defined by

$$\mu_b = \frac{J}{qn_a \mathcal{E}}$$

$$= \frac{qL}{\sqrt{2\pi m^* kT}} \exp\left(-\frac{E_B}{kT}\right) \tag{4}$$

with $\mathcal{E}$ being the electric field. The total mobility across the grain given by
Mathiessen's law becomes

$$\frac{1}{\mu} = \frac{1}{\mu_b} + \frac{1}{\mu_g} \tag{5}$$

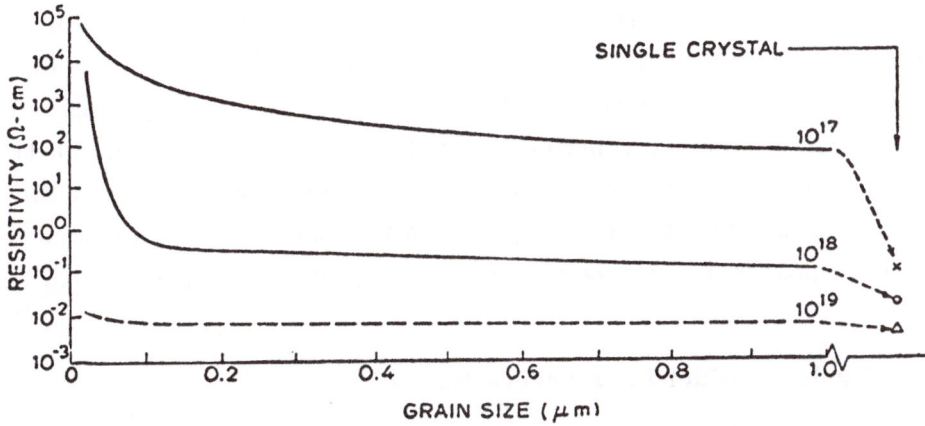

Fig. 8. Simulation of polysilicon resistivity as a function of grain size for three different doping levels.

where μ_g is the mobility caused by the scattering of carriers in the normal lattice. The grain size is involved in both equations (2) and (3). The variation of resistivity as a function of grain size is shown in Figure 8 — the resistivity is most sensitive to the grain size for small grain polysilicon and medium concentrations.

Recently, mobility models including the effects of grain boundary scattering as well as the tunneling component of current at grain boundaries have been proposed by Mandurah et al. [16] and Lu et al. [17]. These new models improve the accuracy for prediction of resistivity at lower temperatures (e.g. liquid nitrogen temperature) at the expense of increased computation time. Although these model improvements offer potential advantages for certain technologies and applications, the present simulator implementation uses only the first-order model.

The trapping model predicts the general behavior of mobility and carrier concentration rather accurately at room temperature over a wide range of doping concentrations. However, it has a serious drawback at high concentrations, where trapping becomes a relatively minor effect. The model does not predict any difference in resistivity between phosphorus- and arsenic-doped polysilicon as observed experimentally [18]. It also does not explain the reversible change of resistivity value when polysilicon is annealed at different temperatures [19]. The impurity segregation model to be discussed below has been proposed to overcome the deficiency of the trapping model.

4. Physical Model Parameters for Polycrystalline Silicon

The previous section has discussed the electrical conduction mechanism in polysilicon. This model utilizes physical parameters — average grain size and doping concentration — which are directly related to the fabrication process. In this section the physical models for grain growth and impurity segregation are discussed, with emphasis given to the underlying coefficients needed to model these phenomena over a typical range of process conditions.

4.1 Grain Growth

The average grain size of polysilicon increases during high temperature processing. This grain growth is believed to be controlled by the silicon self-diffusion across grain boundaries [20]. When the polysilicon is doped, the grain growth rate can be altered as a function of the impurity doping concentration. A model has been developed to determine the average grain size after a given set of conditions. (A detailed discussion of the grain growth model has been published elsewhere [21].)

The effect of dopant concentration on the grain growth rate is summarized in Figure 9. Impurities have two opposing effects on the grain boundary migration rate. On the one hand, they increase the vacancy concentration [22] by shifting the Fermi level, increasing the grain boundary migration rate. On the other hand, the impurities can retard the grain boundary migration due to segregation at grain boundaries and cluster formation in

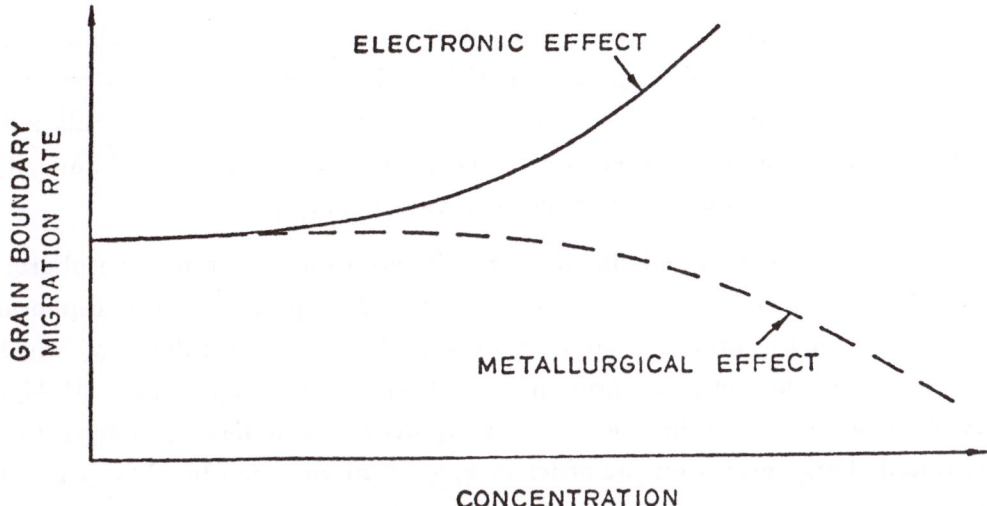

Fig. 9. Effects of dopants on the grain boundary migration rate.

the grains. The net effect of impurity doping on the grain growth depends on the combination of these two effects.

Experimentally, it has been found that arsenic and phosphorus enhance grain growth in polysilicon in varying degrees, whereas boron has little effect on grain growth [21]. Arsenic-doped polysilicon, for example, shows an initial increase in grain size after a rise of concentration beyond 1×10^{19} atoms/cm^3; however this enhancement reaches a maximum and drops at higher concentrations. This is consistent with the experimental results reported by Solmi et al. [23].

The active concentration in the polysilicon layer changes with the grain size and the amount of impurity out-diffused into the adjacent layers at each time step. The increment of grain size ΔL^i at the ith time step, Δt, has been derived in the grain growth model [21] and is given by

$$\Delta L^i = \left(\frac{ab^2}{kT}\right)\left(\frac{\lambda^i D_g^i}{L^i}\right)\Delta t \tag{6}$$

where a and b are geometric and physical factors, T is the annealing temperature, L^i is the average grain size, λ^i is grain-boundary energy, and D_g^i is the silicon self-diffusivity along a boundary, which is related to its bulk self-diffusivity. All the parameters with superscript i are a function of impurity segregation at grain boundaries and can vary during processing.

The experimental and simulated results of grain growth for phosphorus-doped polysilicon are shown in Figure 10. The presence of phosphorus changes both the grain-boundary energy and silicon self-diffusivity, which leads to a change in grain-growth activation energy (Figure 11). At high concentrations, the grain size increases rapidly as the annealing temperature is raised. Large grains on the order of several microns can be obtained with heavy phosphorus doping.

4.2 Impurity Segregation

Impurities can segregate at both the interface and grain boundaries [19], where they become electrically inactive. Therefore, the concentration of mobile carriers in the polysilicon layer is lower than that in single-crystal silicon for a given doping level. The theory of thermal-equilibrium segregation at grain boundaries for metal systems has been applied to describe both the grain boundary [24] and interface segregation. The ratio of concentration in the grain, N_g, to the total, N_t, is given by

$$\frac{N_g}{N_t} = \left[\frac{AQ_s}{LN_{Si}}\exp\left(\frac{G_a}{kT}\right) + 1\right]^{-1} \tag{7}$$

where N_{Si} is the atomic density of silicon crystal, and Q_s is the density of segregation sites at the grain boundary. This value is calculated to be 2.64×10^{15}sites/cm^2 using the reported segregation data of M. Mandurah et al.[19]. The entropy factor A (a dimensionless parameter) changes with

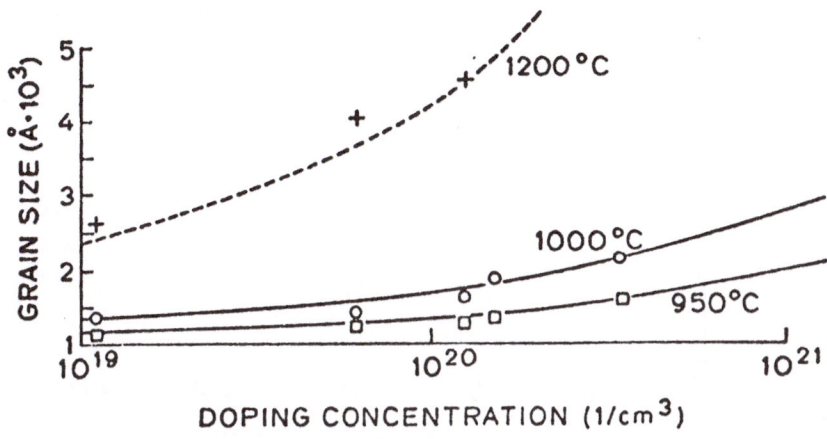

Fig. 10. Variation of average grain size in polysilicon thin layer after annealing at three temperatures for 30 minutes for different phosphorus, concentration both experimental and simulated.

both the dopant type and its concentration. The values reported by M.
Mandurah et al. are 3.02 for arsenic and 2.46 for phosphorus, both at a
concentration of $2.0 \times 10^{19}/cm^3$. The activation energy, G_a, has been determined to be 0.456 eV for both arsenic and phosphorus. Boron does not
segregate significantly at grain boundaries [25]. Figure 12 shows the simulated percentage of impurity segregation at grain boundaries at thermal equilibrium as a function of grain size for arsenic and phosphorus at 1000°C. The
figure shows that arsenic segregates more than phosphorus, and for a given
grain size, that about 30 to 40% of the impurities are segregated at the grain
boundaries. Similarly, the interface segregation can be modeled by equation
(7). The model assumes that thermal equilibrium is immediately achieved
upon annealing. In reality, it is only approached after a period of time such

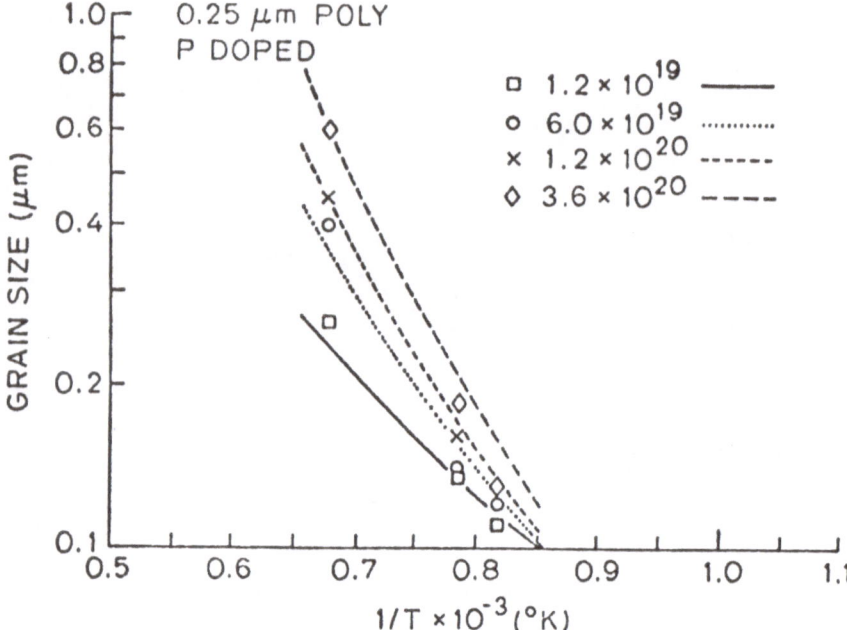

Fig. 11. Change of average grain size as a function of reciprocal of annealing temperature for different phosphorus concentrations. The experimental data are the same as those in Figure 10.

that the impurity diffusion length is approximately half of the average grain
size. For more accurate modeling of the properties, a non-thermal equilibrium
model should be incorporated.

To determine the concentration of active dopants in the polysilicon layer
after grain boundary and interface segregation, an additional condition —
the conservation of total dose, D_{total}, in the multilayer structure — must be
satisfied at all times. The total dose, D_{total}, is reduced by the fraction lost
due to out-diffusion, $D_{out.diff}$, from the polysilicon layer into the adjacent
layers, and the remainder is partitioned as follows:

$$D_{total} - D_{out.diff} = D_g + D_{gb} + D_{in} \tag{8}$$

where D_{in} is the dose segregated at interface, D_{gb} is the dose segregated at

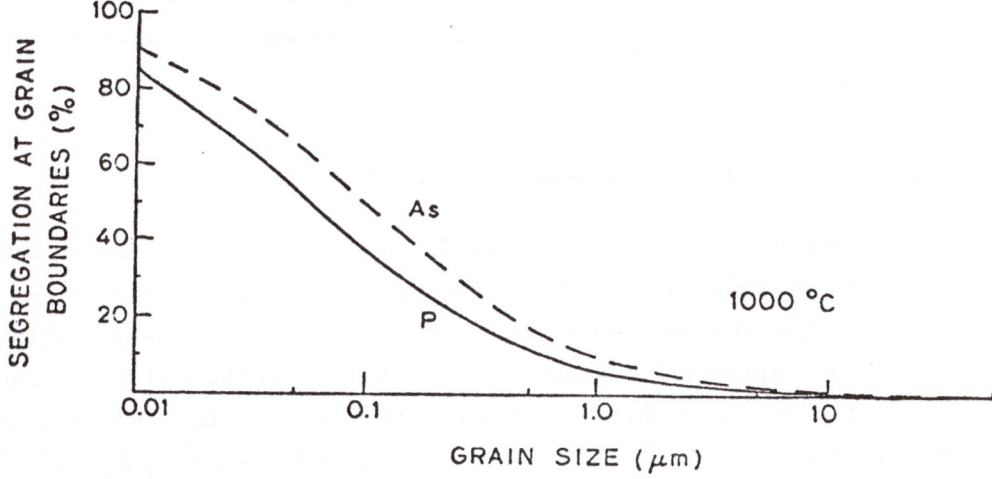

Fig. 12. Percentage of impurity segregation at grain boundaries as a function of grain size for arsenic and phosphorus annealed at 1000°C at thermal equilibrium.

grain boundaries, and D_g is the dose in the grain. The out-diffused dose is calculated by the diffusion algorithm for each time step. By solving equations (7) and (8) simultaneously for interface and grain boundary segregation, the active dose in the polysilicon layer at a given time step can be determined. This active concentration is used not only to determine the electrical properties of the polysilicon layer, such as resistivity, but also to determine its metallurgical properties, such as grain growth and oxidation rate.

5. Examples of Simulation

The following examples show simulated and experimental results for samples with polysilicon multilayer structures. The first examples consider thermal processing in which ion implantation is used to control precisely the amount of impurities in the films. These examples illustrate applications where the results are expected to track process variations, given initial grain size data. An example of laser annealing of LPCVD polysilicon is also presented. The objective of this annealing experiment is to show the potential of combining empirical and calculated results to enhance physical understanding.

5.1 Thermally Processed Polysilicon

An example of the simulation using the conduction model combined with the segregation and grain growth models shows the change of resistivity of a polysilicon layer as a function of phosphorus concentration. The polysilicon layers in this experiment were deposited on 2000 Å of thermal oxide using APCVD at 776°C to a thickness of 0.25 μm and implanted with different doses of phosphorus. The layers were subsequently covered with 1000 Å of Silox oxide deposited at a low temperature to prevent the escape of impurities during thermal annealing. The operation was performed at 1000°C for one hour in a nitrogen ambient. The simulated grain size and the amount of segregated dose for the samples after annealing are shown in Figure 13. The simulated increment in grain size for the more heavily doped polysilicon layers and the resulting effect on carrier trapping and impurity segregation are correctly modeled. Figure 14 shows the resistivity, the grain size, and percentage of impurity segregation at the interface for a second set of experiments carried out with polysilicon layers of three thicknesses: 0.25, 0.5, and 0.75 μm. All samples were doped with phosphorus by ion implantation to an average concentration of $1.2 \times 10^{19}/cm^3$ and subsequently covered with Silox oxide and annealed at 950°C for three hours. The upper curve shows the change of grain size with the polysilicon thickness both simulated and measured. The grains in the thicker polysilicon layer are slightly larger due to the smaller ratio of oxide/polysilicon interface area relative to the grain boundary area. However, a more dramatic effect of the thickness variation is observed in the change in interface segregation, which reduces from 18% to less than 5% as polysilicon thickness increases. This interface segregation is reflected strongly in the change of resistivity as a function of polysilicon thickness, shown in the lowest curve in Figure 14.

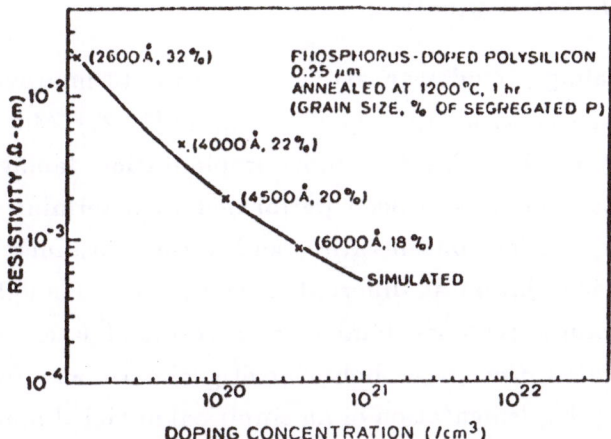

Fig. 13. Change of resistivity in polysilicon layer as a function of phosphorus concentration as determined by experiment and simulation. The simulation model includes the effects of grain growth, interface, and grain boundary segregation.

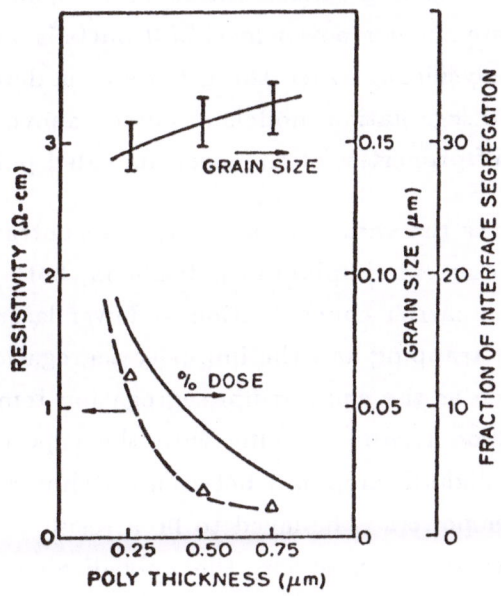

Fig. 14. The change of resistivity, grain size, and fraction of interface segregation of polysilicon layers for three different layer thicknesses after annealing at 950°C for three hours. All the polysilicon layers were doped to the same phosphorus concentration of 1.2×10^{19} by ion implantation.

5.2 CW Laser Annealing

Laser annealing provides an attractive means to improve the electrical and structural properties of the polysilicon layers [26,27]. Moreover, laser annealing is a very efficient way to remove implantation damage and increase grain size. Experiments have been performed to determine the grain size, carrier mobility, carrier concentration and sheet resistance for CW laser-annealed polysilicon layers at different laser powers. Though a theoretical model of the grain growth mechanism as a result of laser annealing is extremely complicated, the simple behavior of grain size as a function of laser power allows easy implementation of an empirical model. Figure 15 shows the variation of grain size of 0.5 μm LPCVD polysilicon on a 1 μm thermal oxide as a function of CW laser power under the following processing conditions: scan speed 50 cm/s, spot size 70 μm, line space 20 μm, substrate temperature 500°C. In this particular experiment, there is no apparent variation of grain size as a function of doping concentration and dopant species. The sharp increase of grain size above a power level of 9 watts is a result of melting and regrowth of the polysilicon. Once the grain size is determined empirically, the conduction and segregation models discussed above should be applicable to the simulation of properties of the laser-annealed polycrystalline silicon.

Figure 16 shows the variation of carrier concentration as a function of laser power for the arsenic-implanted polysilicon, both measured and simulated. The reduced carrier concentration at lower laser power is attributed to both the carrier trapping and the impurity segregation. The silicon melting point was taken as the equilibrium segregation temperature for simulation. The simulation results coincide with the experimental data at high laser powers. The slight discrepancy between the simulated and experimental results at low laser powers is believed to be a result of the non-equilibrium segregation. At these power levels, the polysilicon layer is only partially melted and the thermal cycle is too short to establish equilibrium segregation.

Figure 17 shows the change of mobility for the same samples. Curve

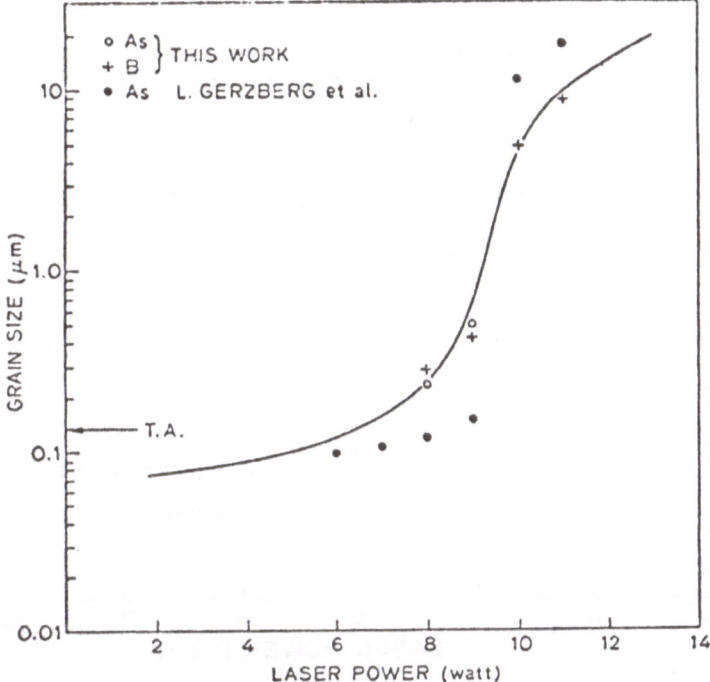

Fig. 15. Variation of grain size of 0.5 μm polysilicon on 1 μm thermal oxide as a function of CW laser power. The data by Gerzberg is from reference [23].

A shows the simulated mobility using the models discussed previously. The simulated mobilitiy is overestimated for lower powers. It is believed that the low power annealing does not effectively remove most of the implantation damage, therefore the mobility is less, due to the defect scattering centers. No attempt was made to study the effects of laser annealing on damage density due to ion implantation. A correction factor C_μ multiplies the mobility plotted in Figure 17. The corrected mobility is plotted as Curve B. The resistivity as a function of both simulated and measured laser power is shown in Figure 18. For comparison, the experimental data for thermally-annealed polysilicon is shown in the same figure. The thermally annealed samples were prepared under the same conditions as in the previous experiments, have the same arsenic concentrations and were annealed at 1000°C for one hour. The

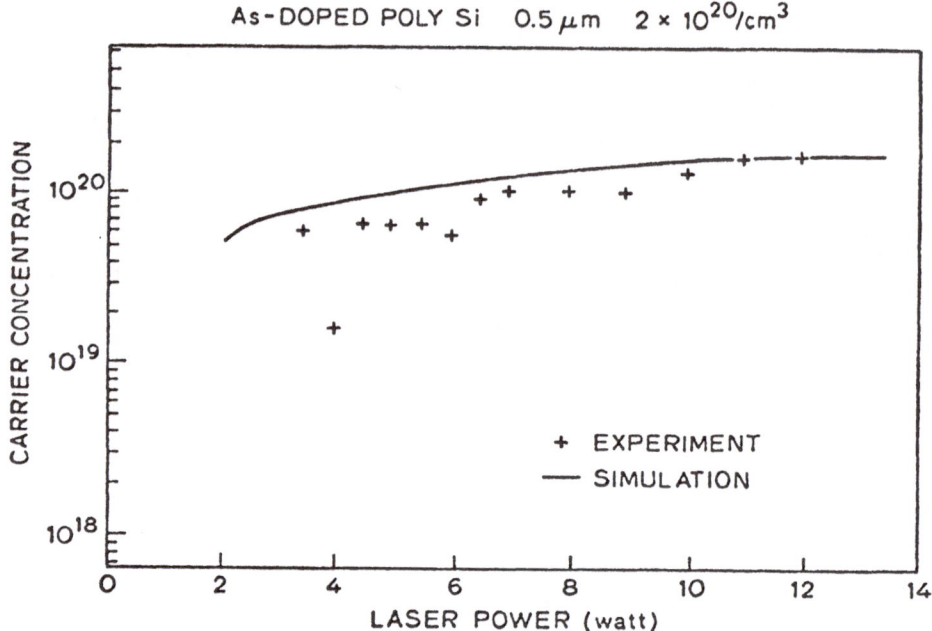

Fig. 16. Variation of carrier concentration as a function of laser power, for arsenic-implanted polysilicon from both experiments and simulation.

resistivity is comparable to that of samples annealed with 6 watts of laser power.

It should be emphasized that the laser annealing model is not based entirely on a complete understanding of the physical processes involved. However, the simulation can be a useful tool when used in conjunction with supporting experiments. In particular, the simulation of segregation and conductivity related to grain size has revealed two important facts. First, the change in grain size with annealing laser power (based purely on an empirical curve) allows one to determine the conditions at which segregated impurities become a small fraction of the total. Second, given the grain size and doping concentration, the simulation shows that at low laser powers, the behavior of both the impurity segregation and carrier mobility deviates from the thermal

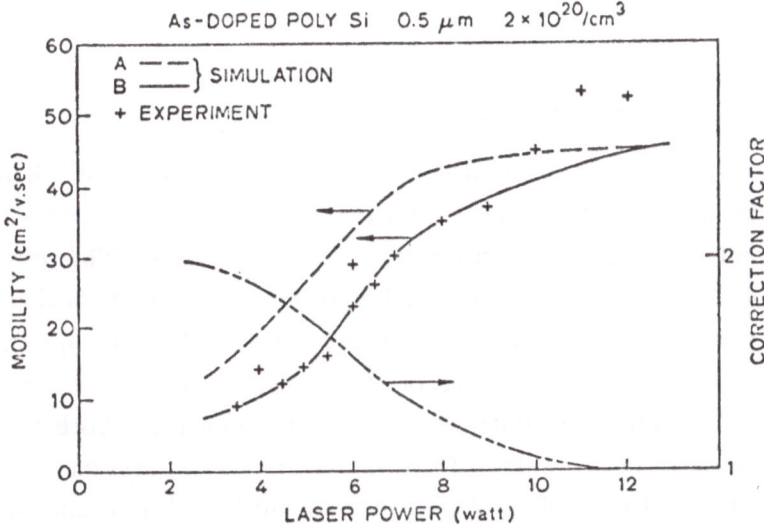

Fig. 17. Change of electron mobility in polysilicon as a function of laser annealing power for the same samples as in Figure 12. Also plotted is the mobility modification factor used in the simulation.

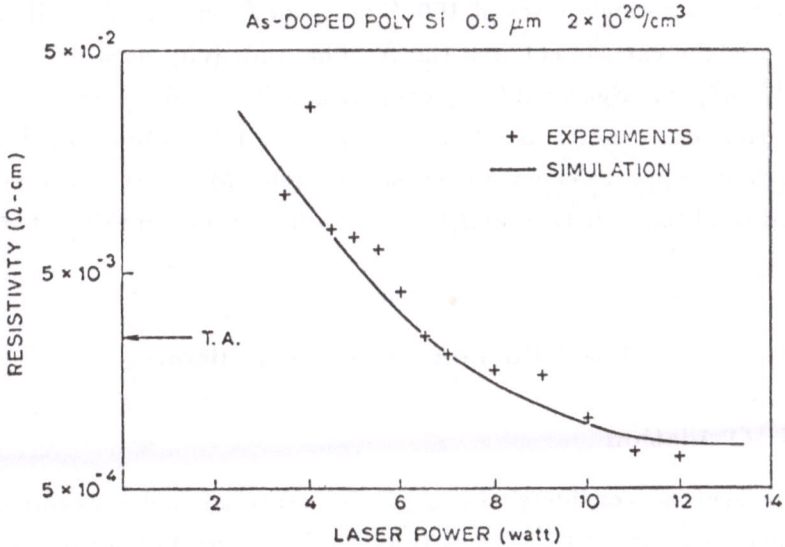

Fig. 18. Resistivity as a function of laser power from the same set of samples as in Figure 12.

equilibrium models. The present model can be used to estimate the effect of such deviations.

5.3 Diffusion Through Thin Oxides

An interesting application of the multilayer diffusion algorithm is the prediction of impurity penetration from polysilicon electrodes through thin gate oxides during high-temperature processing steps. The present program assumes that the diffusivity in oxide is independent of local impurity concentration. This simplified model can be tested by comparing simulated and experimental results. Figure 19 shows the simulated impurity profiles of diffusion from a phosphorus-doped polysilicon layer both on a thin oxide as well as the silicon substrate itself. The original structure consisted of a 3000 Å layer of polysilicon doped to $2.2 \times 10^{20}/cm^3$ and annealed at 1200°C for 24 minutes in a nitrogen ambient. Figure 20 shows the simulated junction depth as a function of annealing time (at 1150°C) for polysilicon on the thin oxide in a hydrogen ambient. The experimental data reported by Shimakura et al. [28] are shown in the same figure for comparison. In the simulation, oxide thicknesses of 100 Å and 120 Å were used, while the oxide thickness in the experiment was 120 Å. The diffusivity used in simulation is $7 \times 10^{-17} cm^2/sec$. Reasonable agreement can be obtained over only part of the experimental conditions. This indicates that the diffusivity is indeed a function of phosphorus concentration in the oxide. More detailed investigation is needed to obtain a better model for the diffusion of impurities through the oxide.

6. Oxidation-Related Phenomena in Polysilicon

6.1 Oxidation

This section considers the oxidation rate of polycrystalline silicon. Experimental evidence [29] shows that the linear oxidation rate of undoped polysilicon has a value between that of the slowest oxidizing orientation (100) and the fastest oxidizing orientation (111) of intrinsic single-crystal silicon.

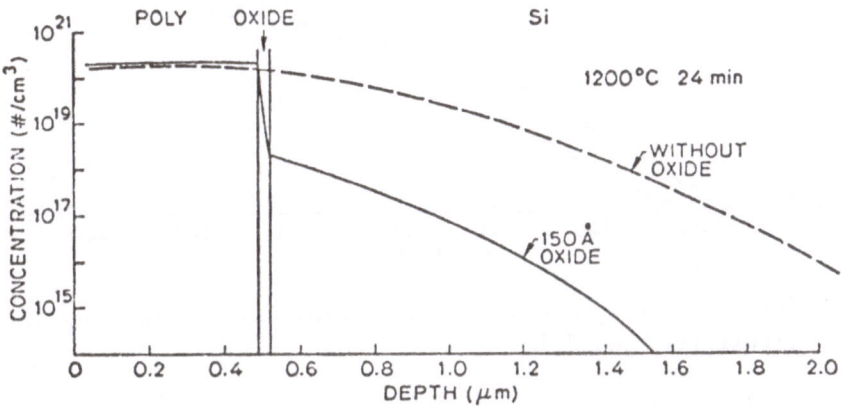

Fig. 19. Simulated profiles of phosphorus diffusion from a doped polysilicon layer both on thin oxide and on silicon substrate.

This seems to suggest that the linear oxidation rate of polysilicon is simply an average over that of its individual grains with different orientations. Grain boundaries may have preferential oxidation, but they apparently do not change the overall oxidation rate dramatically. This would be the case as long as the reaction rate of the oxidizing species at the silicon interface is faster than the diffusion rate along grain boundaries. The linear oxidation rate (B/A) of undoped polysilicon can be expressed by:

$$\left(\frac{B}{A}\right)_{undoped} = \left[k_1\left(\frac{B}{A}\right)_{(100)} + k_2\left(\frac{B}{A}\right)_{(111)}\right] \tag{9}$$

where k_1 and k_2 are constants. Since the parabolic oxidation rate is determined by the diffusion of oxidizing species through oxide, it seems reasonable to assume that it does not change with the silicon crystalline structure.

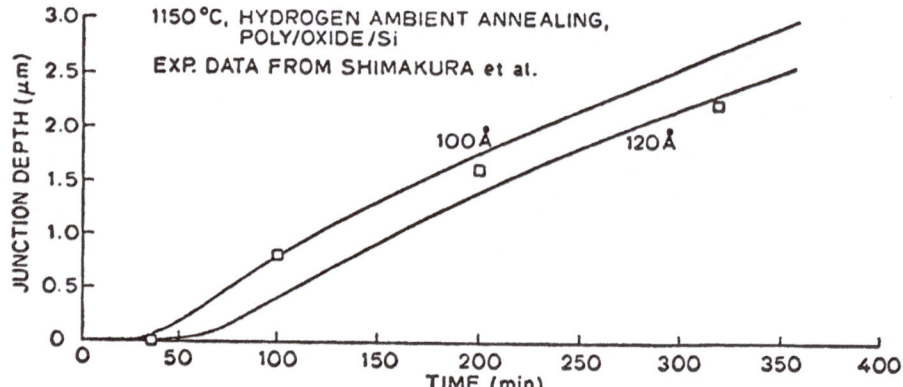

Fig. 20. Simulated junction depths as a function of annealing time for structures 1 and 2 in Fig. 3.

However, the impurity enhancement of the linear oxidation rate in polysilicon may be different from that of similarily doped single-crystal silicon having the same doping concentration. In the present model, the linear oxidation rate of doped polysilicon is given by that of undoped polysilicon multiplied by an enhancement factor:

$$\left(\frac{B}{A}\right)_{doped} = \left(\frac{B}{A}\right)_{undoped} \times f(T, \frac{C_v}{C_v^i}) \tag{10}$$

The enhancement factor $f(T, C_v/C_v^i)$ for single-crystal silicon proposed by C. Ho et al. [22] is:

$$f(T, \frac{C_v}{C_v^i}) = 1 + 2.62 \times 10^3 \exp\left(-\frac{1.1}{kT}\right)\left(\frac{C_v}{C_v^i} - 1\right) \tag{11}$$

where C_v and C_v^i are the vacancy concentration of the doped silicon and intrinsic silicon respectively. In the case of doped polysilicon, the vacancy concentration C_v must be determined using the electrically active concentration in polysilicon layer. The model implemented in the simulator determines

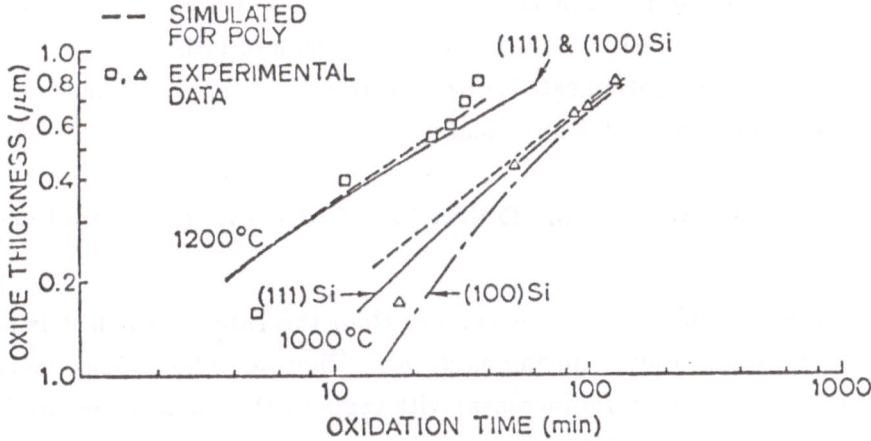

Fig. 21. Oxide thickness vs. oxidation time for a polysilicon layer with phosphorus concentration $3.3 \times 10^{20}/cm^3$ in wet oxidation ambient at 1000°C and 1200°C. Also shown for comparison are the curves for intrinsic single-crystal silicon.

the enhancement factor of the oxidation rate from the active concentration at each time step. Since this active concentration is a strong function of grain size and the polysilicon layer thickness, the oxidation rate is therefore also a function of these parameters. The dopant conservation across multilayer structures, which must be satisfied at all times, has an important influence on the oxidation rate.

The simulated and experimental results of oxide thickness vs. the oxidation time are shown in Figure 21. In this experiment, the polysilicon of 3000 Å was deposited by APCVD on thermal oxide and doped with phosphorus by ion implantation to an average concentration of $3.3 \times 10^{20}/cm^3$. The samples were subjected to wet oxidation at 1000°C and 1200°C. For comparison, the oxidation characteristics of intrinsic single-crystal silicon of both (100) and (111) orientations are plotted in the same figure. It can be seen that at high temperatures both the orientation and doping effects on the oxide thickness are small because the oxidation occurs primarily in the parabolic region. Here the oxidation rate is controlled by the diffusion of oxidizing species through the oxide layer. However, at lower temperatures, the doped polysilicon oxidizes faster than the intrinsic silicon of either orientation. This difference in the oxidation rate is often utilized to obtain a thicker oxide layer from polysilicon for better electrical isolation [9].

6.2   Dependence of Oxidation Rate on the Thickness of Polysilicon

The phosphorus out-flux is smaller than the rate at which it is pushed into the polysilicon layer during oxidation. Thus the phosphorus concentration in the polysilicon layer increases with reduced thickness during oxidation. As a result, the linear oxidation rate increases as predicted by equations (10) and (11). The conservation of dose in the multilayer structure determines the doping concentration, as described by equation (8). This concentration in turn effects the enhancement of the oxidation rate. The oxidation rate, then, depends on the initial thickness of polysilicon. Figure 22 shows the simulation of the oxide thickness for four initial thicknesses of polysilicon as a function of initial phosphorus doping. For comparison, the oxide thickness for single crystal silicon is plotted; the linear oxidation rate is normalized to that of polysilicon by equation (9). In all cases, the oxide thickness increases at higher doping concentrations. However, for a given concentration, the average oxide growth rate is greater for thinner polysilicon layers, due to the impurity pile-up effect just discussed. The experimental evidence for this thickness dependence has been reported in references [29,30].

6.3   Oxidation Enhanced Diffusion

It is well known that impurities diffuse faster when the silicon substrate is exposed to an oxidizing ambient [31]. The enhancement of the diffusion constant, known as the oxidation-enhanced diffuson (OED), is related to the oxidation rate. It is believed that the silicon interstitials generated at the silicon/oxide interface are responsible for the enhanced diffusion. It is reasonable to expect that the oxidation of the polysilicon/bulk structure induces

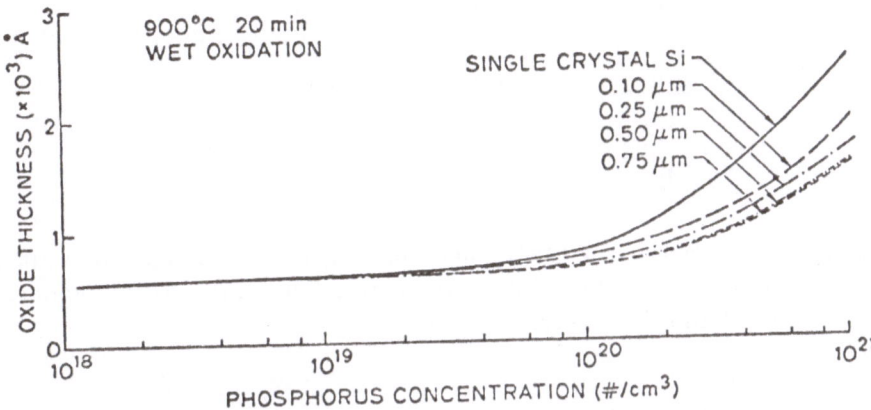

Fig. 22. Results of simulated oxide thicknesses as a function of doping concentration in polysilicon layer with four different polysilicon thicknesses.

the same OED effect. Figure 23 shows a comparative illustration of the OED effect in polysilicon/silicon and single crystal silicon structures. In the polysilicon/silicon structure, the silicon interstitials generated at the oxidizing surface must penetrate through the polysilicon layer into the silicon layer to promote the enhancement effect of diffusion. However, the extent to which the silicon interstitials interact with the defects in the polysilicon layer determines the concentration of silicon interstitials in the bulk silicon. Therefore, consideration of the kinetics of point defects in the polysilicon/bulk structure is essential in predicting bulk junction depths, as in the formation of buried contacts and polysilicon emitters. This section discusses a proposed model for the OED effect in such structures.

In single-crystal silicon, the enhanced diffusivity D_d can be described by

OXIDATION ENHANCED DIFFUSION

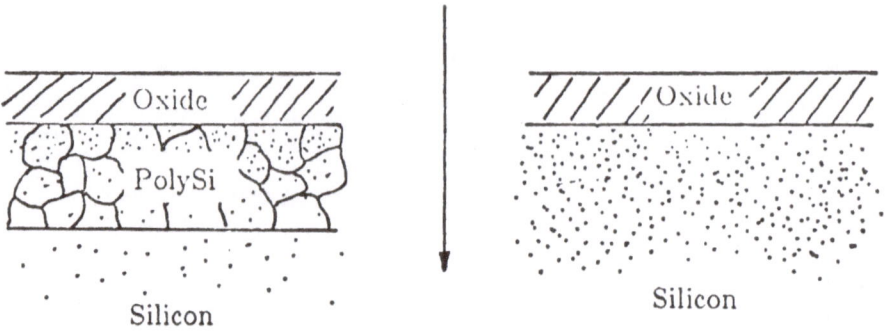

Fig. 23. Comparison of OED effect in poly/bulk silicon and bulk silicon structures. The dots represent silicon interstitials generated at oxidizing interface.

the equation below [31]:

$$D_d = D_0 + A \times \left(\frac{dx}{dt}\right)^{0.4} \tag{12}$$

where D_0, x, and t represent the diffusion constant without oxidation, the oxide thickness, and the oxidation time, respectively, and where A is a fitted process parameter. As a first order approximation, it may be assumed that the rate of interaction between silicon interstitials and grain boundaries is independent of the nature of the grain boundary and temperature. The rate of change of the interstitial concentration $[I]$ along the direction perpendicular to the surface, nz, can now be described by:

$$\frac{d[I]}{[I]} = -B(L) \times dz \tag{13}$$

where $B(L)$ is the characteristic constant of the interstitial-to-grain boundary interaction, which is a function of grain size. It is determined experimentally that a simple function, $B(L) = 1/L$, is sufficient to characterize the interstitial/grain boundary interaction. The interstitial concentration at the polysilicon/silicon interface, $[I(z_p)]$, is therefore given by

$$[I(z_p)] = [I(0)] \exp\left(-B(L)z_p\right) \tag{14}$$

where z_p is the polysilicon thickness. Since the OED effect is directly proportional to the local interstitial concentration, the increment of diffusivity due to OED must be modified by the factor $[I(z)]/[I(0)]$. The diffusivity during oxidation in the bulk silicon substrate covered by a thin layer of polysilicon with thickness z_p then becomes:

$$D_d = D_0 + A \exp\left(-B(L)z_p\right)\left(\frac{dx}{dt}\right)^{0.4} \tag{15}$$

It should be pointed out that as the average grain size L and the thickness of the polysilicon layer change during oxidation the diffusivity changes accordingly. The results of junction depth as a function of processing time both from experiments and simulation are shown in Figure 24. The samples are phosphorus doped (concentration $3.3 \times 10^{21}/cm^3$) polysilicon with thickness 0.31 μm. They are patterned so that some regions are exposed to the oxidizing ambient while others are protected by a 1000 Å nitride layer at the processing temperature 1000°C. Curve A is the result of simulation of the non-oxidizing case. The simulation obeys the characteristic square-root time dependence. Curve B shows the simulation for the oxidized polysilicon layer without taking into account the OED effect. The simulation shows a slight increase in the junction depth compared to that of the non-oxidizing ambient. This is a result of enhanced impurity concentration due to the motion of the oxidizing interface. It is apparent that this effect is not enough to explain the much deeper junction measured. When the OED effect is considered, the result of simulation is illustrated by Curve C, which agrees well with the measured junction depths [32].

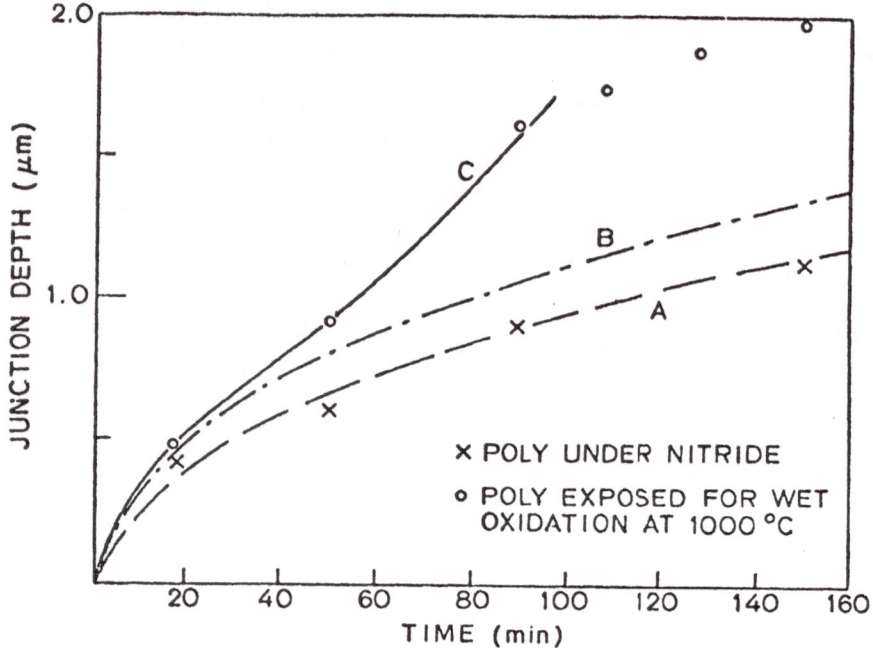

Fig. 24. The depths of junctions formed by the diffusion from doped polysilicon both in inert and oxidizing ambients at 1000°C for different times. The simulated curves B and C for the oxidizing ambient are for those with and without the OED effect.

In this section, it has been demonstrated that the OED effect is indeed present in the polysilicon/bulk silicon structure, and can be characterized by simply considering the interstitial/grain boundary interaction mechanisms.

7. Conclusion

This paper discusses a new process simulation model, which has been implemented into SUPREM. These improvements facilitate the simulation of design processes utilizing polysilicon. Physical models such as grain growth, OED from polysilicon/silicon structures, and oxidation of doped polysilicon have been characterized by experiments. These results agree with a polysilicon model which merges the diffusion/oxidation algorithm with new dopant conservation and multilayer impurity segregation models as needed for polysilicon structures. Within the polysilicon layer, the models for electrical conduction, grain boundary segregation, arsenic clustering, and phosphorus precipitation are integrated together numerically in the time domain. Thus the electrical and metallurgical properties of the polysilicon layer can be determined. A semi-empirical model for laser annealing of polysilicon is also included. It is useful in separating the effects of carrier trapping, impurity segregation, and carrier scattering due to the ion implantation defects.

However, there are several areas which have not been considered to date. For example, the diffusion of impurities through a thin oxide layer from polysilicon remains an area for future work. The kinetics of the phase transformation at high concentrations of impurities in the oxide and the chemical interaction of impurities with the oxide itself must be considered. The grain growth mechanism in polysilicon during oxidation is another area, which may involve quite different kinetics from the grain growth mechanisms without oxidation. Also, for practical purposes, the polysilicon layers are most frequently doped by thermal predeposition processes — a difficult modeling task indeed. A more sophisticated model of conduction in polysilicon including tunneling current is needed to provide an improved simulation of the temperature coefficient of resistivity and the resistivity at lower temperatures.

It is clear that much remains to be considered in the process modeling of polysilicon structures. The model discussed here provides an initial framework and several of the key effects and interactions have been demonstrated using both simulations and experiments. Moreover, it appears hopeful that models will be developed to enhance the basic capabilities described here.

8. Acknowledgements

This work is funded under support from DARPA contract MDA 903-79-C-0257. Initial support from Army Research Office under contract DAAG-29-80-K-0013 is also acknowledged. Helpful discussion and comments from

B. Swaminathan, M. Rivier and Y. Kwark are gratefully acknowledge
Finally, the editorial assistance of D. Ward, J. Rood, and D. Lopez is muc
appreciated.

9. References

1. V. L. Rideout, "One device cells for dynamic RAM: A tutorial," *IEE Trans. Electron Devices*, vol. ED-26, no. 6, p. 837 (1979)

2. II. Yamanaka, T. Wada, O. Kudoh and M. Sakamoto, "Polysilico interconnection technology for IC devices," *J. Electrochem. Soc.*, vc 126, no. 8, p. 1415 (1979)

3. W. Hunter, L. Ephrath, W. Grobman, C. Osburn, B. Crowder, / Cramer, and II. Luhn, "1 μm MOSFET VLSI technology: part V — single-level polysilicon technology using electron-beam lithography *IEEE Trans. Electron Devices*, vol. 26, no. 4, p. 353 (1979)

4. L. Gerzberg, J. Meindl, "Optimization, performance and limitation (monolithic polysilicon distributed RC devices," IEDM Techn. Diges p. 509 (1979)

5. J. Graul, A. Glasl and II. Murrmann, "IIigh performance transisto with arsenic implanted polysil emitters," *IEEE J. Solid-State Circuit* vol. SC-11 , no. 4, p. 491 (1976)

6. K. Tsukamoto, Y. Akasaka, K. Horie, "Arsenic implantation int polycrsytalline silicon and diffusion into silicon substrate," *J. App Phys.*, vol. 48,no. 5, p. 1815 (1979)

7. J. Barnes, J. DeBlasi and B. Deal, "Low temperature differential oxid tion for double polysilicon VLSI devices," *J. Electrochem. Soc.*, , vo 126 no. 10, p. 1779 (1979)

8. D. A. Antoniadis, S. E. Hansen, and R. W. Dutton, "SUPREM II — A Program for IC Process Modeling and Simulation", *Stanford Electronics Laboratory Technical Report No. 5019-2*, Stanford Electronics Lab., June 1978

9. H. Nakashiba, I. Ishida, K. Aomuraand T. Nakamura, "An advanced PSA techbology for high speed bipolar LSI," *IEEE J. Solid-State Circuits*, vol. SC-15, no. 4, p. 455 (1980)

10. L. Mei, B. Swanithan and R. W. Dutton, "Process modeling of multilayer structures involving polysilicon," IEDM Techn. Digest, p. 203 (1980)

11. Y. Tsunoda, "Defects generation at silicon-undoped polysilicon interface in oxidation processes," Jpn. J. Appl. Phys., vol.19 , no. 1, p. 209 (1980)

12. A. Armigliato, M, Servidori, S. Solme and A. Zani, "Role of a polysilicon layer in the reduction of lattice defects associated with phosphorus predeposition in Si," J. of Materials Science, vol. 15, p. 1124 (1980)

13. J. Y. W. Seto, "The electrical properties of polycrystalline silicon," *J. Appl. Phys.*, vol. 46, no. 12, p. 5247 (1975)

14. G. Baccarani, B. Ricco and G. Spadini, "Transport properties of polycrystalline silicon films," *J. Appl. Phys.*, vol. 49, no. 11, p. 5565 (1978)

15. C. Seager and T. Castner, "Zero bias resistance of grain boundaries in neutron transmutation-doped polycrystalline silicon," *J. Appl. Phys.*, vol. 49, no. 7,p. 3879 (1978)

16. M. Mandurah, K. Saraswat and T. Kamins, "A model for conduction in polycrystalline silicon part I: Theory, part II: Comparison of theory and experiment," *IEEE Trans. Electron Devices*, , vol. ED-28, no. 10, p. 1163 and p. 1171 (1981)

17. N. Lu, L. Gerzberg, C. Lu and J. Meindl, "Modeling and optimization of monolithic polysilicon resistors," *IEEE Trans. Electron Devices,* vol. ED-28, no. 7, p. 818 (1981)

18. Y. Wada and S. Nishimatsu, "Resistivity lowering limitations o heavily doped polysilicon," Denki Kagaku, vol. 47, no. 2, p. 118 (1979)

19. M. Mandurah, K. C. Saraswat, C. R. Helms and T. I. Kamins, "Dopant segregation in polysilicon," *J. Appl. Phys.,* vol. 51, no. 11, p.5755 (1981)

20. Y. Wada and S. Nisimatsu, "Grain growth in polycrsytalline silicon," *J. Electrochem. Soc.,* vol. 15, p. 1499 (1978)

21. L. Mei, M. Rivier, Y. Kwart and R. Dutton, "Grain growth mechanism in polycrsytalline silicon," Proc. 4th Internl. Symp. on Silicon Materials and Technol., vol. 81-5, p. 1007 (1981)

22. C. Ho and J. Plummer, "Si/oxide interface oxidation kinetics: a physical model for the influence of high substrate doping levels, part I: Theory , Part II: Comparison with experiment and discussion," *J. Electrochem. Soc.,* vol. 126, no. 9, p. 1519 and p. 1523 (1979)

23. S. Solmi, M. Severi, R. Angelucci, L. Baldi and R. Bilenchi, "Electrical properties of thermally and laser annealed heavily doped with arsenic and phosphorus," Extended Abstract, Oct., (1981)

24. M. Mandurah, K. C. Saraswat and T. Kamins, "Arsenic segregation in polycrystalline silicon," *J. Appl. Phys.,* vol. 36, no. 8, p. 683 (1980)

25. N. Lu, L. Gerzberg, J. Meindl, "A quantitative model of the effect of grain size on the resistivity of polycrystalline silicon," IEEE Electron Devices Lett., vol. EDL-1, no. 1, p. 38 (1980)

26. S. Onga, S. Kohyama, K. Shibata, Y. Nagakubo, II. Iizuk, "Resistivity reduction of polycrystalline silicon films by laser annealing," Proc. 11th Conf. on Solid State Devices, Tokyo, p. 133 (1979)

27. L. Gerzberg, A. Gat, K. Lee, J. Gibbons, J. Peng, T. Magee, V. Deline and C. Evans, "Effect of laser power level in CW laser annealing of polycrystalline silicon," Extended Abstracts, vol. 80-2, Electrochem. Soc., p. 1053 (1980)

28. K. Shimakura, T. Suzuki and Y. Yadoiwa, "Boron and phosphorus diffusion through an SiO_2 layer from a doped polysilicon source under various drive-in ambients," Solid-State Electronics, vol. 18, p. 991 (1975)

29. T. Kamins, "Resistivity of LPCVD polycrystaline silicon films," J. Electrochem. Soc., vol. 126, no. 5, p. 838 (1979)

30. H. Sunami, M. Koyanagi, N. Hashimoto, "Intermediate oxide formation in double-polysilicon gate MOS structure," J. Electrochem. Soc., vol. 127, no. 11, p. 2499 (1980)

31. A. Lin, D. Antoniadis and R. W. Dutton, "The oxidation rate dependence of oxidation-enhanced-diffusion of boron and phosphorus in silicon," J. Electrochem. Soc., vol. 128, no. 5, p. 1131 (1981)

32. B. Swaminathan, L. Mei, A. M. Lin and R. W. Dutton, "Enhanced diffusion in the single crystal silicon substrate during the oxidation of a doped polysilicon doping source," presented in the IEEE Interface Specialist Conf., Ft. Lauderdale, Fl. Dec., (1980)

TWO-DIMENSIONAL PROCESS SIMULATION - SUPRA

Michael Kump and Robert W. Dutton

Integrated Circuits Laboratory
Stanford University
Stanford, California 94305

1 INTRODUCTION

The purpose of this chapter is to discuss two-dimensional process modeling and to show applications in IC technology design. In the previous discussion of one-dimensional process modeling, the emphasis was placed on the kinetics of multilayer structures. Indeed, current technology uses oxide-nitride, polysilicon, and silicide layers frequently. Of particular interest is the interaction of these materials with the bulk and the electrical impact on one dimensional profiles. For example, channel and field threshold voltages, source resistance of buried contacts, and gate sheet resistance are typical design values of concern. Yet in contrast to these one-dimensional parameters, there are many two-dimensional quantities which are integral to device design and require detailed approximations of two-dimensional profiles. For example, source-bulk sidewall capacitance and breakdown are two parameters which depend critically on both 2D technology and device properties.

In this chapter the process modeling of complete technology cross-sections will be considered as the driving force for the discussion. Toward this end, Fig. 1 shows a hypothetical 2μm NMOS process cross-section

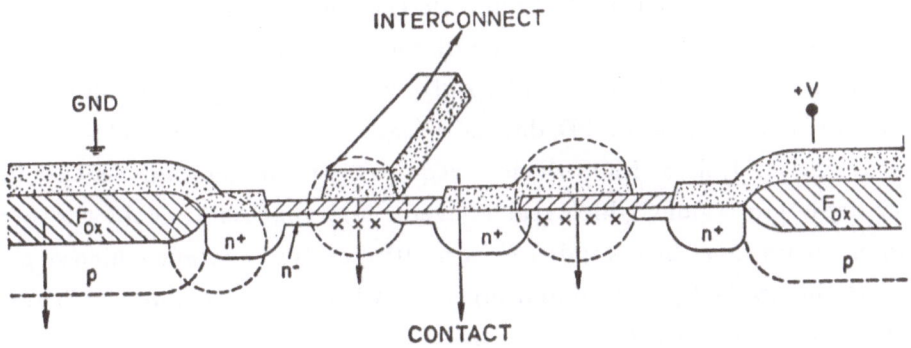

Fig. 1. An NMOS cross-section showing enhancement/depletion devices and LOCOS field oxidation regions.

with enhancement/depletion devices and a LOCOS field isolation. Several features can be identified pertinent to 2D process modeling.

The technology as depicted in Fig. 1 shows a surface topography which is nonplanar due to the local oxidation (LOCOS) of field regions, and multiple functions for the polysilicon interconnection are also shown. Just below the oxide surface, there are junction regions which reflect the coupled effects of oxidation and diffusion. Moreover, the deep n^+ junction results from a multilayer diffusion involving a polysilicon dopant source and the poly-bulk interface. Finally, in and near the channel regions of both enhancement and depletion devices the topology, while reasonably planar, involves design requirements for tight dimensional control of vertical and lateral junctions. Denoted on the figure are a number of one-dimensional arrows as well as a few enclosed regions. While the arrows indicate regions for 1D modeling, the encircled areas require 2D models.

One might infer from the limited number of highlighted 2D sections that only a few regions require detailed analysis. Two key arguments are now presented to dispel this conclusion. First, the lateral extent of 2D process effects is currently a topic of ongoing research [1,2]. At near- and sub-micron device dimensions the extent of point defect kinetics – for example oxidation enhanced diffusion – can extend the diameter of the circled areas by factors of 2-3 times. Second, the objective of process modeling is not for its own sake but as a vehicle for 2D device design. As will become clear shortly, the device analysis of MOS devices requires 2D Poisson analysis extending over depletion regions which reach several microns. As a result, the circled regions again must be extended by a multiplicative constant which depends directly on applied potential and inversely with substrate doping. Examples to be shown later clearly indicate these facts.

With this dual objective of process/device design in mind, let us proceed in considering the approach of this chapter. Because of the device design implications, and the tight geometry coupling of scaled devices, a major factor affecting process modeling is the design specification process itself. Layers must be grown, deposited, and etched many times in the process of achieving single species doping such as the field isolation or source-drain junction regions. The specification of these initial conditions are frequently more critical to the final device outcome than the subsequent ion implant or oxidation steps. Tapered poly gates and trenched regions for LOCOS are examples of highly pattern sensitive inputs. The next section begins by considering two generic cross sections – that of the field isolation and gate regions for MOS devices. As can be seen from Fig. 2(a), both structures involve pattern-sensitive specification for the subsequent ion implantations. However, the objectives of oxidation and diffusion differ radically in the two structures as shown in Fig. 2(b). Here one sees that in the field region it is desirable to grow a thick thermal oxide layer and thereby redistribute the boron by a non-negligible amount. Although novel isolation techniques are under investigation [3], lateral impurity diffusion is a natural consequence. By contrast, the source-drain regions are intended to be shallow to reduce a variety of short-channel effects. Hence, minimum diffusion is desired while surface passivation must simultaneously be maintained.

The final point to be made in Fig. 2(c) is the differences in analysis techniques used to model these generic structures. The LOCOS techniques

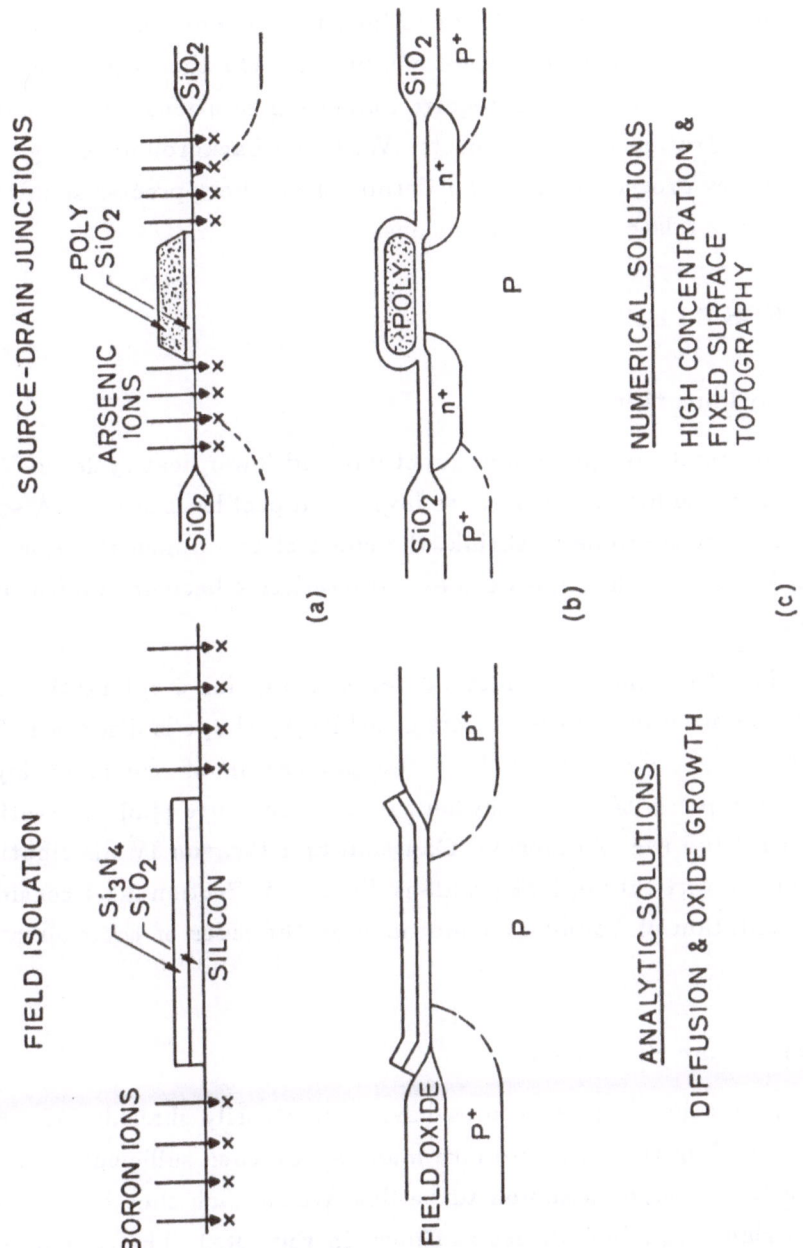

Fig. 2. Generic cross-sections of MOS field isolation and gate regions.

tend not to occur simultaneously with high concentration diffusion. Hence, an analytical technique is exploited in the field region because of its computational efficiency and adaptability to incorporate empirical data for the oxide shape as a function of process conditions. On the other hand, the shallow junction regions indeed require numerical solutions resulting from the high concentration diffusion effects. With the background of Fig. 2 in mind, let us now proceed to consider details of all these process simulation concerns as they relate to two dimensions.

2 PROCESS STEPS

2.1 Ion Implantation

With the trend toward shallow junctions and lower heat cycles in VLSI technologies, the ability to calculate as-implanted profiles is crucial. Also, as lateral dimensions continue to shrink, the effect of two-dimensional profiles near a mask edge on the final device characteristics becomes increasingly more significant.

In section 2.1.1 the basic method for treating ion implantation into structures consisting of multiple layers of arbitrary shape is discussed. The scope of this work does not include the patterning of the mask layers although this topic is treated elsewhere in the literature [4,5]. In sections 2.1.2 and 2.1.3 the use of binormal Gaussian and Pearson IV distributions to represent the vertical implant profile is discussed. Section 2.1.4 considers the implementation of boundary conditions at the sides of the simulation space.

2.1.1 Nonplanar Structures

Ion implantation into structures having arbitrarily shaped surfaces is handled by dividing the structure into many slices, each sufficiently narrow that the surfaces can be assumed to be flat within each slice between the left x_k and right x_{k+1} boundaries as shown in Fig. 3(a). The contribution from implantation within each segment is found, and then all contributions are superposed to find the total implantation profile.

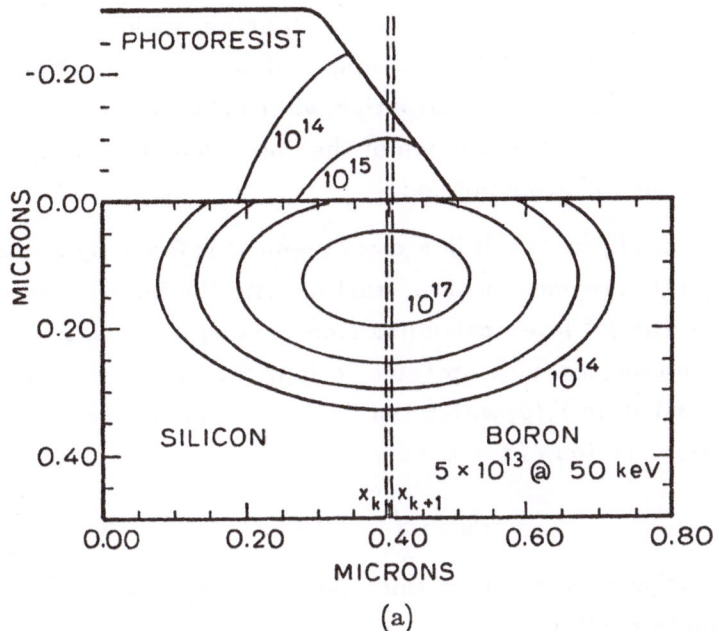

(a)

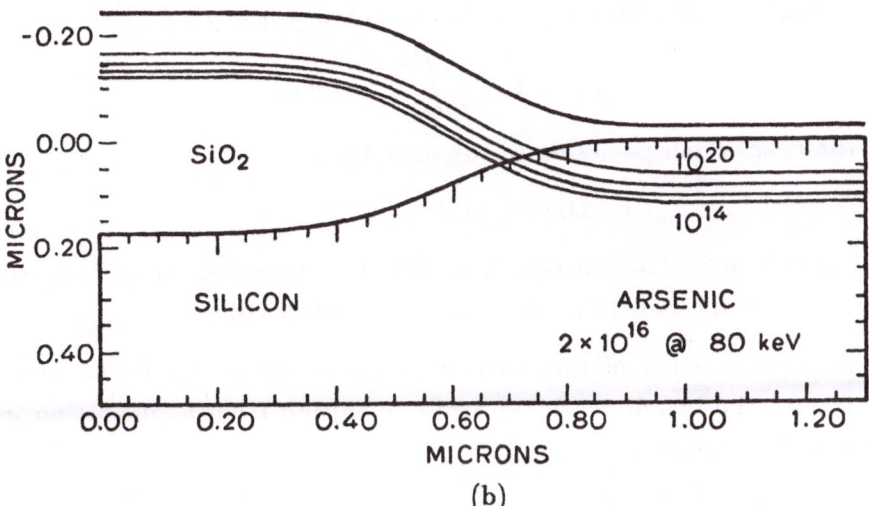

(b)

Fig. 3. (a) The boron implantation profile from a single segment through a sloped photoresist mask edge. (b) The complete arsenic implant profile through a semi-recessed oxide layer.

310

Each segment may contain several layers of material. Depending upon the model used, the vertical distribution function within each layer may be a Gaussian or Pearson IV distribution, as described later. The parameters describing the shape of the distribution will be characteristic of the material composing the layer. The fraction of the dose allocated within each layer is found using the following method.

The vertical profile within a given segment is found by treating each of the layers in that segment in turn, starting with the topmost layer. Consider for example the jth layer, extending from $y = t_j$ down to $y = b_j$. We first use interpolation in a look-up table to find the parameters describing the vertical distribution $Y_j(y)$ within that layer. The area under the distribution function from the top of the layer is

$$A_j = \int_{t_j}^{\infty} Y_j(y)dy. \tag{1}$$

If the impurity dose entering this layer is D_{IN_j}, then the peak of the distribution in cm^{-3} is

$$I_{max_j} = \frac{D_{IN_j}}{A_j}. \tag{2}$$

The impurity dose which is actually deposited in the jth layer is

$$D_j = I_{max_j} \int_{t_j}^{b_j} Y_j(y)dy, \tag{3}$$

and the dose which passes on to the next layer is

$$D_{IN_{j+1}} = D_{IN_j} - D_j. \tag{4}$$

The procedure outlined in equations (1)-(4) is repeated for each successive layer until all the implanted dose has been allocated.

A generalization of the method proposed by Runge [6] is then used to estimate the two-dimensional profile resulting from implantation in the segment. The result is

$$I_{jk}(x, y) = \frac{I_{max_j}}{2} Y_j(y)$$
$$\cdot \left[\text{erfc}\left(\frac{x - x_{k+1}}{\sqrt{2}\Delta x_j}\right) - \text{erfc}\left(\frac{x - x_k}{\sqrt{2}\Delta x_j}\right) \right] \quad t_j \le y \le b_j \tag{5}$$

where

$$\mathrm{erfc}(x) = \frac{2}{\sqrt{\pi}} \int_x^\infty \exp(-u^2)du. \qquad (6)$$

Δx_j is the lateral standard deviation characteristic of the material within the jth layer, and x_k and x_{k+1} are the left and right boundaries of the segment, respectively. The total implanted profile is the superposition of the contribution from each layer of each segment given by

$$I(x,y) = \sum_k \sum_j I_{jk}(x,y). \qquad (7)$$

Fig. 3(a) is an example of boron implantation into silicon using a photoresist mask with a sloped edge. Concentration contours resulting from implantation through a single mask segment are plotted both in the photoresist and in the silicon. Fig. 3(b) shows the complete profile for an arsenic implantation through a semi-recessed oxide which was calculated using this segmentation and superposition technique. This profile is typical of the arsenic implantation used to form the source-drain regions indicated in Figures 1 and 2.

2.1.2 Gaussian Distributions

The simplest description of the vertical distribution of an implanted impurity is the Gaussian distribution

$$Y(y) = \exp\left[\frac{-(y - R_p)^2}{2\Delta R_p^2}\right] \qquad (8)$$

where R_p and ΔR_p are the projected range and vertical standard deviation, respectively. However, experimental distributions of some ions, such as boron and arsenic, are found to be asymmetric even in amorphous targets. For cases where the asymmetry is not excessive, Gibbons and Mylroie [7] have shown that the use of the third central moment in addition to R_p and ΔR_p is enough to construct accurate distributions. In these cases, the distribution can be represented by two half-Gaussian profiles, joined together at a modal range R_m. The distribution is given by

$$
Y(y) = \begin{cases} \exp\left[\dfrac{-(y-R_m)^2}{2\Delta R_{p1}{}^2}\right], & y \leq R_m \\[3mm] \exp\left[\dfrac{-(y-R_m)^2}{2\Delta R_{p2}{}^2}\right], & R_m \leq y \end{cases}
\tag{9}
$$

where the modal range R_m is given in terms of the projected range R_p and the shallow and deep vertical standard deviations ΔR_{p1} and ΔR_{p2} by

$$
R_m = R_p - 0.8(\Delta R_{p2} - \Delta R_{p1}).
\tag{10}
$$

This distribution is suitable for phosphorus and arsenic, but boron is more accurately represented by a Pearson type IV distribution described in the next section.

2.1.3 Pearson IV Distribution

Hofker *et al.* [8] found that Pearson IV distributions are especially well suited to describe implanted boron profiles in silicon. Ryssel *et al.* [9] found that this is also true for boron in SiO_2 and Si_3N_4 . The Pearson type IV distribution is given by

$$Y(y) = K \left[b_0 + b_1(y - R_p) + b_2(y - R_p)^2 \right]^{1/(2b_2)}$$

$$\cdot \exp \left[-\frac{(b1/b2) + 2a}{\sqrt{4b_2b_0 - b_1^2}} \arctan \frac{2b_2(y - R_p) + b_1}{\sqrt{4b_2b_0 - b_1^2}} \right] \tag{11a}$$

where the third and fourth moment ratios are

$$\mu_i = \int_{-\infty}^{\infty} (y - R_p)^i Y(y) dy, \qquad i = 3, 4 \tag{11b}$$

$$\gamma = \frac{\mu_3}{\Delta R_p^3} \tag{11c}$$

$$\beta = \frac{\mu_4}{\Delta R_p^4} \tag{11d}$$

and

$$a = -\frac{\Delta R_p \gamma (\beta + 3)}{A} \tag{11e}$$

$$b_0 = -\frac{\Delta R_p^2 (4\beta - 3\gamma^2)}{A} \tag{11f}$$

$$b_1 = a \tag{11g}$$

$$b_2 = -\frac{2\beta - 3\gamma^2 - 6}{A} \tag{11h}$$

$$A = 10\beta - 12\gamma^2 - 18 \tag{11i}$$

$$K = b_0^{-1/(2b_2)}. \tag{11j}$$

Data for the third and fourth moment ratios γ and β come from the work of Ryssel et al. [9]. The ratios were found by fitting the Pearson IV distribution to measured boron profiles. These empirically determined moments agreed well with the theoretical values for boron in SiO_2 and Si_3N_4 ; however, the third moment μ_3 for boron in silicon differs substantially from the theoretical value. It was suggested that this discrepancy may be due to residual channeling. The empirically determined fourth moment ratio is [9]

$$\beta = 1.5 \left(\frac{48 + 39\gamma^2 + 6(\gamma^2 + 4)^{3/2}}{32 - \gamma^2} \right). \tag{12}$$

2.1.4 Reflection Symmetry

For most applications it is convenient to impose reflection symmetry at the left and right edges of the simulation space. This boundary condition is appropriate if the edge of the region to be simulated occurs at the middle of a mask opening or is far enough from a mask edge that the structure is no longer changing in the horizontal direction.

This symmetry boundary condition can be imposed by assuming the existence of fictitious (or image) sources of implanted ions located at a symmetric position on the other side of both the left and right boundaries. Thus, if the function describing the contribution of implanted ions from the jth layer between x_k and x_{k+1} is

$$I_{jk}(x, y) = \frac{I_{max_j}}{2} Y_j(y) \left[\text{erfc} \left(\frac{x - x_{k+1}}{\sqrt{2}\Delta x_j} \right) - \text{erfc} \left(\frac{x - x_k}{\sqrt{2}\Delta x_j} \right) \right], \quad (13)$$

then the contribution from the image source across the left boundary at $x = 0$ is

$$I_{jk}(x, y) = \frac{I_{max_j}}{2} Y_j(y) \left[\text{erfc} \left(\frac{x + x_{k+1}}{\sqrt{2}\Delta x_j} \right) - \text{erfc} \left(\frac{x + x_k}{\sqrt{2}\Delta x_j} \right) \right], \quad (14)$$

and the contribution from the image source across the right boundary at $x = L$ is

$$I_{jk}(x, y) = \frac{I_{max_j}}{2} Y_j(y) \left[\text{erfc} \left(\frac{x + L - x_{k+1}}{\sqrt{2}\Delta x_j} \right) - \text{erfc} \left(\frac{x + L - x_k}{\sqrt{2}\Delta x_j} \right) \right]. \quad (15)$$

Image sources farther than several times the lateral straggle Δx_j on the other side of a boundary make an insignificant contribution to the implant profile within the simulation region and need not be considered.

2.2 Low Concentration Diffusion

If the impurity diffusion coefficient is constant during high temperature processes, as with low concentration boron diffusion, the diffusion equation $\frac{\partial N}{\partial t} = \vec{\nabla} \cdot \left(D \vec{\nabla} N \right)$ becomes linear. In this case, a closed form analytic solution to the diffusion equation can be used. The analytic solution assumes planar boundaries, however structures with nonplanar boundaries can be treated by applying the same segmentation and superposition method used

for implantation. The solution for an inert drive-in is considered in Section
2.2.1. When oxidation occurs along with the impurity diffusion, an additional flux should be considered at the SiO_2 interface to account for moving
boundary and segregation effects, as will be discussed in Section 2.2.2.

2.2.1 Inert Drive-In

The solution for a Gaussian implant within the segment x_k to x_{k+1}
followed by an inert drive-in is given by [10]

$$N_i(x, y, t) = N_{ix}(x, t) N_{iy}(y, t) \tag{16a}$$

$$N_{ix}(x, t) = \frac{1}{2} \left\{ \text{erfc} \left(\frac{x - x_{k+1}}{\sqrt{2\Delta x^2 + 4Dt}} \right) - \text{erfc} \left(\frac{x - x_k}{\sqrt{2\Delta x^2 + 4Dt}} \right) \right\} \tag{16b}$$

$$N_{iy}(y, t) = \frac{I_{max}\Delta R_p}{2\sqrt{2Dt + \Delta R_p^2}} \{\Omega(y, t) + \Omega(-y, t)\} \tag{16c}$$

$$\Omega(y, t) = \exp\left[\frac{-(y - R_p - d_k)^2}{4Dt + 2\Delta R_p^2} \right]$$
$$\cdot \left\{ 2 - \text{erfc} \left(\frac{y \Delta R_p^2 + 2(R_p - d_k)Dt}{\Delta R_p \sqrt{2Dt(4Dt + 2\Delta R_p^2)}} \right) \right\} \tag{16d}$$

where the subscript i refers to inert ambient conditions. D is the impurity
diffusivity, t is the time of the diffusion step, and d_k is the effective silicon
thickness of the layers above the substrate. The SiO_2 interface is assumed
to be a reflecting boundary, and diffusion in the oxide layer is ignored during
the analytic calculations. The profile resulting from an inert drive-in of
the implant through one mask segment is shown in Fig. 4 (broken lines).
Once again, the complete drive-in profile is constructed by superposing the
solution for each mask segment.

2.2.2 Moving Boundary Diffusion

Huang and Welliver [11] approximated the one-dimensional profile following oxidation by adding a correction factor to the inert drive-in concentrations:

$$N_o(y, t) = N_i(y, t) + N_c(y, t) \tag{17}$$

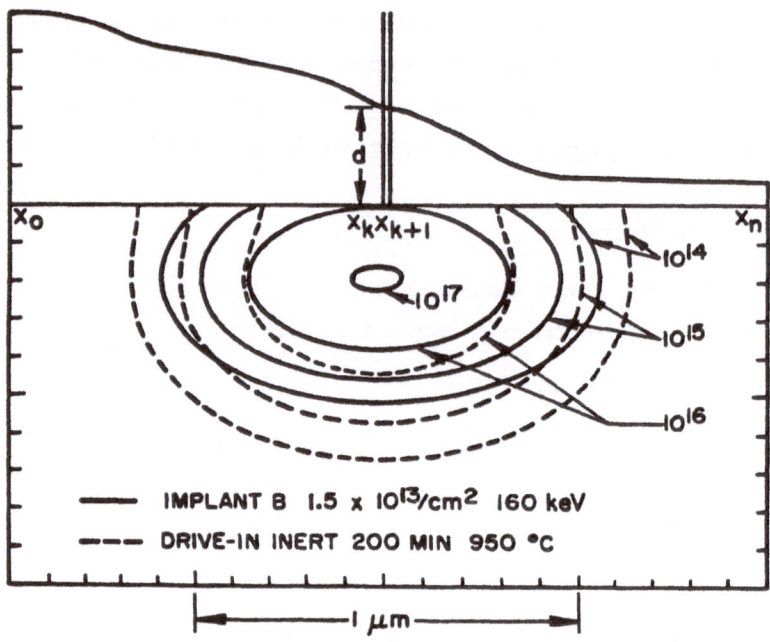

Fig. 4. The implantation (solid line) and inert drive-in (broken line) profiles through a single mask segment.

where the correction factor $N_c(y, t)$ was expressed by a term involving a complementary error function. In this approach the correction function basically subtracts off an appropriate fraction of dopant to account for segregation and out diffusion into the oxide. A boundary condition which properly determines the correction term has been found, and an extensive discussion of this approach for the one-dimensional case and comparison of analytical and numerical results have been presented elsewhere [10]. For low concentration diffusion this analytic method gives agreement typically better than 10% with numerical calculations [12], and it has a computation-time advantage of more than a factor of ten.

A quasi two-dimensional profile has been obtained by assuming that the oxide layer grows only in the vertical direction during a semi-recessed oxidation [10]. However, the effect of lateral oxide growth may not be negligible in a case such as fully recessed oxidation. Therefore the following generalized boundary condition is proposed in this work, which can account for not only vertical but also lateral oxidation:

$$D\vec{\nabla}N_o(x,y,t)\Big|_{x_f,y_f} = \left(\frac{1}{m} - \alpha\right)N_o(x_f,y_f,t)\,g(x_f,y_f,t)\,\hat{n}(x_f,y_f) \qquad (18)$$

where the coefficient m is the equilibrium segregation factor and α is the volumetric ratio of silicon consumed in forming one unit of oxide (0.44). The quantity $g(x_f,y_f,t)$ is the oxide growth rate, $\hat{n}(x_f,y_f)$ is a unit vector normal to the SiO_2 interface, and (x_f,y_f) is the final interface point closest to (x,y). Provided that $N_o(x,y,t)$ can be separated in the x and y variables, the above boundary condition is also separable in the vertical and lateral directions. Therefore the vertical correction technique can also be applied in the lateral direction. In practical processes, inert drive-in steps without a moving boundary must be considered in addition to oxidation steps. The present correction method allows such combined high-temperature steps as well as lateral oxidation. The resulting 2D profile is given by

$$N_o(x,y,t) = N_{ox}(x,t)N_{oy}(y,t) \qquad (19)$$

$$N_{ox}(x,t) = N_{ix}(x,t)\left\{1 + A_x\mathrm{erfc}\left(\frac{x - x_f}{2\sqrt{Dt_{ox}}}\right)\right\} \qquad (20a)$$

$$N_{oy}(y,t) = N_{iy}(y,t)\left\{1 + A_y\mathrm{erfc}\left(\frac{y - y_f}{2\sqrt{Dt_{ox}}}\right)\right\} \qquad (20b)$$

where the subscripts o and i indicate oxidizing and inert conditions, respectively. The Dt terms implicit in N_{ix} and N_{iy} represent the diffusivity-time product for all high temperature steps. The correction terms involve only the moving boundary and segregation effects, hence Dt_{ox} represents the diffusivity-time product for oxidation steps only.

The coefficients A_x and A_y are obtained from (20) and the boundary condition (18):

$$A_z = -\frac{((1/m) - \alpha)g_z N_{iz}(z_f,t) - D\frac{\partial N_{iz}(z,t)}{\partial z}\Big|_{z=z_f}}{(((1/m) - \alpha)g_z + (D/\sqrt{\pi Dt_{ox}}))N_{iz}(z_f,t) - D\frac{\partial N_{iz}(z,t)}{\partial z}\Big|_{z=z_f}} \qquad (21)$$

where the variable z represents either the x or y variable. Again g_z is the oxide growth rate in either the vertical or lateral direction.

When oxidation occurs across the entire surface of the simulation, the shape of the oxide is determined by applying the Deal-Grove relationship [13] at every horizontal location. When a local oxidation is performed by masking a portion of the surface by nitride, the oxide shape is approximated by an empirically determined shape function [14]. Work is underway to model the oxide shape more accurately by simulating the diffusion of the oxidizing species along with the growth and viscous flow of the oxide film [15].

2.3 High Concentration Diffusion

If the impurity concentration is sufficiently large compared to the intrinsic electron concentration at the diffusion temperature, then diffusivity will no longer be constant throughout the simulation space. The diffusion equation then becomes nonlinear and must be solved numerically. Also, there may be several impurities present simultaneously which can interact by way of electric field or Fermi level effects, and impurity clustering can become important.

Section 2.3.1 discusses the general formulation of the diffusion equation which accounts for these high concentration effects. Sections 2.3.2 and 2.3.3 describe the clustering model and the phosphorus diffusion models, respectively. Finally, the numerical solution of the nonlinear diffusion equation is considered in section 2.3.4.

2.3.1 Diffusion Equation Formulation

The diffusion equation for each impurity species is [16]

$$\frac{\partial C}{\partial t} = -\vec{\nabla} \cdot \vec{\jmath} \tag{22}$$

$$\mathbf{J} = -D\vec{\nabla}N \mp \frac{q}{kT}(DN\vec{\nabla}\phi). \tag{23}$$

where the sign of the field term is negative for donors and positive for acceptors. C and N are the total and the active dopant concentrations, respectively. The electrostatic potential ϕ is given by

$$\phi = \frac{kT}{q} \ln(\frac{n}{n_i}) \tag{24}$$

where n and n_i are the electron and intrinsic electron concentrations respectively. The electron concentration n is found by assuming that the sample remains quasi-neutral throughout the diffusion and that any charge due to charged vacancies or impurity clusters is negligible. Then [17]

$$n = \frac{U + \sqrt{U^2 + 4n_i^2}}{2} \tag{25}$$

where U is the algebraic sum of the electrically active concentration of each impurity species present given by

$$U = \sum_{donors} N_j - \sum_{acceptors} N_j. \tag{26}$$

Combining equations (23) - (25) we can write the diffusion equation solely in terms of impurity concentrations as

$$\mathbf{J} = -D\vec{\nabla}N \mp \frac{DN}{\sqrt{U^2 + 4n_i^2}} \vec{\nabla}U \tag{27}$$

where the minus sign applies for donors and the plus sign for acceptors. Assuming there is a differentiable relation between the electrically active concentration N and the total chemical concentration C such as that described in the next section, the expression for flux can be written as

$$\mathbf{J} = -D^*\vec{\nabla}C \mp \frac{DN}{\sqrt{U^2 + 4n_i^2}} \vec{\nabla}U \tag{28}$$

where

$$D^* \equiv D\frac{dN}{dC}. \tag{29}$$

D^* is often referred to as the effective diffusivity and will in general be less than the true diffusivity D since only a portion of the total impurity concentration is mobile at high concentrations due to clustering. Equation (28) is the form of the flux expression that is most convenient for numerical implementation, as will be shown in section 2.3.4.

It is assumed that dopants diffuse primarily via interaction with vacancies in various charge states. The concentration-dependent diffusion constants are given by [18]

$$D = D^+\left(\frac{n_i}{n}\right) + D^x + D^-\left(\frac{n}{n_i}\right) + D^=\left(\frac{n}{n_i}\right)^2 \tag{30}$$

where D^+, D^x, D^-, and $D^=$ are the diffusivities of the dopant ion when it is associated with a positive, neutral, negative, or doubly negative vacancy, respectively.

2.3.2 Clustering

At high concentrations, inactive complexes of impurity ions are formed. These clusters are assumed to be immobile and electrically inactive. There is ongoing discussion concerning the number of ions in a cluster, the charge state of the clusters, and the nature of other particles that may be associated with clustered ions such as vacancies and electrons [19–25]. Regardless of the exact clustering mechanism, the results of most clustering studies can be fit with the relation

$$C = N + rK(T)N^r \tag{31}$$

between the total (C) and active (N) impurity concentrations. In this expression r is the number of atoms in a cluster and $K(T)$ is the temperature-dependent equilibrium constant.

The clustering data used in the program have been gathered from a number of different sources. The results for arsenic indicate a cluster size of between 2 and 4 [19–21,23,25]. For phosphorus, the data of Fair and Tsai [22] can be fit with a temperature-independent equilibrium constant and a cluster size of three. For boron the only data available is that of Ryssel et al. [24]. Here only the active boron concentration limit is given, so a cluster size and equilibrium constant have been picked arbitrarily to reproduce the observed active concentration limit.

2.3.3 Phosphorus

The models described previously for concentration-dependent diffusivity and clustering adequately predict the diffusion profiles for boron and arsenic. For phosphorus, however, they are not sufficient to predict the kink formation [26] and "base push" effect [27] commonly observed during high concentration phosphorus diffusions. A model for the diffusive migration of phosphorus has been presented by Fair and Tsai [22]. According to this model the physical explanation of these "anomalous" effects lies in the enhancement of vacancy concentrations in the silicon caused by the dissociation of the phosphorus-doubly ionized vacancy pairs that flow from the region of high phosphorus concentration. A typical high concentration phosphorus profile is composed of three regions:

(I)   The high concentration region in which phosphorus diffusion is assumed to take place by interaction with neutral and with doubly negative vacancies. In this region the diffusivity is given by

$$D_{high} = D^x + D^= \left(\frac{n}{n_i}\right)^2 \qquad (32)$$

where D^x, $D^=$, n, and n_i are as in section 2.3.1. The end of the high concentration region is assumed to lie at the point where the Fermi level drops below .11 eV from the conduction band, which is identified as the energy level for doubly negative vacancies. This occurs at a characteristic electron concentration n_e given by

$$n_e = 4.65 \times 10^{21} \exp\left(-\frac{0.39\text{eV}}{kT}\right). \qquad (33)$$

(II)   The intermediate region in which it has been observed by means of Boltzmann-Matano analysis [22] that the phosphorus diffusivity is proportional to $(n_i/n)^2$. This empirical observation is used to determine the diffusivity in the intermediate region and to locate the beginning of the tail region. The diffusivity calculated at the edge of the high concentration region, at the point where $n = n_e$, is multiplied by $(n_i/n)^2$ as the distance from the high concetration region increases, until the point where the value D_{tail} is reached. At that point, the tail region is assumed to begin.

(III)   The tail region where the diffusivity is enhanced relative to the intrinsic value due to supersaturation of the silicon lattice by vacancies resulting from the phosphorus-double negative vacancy pair dissociation. In this region the diffusivity is given by

$$D_{tail} = D^x + D^- \left[\frac{N_{peak}^3}{n_i n_e^2} \right] \left[1 + \exp \left(\frac{0.3\text{eV}}{kT} \right) \right] \qquad (34)$$

where N_{peak} is the peak electrically active phosphorus concentration occuring within the silicon substrate. Since the model predicts that there exists a vacancy supersaturation in the bulk, any other impurity elements present in the tail region of the phosphorus profile should also experience a diffusivity enhancement equal to that of phosphorus. Thus, the maximum diffusivity enhancement factor is given by

$$E_{max} = \frac{D_{tail}}{D^x} \qquad (35)$$

where D_{tail} and D^x are the phosphorus tail region and neutral vacancy diffusivities respectively. The vacancy supersaturation is assumed to decay exponentially from the point of peak phosphorus concentration. Thus the enhancement factor as a function of position is

$$E(x, y) = E_{max} \exp \left(\frac{|x - x_{peak}|}{L_x} \right) \exp \left(\frac{|y - y_{peak}|}{L_y} \right). \qquad (36)$$

L_x and L_y are the horizontal and vertical vacancy diffusion lengths and x_{peak} and y_{peak} are the horizontal and vertical locations of the peak phosphorus concentration, respectively. $E(x, y)$ multiplies the diffusivity of any other impurity in the tail region of the phosphorus profile.

The modeling of phosphorus diffusion phenomena is indeed complex. Fahey [28] has recently pointed out that values of electron mobility used to extract diffusivity can alter substantially the coefficients used in equations (32)-(35). In fact, there is evidence that the role of electrical activity and clustering phenomena involving phosphorus may differ from those for arsenic and thereby provide alternative models to the one discussed here. Finally, the two-dimensional spatial dependence reflects still another unknown for 2D modeling. Innovation and novel experimental techniques are clearly needed to resolve the detailed kinetics of 2D diffusion – especially for phosphorus.

2.3.4 Finite Difference Method

The numerical solution of the diffusion equation is formulated by approximating the continuous concentration profile with its values on a network of nodes within the device boundaries. The grid structure chosen for this work is rectangular with nonuniform spacing in both spatial directions. A typical rectangular grid is illustrated in Fig. 5, where the nodes occur at the intersections of horizontal and vertical lines.

Fig. 6 shows a node surrounded by its four nearest neighbors and defines the locations and concentrations associated with these nodes. The finite difference approximation at the center node is derived by integrating the diffusion equation (22&28) over S [29]. In analogy to the method used by Greenfield [30], the resulting finite difference equation is

$$\frac{\partial C_0}{\partial t} = \sum_{m=1}^{4} [B_m(C_m - C_0) \pm E_m(U_m - U_0)] \tag{37}$$

$$B_m = \frac{L_m D_m^*}{h_m A} \tag{38}$$

$$E_m = \frac{L_m D_m N_m}{h_m A \sqrt{U_m^2 + 4n_i^2}} \tag{39}$$

$$L_m = \int_{l_m} dl \tag{40}$$

$$A = \sum_{m=1}^{4} A_m \tag{41}$$

where the plus sign in (37) applies for donors and the minus sign for acceptors. A_m is the area of S_m when the center node does not lie on the boundary of a layer. When the center node does lie on a boundary, L_m and A_m are modified to account for the possibly nonplanar boundaries [30].

For approximating the time derivative $\partial C_0/\partial t$, it has been found that a three-level time discretization scheme is advantageous for treating nonlinear parabolic equations such as the one considered here [31]. We thus make the approximation

$$\frac{\partial C_0}{\partial t} \approx \frac{C_0^{n+1} - C_0^{n-1}}{k} \tag{42}$$

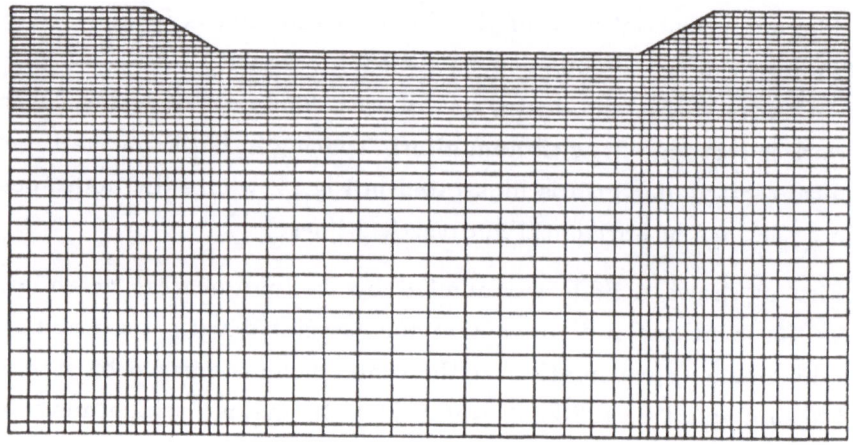

Fig. 5. A typical nonuniform rectangular grid used to discretize the diffusion equation. The nodes occur at the intersections of the horizontal and vertical lines.

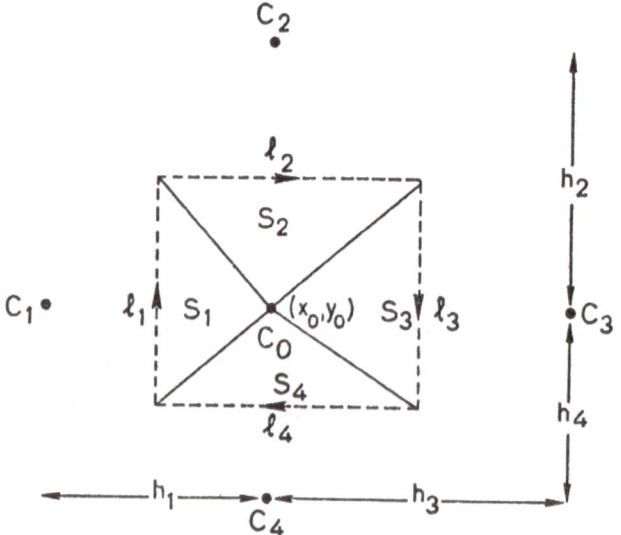

Fig. 6. A node and its four nearest neighbors with their associated locations and concentration values. The l_m are perpendicular bisectors of the lines joining the center node to the outside nodes and define a rectangle S with triangular subregions S_m.

where k is the value of the time step and the superscripts $n + 1$ and $n - 1$ denote values at the future and previous time levels, respectively. Evaluating the spatial differences at the current time level n and solving for the concentration at the future time level, we have

$$C_0^{n+1} = C_0^{n-1} + k \sum_{m=1}^{4} [B_m^n(C_m^n - C_0^n) \pm E_m^n(U_m^n - U_0^n)]. \tag{43}$$

This expression is called the "leapfrog" method and could be used for calculating concentrations at the future time level in terms of known concentrations. This method, however, is unconditionally unstable when applied to the diffusion equation [31]. That is, repeated application of the above formula results in unbounded growth of truncation and round-off errors.

A modified formula proposed by DuFort and Frankel [32] is found by defining

$$R_m^n = U_m^n \mp N_m^n, \qquad 0 \leq m \leq 4 \tag{44}$$

where the minus sign applies for donors and the plus sign for acceptors, and by then replacing C_0^n and N_0^n by their time averages

$$C_0^n \approx \frac{C_0^{n+1} + C_0^{n-1}}{2} \tag{45a}$$

$$N_0^n \approx \frac{N_0^{n+1} + N_0^{n-1}}{2}. \tag{45b}$$

Substituting (44) and (45) into (43) and once again solving for the concentration at the future time level, we have

$$C_0^{n+1} = \frac{(1 - F_1^n)C_0^{n-1} + \sum_{m=1}^{4} [F_2^n C_m^n \pm F_3^n(R_m^n - R_0^n)]}{(1 + F_1^n)} \tag{46a}$$

$$F_1^n = k \sum_{m=1}^{4} \frac{L_m D_m^{*n}}{2h_m A} \left(1 + \frac{N_m^n}{\sqrt{U_m^{n^2} + 4n_i^2}}\right) \tag{46b}$$

$$F_2^n = k \frac{L_m D_m^{*n}}{h_m A} \left(1 + \frac{N_m^n}{\sqrt{U_m^{n^2} + 4n_i^2}}\right) \tag{46c}$$

$$F_3^n = k \sum_{m=1}^{4} \frac{L_m D_m^n}{h_m A} \frac{N_m^n}{\sqrt{U_m^{n^2} + 4n_i^2}}. \tag{46d}$$

By contrast to the leapfrog method, this method is unconditionally stable for the linear diffusion equation [31]. That is, truncation and round-off errors will remain bounded independent of the grid and time step sizes. Although unconditional stability is not guaranteed when this method is applied to the nonlinear diffusion equation, this approach has proved to be stable for all grid sizes and time steps which yield acceptable accuracy.

3 APPLICATIONS

To show the use of the program as a practical design tool, several examples of its use are now given. The first two examples (sections 3.1 and 3.2) consider some of the parasitic circuit elements including sidewall junction capacitance, avalanche breakdown, and field threshold voltage. The final example (section 3.3) illustrates the generality of the program by using it to extract polysilicon grain boundary diffusion constants.

3.1 Breakdown and Capacitance Test Structures

3.1.1 Fabrication

In order to verify the process simulation results, the test structures shown in Fig. 7 have been fabricated to study avalanche junction breakdown and perimeter junction capacitance. These structures correspond to the p-n^+ junction edge indicated in Fig. 1. The device for measuring breakdown voltage (marked B) is a circular structure designed to avoid corner break-down. Capacitance structures with various finger widths (marked C) are used to vary the ratio between perimeter capacitance and area capacitance—finger widths range from 3 to 20 microns. The area and perimeter junction capacitances are obtained from the total measured capacitance by solving the following equation as a function of finger width:

$$C_T = AC_a + PC_{per} \tag{47}$$

where C_T, C_a, and C_{per} are total, area, and perimeter capacitances, respectively. A and P are the area and perimeter as designed on the mask.

Fig. 7. A picture of the test structure. Breakdown (B) and capacitance (C) devices are indicated.

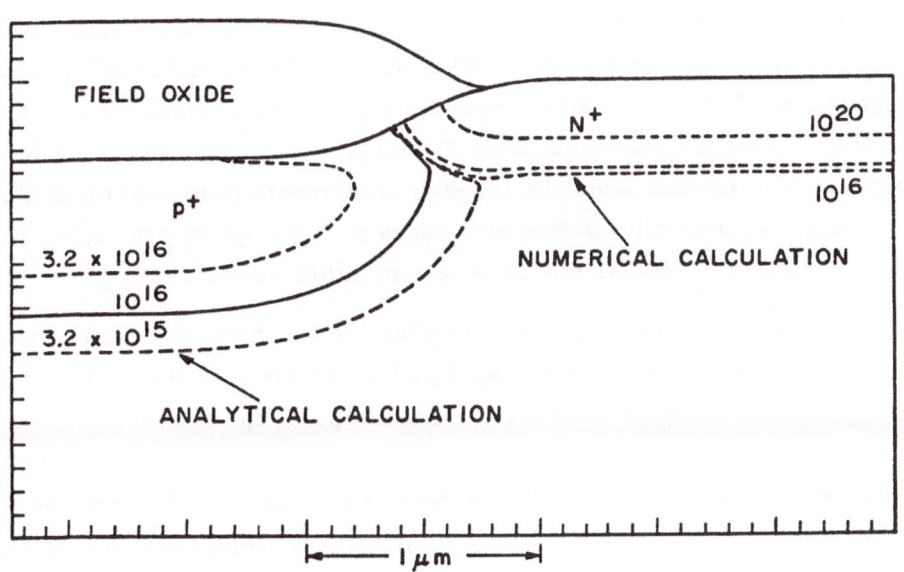

Fig. 8. A simulated cross section of the test structure used for the study of breakdown and capacitance.

The process schedule utilized is described briefly as follows. A gate oxide (740Å) is grown on a lightly doped <100> p-type substrate (10Ω · cm). Boron is implanted into the channel-stop regions using a photoresist mask to protect the active device regions. A semi-recessed oxide (0.8μm) is grown using a nitride layer (800Å) to mask the oxidation. After polysilicon gate definition, arsenic is implanted into the source/drain regions (dose $6.0x10^{15}/cm^2$ at 100 keV) and the structure is annealed for 30 min at 1000°C.

Boron field implant dose and local oxidation time are two process parameters which affect perimeter capacitance and breakdown voltage. To study this sensitivity, the field implantation dose was varied over the range of $0.5x10^{13}$ to $5.5x10^{13}/cm^2$ with an energy of 160 keV. The local oxidation time was varied from 2.5 to 3.5 hours at 1000°C. These values were chosen as a practical spread about a standard NMOS process.

3.1.2 Simulation

As described previously, the process simulation program consists of two main parts—an anlaytical solution which provides a reference frame and numerical calculations for solving the nonlinear diffusion equation. Analytical solutions are used to model changes in the overall device structure (i.e. deposition/etch and oxide growth), ion implantation, and boron diffusion. The process sequence through the arsenic implantation is simulated using the analytic solution subprogram. Subsequent processing steps are then simulated numerically by a second subprogram.

The physical regions of primary application for the analytic and numerical subprograms are denoted in Fig. 8, which shows a section of the structure considered in the present work. The shape of the semi-recessed oxide is obtained by an empirical formulation [14]. The boron profile beneath the bird's beak is obtained from the analytic model described in Section 2.2. The arsenic profile has been calculated numerically using the concentration-dependent diffusion model outlined in Section 2.3. The typical computer execution time of the analytic part with 50x50 points is about 1 minute using a Hewlett-Packard minicomputer (HP 1000F) for each process step.

The numerical calculation using the same grid takes about 10 minutes for the arsenic profile calculation.

The composite 2D impurity profile ouput is used as input to a 2D device analysis program [30]. The device analysis imposes a serious limit on computation speed since the total structure must be considered numerically. The analytic part of the process simulator requires no grid points, while the numerical process calculation is applied only to the final arsenic annealing step. Thus in choosing grid for device analysis, the analytical process modeling poses no limitation since doping values at all spatial points can easily be computed by the given analytic expressions. However the device analysis grid poses greater constraints on carrier distributions than does the grid used for the numerical solution of the impurity diffusion. Hence it is most convenient to choose the device analysis grid as the reference frame for the numerical process simulation. Moreover this has the advantage that no interpolation is required in transferring the 2D impurity concentrations.

3.1.3 Experiment and Simulation Results

The breakdown voltage under the locally oxidized region is governed by the integrated net charge between the p and n^+ depletion edges. Electric field lines and the depletion region edge of the device in Fig. 8 biased near its breakdown voltage are plotted in Fig. 9. The maximum impurity concentration gradient and hence the first point of breakdown occurs along the E field line between points A and B. A one-dimensional plot of the impurity distribution along this path is shown in Fig. 10. The profile shape is nearly linear at the metallugical junction but slightly concave further into the boron field implant region.

One-dimensional theory predicts that breakdown voltage varies as $N_B^{-\frac{3}{4}}$ for an abrupt junction and as $N_B^{-\frac{2}{5}}$ for a linear-graded junction [33], where N_B is the background concentration. Since the impurity concentration in the field region is directly proportional to the field implantation dose, an analogous power dependence of breakdown voltage on dose is expected. The results of both experiment and simulation in Fig. 11 show a power

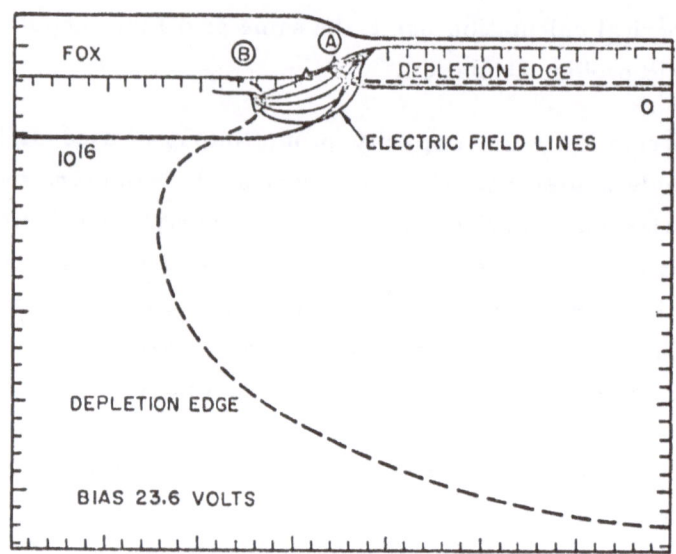

Fig. 9. The electric field lines and depletion region edge of the test structure biased at breakdown voltage. The marks (▲) show the two
points between which breakdown occurred.

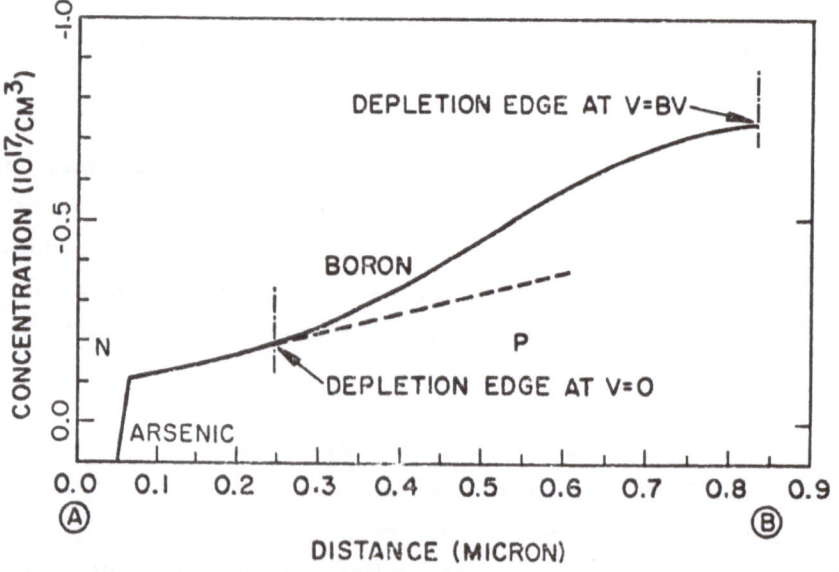

Fig. 10. One-dimensional impurity profile along the electric field line (A-B)
in Fig. 9.

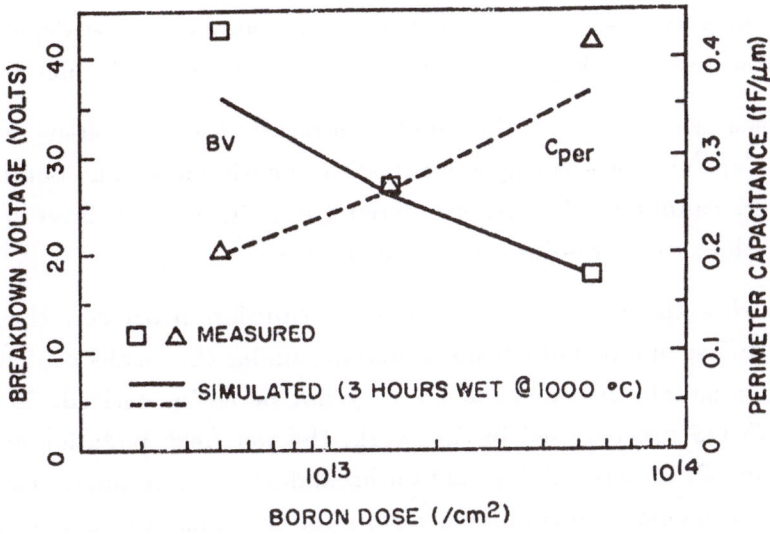

Fig. 11. Breakdown voltage and perimeter capacitance versus boron channel-stop implantation dose at 160 keV.

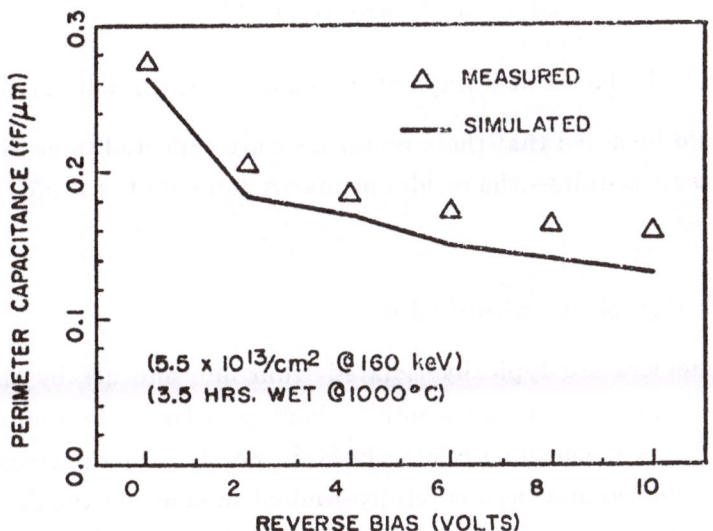

Fig. 12. Perimeter junction capacitance versus reverse junction bias.

dependence of $-\frac{1}{3}$, which implies a quasi-linearity of the impurity distribution. The perimeter capacitance at zero bias shows a power dependece of $\frac{1}{3}$ on boron dose, which also corresponds to a linear-graded junction.

As shown in Fig. 12, the simulation results for C_{per} versus reverse bias voltage are also in good agreement with experiments. The doping profile, however, cannot be directly extracted from the C_{per}-V curve due to the complexity of the two-dimensional structure.

During the high temperature redistribution processes, the impurity atoms diffuse in a two-dimensional manner under the locally oxidized region, so that a simple time dependence may not be easily derived. In the standard NMOS process used in this work, the wet field oxidation cycle is the dominant Dt in determining junction breakdown and perimeter capacitance. The experimental and simulated results for the capacitance and breakdown as a funtion of field oxidation time are shown in Fig. 13. Approximate power law relationships have been extracted as expressed in the following equations:

$$BV \quad \alpha \quad (dose)^{-\frac{1}{3}} (t_{ox})^{.75} \tag{48a}$$
$$C_{per} \quad \alpha \quad (dose)^{\frac{1}{3}} (t_{ox})^{-.75} \tag{48b}$$

where *dose* is the boron field implant dose and t_{ox} is the wet oxidation time.

It should be noted that these power laws are expected to be different for other process schedules—the results primarily indicate the trends in process dependencies.

3.2 Field Threshold Simulation

MOS devices are typically isolated from one another by thick oxide regions and substrate doping profiles which give large threshold voltages. These field regions consume relatively large portions of the silicon surface but have previously not been carefully studied to consider the design trade-offs. The present work considers the design of this barrier region with the objective of minimizing chip area while maintaining high field threshold voltage and low leakage current.

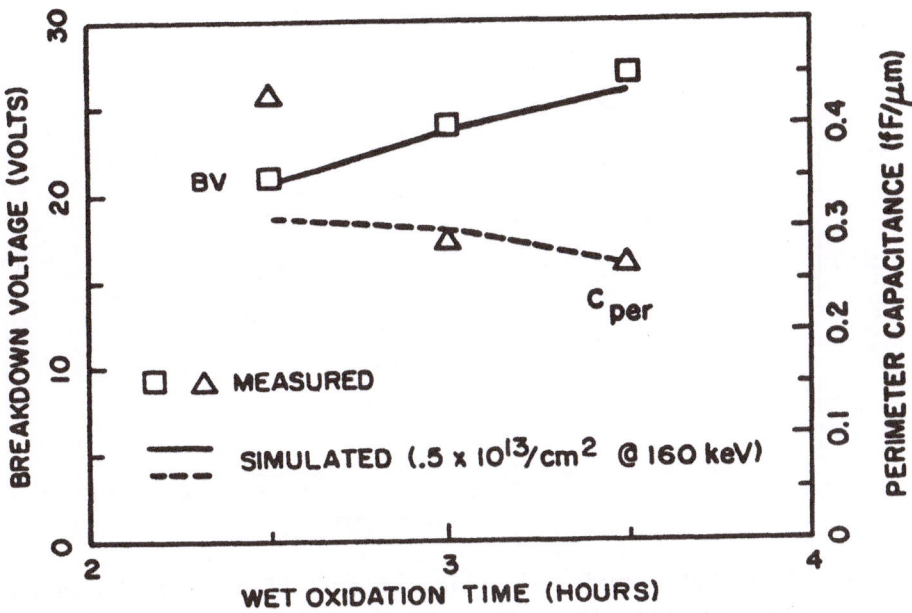

Fig. 13. Breakdown voltage and perimeter capacitance versus wet oxidation time of the field region.

The structure considered is a polysilicon layer passing over a barrier region abutted by the channel regions of two buried channel transistors (Fig. 14). This device was chosen for the study because it was expected to have the largest leakage of any structure on the chip. The structure was fabricated on a lightly doped <100> p-type substrate by first growing 500Å of oxide. Boron was then implanted into the channel-stop region (3.5×10^{12} at 110 keV) using a photoresist mask to protect the active device regions. A semi-recessed oxide (1.1 μm) was grown using a nitride layer (700 Å) to mask the oxidation. After etching of the original pad oxide and growth of the thin gate oxide, arsenic was implanted into the buried channel transistor regions, the polysilicon gate was deposited, and the arsenic was annealed. The field region oxidation and diffusion was simulated using the analytical calculations outlined in section 2.2.2, and the arsenic diffusion simulation was performed numerically as in section 2.3.

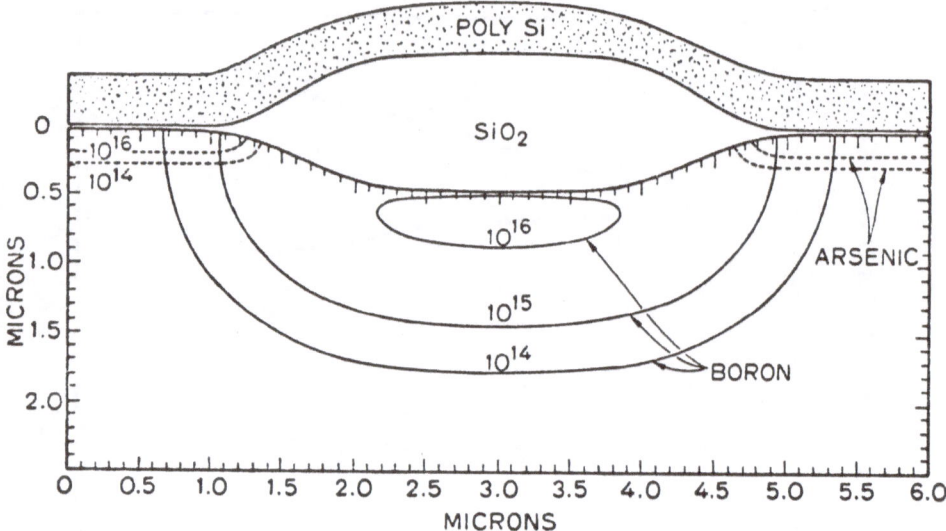

Fig. 14. Channel-stop region showing contours of constant boron concentration beneath the thick oxide and arsenic contours in the neighboring gate regions.

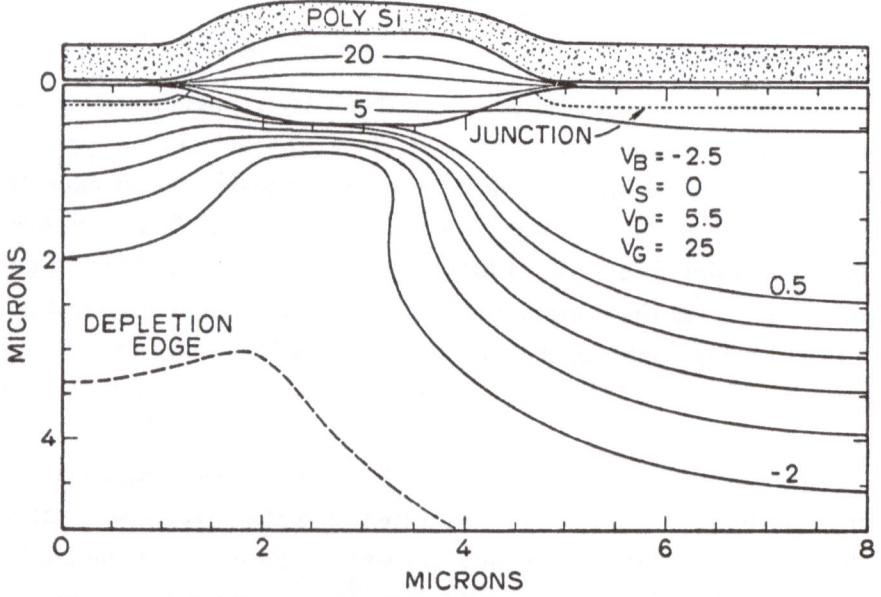

Fig. 15. Contours of constant electrostatic potential when the channel-stop region is biased near threshold.

A typical solution of Poisson's equation on this structure is shown in Fig. 15, where potential contours and the depletion region edge are shown when one of the n-type regions and the polysilicon gate are biased so that the structure is near threshold. In a case such as this where conduction is limited by a potential barrier near the source, the current flow can be calculated approximately from the solution of Poisson's equation [30]. Using this method, subthreshold characteristics and threshold voltage can be readily extracted.

The present study has dealt with finding the oxide thickness which maximizes the field threshold. For field structures where the separation between n^+ regions is many microns, one-dimensional theory is applicable and indicates that field threshold can always be increased by increasing the thickness of the field oxide [33]. However when the separation between adjacent n^+ regions is only two or three microns as in Figures 15 and 16 and a significant drain voltage is applied, it is found that increasing the field oxide above a certain thickness actually decreases the threshold voltage. Thus, a field oxide thickness which maximizes the threshold voltage exists for each channel length and n^+ bias condition. Further work is needed to determine these optimum oxide thicknesses and also to optimize the boron doping profile subject to the tradeoffs with avalanche breakdown and junction capacitance as indicated in Section 3.1.

3.3 Polysilicon Grain Boundary Diffusion

Polysilicon has a wide and growing range of application in VLSI circuits. Polycrystalline silicon is currently used, for example, as high-value resistors in memory circuits and to form "buried" contacts in self-aligned circuits and may in the future be used as a material in which active devices are fabricated. In each of these applications it is important to know the details of impurity diffusion within the polysilicon. Diffusion in polysilicon is complicated, however, because of the impurity segregation to the grain boundaries and the rapid diffusion along them occuring in parallel with the slower diffusion in the interior of the grain. Because of the two-dimensional nature of diffusion in poly, it is difficult to determine the grain boundary diffusivity experimentally.

Our approach was to match experimantal profile measurements with 2D simulation results in which all simulation parameters are known except the grain boundary diffusivity. The idealized polysilicon structure used in the simulation assumes that the poly layer consists of rectangular grains separated by vertical grain boundaries. Due to the assumed symmetry, only half of the grain and half of the grain boundary need be simulated, as shown in Fig. 16. The boundary region is considered by the program to be a very thin strip of thermal oxide, but the parameters of this region (segregation constant and diffusivity) can be adjusted to be typical of the grain boundary region. Values for grain boundary width and arsenic segregation coefficient come from the work of Mandurah [34]. The average grain size was measured experimentally, and the diffusivity typical of single crystal silicon was used in the interior of the grain [18]. Following an arsenic implantation and anneal, the arsenic has segregated to the boundary, diffused rapidly down the boundary, and subsequently diffused to a limited extent back into the grain as shown in Fig. 16.

Experimentally, a 1 μm thick layer of polysilicon was deposited onto a thermally grown SiO_2 layer. Arsenic was implanted and annealed for 6 hours at 800°C. The resulting arsenic profile was measured with the Rutherford backscattering technique to yield the profile shown in Fig. 17.

In order to compare the 2D simulation to the experimental profile, the 2D simulation has been averaged horizontally at every vertical location to produce a vertical profile comparable to the backscattering measurements. The value of the grain boundary diffusivity was extracted by matching the simulation to the measured profile as in Fig. 17. Determining the grain boundary diffusivity at several temperatures yields the Arrhenius relation

$$D_{As-grainboundary} = 1.3 \times 10^5 \exp\left(\frac{-3.9\text{eV}}{kT}\right) \qquad \text{cm}^2/\text{sec.} \qquad (49)$$

Using this value for grain boundary diffusivity, we can estimate the amount of lateral diffusion that might be expected from a contact region in

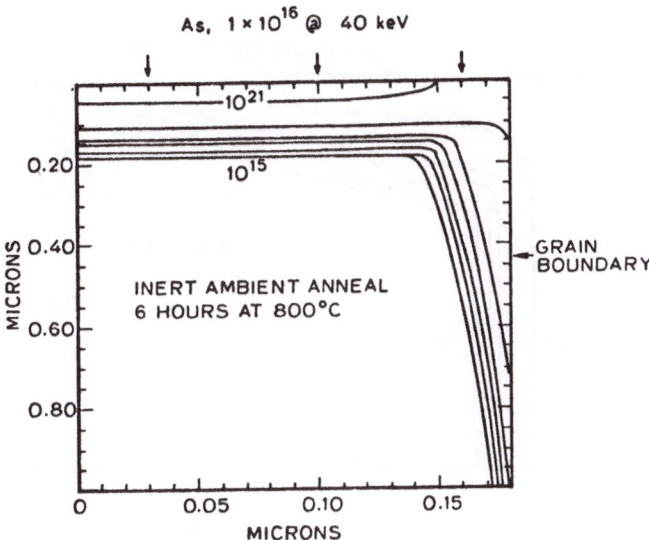

Fig. 16. Cross-sectional view of simulated arsenic contours in a single grain of polysilicon following implantation and diffusion. The grain boundary passes vertically along the right hand side.

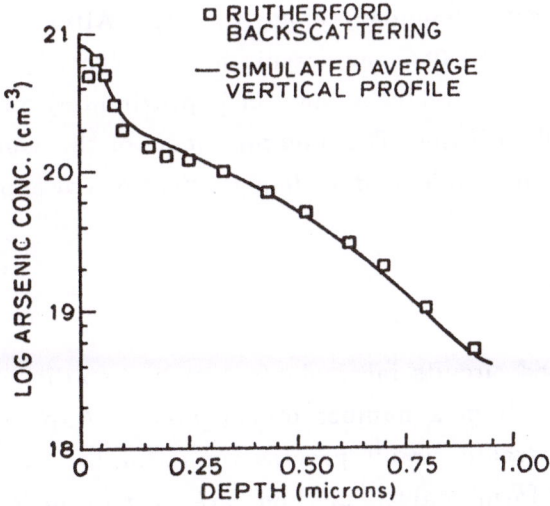

Fig. 17. Vertical profile of average arsenic concentration versus depth through the polysilicon layer.

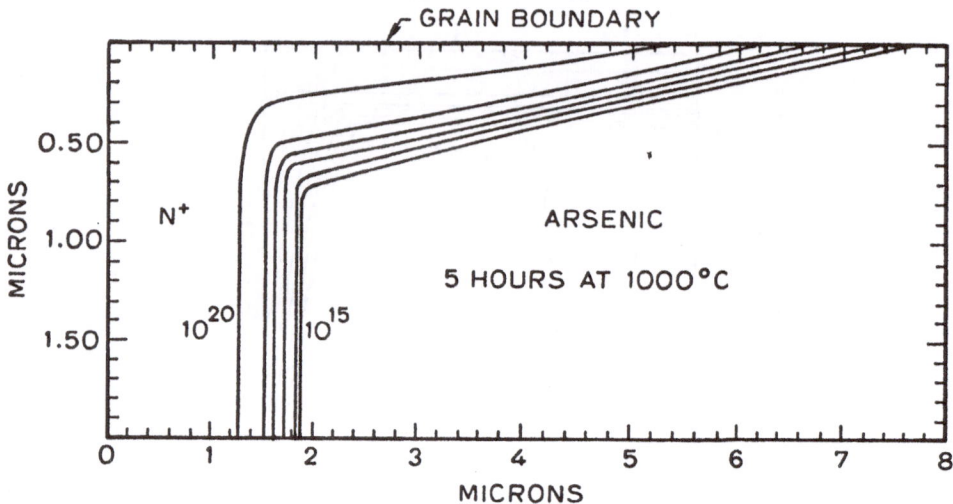

Fig. 18. Top view of lateral arsenic diffusion from a highly doped region along a grain boundary.

a polysilicon resistor or transistor structure. The structure shown in Fig. 18 is the top view of a polysilicon layer with a highly doped arsenic region on the left and a grain boundary along the top. After an inert ambient anneal for 5 hours at 1000°C the arsenic has segregated to the boundary and diffused along it to produce a finger of approximately 5 microns length. This compares well with the EBIC measurements of the same structure by Johnson *et al.* [35] that indicate dark finger extensions of approximately the same length.

4 SUMMARY

VLSI integrated circuits increasingly require two-dimensional process and device simulation in a number of key areas. A practical approach to a complete integrated circuit process simulator has been described in this paper. Closed-form analytic methods are used to model implantation, oxidation, and low concentration diffusion because of their accuracy, adaptability to include empirical results, and computational speed advantages.

More general, but more time consuming, numerical calculations are available to treat high concentration impurity diffusion on structures having arbitrarily shaped boundaries. The 2D output of the process simulator is accepted directly by a device simulation program using the same grid and thus avoiding the loss of accuracy incurred when interpolating onto a new grid. Although no direct experimental verification of the 2D impurity profiles is possible at the present time, the ability to accurately predict device electrical characteristics has been demonstrated, which is the most tangible goal of process simulation.

5 ACKNOWLEDGEMENTS

The authors wish to thank Daeje Chin for his work on the analytic oxidation/diffusion methods, Jim Greenfield for the use of his device simulation program, Balaji Swaminathan for his measurements of diffusion profiles in polysilicon, and Stephen Hansen for many helpful discussions. This work was supported under ARO contract number DAAG29-80-K-0013.

6 REFERENCES

1. A. M. Lin, R. W. Dutton, and D. A. Antoniadis, "The Lateral Effect of Oxidation on Boron Diffusion in <100> Silicon," *Applied Phys. Lett.*, vol. **35**, pp. 799–801, Nov. 1979.

2. M. Hamasaki, J. Suzuki, and T. F. Shimada, "Lateral Effect of Oxidation Enhanced Diffusion in Silicon," in *IEEE Int. Electron Devices Meet., Dig. Tech. Papers*, pp. 542–545, Dec. 1981.

3. K. Chiu, J. Moll, and J. Manoliu, "A Bird's Beak Free Local Oxidation Technology Feasible for VLSI Circuits Fabrication," *IEEE Trans. Electron Devices*, vol. **ED-29**, pp. 536–540, Apr. 1982.

4. A. R. Neureuther, C. H. Ting, and C-Y Liu, "Application of Line-Edge Profile Simulation to Thin-Film Deposition Processes," *IEEE Trans. Electron Devices*, vol. **ED-27**, pp. 1449–1455, Aug. 1980.

5. W. G. Oldham, A. R. Neureuther, C. Sung, J. L. Reynolds, and S. N. Nandgaonkar, "A General Simulator for VLSI Lithography and Etching Processes: Part II – Application to Deposition and Etching," *IEEE Trans. Electron Devices*, vol. **ED-27**, pp. 1455–1459, Aug. 1980.

6. H. Runge, "Distribution of Implanted Ions under Arbitrarily Shaped Mask Edges," *Phys. Stat. Sol. (a),* vol. **39**, pp. 595–599, 1977.

7. J. F. Gibbons and S. Mylroie, "Estimation of Impurity Profiles in Ion-implanted Amorphous Targets using Joined Half-Gaussian Distributions," *Applied Phys. Lett.,* vol. **22**, pp. 568–569, June 1973.

8. W. K. Hofker, "Implantation of Boron in Silicon," *Phillips Res. Reports,* suppl. 8, 1975.

9. H. Ryssel, K. Haberger, K. Hoffmann, G. Prinke, R. Dümcke, and A. Sachs, "Simulation of Doping Processes," *IEEE Trans. Electron Devices,* vol. **ED-27**, pp. 1484–1492, Aug. 1980.

10. H. G. Lee, R. W. Dutton, and D. A. Antoniadis, "On Redistribution of Boron during Thermal Oxidation of Silicon," *J. Electrochem. Soc.,* vol. **126**, pp. 2001–2007, 1979.

11. J. Huang and L. Welliver, "On the Redistribution of Boron in the Diffused Layer during Thermal Oxidation," *J. Electrochem. Soc.,* vol. **117**, pp. 1577–1580, 1970.

12. D. A. Antoniadis, S. E. Hansen, and R. W. Dutton, "SUPREM II - A Program for IC Process Modeling and Simulation," *Stanford Electronics Laboratory Technical Report No. 5019-2,* Stanford Electronics Lab., June 1978.

13. B. E. Deal and A. S. Grove, "General Relationship for the Thermal Oxidation of Silicon," *J. Appl. Phys.,* vol. **36**, pp. 3770–3778, Dec. 1965.

14. Hee-Gook Lee, "Two-dimensional Impurity Diffusion Studies: Process Models and Test Structures for Low-Concentration Boron Diffusion," *Stanford Electronics Laboratory Technical Report No. G201-8,* Stanford Electronics Lab., Aug. 1980.

15. D. Chin, paper to be presented at Joint SIAM/IEEE Conference on Numerical Simulation of VLSI Devices, Boston, Nov. 1982.

16. R. Tielert, "Two-dimensional Numerical Simulation of Impurity Redistribution in VLSI Processes," *IEEE Trans. Electron Devices,* vol. **ED-27**, pp. 1479-1483, Aug. 1980.

17. A. S. Grove, *Physics and Technology of Semiconductor Devices,* New York: Wiley, Chap. 6, 1978.

18. R. B. Fair, *Impurity Doping Processes in Silicon,* (F.F.Y. Wang ed.), Amsterdam: North Holland Publishing Co., 1981.

19. R. O. Schwenker, E. S. Pan, and R. F. Lever, "Arsenic Clustering in Silicon," *J. Appl. Phys.,* vol. **42**, pp. 3195–3200, 1971.

20. R. B. Fair and G. R. Weber, "Effect of Complex Formation on Diffusion of Arsenic in Silicon," *J. Appl. Phys.,* vol. **44**, pp. 273–279, Jan. 1973.

21. R. B. Fair and J. C. C. Tsai, "The Diffusion of Ion-Implanted Arsenic in Silicon," *J. Electrochem. Soc.,* vol. **122**, Dec. 1975.

22. R. B. Fair and J. C. C. Tsai, "A Quantitative Model for the Diffusion of Phosphorus in Silicon and the Emitter Dip Effect," *J. Electrochem. Soc.,* vol. **124**, pp. 1107–1118, July 1977.

23. K. Murota, E. Arai, K. Kobayashi, and K. Kudo, "Relationship between Total Arsenic and Electrically Active Arsenic Concentrations in Silicon Produced by the Diffusion Processes," *J. Appl. Phys.,* vol. **50**, pp. 804–808, Feb. 1979.

24. H. Ryssel, K. Müller, K. Haberger, R. Henkelmann, and F. Jahnel, "High Concentration Effects of Ion Implanted Boron in Silicon," *Applied Phys.,* vol. **22**, pp. 1–4, 1980.

25. M. Y. Tsai, F. F. Morehead, J. E. E. Baglin, and A. E. Michel, "Shallow Junctions by High-Dose As Implants in Si: Experiments and Modeling," *J. Appl. Phys.,* vol. **51**, pp. 3230–3235, June 1980.

26. F. N. Schwettmann and D. L. Kendall, "On the Nature of the Kink in the Carrier Profile for Phosphorus-Diffused Layers in Silicon," *Applied Phys. Lett.,* vol. **21**, pp. 2–4, July 1972.

27. S. M. Hu, "Diffusion in Silicon and Germanium," in *Atomic Diffusion in Semiconductors,* D. Shaw, editor, chap. 5, London: Plenum Publishing Corp., 1972.

28. R. W. Dutton, P. Fahey, K. Doganis, L. Mei, and H. G. Lee, "Computer-Aided Process Modeling for Design and Process Control," Symposium on Silicon Processing, San Jose, California, Jan. 19-22, 1982.

29. R. S. Varga, *Matrix Iterative Analysis,* Englewood Cliffs: Prentice-Hall, Chap. 6, 1962.

30. J. A. Greenfield and R. W. Dutton, "Nonplanar VLSI Device Analysis Using the Solution of Poisson's Equation," *IEEE Trans. Electron Devices,* vol. **ED-27**, pp. 1520–1532, Aug. 1980.

31. A. R. Mitchell, *Computational Methods in Partial Differential Equations,* New York: Wiley, Chap. 2, 1969.

32. E. C. Du Fort and S. P. Frankel, "Stability Conditions in the Numerical Treatment of Parabolic Differential Equations," *M.T.A.C.,* vol. **7**, pp. 135–152, 1953.

33. S. M. Sze, *Physics of Semiconductor Devices,* New York: Wiley-Interscience, 1969.

34. Mohammad Mahmoud Mandurah, "The Physical and Electrical Properties of Polycrystalline-Silicon," *Stanford Electronics Laboratory Technical Report No. G503-2,* Stanford Electronics Lab., June 1981.

35. N. M. Johnson, D. K. Biegelsen, and M. D. Moyer, "Grain Boundaries in p-n Junction Diodes Fabricated in Laser-recrystallized Silicon Thin Films," *Applied Phys. Lett.,* vol. **38**, June 1981.

NUMERICAL SIMULATION OF IMPURITY
REDISTRIBUTION NEAR MASK EDGES

R. Tielert

Siemens AG, Research Laboratories, Otto-Hahn-Ring 6, D-8000 München 83, W. Germany

1. INTRODUCTION

Two-dimensional device analysis programs have become a commonly used tool for the development of integrated circuits. The two-dimensional doping distributions being crucial input data of those programs, are usually approximated on the basis of one-dimensional process models or one-dimensional profile measurements. Lateral spreading of the doping distributions, however, is no longer a second order effect in scaled down devices and needs accurate determination in order to enable reliable prediction of the device's electrical performance.

Recent advances in the field of two-dimensional process simulation have been reviewed in (1). Adequate mathematical tools have been developed for solving the equations of the impurity transport, including the effect of nonlinear diffusion and nonuniformly moving boundary (2-8). The underlying physical models, however, still need further development in order to arrive at a status of process simulation comparable to that of device simulations.

The goal of this paper is to provide the numerical techniques necessary for the modeling of two-dimensional doping distributions. Equations are given in discretized form enabling also the nonspecialist to develop his own program. This includes the models of nonlinear diffusion mechanisms and the oxidation related phenomena. A strategy of combining different solution algorithms corresponding to a different degree of complexity of the applicational area is demonstrated by the simulation program LADIS, which has been developed by the author.

2. DIFFUSION MODELING - INERT AMBIENT

2.1 Transport Equation

The two-dimensional simulation of the redistribution of impurities is a mathematical extension of the principles that are already established in the one-dimensional simulation. Basis is the transport equation which in a complete form includes both the diffusion and the field induced flux of impurities:

$$\frac{\partial}{\partial x}(D\ \frac{\partial N}{\partial x}) + \frac{\partial}{\partial y}(D\ \frac{\partial N}{\partial y})$$

$$\pm\left(\frac{\partial}{\partial x}(D\ N\ \frac{\partial \Psi}{\partial x}) + \frac{\partial \Psi}{\partial y}(D\ N\ \frac{\partial \Psi}{\partial y})\right) = \frac{\partial C}{\partial t} \tag{1}$$

The sign of the field term corresponds to the polarity of the ions under consideration. In this equation

$$\Psi = \ln \frac{n}{n_i} \tag{2}$$

denotes the normalized electrical potential, N the concentration of the electrical active doping centers, and C the total doping concentration. The directions x and y are defined by Fig. 1. The simulation area is a plane that is perpendicular to the silicon surface and perpendicular to a mask edge which is the origin of the two-dimensional problem.

Eq. (1) relies on the assumption that only the active atoms N are mobile whereas the clustered or precipitated impurity atoms contained in C are assumed to be immobile. A relation $N = f(C)$ is necessary and the potential Ψ has to be known before eq. (1) can be solved. Additionally, since D is a constant only at low impurity concentrations, a model $D = f(C,n,x,y)$ has to be derived.

This paper is mainly concerned with the numerical treatment of the problem. Thus, the underlying physical

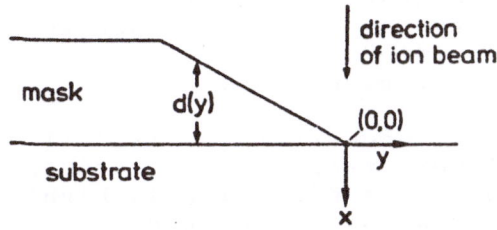

FIG. 1   Definition of coordinates.

models of impurity redistribution are shortly reiterated as far as they are implemented in the program discussed in the following sections.

2.2 Physical Models

The diffusion coefficient is modeled according to
SUPREM (9) using

$$D = \frac{D_i (1 + \beta \frac{n}{n_i})}{(1 + \beta)} \tag{3}$$

where D_i is known as intrinsic diffusion constant and
β is an empirical constant depending on the special
dopant. This relation has been used for arsenic and
phosphorus. In the case of boron the factor $\frac{n}{n_i}$ must be
replaced by $\frac{p}{n_i}$ accounting for the fact that boron
diffusion is enhanced by the positively charged vacancies instead of the negatively charged vacancies which
apply in the case of donors. An additional space de-
pendency is assumed if oxidizing border conditions are
applied. This will be discussed in section 3. The
simulation of phosphorus has been confined to low and
medium concentrations where eq. (3) yields acceptable
agreement with experiments. At high concentrations,
however, the model still needs further development.

Electron density n is an important variable used
in the models. It is simply derived by

$$n = \frac{1}{2}\left((N_D - N_A) + \sqrt{(N_D - N_A)^2 + 4n_i^2}\right) \tag{4}$$

Relation (4) holds at typical high temperature conditions where diffusion is modeled, the assumed thermal

equilibrium and local neutrality being justified there. In case of low impurity concentrations $n \simeq n_i$ holds. There, constant diffusivity may be assumed and the field term in eq. (1) may be omitted, in order to simplify the problem and save computation time.

At the high impurity concentrations typical of source/drain regions of MOS transistors or emitter regions of bipolar transistors diffusion is highly non-linear due to the concentration dependence described by eq. (3) and other effects like solubility limitation, clustering and precipitation of impurities. A number of models have been proposed, refering to the high concentrations anneal of arsenic, which assume 2 or 4 arsenic atoms forming a cluster (10,11). A relation based on the law of mass action has been proposed

$$C = N + mk_{eq}N^m \tag{5}$$

with k_{eq} being the equilibrium constant of the clustering reaction and m being the number of As-atoms per cluster.

Other authors have proposed arsenic-vacancy clusters (12) and electrically charged clusters (13). The latter assumption has been investigated thoroughly in (14), concluding some important consequences for diffusion modeling:

a) A solubility limit of active doping centers is explained by the fact that charge neutrality of charged clusters and free electrons must be maintained. This solubility limit is given by

$$N_{max} = \left((m - 1)k_{eq}\right)^{-\frac{1}{m}} \tag{6}$$

b) At high temperature electron density is not limited by N_{max} but increases monotonically with total concentration C, i.e.

$$n = N + \frac{m - 1}{m} (C - N) \tag{7}$$

c) Clusters loose their charge when wafers are cooled to room temperature. Thus, a lower electron density proportional to N is measured at laboratory conditions.

d) The clustering formula relating C to N is given

by

$$C = N + \frac{mk_{eq}N^{m+1}}{1 - (m-1)k_{eq}N^{m}} \quad . \tag{8}$$

Fig. 2 displays the temperature dependence of N_{max}.
Fig. 3 shows a comparison of eq. (8) (curve 2) and eq.
(5) (curve 1) fitted to experimental data. The improved
agreement of the model using charged clusters is apparent. Therefore, this model has been adopted in this paper using m = 3 for the simulations of high concentration arsenic profiles.

In the case of high concentrations of boron a
reasonable agreement with experimental profiles is obtained if eq. (5) is used together with a clustersize
m ≥ 12. In this case uncharged clusters are assumed,
with n being proportional to the electrical active
atoms N. The large number m results from the very abrupt onset of the solubility limit of active boron,
which is probably due to a precipitation mechanism.
Since the specific mechanism is yet not well understood
high concentration boron profiles are at present difficult to predict and depend on the narrow range of validity of the empirical modeling parameters.

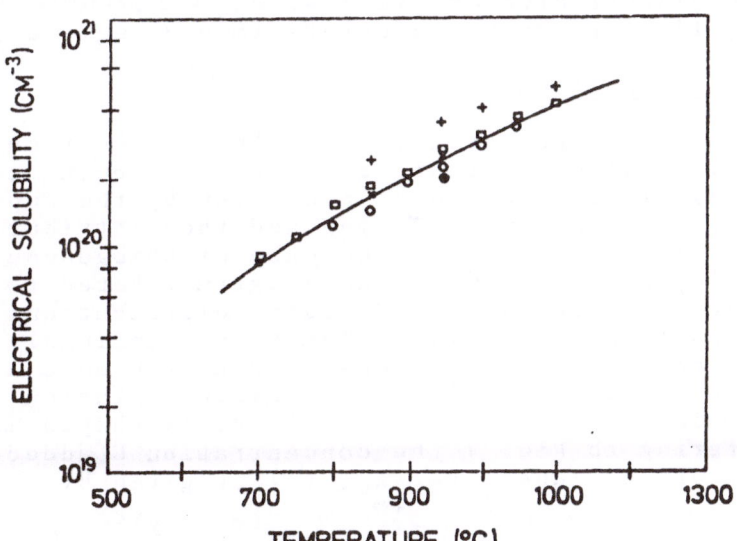

FIG. 2   Electrical solubility (N_{max}) vs. temperature
(14). The solid curve fitting published experimental
data is given by:
N_{max} = 1.896·10^{22} exp(-0.453 eV/kT) cm^{-3} .

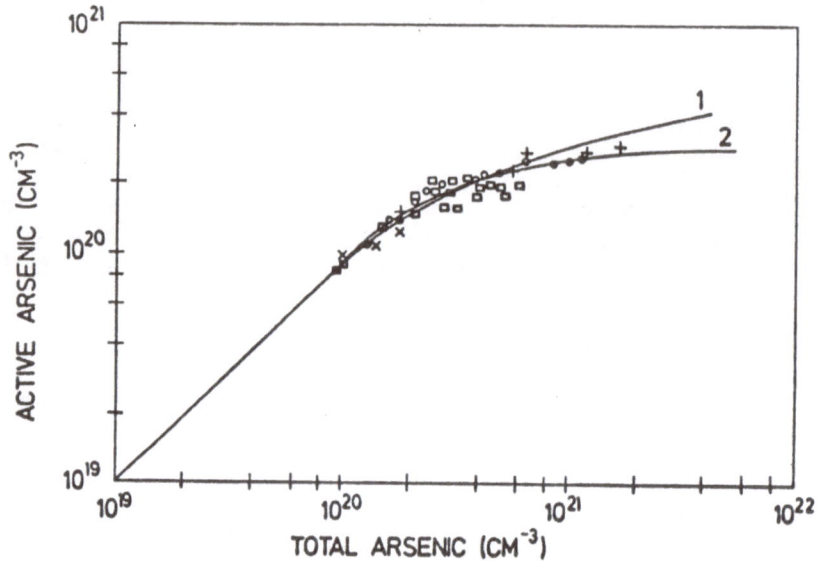

FIG. 3   Total As-concentration (C) vs. electrically active As-concentration (N) for T = 1000°C. Curve 1 holds for neutral clusters (m = 4, k_{eq} = 2.22·10^{-62} cm^9). Curve 2 holds for charged clusters (m = 3, eq. (6)).

2.3 Initial Conditions

Solution of the partial differential equation (1) requires initial conditions, i.e. initial doping concentrations. In general these are given by the superposition of the substrate doping and the distributions due to ion implantation. In the case of homogeneous implantation without a mask the program refered to in this paper generates simple Gaussian distributions based on published data (15). The two-dimensional distributions of implanted ions near a mask edge are calculated according to (16). A modified solution (17) is used which accounts for an arbitrarily shaped mask edge. Refering to Fig. 1 the concentration C added to any point of the substrate (x_0, y_0) is given by

$$C(x_0,y_0) = \frac{C_\square}{2\Pi\Delta R_p \Delta R_{py}} \int_{-\infty}^{+\infty} exp -(\frac{(y_0 - y)^2}{2\Delta R_{py}^2}$$

$$+ \frac{(x_0 + d(y) - R_p)^2}{2\Delta R_p^2}) \ dy \qquad (9)$$

Therein, $C_\square$ denotes the ion dose, R_p the projected
range, ΔR_p the projected standard deviation, and ΔR_{py}
the lateral projected standard deviation. Data for R_p,
ΔR_p, and ΔR_{py} published in (15) are used.

Integration of (9) is performed numerically by
Simpson's rule. Therefore, the infinite integral is
approximated by a finite integral with integration
limits chosen several ΔR_{py} wider than the simulated
area. In the case of an abrupt mask edge (e.g. at y = 0)
the discontinuity of the integrand can cause integration
errors. By partitioning the integral into two this is
avoided. Equal stopping power of mask and substrate is
assumed in this paper. A different stopping power, however, can approximately be accounted for by using an
effective mask thickness

$$d(y) = \frac{d_{mask}(y) \, R_{p,Si}}{R_{p,mask}} \quad . \tag{9a}$$

Although eq. (9) can be solved analytically (18) for
some typical functions describing the slope of the mask
edge, the more time-consuming numerical integration has
been chosen in this paper. This more general solution
is required if arbitrarily shaped oxide layers due to
local oxidation are used as an implantation mask.
Taking the original silicon surface as a reference line,
also negative mask thicknesses can be applied. This is
very useful in a sequence of processing steps where the
silicon surface is not planar due to local oxidation
and etching.

2.4 Discretization of Transport Equation

The numerical solution of the partial differential
equation (1) requires a discretization of the equation
in both space and time. The continuous equation is
approximated on a grid of $I \cdot J$ meshpoints, where I and
J are numbers of discretization steps in y- and x-direction, respectively. This results in a set of $I \cdot J$
ordinary algebraic equations which have to be solved
for the unknown values $C_{i,j}^{n+1}$, where n denotes the time
discretization step and i, j are the indices of the
matrix.

The Crank-Nicolson equation (19) has been chosen
in this paper as an adequate analog of the parabolic
type of equations typical of the diffusion problem. In
the Crank-Nicolson representation the time discretiza-

tion is performed at the time step $n + \frac{1}{2}$, i.e. the function is averaged between the known and next unknown time step. Advantage of the Crank-Nicolson analog is a very good stability of the integration algorithm regardless of the size of Δx and Δt which are the discretization steps in space and time, respectively. A disadvantage, however, is that the unknowns are given implicitly, i.e. a simultaneous or iterative solution of the whole set of difference equations is necessary. This means a greater computational overhead per time step as compared to explicit integration techniques. On the other hand this disadvantage can sometimes be compensated in the implicit scheme by choosing larger time steps, depending on the accuracy required.

Two-dimensional discretization of the parabolic equation yields a 5-point representation as shown in Table 1. Using an equally spaced grid the spatial discretization is performed straightforward by replacing the differentials by finite differences. Coefficients are averaged between the meshes. Eq. (1) can be thus transformed into the difference equation (10) displayed in Table. 2.

2.5 Boundary Conditions

If diffusion is performed in inert ambient very simple boundary conditions apply. The posterior boundary of the simulated area must always be chosen apart from the range of interest to assure zero flux and zero field strength. The boundaries at the left and right side must be symmetry lines, i.e. reflecting boundaries with flux and field strength being zero. At the Si/SiO_2 interface flux and field strength are assumed to be zero, too, consequent of negligible diffusion in oxide.

Boundary conditions are commonly accounted for in the numerical representation by virtual points outside the simulation grid. For the simple reflecting border conditions the concentrations at the virtual points are identical to those at the border mesh. Thus, the virtual points need no extra storage and can be simply generated by the computer algorithm.

In a sequence of processing steps the silicon surface is not planar if local oxidation is performed. Modeling a nonplanar surface or Si/SiO_2 interface in a rectangular grid results in a boundary that crosses arbitrarily the discretization meshes. In inert ambient reflecting boundary conditions apply at this line.

BLE 1

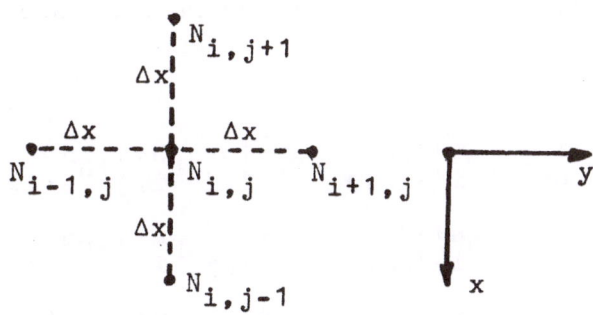

Spatial discretization:

$$\frac{\partial}{\partial y}\left(D\,\frac{\partial N}{\partial y}\right) \;\rightarrow\; \frac{1}{\Delta x}\{D_{i+1/2,j}\,\frac{1}{\Delta x}\,(N_{i+1,j}-N_{i,j})$$

$$-D_{i-1/2,j}\,\frac{1}{\Delta x}\,(N_{i,j}-N_{i-1,j})\}$$

$$\frac{\partial}{\partial y}\left(D\,N\,\frac{\partial \Psi}{\partial y}\right) \;\rightarrow\; \frac{1}{\Delta x}\{D_{i+1/2,j}\,N_{i+1/2,j}\,\frac{1}{\Delta x}\,(\Psi_{i+1,j}-\Psi_{i,j})$$

$$-D_{i-1/2,j}\,N_{i-1/2,j}\,\frac{1}{\Delta x}\,(\Psi_{i,j}-\Psi_{i-1,j})\}$$

Spatial averages:

$$D_{i+1/2,j} = \frac{1}{2}\,(D_{i,j}+D_{i+1,j})$$

Spatial and time averages:

$$D_{i+1/2,j}^{n+1/2} = \frac{1}{4}\,(D_{i,j}^{n+1}+D_{i+1,j}^{n+1}+D_{i,j}^{n}+D_{i+1,j}^{n})$$

TABLE 2   Analog of transport equations (eq. (10)).

$$0 = f_{i,j} =$$

$$\frac{1}{2\Delta x^2} \{ D_{i+1/2,j}^{n+1/2} \ (N_{i+1,j}^{n+1} - N_{i,j}^{n+1} + N_{i+1,j}^{n} - N_{i,j}^{n})$$

$$+ D_{i-1/2,j}^{n+1/2} \ (N_{i-1,j}^{n+1} - N_{i,j}^{n+1} + N_{i-1,j}^{n} - N_{i,j}^{n})$$

$$+ D_{i,j+1/2}^{n+1/2} \ (N_{i,j+1}^{n+1} - N_{i,j}^{n+1} + N_{i,j+1}^{n} - N_{i,j}^{n})$$

$$+ D_{i,j-1/2}^{n+1/2} \ (N_{i,j-1}^{n+1} - N_{i,j}^{n+1} + N_{i,j-1}^{n} - N_{i,j}^{n}) \}$$

$$\pm \frac{1}{2\Delta x^2} \{ D_{i+1/2,j}^{n+1/2} \ N_{i+1/2,j}^{n+1/2} \ (\psi_{i+1,j}^{n+1} - \psi_{i,j}^{n+1} + \psi_{i+1,j}^{n} - \psi_{i,j}^{n})$$

$$+ D_{i-1/2,j}^{n+1/2} \ N_{i-1/2,j}^{n+1/2} \ (\psi_{i-1,j}^{n+1} - \psi_{i,j}^{n+1} + \psi_{i-1,j}^{n} - \psi_{i,j}^{n})$$

$$+ D_{i,j+1/2}^{n+1/2} \ N_{i,j+1/2}^{n+1/2} \ (\psi_{i,j+1}^{n+1} - \psi_{i,j}^{n+1} + \psi_{i,j+1}^{n} - \psi_{i,j}^{n})$$

$$+ D_{i,j-1/2}^{n+1/2} \ N_{i,j-1/2}^{n+1/2} \ (\psi_{i,j-1}^{n+1} - \psi_{i,j}^{n+1} + \psi_{i,j-1}^{n} - \psi_{i,j}^{n}) \}$$

$$- \frac{1}{\Delta t} \ (C_{i,j}^{n+1} - C_{i,j}^{n})$$

Again, this can be easily accounted for by using virtual points as discussed before. The simulation area is now confined to the grid points lying on the silicon side of that border line and points lying beyond that line are taken as virtual points. Another application of virtual points is modeling the predeposition of impurities, where the virtual points, however, have a fixed value corresponding to the surface concentration assumed.

2.6 Nonlinear SOR Iteration

In general the discretized equation (10) is a nonlinear function $f_{i,j} = f(C_{i,j})$ of the unknowns. Its solution thus requires successive approximations of the function in an iteration procedure, where the function $f_{i,j}$ is repeatedly linearized. A suited method is the Newton iteration scheme which is based on the linear expansion of the function $f_{i,j} = 0$.

$$0 = f^0_{i,j} + \left. \frac{\partial f_{i,j}}{\partial c_{i,j}} \right|_{f_{i,j} = f^0_{i,j}} \Delta c_{i,j} \qquad . \qquad (11)$$

Starting the iteration at a given $C_{i,j}$ the succeeding better approximation is given by $C_{i,j} + \Delta C_{i,j}$ with

$$\Delta C_{i,j} = \frac{-f^0_{i,j}}{\left. \frac{\partial f_{i,j}}{\partial c_{i,j}} \right|_{f_{i,j} = f^0_{i,j}}} \qquad . \qquad (11a)$$

This procedure is continued until $\Delta C_{i,j}$ does not exceed the required accuracy limit.

Applying this technique to the system of equations (10) makes a repeated simultaneous solution of approx. 3000 linearized equations necessary at each time step, which results in the great computational overhead of a direct solution technique. In this paper an iterative technique is used for solving the simultaneous equations, known as the Gauß-Seidel point relaxation method. It evaluates $C_{i,j}$ in equation (10) separately at each point of the simulation grid by using the results of the preceding iteration. After repeated scanning through all meshes convergence to the true solution is obtained. Newton iterations are performed at each point separately, the required number of iterations depending on the nonlinearity of the function at that point. In a typical application only a small fraction of the points exhibits strong nonlinearities. Thus computation time is saved in contrast to a direct solution technique in which all points undergo the same number of Newton iterations related to the strongest nonlinearity of the system.

Further improvement of computation time is achieved by the use of an overrelaxation factor in the Gauß-Seidel procedure. This factor r modifies the result of the actual iteration m according to

$$c^m_{i,j} = c^{m-1}_{i,j} + r \, (c^m_{i,j} - c^{m-1}_{i,j}) \qquad . \qquad (12)$$

Convergence is drastically improved if an optimal parameter r is applied. Since r cannot in general be predicted it is optimized empirically by the computer algorithm in the course of iterations. The procedure
starts with r = 1 and increases r by 0.1 or less in the
succeeding iterations until the residuals monitored
change sign, which is an indicator of an oscillating
solution. Oscillations with an amplitude being smaller
than the correction of succeeding solutions assure a
rapid and stable convergence. Small oscillations are
thus maintained by decreasing or increasing again the
factor r by adequate steps depending on the size of
oscillations.

A further augmentation of the numerical procedure
is necessary if simultaneous diffusion of multiple
species of impurities has to be simulated. Due to the
coupling of the transport equation via the free electron density the equations of the different species
must be solved simultaneously.

In this paper simultaneous diffusion of up to
three dopants will be considered. Hence, in the successive overrelation (SOR) procedure at each point a sub-
set of three equations is solved. Analogous to (11) we
can write:

$$f_B = - \frac{\partial f_B}{\partial C_B} \Delta C_B - \frac{\partial f_B}{\partial C_{As}} \Delta C_{As} - \frac{\partial f_B}{\partial C_P} \Delta C_P$$

$$f_{AS} = - \frac{\partial f_{As}}{\partial C_B} \Delta C_B - \frac{\partial f_{As}}{\partial C_{As}} \Delta C_{As} - \frac{\partial f_{As}}{\partial C_P} \Delta C_P$$

$$f_P = - \frac{\partial f_P}{\partial C_B} \Delta C_B - \frac{\partial f_P}{\partial C_{As}} \Delta C_{As} - \frac{\partial f_P}{\partial C_P} \Delta C_P$$

In this set of equations the indices B, As, P denote
the dopants boron, arsenic and phosphorus, respectively.

2.7 Simulation Results

The simulation program LADIS solves eq. (10) on an
equally spaced grid of 64 x 46 meshes, including the
cooperative effect of the diffusion of up to three
species. Clustering and solubility limit of active impurities are accounted for.

The program starts by reading a file in which the user defines a sequence of processing steps by using keywords. These stand for substrate and grid definition, implantation, predeposition, drive-in, oxidation, oxide deposition (etching), mask definition, plotting, and saving results on a file. An additional keyword defines three different levels of complexity of the diffusion model – constant diffusivity and no interaction – concentration dependent coefficients and no interaction – inclusion of all effects mentioned. This feature helps to save computation time when simple conditions are sufficient. Furthermore, the program performs one-dimensional simulations unless a two-dimensional mask is defined in the sequence of simulation steps.

A linear diffusion model holds for the first example. It starts with the distribution of boron implanted near a step mask. The corresponding contours of equal concentrations are plotted in Fig. 4. A following anneal

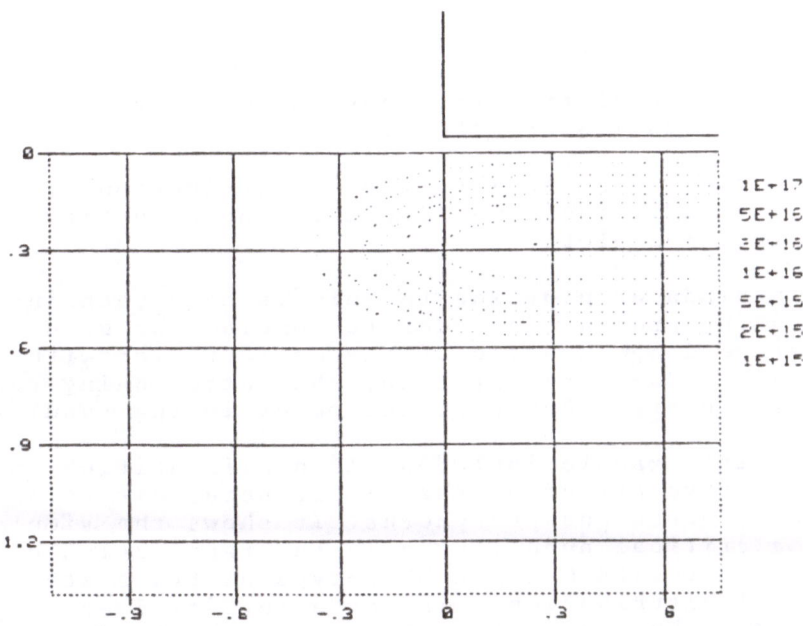

FIG. 4   Concentration of boron implanted near a step mask. (Vertical scale: Depth coordinate x in microns. Horizontal scale: Lateral coordinate y in microns. Parameters of the contourlines (in cm^{-3}) are listed on the right)

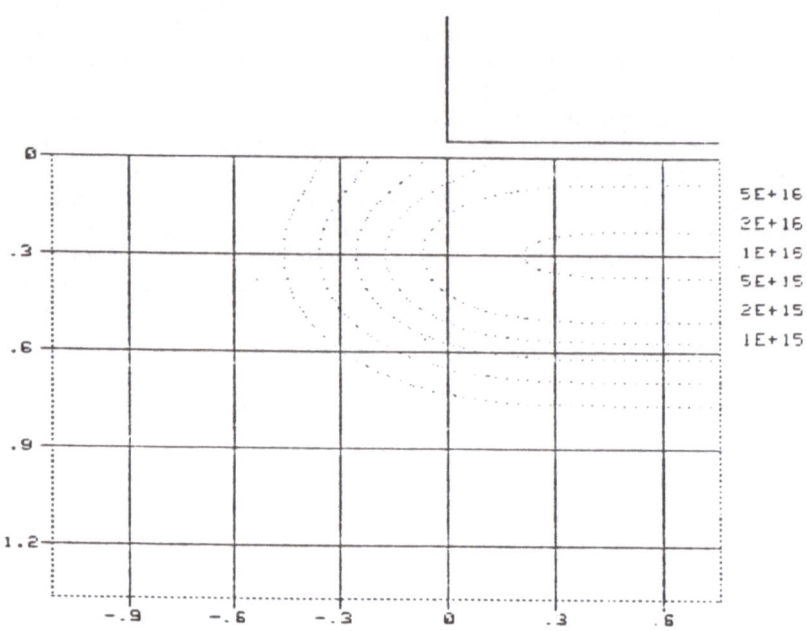

FIG. 5   Boron distribution corresponding to Fig. 4
annealed for 120 min at 1000°C.

at inert conditions results in the distribution displayed in Fig. 5. The same result is shown in Fig. 6
as a perspective plot.

     Both kinds of plotting techniques have been in-
cluded in the program in order to combine the good
quantitative documentation of the contour plot with the
more detailed 3d-representation, the latter being very
helpful if multiple pn-junctions occur in the results.

     The next example includes all nonlinearities and
the cooperative effect of the simultaneous diffusion
of arsenic, boron and phosphorus. It shows the simulation of a critical anneal step in the fabrication of
integrated circuits using a DMOS-type of transistor.
The initial distributions displayed in Fig. 7 and Fig. 8
result from a homogeneous implant and anneal of phosphorus, followed by an implant of arsenic and boron
which is aligned by the tapered edge of a polysilicon
gate (Fig. 7).

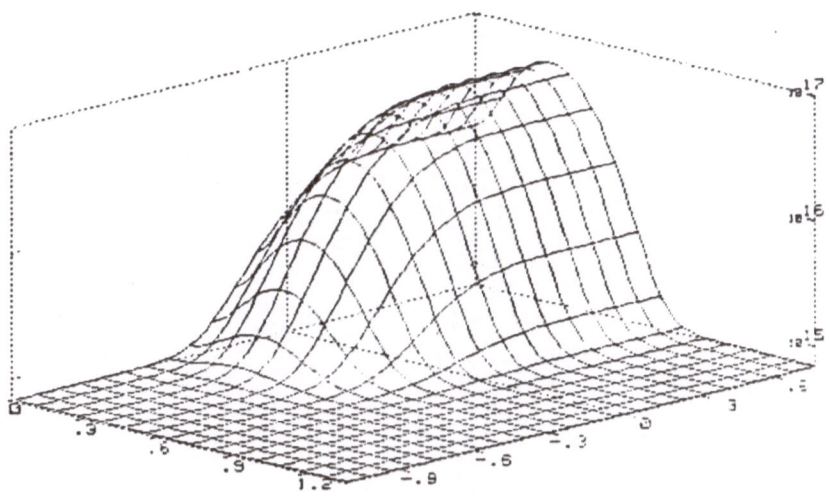

FIG. 6   3D-representation of Fig. 5.

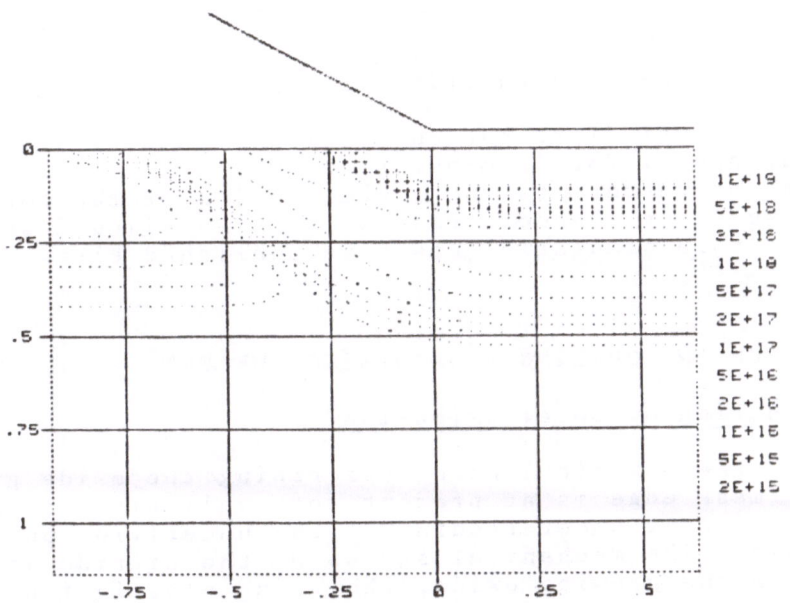

FIG. 7   Net total donor and acceptor concentrations
in the source region of a DIMOS transistor prior to the
activation of the implanted arsenic and boron profiles.

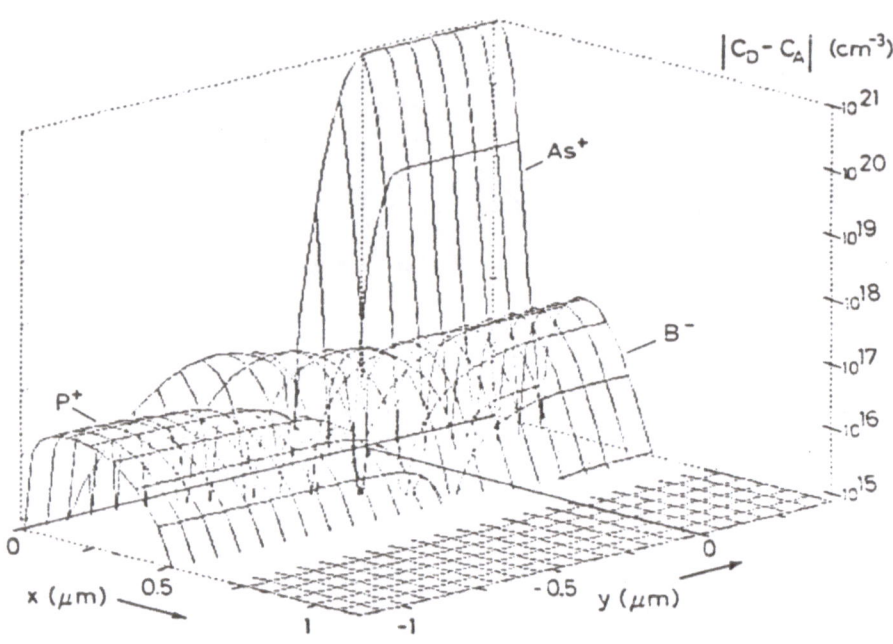

FIG. 8   3D-representation of Fig. 7.

The final distribution displayed in Fig. 9 results from an anneal for 20 minutes at 1000°C. The distribution shown represents the source region of the corresponding short channel transistor. A detailed discussion of this simulation has been published elsewhere (2).

3. DIFFUSION MODELING - OXIDIZING AMBIENT

3.1 Modeling of Local Oxidation

A true numerical model describing the oxide growth near a mask edge is at present not available. To arrive there a closer understanding of the underlying physical mechanism like mechanical stress of the nitride mask layer on the growing oxide, the visco-elastic flow of oxide under the mask, and oxygen transport at these conditions will be necessary. Experimental data showing the effect of nitride layer thickness, pad oxide thickness, oxidizing ambient, and temperature on the shape of the grown "bird's beak" have been published (21,22).

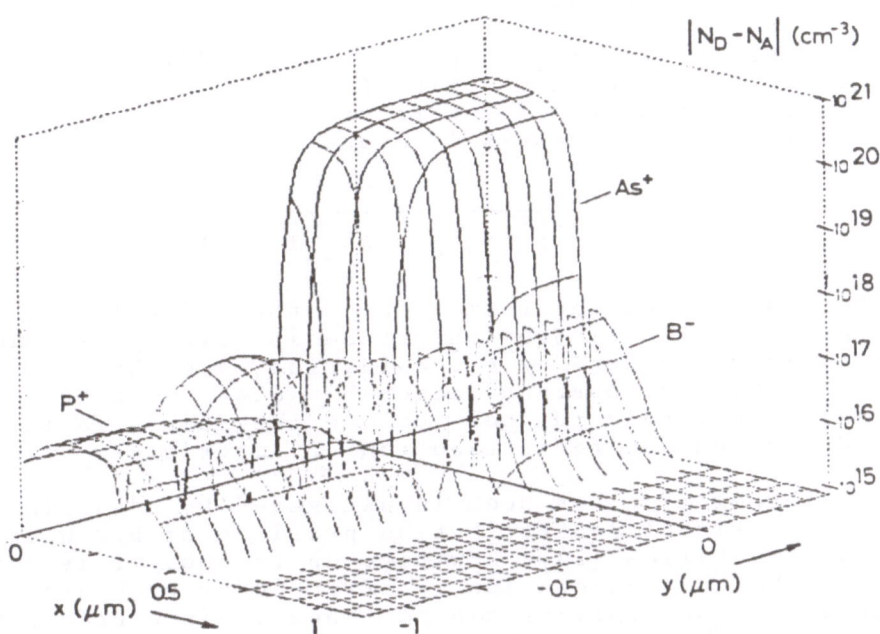

FIG. 9   Net active concentrations corresponding to
Fig. 8 after a drive-in for 20 min at 1000°C.

On the basis of such data parametrical descriptions of
the oxide growth have been derived (8), suited to fit
the data of a specific technology.

    In this paper oxide thickness as a function of
lateral dimension and time is approximated after (5) by

$$X_{ox}(y,t) = X(t) \frac{1}{2} \left\{ 1 + erf(\frac{\sqrt{2}}{\gamma} \frac{y}{X(t)}) \right\} \tag{13}$$

where γ is an arbitrary parameter fitting the final
slope of the oxide edge, and X(t) is the one-dimensional oxide thickness that is obtained with no masking
layer present. X(t) is calculated using Grove's model

$$X(t) = -\frac{A}{2} + \sqrt{\frac{A^2}{4} + B(t + t_o)} \tag{14}$$

with

$$t_o = \frac{1}{B}(X_o^2 + AX_o) .$$

Herein, B/A is the so-called linear rate constant, and B
is the parabolic rate constant; X_o denotes the initial
oxide layer thickness.

A useful simulation program must allow for the
modeling of a sequence of different oxidation steps.
Therefore, a differential form of (14) is used (9) which
accounts for the changes in A and B in successive oxidation steps

$$\Delta X(t) = -\frac{A}{2} - \sqrt{\frac{A^2}{4} + X_o^2 + AX_o + B\,\Delta t} - X_o \quad . \tag{15}$$

Having described the formation of the oxide layer, the
movement of the Si/SiO_2 interface is given by the amount
of silicon consumed by the oxidation, i.e. by multiplying X_{ox} by the constant factor b = 0.44. A reliable
simulation should further include the main effects of
oxidation on the redistribution of impurities.

Segregation has been investigated by measuring the
one-dimensional redistribution profiles of boron and
other impurities (22), however, up to now, it is not
clear whether segregation is affected by mechanical
forces in the range of a bird's beak. Since no other
data are available at present the one-dimensional segregation coefficients will be used in this paper.

Oxidation enhanced diffusion (OED) is another
important factor which should be accounted for in the
simulation since it can increase the effective diffusion coefficient by more than a factor of three (23).
The origin of OED has been attributed to the generation
of silicon self-interstitials (24), their generation
rate being proportional to the oxidation rate. Assuming
a quasi-equilibrium of self-interstitials and vacancies,
the diffusion coefficient at the Si/SiO_2 interface can
be calculated by an empirical formula (23):

$$D = D_i + \delta_{OED} \left(\frac{\partial X_{ox}}{\partial t}\right)^{0.3} \exp(-2.08/kT) \quad . \tag{16}$$

Depth dependence of OED, explained by the recombination
of self-interstitials and vacancies in the bulk silicon,
was found to have a typical decay constant of 25 μm.
Hence, it appears to be negligible in VLSI applications,
where profile depths less than 2 or 3 microns are
typical. A more precise simulation, allowing also for
deep profiles, should resolve the two-dimensional contours of the self-interstitial density, i.e. solve the

corresponding diffusion equation with appropriate boundary conditions. In this paper, however, eq. (16) will be used, neglecting the depth dependence of the OED.

3.2 Coordinate Transformation

The difficulties of modeling a nonplanar moving boundary in a fixed rectangular grid can be overcome by the use of a transformation technique (5). It maps the simulation grid to the time-depending contour of the Si/SiO_2 interface, which has the advantage that boundary conditions for oxidation apply on one single line at the top border of the grid. The method will be reiterated in this paragraph in order to derive the discretized equations used in this paper.

The transformation conserves the lateral coordinate y, but defines a new depth coordinate

$$\xi = x - b\, X_{ox}(y,t) \quad . \tag{17}$$

Applying this transformation to the transport equation is straightforward. In the following, however, the field term in eq. (1) will be omitted in order to reduce the computational effort. This simplification means that simulations must be confined to concentrations below the intrinsic density at diffusion temperatures, which, however, is not crucial to this paper since the applications envisaged meet this requirement.

In eq. (1) the concentrations $C(x,y)$ are replaced by $C(\xi,y)$. Hence, the derivates must be altered as demonstrated in the examples given in Table 3. Substitution of the derivates and rearranging the terms yield the diffusion equation (18) in the transformed plane (see Table 3). Discretization of this equation requires a 9-point formula due to the occurrence of cross derivatives. Table 4 displays the additional terms required. The complete discretized diffusion equation (19), displayed in Table 5, is obtained by adding the terms previously discussed.

TABLE 3

<u>Substitution of derivatives in transformed plane</u>:

$$\frac{\partial C}{\partial x} = \frac{\partial C(\xi,y)}{\partial \xi} \quad \frac{\partial \xi}{\partial x} = \frac{\partial C(\xi,y)}{\partial \xi}$$

$$\frac{\partial C}{\partial y} = \frac{\partial C(\xi,y)}{\partial y} + \frac{\partial C(\xi,y)}{\partial \xi} \quad \frac{\partial \xi}{\partial y}$$

$$= \frac{\partial C(\xi,y)}{\partial y} - b \frac{\partial Xox(y,t)}{\partial y} \frac{\partial C(\xi,y)}{\partial \xi}$$

$$\frac{\partial C}{\partial t} = \frac{\partial C(\xi,y)}{\partial \xi} \frac{\partial \xi}{\partial t} + \frac{\partial C(\xi,y)}{\partial t}$$

$$= \frac{\partial C(\xi,y)}{\partial t} - b \frac{\partial Xox}{\partial t} \frac{\partial C(\xi,y)}{\partial \xi}$$

<u>Diffusion equation in transformed plane</u>: (eq. (18))

$$\{1+(b \frac{\partial Xox}{\partial y})^2\} \frac{\partial}{\partial \xi} (D \frac{\partial C}{\partial \xi}) + \frac{\partial}{\partial y} (D \frac{\partial C}{\partial y})$$

$$+b (\frac{\partial Xox}{\partial t} - D \frac{\partial^2 Xox}{\partial y^2}) \frac{\partial C}{\partial \xi}$$

$$-b \frac{\partial Xox}{\partial y} \{\frac{\partial}{\partial \xi} (D \frac{\partial C}{\partial y}) + \frac{\partial}{\partial y} (D \frac{\partial C}{\partial \xi})\} = \frac{\partial C}{\partial t}$$

TABLE 4   Spatial discretization.

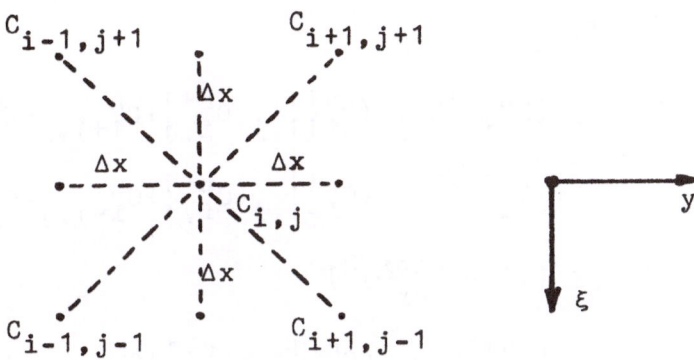

$$\frac{\partial}{\partial \xi} \left(D \; \frac{\partial C}{\partial y} \right) + \frac{\partial}{\partial y} \left(D \; \frac{\partial C}{\partial \xi} \right) \rightarrow$$

$$\frac{1}{8\Delta x^2} \{ D_{i,j-1}^{n+1/2} \; (C_{i+1,j-1}^{n+1} - C_{i-1,j-1}^{n+1} + C_{i+1,j-1}^{n} - C_{i-1,j-1}^{n})$$

$$+ D_{i,j+1}^{n+1/2} \; (C_{i-1,j+1}^{n+1} - C_{i+1,j+1}^{n+1} + C_{i-1,j+1}^{n} - C_{i+1,j+1}^{n})$$

$$+ D_{i+1,j}^{n+1/2} \; (C_{i+1,j-1}^{n+1} - C_{i+1,j+1}^{n+1} + C_{i+1,j-1}^{n} - C_{i+1,j+1}^{n})$$

$$+ D_{i-1,j}^{n+1/2} \; (C_{i-1,j+1}^{n+1} - C_{i-1,j-1}^{n+1} + C_{i-1,j+1}^{n} - C_{i-1,j-1}^{n}) \}$$

$$D \; \frac{\partial C}{\partial \xi} \rightarrow \frac{1}{4\Delta x} D_{i}^{n+1/2} \; (C_{i,j-1}^{n+1} - C_{i,j+1}^{n+1} + C_{i,j-1}^{n} - C_{i,j+1}^{n})$$

TABLE 5   Discretized diffusion equation (eq. (19)).

$0 = f_{i,j} =$

$\frac{1}{2\Delta x^2} \{ D_{i+1/2,j}^{n+1/2} (C_{i+1,j}^{n+1} - C_{i,j}^{n+1} + C_{i+1,j}^n - C_{i,j}^n)$

$+ D_{i-1/2,j}^{n+1/2} (C_{i-1,j}^{n+1} - C_{i,j}^{n+1} + C_{i-1,j}^n - C_{i,j}^n) \}$

$+ \frac{1}{2\Delta x^2} \{ 1 + (b \frac{\partial Xox}{\partial y})^2 \}$

$\{ D_{i,j-1/2}^{n+1/2} (C_{i,j-1}^{n+1} - C_{i,j}^{n+1} + C_{i,j-1}^n - C_{i,j}^n)$

$+ D_{i,j+1/2}^{n+1/2} (C_{i,j+1}^{n+1} - C_{i,j}^{n+1} + C_{i,j+1}^n - C_{i,j}^n) \}$

$- \frac{1}{8\Delta x^2} b \frac{\partial Xox}{\partial y}$

$\{ D_{i,j-1}^{n+1/2} (C_{i+1,j-1}^{n+1} - C_{i-1,j-1}^{n+1} + C_{i+1,j-1}^n - C_{i-1,j-1}^n)$

$+ D_{i,j+1}^{n+1/2} (C_{i-1,j+1}^{n+1} - C_{i+1,j+1}^{n+1} + C_{i-1,j+1}^n - C_{i+1,j+1}^n)$

$+ D_{i+1,j}^{n+1/2} (C_{i+1,j-1}^{n+1} - C_{i+1,j+1}^{n+1} + C_{i+1,j-1}^n - C_{i+1,j+1}^n)$

$+ D_{i-1,j}^{n+1/2} (C_{i-1,j+1}^{n+1} - C_{i-1,j-1}^{n+1} + C_{i-1,j+1}^n - C_{i-1,j-1}^n) \}$

$+ \frac{1}{4\Delta x} b (\frac{\partial Xox}{\partial t} - D_{i,j}^{n+1/2} \frac{\partial^2 Xox}{\partial y^2})$

$(C_{i,j-1}^{n+1} - C_{i,j+1}^{n+1} + C_{i,j-1}^n - C_{i,j+1}^n)$

$- \frac{1}{\Delta t} (C_{i,j}^{n+1} - C_{i,j}^n)$

3.3 Boundary Conditions

The moving boundary induces a flux of impurities depending on the segregation coefficient k, the surface concentration, the velocity v_{ox}, and the silicon volume change b at the interface

$$F = (1 - \frac{k}{b}) \, v_{ox} \, C \quad , \tag{20}$$

where k is defined as the ratio of the impurity concentration in oxide to that in silicon. This flux is balanced by the net flux of impurities diffusing through the interface. In this paper diffusion in oxide will be neglected, which makes the simulation much easier and which is, on the other hand, a good approximation for the impurities considered, i.e. for boron, arsenic and phosphorus.

Denoting the impurity gradient normal to the interface by $\partial C/\partial n$, the two-dimensional boundary condition at the interface line reads (25)

$$D \frac{\partial C}{\partial n} + (b - k) \left(\frac{\partial X_{ox}}{\partial t}\right)_n C = 0 \quad . \tag{21}$$

Defining an angle α by the positive y-direction and the contour of the interface line eq. (21) can be expressed in x- and y-coordinates:

$$D \left(\frac{\partial C}{\partial x} \cos\alpha - \frac{\partial C}{\partial y} \sin\alpha\right) + (b - k) \, \dot{X}_{ox} \cos\alpha \, C = 0 \quad . \tag{22}$$

Using further

$$tg\alpha = b \, X'_{ox}$$

and

$$\cos\alpha = (1 + (b \, X'_{ox})^2)^{-\frac{1}{2}}$$

and transforming (22) into the ξ- and y-plane yields:

$$D \left(1 + (b \, X'_{ox})^2\right) \frac{\partial C}{\partial \xi} - D \, b \, X'_{ox} \frac{\partial C}{\partial y} + (b - k) \, \dot{X}_{ox} \, C = 0 \tag{23}$$

This equation is discretized as shown in Table 6. It is finally used to calculate the virtual points R_i^{n+1}

TABLE 6

Discretization of boundary equation:

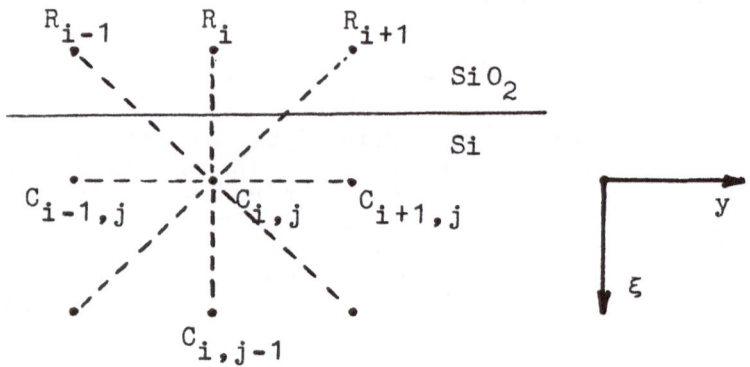

$$D \frac{\partial C}{\partial \xi} \rightarrow \frac{1}{2\Delta x} D_{i,j}^{n+1/2} (C_{i,j}^{n+1}-R_i^{n+1}+C_{i,j}^n-R_{i,j}^n)$$

$$D \frac{\partial C}{\partial y} \rightarrow \frac{1}{4\Delta x} D_{i,j}^{n+1/2} (C_{i+1,j}^{n+1}-C_{i-1,j}^{n+1}+C_{i+1,j}^n-C_{i-1,j}^n)$$

$$C \rightarrow \frac{1}{2} (C_{i,j}^{n+1}+C_{i,j}^n)$$

Virtual points:

$$R_i^{n+1} = \{1+(b \frac{\partial Xox}{\partial y})^2\}^{1/2} (C_{i,j}^{n+1}+C_{i,j}^n-R_i^n)$$

$$- \frac{1}{2}b \frac{\partial Xox}{\partial y} (C_{i+1,j}^{n+1}-C_{i-1,j}^{n+1}+C_{i+1,j}^n-C_{i-1,j}^n)$$

$$+ \frac{\Delta x}{D_{i,j}^{n+1/2}} (b-k) \frac{\partial Xox}{\partial t} (C_{i,j}^{n+1}+C_{i,j}^n)$$

required in the numerical solution of the transport equation.

3.4 Simulation Results

A sequence of different oxidation steps can be
simulated by LADIS. Local oxidation is modeled on the
basis of eq. (13) using a pad oxide defined by the
preceding simulation step. Assuming a mask edge at
y = 0, the shape of the bird's beak is calculated depending on the lateral oxidation parameter γ.

The first exampe, displayed in Fig. 10, demonstrates the effect of segregation on the redistribution
process. This simulation should be compared to the
inert anneal which has been shown in Fig. 5. It uses
the same initial distribution of boron as displayed in
Fig. 4 and also the same diffusion coefficient, anneal
time, and temperature. This simulation shows a very

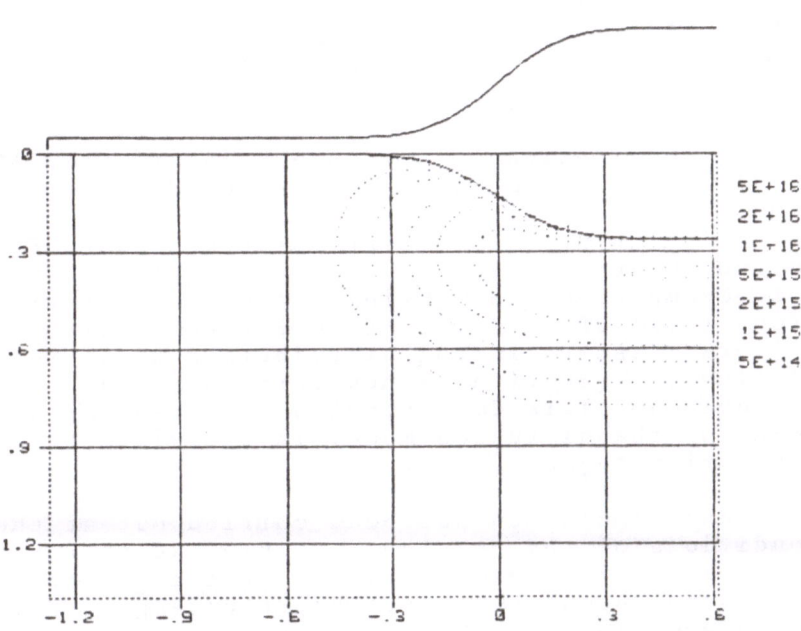

FIG. 10   Local oxidation in wet ambient for 120 min
at 1000°C assuming constant diffusivity. For the initial distribution see Fig. 4. Lateral oxidation parameter $\gamma = 0.5$

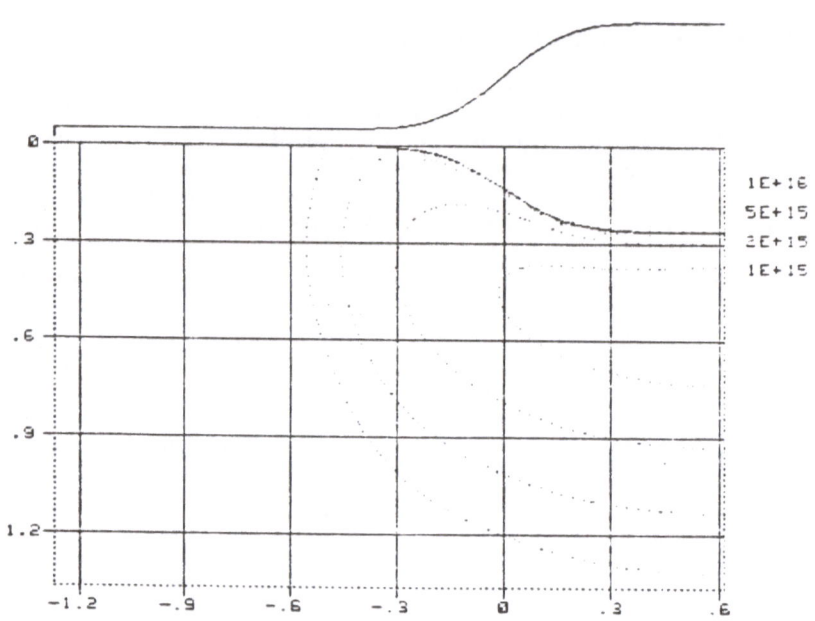

FIG. 11   Local oxidation corresponding to Fig. 10
with oxidation enhanced diffusion included.

similar diffusion profile except for the amount of
silicon and impurities consumed by the moving boundary.
This same example, but now using the more realistic
model (eq. (16)) of oxidation-rate-dependent diffusion
coefficients, results in the distribution shown in
Fig. 11. The distribution is now very different from
the previous one. This demonstrates the need for a
refinement of the diffusion model in the vicinity of
local oxidation edges.

    The next simulation has been carried out at the
same conditions of the previous example except for the
missing implantation mask, i.e. the initial distribution is a homogeneous implant. The resulting distribution, displayed in Fig. 12, gives an idea of the local
variation of the diffusion coefficient.

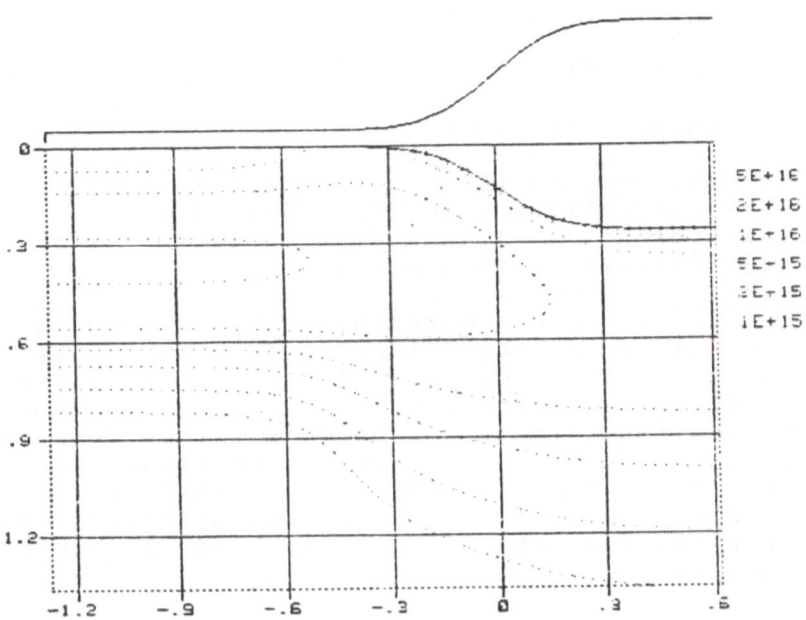

FIG. 12   Local oxidation performed on a homogeneous
implant of boron, showing the effect of OED.

    The next example consists of a sequence of simula-
tion steps typical of the fabrication of today's MOS
integrated circuits:

1.   Pad oxide growth, 50 nm.

2.   Si-nitride deposition, and etching.

3.   Field implant, boron (similar to Fig. 4).

4.   Local oxidation, wet ambient, 900°C (similar
     to Fig. 11).

5.   Inert anneal, 900°C.

6.   Reoxidation of pad oxide, O_2 + HCl ambient,
     900°C (see Fig. 13).

7.   60 nm etching (see Fig. 14).

8.   Gate oxide growth, 900°C (see Fig. 15).

9.   Channel doping, double implant (see Fig. 16).

10. Anneal steps, 900°C.

11. Source/drain implant, arsenic.

12. Anneal steps, 900°C.

Fig. 17 displays the final boron distribution in the cross-cut channel/field doping. Fig. 18 and 19 display the final boron and arsenic distributions in the cross--cut source/drain field doping.

A final example demonstrates the effects of different temperatures at which local oxidation is carried out. Using the same sequence of process steps except for a different temperature at step 4, taking now 1000°C, the resulting final boron distribution, displayed in Fig. 20, shows a more pronounced penetration of the channel stop doping into the channel than in the previous simulation. This is due to the different dependence of oxidation rate and diffusion coefficients on temperature.

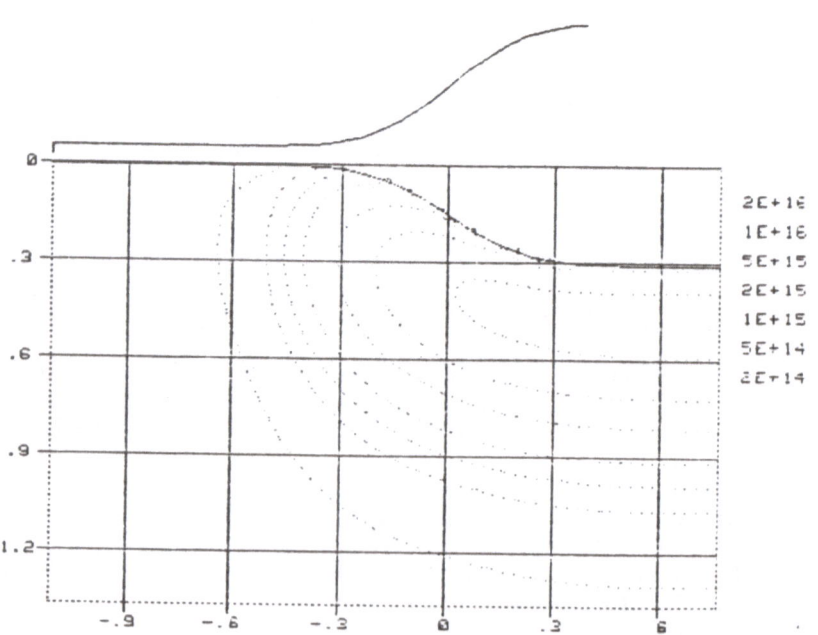

FIG. 13   Field implant after reoxidation of pad oxide.

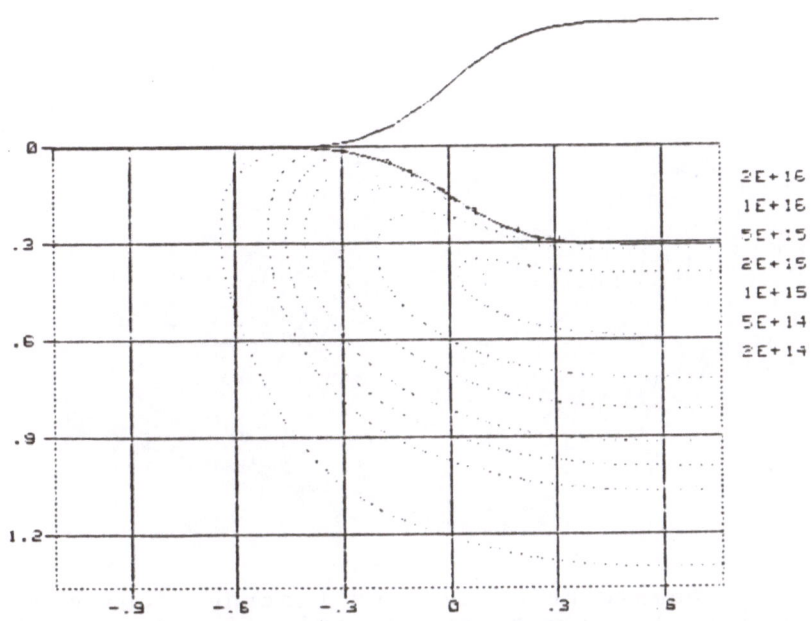

FIG. 14   Field implant after 60 nm etching.

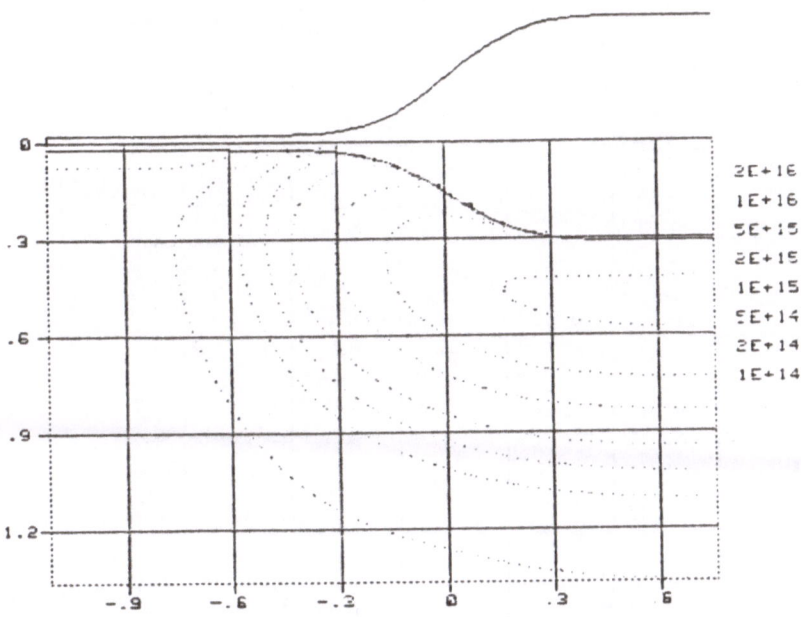

FIG. 15   Field implant after growing gate oxide in
HC1-O_2 ambient, 135 min at 900°C.

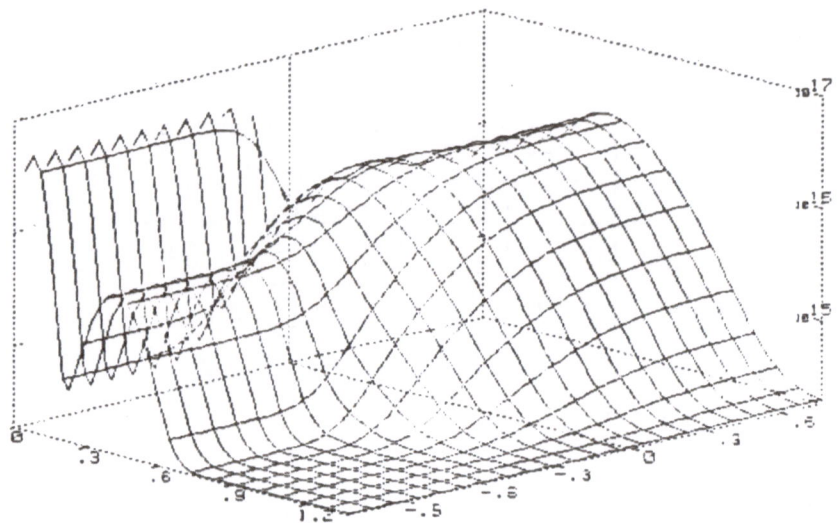

FIG 16.   Double boron implant (channel doping of
MOS-transistor) added to the distribution displayed in
Fig. 15.

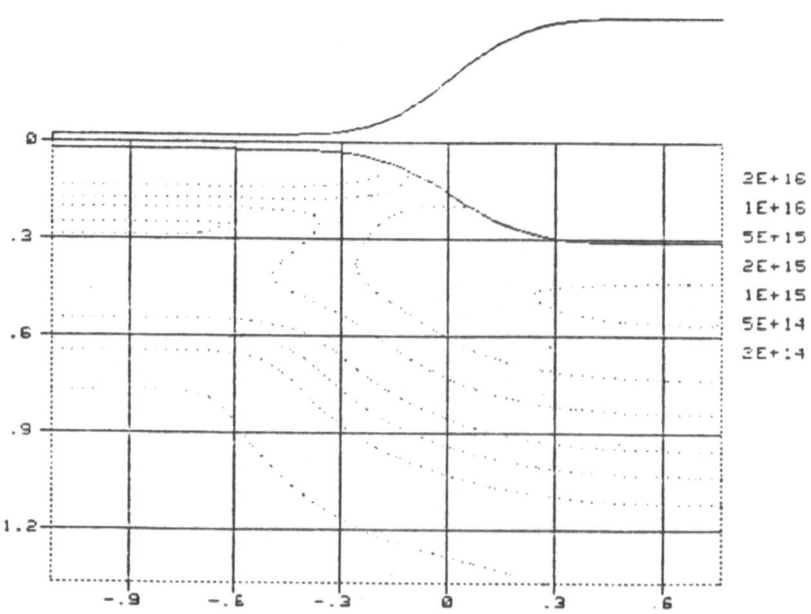

FIG. 17   Field and channel doping near the bird's
beak after final annealing steps, 5 h at 900°C.

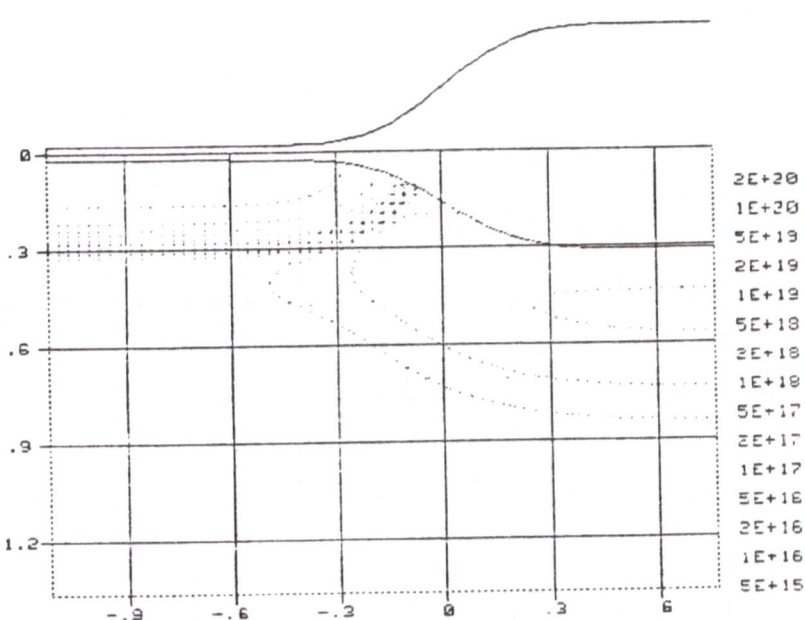

FIG. 18   Arsenic implant after final anneal adding
to the distribution displayed in Fig. 17.

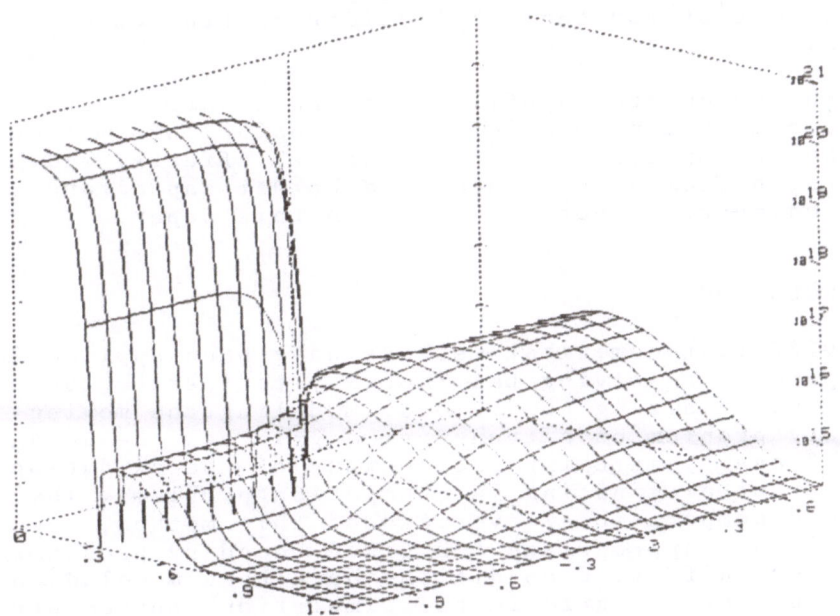

FIG. 19   3D-representation of Fig. 18.

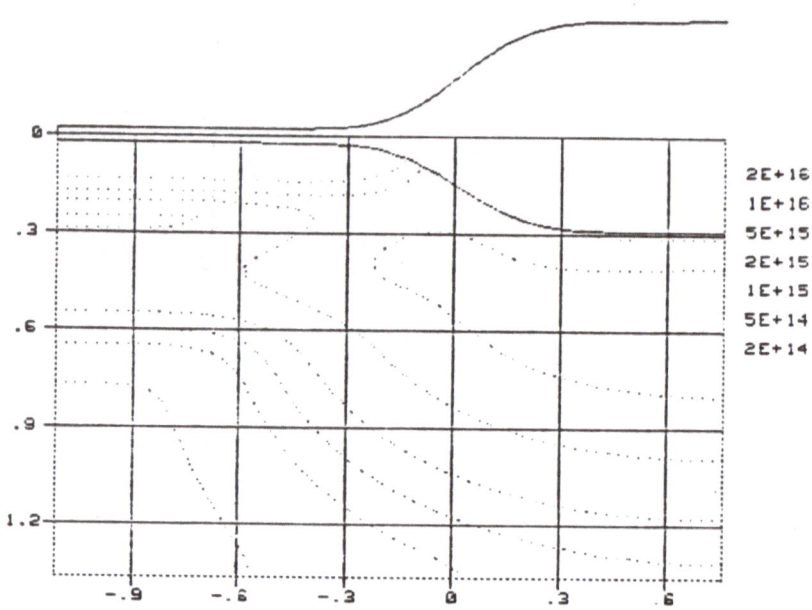

FIG. 20 Field and channel doping comparable to
Fig. 17 except for the local oxidation step using
1000°C.

The calculated profiles are input data of the
device simulation, providing the crucial information
for the optimization of short channel effects, width
effects, bottom capacitances and border capacitances.
This, however, is not discussed in this paper.

4. CONCLUSION

Different numerical methods have been applied in
this paper for solving the impurity redistribution at
inert and oxidizing conditions. Modeling the moving
boundary problem on a transformed grid has the advantage of easily matching the contour of the oxidation
front, thus eliminating interpolation problems. The
enhanced computational overhead of this method, however, is not appropriate for fixed boundary problems.
Thus, inert diffusion conditions have been modeled on a
fixed rectangular grid in the simulation program presented.

A nonplanar interface is accounted for in the fixed rectangular grid by omitting the points lying outside the silicon plane. The original silicon surface is taken as a reference of the interface coordinates and the surface topology in the sequence of simulation steps. This information provides also the oxide thickness present on top of the silicon, which is considered in the modeling of oxidation, oxide deposition, etching and ion implantation.

The simulation program LADIS has been applied to typical doping processes of the MOS technology. Computation time is minimized by adapting the level of complexity of the simulation method for the complexity of the process.

The two-dimensional simulation of diffusion under non-oxidizing conditions appears as an adequate mathematical extension of established one-dimensional models. No extra physical phenomena are involved in the lateral diffusion. Hence, those simulations are presumably as reliable as their one-dimensional counterpart.

The examples of modeling the local oxidation problems, on the other hand, reveal the dominant effect of the local variation of the diffusivity on the resulting distributions. These models are still at their infancy. Their refinement will depend on a further development of the yet rather primitive methods available for directly measuring two-dimensional impurity distributions. An indirect verification of the assumptions involved in the simulations, however, can be achieved by comparing electrical characteristics of test devices and the results obtained by a combined two-dimensional process and device simulation.

ACKNOWLEDGEMENT

The author wishes to thank E. Guerrero for the use of research work on the arsenic cluster model, and he wishes to thank A. Seidl for helpful discussions of the moving boundary problem.

REFERENCES

1. R.W. Dutton, and S.E. Hansen, "Process modeling of integrated circuit device technology", Proc. of the IEEE, vol. 69, no. 10, (1981), pp. 1305-1320

2. R. Tielert, "Two-dimensional numerical simulation of impurity redistribution in VLSI process", IEEE Trans. Electron Devices, vol. ED-27, (August 1980), pp. 1479-1483

3. H. Ryssel, K. Haberger, K. Hoffmann, G. Prinke, R. Dumcke, and A. Sachs, "Simulation of doping processes", IEEE Trans. Electron Devices, vol. ED-27, (August 1980), pp. 1484-1492

4. B.R. Penumalli, "Process simulation in two dimensions", ISSCC Dig. Tech. Papers, (1981), pp. 212-213

5. B.R. Penumalli,"Lateral oxidation and redistribution of dopants", presented at 2nd Int. Conf. on Numerical Analysis of Semiconductor Devices and IC's (NASECODE II), Dublin, Ireland, (17.-19. June 1981)

6. W.D. Murphy, W.F. Hall, and C.D. Maldonado, "Efficient numerical solution of two-dimensional nonlinear diffusion equations with nonuniformly moving boundaries: A versitale tool for VLSI process modeling", presented at 2nd Int. Conf. on Numerical Analysis of Semiconductor Devices and IC's (NASECODE II), Dublin, Ireland, (17.-19. Juni 1981)

7. K. Taniguchi, M. Kashiwagi, and H. Iwai, "Two-dimensional computer simulation models for MOSLSI fabrication processes", IEEE Trans. Electron Devices, vol. ED-28, (May 1981), pp. 574-580

8. H.G. Lee, and R.W. Dutton, "Two-dimensional low concentration boron profiles: Modeling and measurement", IEEE Trans. Electron Devices, vol. ED-28, (October 1981), pp. 1136-1147

9. D.A. Antoniadis, S.E. Hansen, and R.W. Dutton, "SUPREM II - A program for IC process modeling and simulation", Stanford University, Integrated Circuits Laboratory, Tech. Rep. 5019-2, (June 1978)

10. R.O. Schwenker, E.S. Pan, and R.F. Lever, J. Appl. Phys., vol. 42, (1971), p. 3195

11. S.M. Hu, Atomic Diffusion in Semiconductors, edited by D. Shaw, New York: Plenum, (1973), p. 306

12. R.B. Fair, and C.R. Weber, J. Appl. Phys., vol. 1, (1973), p. 273

13. M.Y. Tsai, F.F. Morehead, and J.E.E. Baglin, J. Appl. Phys., vol. 51, (1980), p. 3230

14. E. Guerrero, H. Pötzl, R. Tielert, M. Grasserbauer, and G. Stingeder, "Generalized model for the clustering of As dopants in silicon", J. Electrochem. Soc., vol. 129, (August 1982), pp. 1826-1831

15. J.F. Gibbons, W.S. Johnson, and S.W. Mylroie, Projected range statistics, (Stroudsburg, PA: Dowdon, Hutchinson, and Ross, 1975)

16. S. Furukawa, H. Matsumara, and H. Ishiwara, Japan. J. Appl. Phys., vol. 11, (1972), p. 134

17. H. Runge, Phys. Status Solidi, vol. (a) 39, (1977), p. 595

18. D.C. Chin, M.R. Kump, H.G. Lee, and R.W. Dutton, "Process design using two-dimensional process and device simulators", IEEE Trans. Electron Devices, vol. ED-29, (Feb. 1982), pp. 336-340

19. D.U. von Rosenberg, Methods for the Numerical Solution of Partial Differential Equations, (New York: Elsevier, 1969)

20. E. Bassons, H.N. Yu, and V. Maniscalo, "Topology of silicon structures with recessed SiO_2", J. Electrochem. Soc., vol. 123, (1976), p. 1729

21. J.C.-H. Hui, T.-Y. Chin, S.-W.S. Wong, and W.G. Oldham, "Sealed-interface local oxidation technology", IEEE Trans. Electron Devices, vol. ED-29, (April 1982), pp. 554-561

22. J.W. Colby, and L.E. Katz, "Boron segregation at $Si-SiO_2$ interface as a function of temperature and orientation", J. Electrochem. Soc., vol. 123, (1976), p. 409

23. K. Taniguchi, K. Kurosawa, and M. Kashiwagi, "Oxidation enhanced diffusion of boron and phophorus in (100) silicon", J. Electrochem. Soc., vol. 127, (October 1980), pp. 2243-2248

24. R.B. Fair, "Oxidation, impurity diffusion, and defect growth in silicon - An overview", J. Electrochem. Soc., vol. 128, (June 1981), pp. 1360-1368

25. A. Seidl, private communication (1982)

FEDSS - FINITE ELEMENT DIFFUSION SIMULATION SYSTEM

K. A. Salsburg

International Business Machines, Federal Systems Division
Manassas, Virginia
H. H. Hansen

International Business Machines, General Technology Division
Essex Junction, Vermont

Abstract

An accurate model of the diffusion of impurity atoms into a substrate is necessary to assess the effects of process changes on impurity profiles. The FEDSS program simulates semiconductor processes in two dimensions. The process steps to be modeled include ion implantation, oxidation/drive-in, chemical pre-deposition through the surface, and oxide deposition. The finite-element method transforms the diffusion equation for impurity atoms to a simulation model at a discrete number of points. Direct techniques are used to solve the resulting matrix equations. Pre- and post-processors enable users to rapidly generate new models and analyze results. The impurity distributions resulting from sequences of the process steps and the simulated shape of two-dimensional oxide growth are shown.

SECTION 1 OVERVIEW

One-dimensional programs no longer are adequate in the realm of VSLI technology. To model small devices successfully, and to establish valid design rules, two-dimensional (2-D) capabilities have to be heightened in the fields of process simulation and impurity profiles. Two-dimensional process

simulation programs have been discussed previously [1-4]. However, these have been based primarily on finite difference
techniques and have been limited somewhat in scope. A general
purpose 2-D process simulation program called FEDSS (Finite
Element Diffusion Simulation System) will be discussed in this
paper. The redistribution of arsenic, boron and phosphorus in
one- or two-dimensions is simulated by this program. The composite 2-D impurity profile output from FEDSS is used as input to
the device analysis program, FIELDAY [5].

In this program, the finite element method is used to
approximate numerically the solution to the diffusion equation.
The algorithm utilizes a symbolic and numeric factorization
technique so that the solution of the appropriate sparse matrix
equation is rapid. The finite element method defines the
irregular geometries so that spatially dependent parameters can
be handled readily. In addition, more discretization nodes can
be allocated to regions of physical interest without increasing
the cost of solution.

The primary method of introducing impurities is through ion
implantation, although impurities also can be introduced by
chemical predeposition through the surface. The ion implantation
models include angled-implant distributions, Gaussian, joined
half-Gaussian, and Pearson IV distributions. Implants can be
windowed to place them in a region defined by a mask where
either vertical mask edges or arbitrary edges can be specified.
A Gaussian lateral implant component also can be included in the
models. Where required, exponential tails are added to the
as-implanted distributions.

Once introduced into the oxide/silicon, various impurity
redistribution models are used to determine the appropriate
diffusivity which is then used in the diffusion equation. Where
required, concentration-dependent diffusion phenomena are
accounted for in these models. In the case of arsenic diffusion,
the model includes the interaction with neutral, singly-charged
and doubly-ionized vacancies [6]. The clustering is accounted
for by a two arsenic model with both equilibrium and non-
equilibrium conditions considered.

While a generalized oxidation scheme has not been realized
for this process simulator, some preliminary attempts have been
made which appear to hold promise for simulating the growth of
"birds beak" regions. The method employed is to solve the
diffusion equation for the transport of the oxidizing species
through the oxide and then to change the material property of
the silicon when a sufficient amount of the oxidizing species is
present. A deposited silicon nitride layer acts as a diffusion
barrier in regions where oxidation is not desired.

380

A particularly useful feature is the ability to simulate
low temperature oxide deposition. This often is used where
impurities are introduced after an oxidation step and the redistribution in subsequent temperature cycles is to be studied.

To use the program, the operator first constructs a finite
element mesh describing the problem's geometry. A special
user-interactive program is employed for this step, or previously-
created meshes are used. FEDSS then is run, simulating each
process step and using as an initial condition the results of
the previous process step. At the end of the simulation, the
results are viewed graphically as line plots, contour plots, or
as perspective plots using a user-interactive plotting program.
The specific profiles plotted are done at the user's option.
FEDSS is but one part of an overall system based on the finite
element method. Figure 1 shows an overall flow chart of this
system. Where coupling into the device analysis program is
required, meshes suitable for device calculations are used as
they place a greater dependence on the grid than do process
simulations.

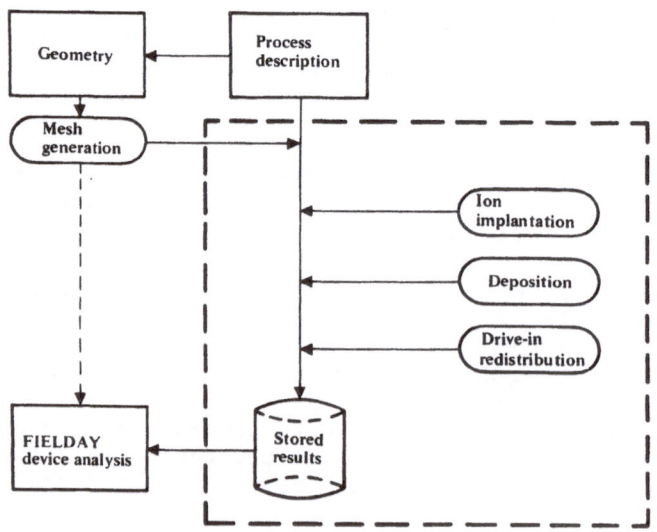

Note: Dashed lines show process modeling activity.

Figure 1.   FEDSS process flow.

SECTION 2   PROCESSING MODELS

There are currently five different types of processing steps that can be modeled by FEDSS: ion implantation, drive-in, oxidation, chemical predeposition through the surface, and low temperature oxide deposition. Drive-ins are identical to oxidations except that drive-ins occur in nitrogen or neutral ambients. Each step in a processing sequence uses the distributions that resulted from the previous step as the starting point for its calculations.

Ion Implantation

Ion implantation is a method of introducing impurities into the silicon or silicon dioxide. Ion implantation steps in FEDSS may be specified in two ways. One way is to specify the implant energy in KeV and the dose in ions/cm. The program uses tables to calculate parameters used in distribution fitting. The other way is to specify the distribution parameters and the dose. The distributions available are a Gaussian distribution, two joined half-Gaussian distributions [7], a Pearson IV distribution, an angled-Gaussian distribution [8], a user-defined distribution, and a Gaussian distribution with a lateral component. The default profile is a two-sided Gaussian for the elements, arsenic, and phosphorous, and a modified Pearson-type IV distribution for boron. Implants can be windowed to place them in a region defined by a mask. The effects of a mask and the mask edge are taken into account. Either vertical mask edges or arbitrary edges can be specified. Where required, exponential tails are added to the as-implanted distributions.

The default implant model for arsenic and phosphorus is the joined half-Gaussian distribution [7,9]. However, in the case of arsenic, it also is possible to enter the parameters for a Pearson IV distribution if a more accurate determination of the implanted profile is needed [10]. The Pearson IV parameters give a profile very near that of a Gaussian.

Drive-In

Solid-state diffusion is the mechanism responsible for impurity migration within the silicon body during high-temperature process steps. Given an initial concentration distribution, the diffusion equation

$$\nabla \cdot (D \cdot \nabla C) = \frac{\partial C}{\partial t} \tag{1}$$

is simulated for a given number of minutes at a given temperature. The diffusivity D, can be concentration dependent. The left side of the equation is replaced by the finite element functional while the right side of the equation is related to a first order, fully-implicit finite difference approximation.

The drive-in step is a high temperature one which redistributes impurities already present according to Equation (1). The user specifies the total step time and the wafer temperature. The program calculates a time step (Δt), determines whether to use an implicit or explicit method, and calculates a value for diffusivity for each element at each step in time. Using this set of diffusivity values, a vector of concentration values is calculated for each node at each step in time. When the total step time is completed, the drive-in step is finished and the resulting concentration values are ready to be used as an initial condition for the next process step.

The specific form of the diffusion equation solved by FEDSS is

$$\nabla \cdot (D_{eff} \cdot \nabla C) = \frac{\partial C}{\partial t} \qquad (2)$$

where C is the impurity concentration being diffused and D_{eff} is the effective diffusivity calculated by various phenomenological impurity redistribution models. D_{eff} is the product of three terms; namely, the intrinsic diffusivity at the process temperature, the electric field enhancement factor, and a term depending on the concentration of neutral and charged vacancies.

Once introduced into the oxide/silicon, various impurity redistribution models are used to determine the appropriate diffusivity which then is used in the diffusion equation. Where required, concentration-dependent diffusion phenomena are accounted for in these models. In the case of arsenic diffusion, the model includes the interaction with neutral, singly-charged and doubly-ionized vacancies [11]. The clustering is accounted for with a two arsenic model considering both equilibrium and non-equilibrium conditions. Three arsenic atoms and an electron together form an alternate clustering model.

The arsenic diffusion model is of some importance in that it is used commonly to form source-drain regions in FET devices and serves as the emitter and subcollector of bipolars. A relatively simple arsenic clustering model based on two arsenic atoms forming a neutral cluster has been incorporated into FEDSS. In this model, the total arsenic flux is assumed to be due to the diffusion of AsV+, AsV, and AsV- bound defect complexes*. Using this assumption, the flux for the total arsenic

---

* This model was developed by F. Morehead.

concentration C_T, including the effect of the electric field on the charged vacancy complexes can be written as:

$$-J_{C_T} = D_{eff} \frac{\partial C_T}{\partial x} \tag{3}$$

in one dimension, where

$$D_{eff} = D_i \left\{ \frac{1 + \beta \frac{n}{n_i} + \gamma \frac{n^2}{n_i}}{1 + \beta + \gamma} \right\} C_A \left\{ \frac{\partial \ln C_A}{\partial C_T} + \frac{\partial \ln n}{\partial C_T} \right\} \tag{4}$$

and C_A is the electrically-active arsenic concentration, n is the electron concentration, n_i is the intrinsic electron concentration at the diffusion temperature [12], and the parameters beta=100 and gamma=4. Thus, for a specific relationship between C_T and C_A, the value of the effective diffusivity, D_{eff}, can be calculated.

The arsenic clustering is assumed to proceed through the following fourth order reaction:

$$2As^+ + 2e^- + V \rightarrow As_2V \tag{5}$$

and

$$K_{eq} = \frac{[As_2\ddot{V}]}{C_A^2 n^2 [v]} \tag{6}$$

using the relationship

$$C_T = C_A + 2[As_2V] \tag{7}$$

the following equation then gives the relationship desired between the total and active arsenic:

$$C_T = C_A + \frac{2C_A^2 n^2}{s^3} \tag{8}$$

where s is a parameter related to the clustering equilibrium constant by

$$s = \sqrt[3]{\frac{1}{K_{eq}[V]}} \tag{9}$$

A value of $s/n_i = 40$ provides a good fit to the data, except in the limit of high concentrations ($C_T > 100 n_i$).

384

The diffusion of boron is assumed to be through neutral and positively-charged vacancies, so the following equation gives the effective concentration-dependent diffusivity:

$$\frac{D}{D_i} = \frac{1 + \beta \frac{p}{n_i}}{1 + \beta} \tag{10}$$

The parameter beta is chosen to be ten [13].

Phosphorous redistribution is treated using the model due to R. B. Fair [14]. The model predicts, with reasonable accuracy, the phosphorous kink formation as well as the base push effect, commonly observed during heavy emitter diffusions in bipolar technology.

A technique commonly used to obtain approximate solutions for Equation (1) is the finite difference method, which approximates derivatives by using differences. The physical domain of interest conceptually is divided into a large number of two-dimensional rectangular cells. Impurity concentrations in the device being studied are calculated at the intersections of the rectangular cell boundaries called nodal points (Figure 2). The density of nodes required in any region of the device is directly-related to the magnitude of the impurity concentration gradient in that region. A problem with the finite difference method is that all nodes in the system must be at the corners of neighboring cells. This constraint forces an increase in the density of nodes in areas otherwise of little interest solely because they are aligned horizontally or vertically with regions containing higher impurity variations. Since the solution's computer time requirements are related to the number of nodes, a more desirable system would be one which allows changes in the density of nodes

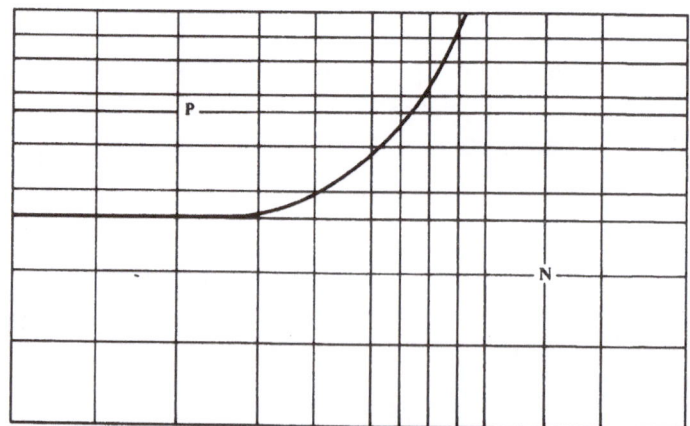

Figure 2. Finite difference mesh.

across the device as impurity variations dictate, yet remains
independent of the geometry of neighboring nodes.

The finite element method, used in FEDSS, allows continuous
and independent variation in cell density resulting in fewer
nodes being required for a given problem. Triangular elements
are used to divide the device domain (Figure 3). An expression
can be obtained for each element which relates the geometry and
diffusivity of the element to the impurity concentration at the
nodes of that element [15]. It can be shown using variational
calculus that the functional representing the right side of
Equation (1) is

$$I(C) = \frac{1}{2} \int_A D|\nabla C|^2 \, dA \qquad (11)$$

while the left side of Equation (1) can be expressed as a first
order backward difference over time

$$\frac{\partial C}{\partial t} \approx \frac{C(t) - C(t-\Delta t)}{\Delta t} = \frac{c_i - c_{i-1}}{\Delta t} \qquad (12)$$

Linear functions of C over the element are used and these
functions for C are substituted into Equation (11), the first
variation taken and set to zero. Using these expressions, the
following matrix equation is obtained:

$$[A]\{c_i\} = [B]\left\{\frac{c_i - c_{i-1}}{\Delta t}\right\} \qquad (13)$$

The element coefficient matrix A is symmetric and its terms are
a function of elemental geometry and diffusivity. The matrix B

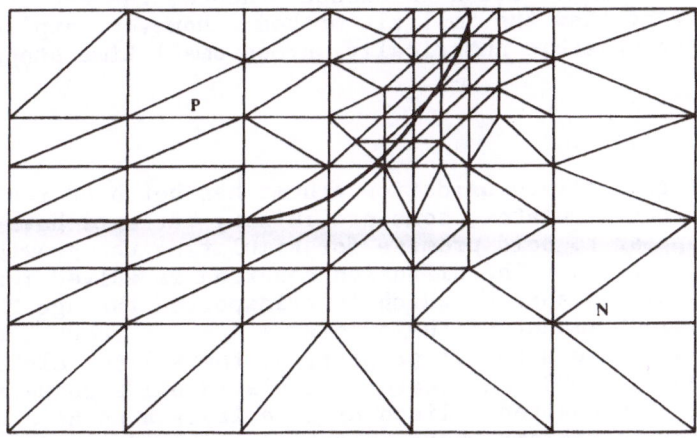

Figure 3. Finite element mesh.

results from relating the number of impurity atoms and concentration at a node with a weighting scheme on an element-by-element basis. The global forms of these matrices then are assembled in the usual way.

The matrix A may be large since its order is equal to the number of nodes being analyzed. The matrix equation must be solved many times, once for each different impurity at each time step. Sophisticated matrix techniques are required to obtain all these solutions in a reasonable amount of computer time. Many research papers have been published on these techniques since solution of large sets of equations of the form of Equation (13) takes up the major cost of computation in semiconductor analysis, linear programming, circuit analysis, and structural analysis [16, 17, 18]. The matrices are sparse since the number of non-zero terms usually is less than 5 percent. A direct matrix solution technique was chosen rather than an iterative technique since the solution time for iterative methods depends on the numerical conditioning of the matrix. FEDSS uses a technique from the Waterloo Sparse Matrix Package (SPARSPAK) [19] which orders the coefficient matrix based on its symbolic structure, i.e., where non-zeroes appear, then numerically solving the matrix equations as a separate step. The symbolic processing is done only once, but the numeric solution is evaluated many times.

If explicit time differencing is used, the following matrix equation is obtained:

$$[A]\{c_{i-1}\} = [B]\left\{\frac{c_i - c_{i-1}}{\Delta t}\right\} \tag{14}$$

To obtain the desired unknowns $\{c_i\}$ with this technique, only matrix vector multiplication is needed. Hence, the explicit method is faster than the implicit method. However, explicit methods can be unstable numerically unless small time steps (Δt) are used.

Oxidation

While a generalized oxidation scheme has not been realized for this process simulator, some preliminary attempts have been made which appear to hold promise for simulating the growth of "birds beak" regions. The diffusion equation is solved for the amount of oxidizing species which is transported through the oxide. When a sufficient amount of the oxidizing species reaches the silicon surface, the material property of the silicon elements is changed. Hence, FEDSS can create a simulated oxide shape as shown in Figure 4. A deposited silicon nitride layer acts as a diffusion barrier in regions where oxidation is not desired.

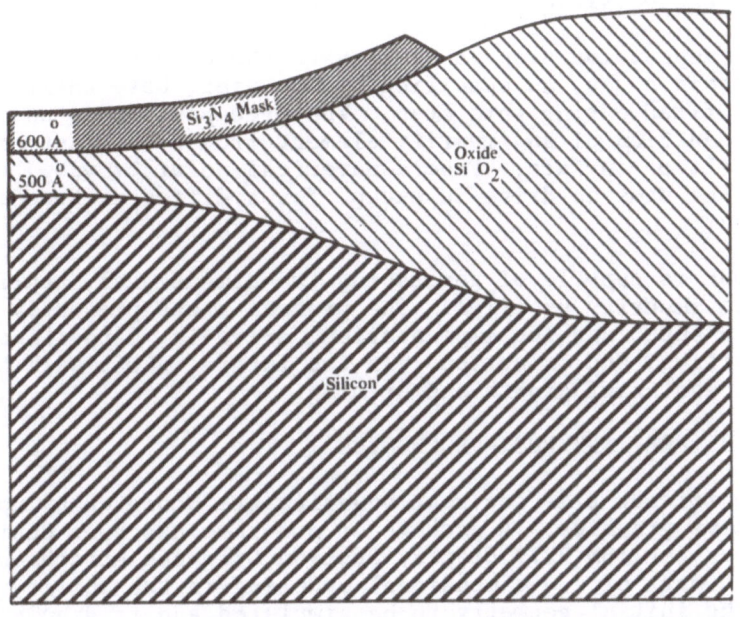

Figure 4. Simulated semi-recessed oxide shape.

The oxidation step is a high temperature step which includes
the redistribution of the impurities present due to diffusion
and evaporation. It involves solving the diffusion equation for
the concentration of oxygen and accounting for the growth of
oxide. The Deal/Grove model presently is being used [20].
During oxidation, in addition to the impurities diffusing,
oxygen (O_2) is diffusing. When a sufficient quantity of O_2
reached the silicon, SiO_2 molecules are formed. This results in
a moving silicon-silicon dioxide interface.

Low Temperature Oxide Deposition

A particularly useful feature is the ability to simulate
low temperature oxide deposition. This is used often where
impurities are introduced after an oxidation step and the redistribution in subsequent temperature cycles is to be studied. In
this step, doped or undoped oxide is deposited on the silicon
surface or on top of any existing oxide. The amount of oxide
deposited is the interior of the rectangle whose vertices are
specified. If the deposited oxide is to be doped, then the
impurity type and concentration is specified.

Chemical Predeposition Through the Surface

The deposition step is similar to the drive-in step except that certain nodes, as specified by the user, have their concentrations fixed during this step. The predeposition step models the introduction of an impurity element from a constant source at the silicon surface. From a numerical point of view, this step is the same as drive-in except for a different boundary condition.

FEDSS is capable of generating two-dimensional, nonlinear diffusion impurity profiles. The process steps presently implemented in FEDSS are ion implantation, drive-in, deposition, predeposition, and oxidation. In the future, FEDSS will be able to model epitaxial growth and etching.

SECTION 3 PRE- AND POST-PROCESSING

Significant time and cost benefits are achieved by using interactive graphics. A package of programs has been designed for use on the IBM 3277 Display Station Graphics Attachment. An important part of two-dimensional modeling is the ability to define the initial geometry to be simulated and to display for interpretation the results of the simulation. Extensive pre- and post-processors are available for use with FEDSS for defining the initial mesh and for displaying the results on graphics display units. These processors considerably enhance the utility of the program.

Model definition consists of generating a finite element mesh and assigning material properties to each element. The FEDSS user has available several options for model definition. One such option is (TRIangular Mesh generator) TRIM. Here, the user specifies the boundary of a region and selects a mathematically-regular grid, such as a rectangular mesh. TRIM generates a conformal map of that mesh onto the user's model. The same mesh may be mapped onto geometrically-similar models; thus, the user need not respecify the mesh generation information but only a few details relative to his model. TRIM generates, displays, and stores the mesh for later use by FEDSS. During mesh display, windowing and node and element numbering may be utilized. Figure 5 shows a mesh, with a typical polysilicon shape, generated by TRIM. Note the varying density of elements over the model. This increases the accuracy of solution at points where the impurity concentration gradients are changing rapidly.

A library of FEDSS models exists for frequently-modeled devices with well-defined structures. With these models, the user supplies various parameters such as oxide thickness and

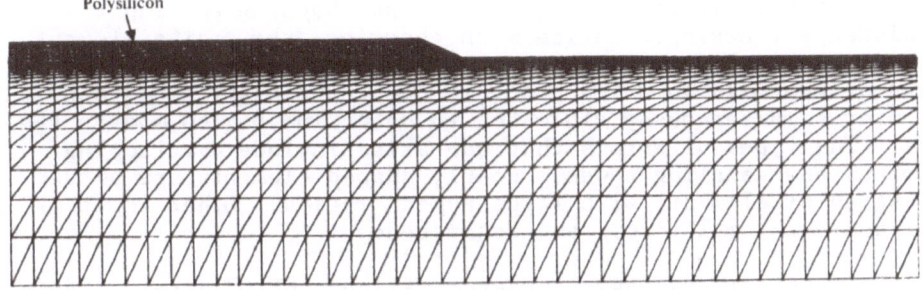

Figure 5. TRIM-generated mesh with a typical polysilicon shape.

junction depth and the model is stretched to reflect the given
parameters. Again, mesh storage and display is possible. The
user then can delete elements or change their material properties.
Figure 6 shows a mesh used for simulating a doubly-diffused
IGFET.

Figure 6. Mesh used for simulating a doubly-diffused IGFET.

Input verification consists of mesh checking and input consistency checking. During mesh checking, the finite element mesh is examined for errors and poorly shaped-elements. The areas of the mesh in which problems occur are highlighted and the user simultaneously can watch to determine how to modify the mesh. The types of errors that can arise include overlapping elements and dangling nodes. Poorly-shaped elements are obtuse triangles or elements with large aspect ratios. Other input is examined for completeness and consistency.

For post-processing, a program called FEMPLOT permits rapid interpretation of the FEDSS analysis with the ability to see the results as they occur. FEMPLOT will display nodal values of impurity concentration with contour plots, line graphs, or perspective plots. Contour plots are a means of displaying nodal data values. Lines of equal value are drawn through points of equal value interpolated along the sides of the elements. Perspective graphs are a means of displaying all the nodal data from a two-dimensional surface of a model.

Model definition using interactive graphics replaces the time-consuming task of meticulously defining every node and element in the finite element mesh, which formerly took 65-70% of the total analysis time. FEMPLOT minimizes the time the FEDSS user spends searching through results on a node-by-node or element-by-element basis. Process designers are able to optimize designs by rapidly viewing simulation results and noting impurity profiles within the device that cannot be measured experimentally. This capability results in fewer design projects terminating after considerable resource expenditure. Therefore, the use of interactive graphics results in higher engineering productivity by reducing development time, lowering development costs, and making more resources available for product optimization.

SECTION 4 PROCESS SIMULATION RESULTS

Boron profiles are calculated using a Pearson IV distribution with an attached exponential tail [9, 10]. The tail is included as part of the default distribution, but since its presence is somewhat dependent on the process, it is possible to leave the tail off of the distribution, if desired. A tail attached at 45% of the peak concentration and with a characteristic length of 120 nm was found appropriate for boron-implanted profiles measured in IBM's laboratory. An example of a boron implant at 40 KeV followed by a neutral ambient drive-in is shown in Figure 7 and is compared with data measured by pulsed C-V techniques. The fit is excellent.

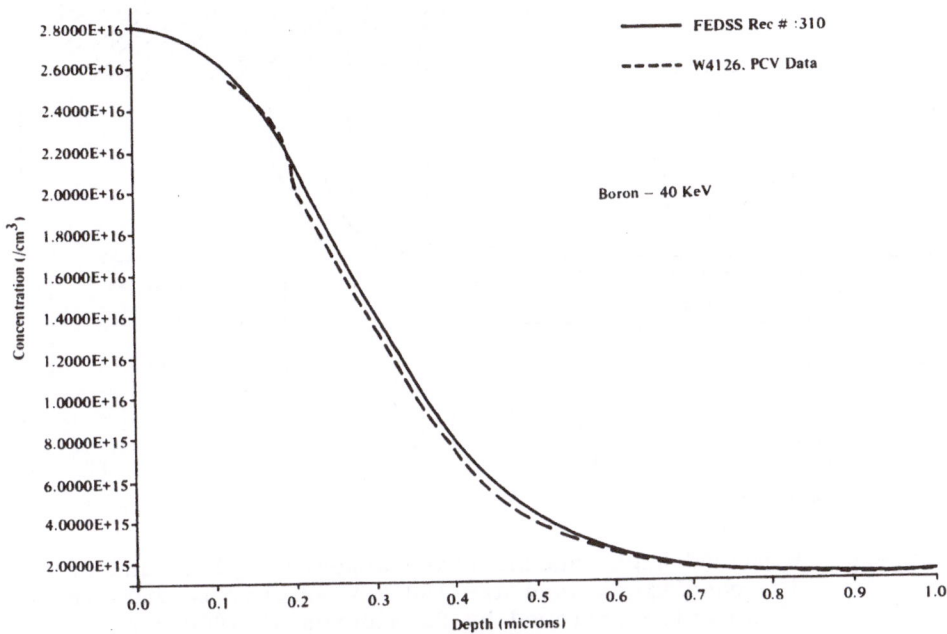

Figure 7. Boron implanted at 40 KeV and driven in compared to pulsed CV measurements.

An arsenic implant and drive-in provides a test for the arsenic model. A 25 nm screen oxide was put into place using the STEP='DEPO' process. Next, the arsenic was implanted through the oxide at 140 keV to a dose of 2.0E16 using the default-joined half-Gaussian model. The neutral ambient drive-in was for 20 minutes at 1000 degrees centigrade. The symmetry of the problem enables one to choose the window edge at x=0.5 microns. The two-dimensional result is shown in Figure 8 using a linear plot for the concentration contours. A line plot of the same result enables the results to be compared against a profile obtained by spreading resistance (the electrically-active arsenic only). The result is shown in Figure 9. The simulation is seen to be accurate -- both results give a junction depth of 0.34 microns. Figure 9 also shows the total arsenic concentration profile.

An example illustrating boron and phosphorus redistribution at low concentrations provides a practical illustration of the use of the program as a 2-D process simulator. Windowed implants and enhanced diffusivity during "oxidation" were used in the simulation of a proposed bulk CMOS process. The simulation includes a phosphorus implant to form the N-well region and a boron implant to form the field region. Of interest is not only

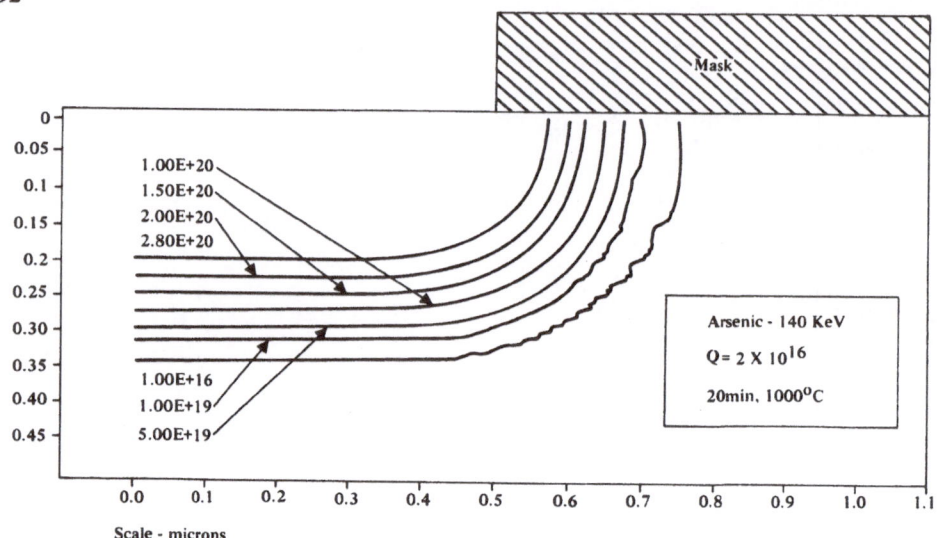

Figure 8.   Windowed high concentration arsenic implant and drive-in.
The conditions used are 140 KeV arsenic to 2E16 through
25 nm oxide followed by 20 minutes at 1000 deg C.

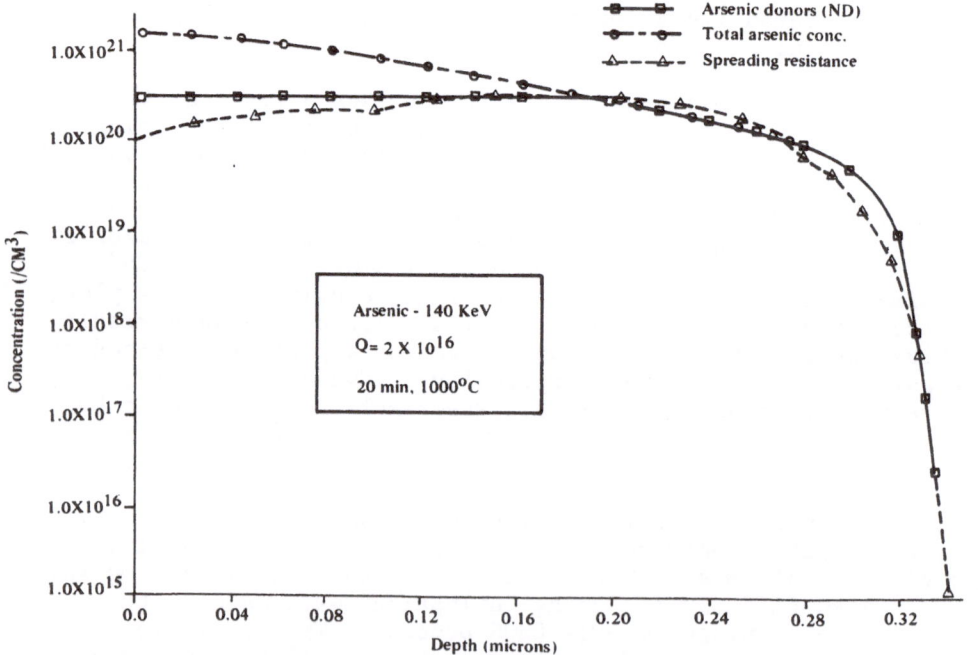

Figure 9.   Line plot of Figure 5 compared to spreading
resistance data.

the junction depth of the N-well region, but mostly the extent of
the lateral penetration of the boron into the N-well. The simulated process is outlined in Table 1. The mask openings given in
Table 1 have the N-well and field mask edges at the same point.
Other possibilities are to examine the effects of various mask
mis-alignments. These results can be used to aid in the development of preliminary ground rules for designers. The results for
a small displacement of the boron mask are shown in Figure 10.

Table 1.   Simulated Bulk CMOS Process in 2-D.

1.   Substrate, concentration 1.0E15.
2.   Screen oxide-deposition of 40 nm SiO_2.
3.   N-Well Ion-implant. Implant phosphorus at 400 KeV to a dose of 5.0E12, using an implant model which includes the lateral scattering. The mask opening extends from 0.0 to 4.4 microns.
4.   N-Well drive-in. Neutral ambient drive-in at 1130o C.
5.   Field ion-implant. Implant boron at 40 KeV. Mask opening from 4.4 to 8.8 microns.
6.   Neutral Ambient drive-ins.   a.   The 950° and 1000° C. drive-ins are done with a boron-enhanced diffusivity to simulate "oxidation".   b.   An additional 1000° C. normal drive-in/anneal using a normal boron diffusivity.

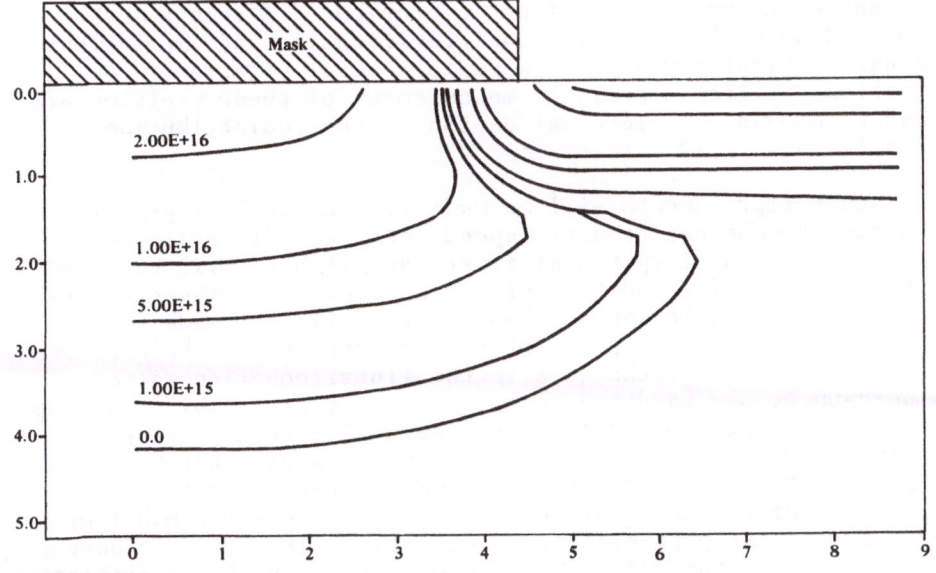

Figure 10.   Windowed phosphorus and Boron implants for a bulk
             CMOS process.

SECTION 5   SUMMARY AND CONCLUSIONS

The capabilities and methods of the FEDSS program have been described in this work. Several specific examples of its application have been presented to illustrate the flexibility of the program. Close correlation of simulation results and experimental data illustrate the accuracy of the model and the credibility of its underlying assumptions and computational methods. Pre-processors speed the creation of new models through mesh generation. Post-processing programs allow rapid examination of the internal profiles of devices and subsequent improvement of process design.

These capabilities form a comprehensive process CAD tool which allows prediction of the characteristics of new processes and rapid reponse to problems affecting device function and reliability. Design frequently can involve much trial and error; it is the objective of process modeling and device simulation to reduce the design time while simultaneously increasing the success rate in achieving well-designed processes. The relative time and cost for simulation makes it very useful since many fabrication runs and device topologies can be considered in the time it takes to create the first working device. The ability to make variations is desirable both for optimizing designs and for targeting initial choices of experiments to be performed. Simulation allows examination of the internal profiles of devices, and the resulting insights often spark innovative solutions to problems. Although much has been done on FEDSS, more model verification and improvement remains, particularly with respect to lateral effects of concentration profiles. Reliable measurement of these profiles are needed to provide complete validation of the redistribution models.

Future improvements will be made to the FEDSS program in the areas of enhanced physics, speed, and user friendliness. The process steps of epitaxial growth and etching will be added to the program. The addition of the interaction between impurity species will be implemented. Improvements will be made in the diffusivity models. Computer times will continue to be decreased by a mixture of new techniques in the linear equation solver, and perhaps by new hardware such as an array processor. The implementation of higher order elements would give better accuracy with less computer time required. A new data base has been designed which will permit better communication between the pre-processing programs, process, device, and circuit modeling programs, and the post-processing programs. The new data base will contain all the run inputs and outputs. The human factors use of the FEDSS system will be improved. Easier mesh generation will be available while more meshes will be found in the pertinent libraries. Suitable methods for displaying and interpreting results will be discovered.

FOOTNOTES

1. D. D. Warner and C. L. Wilson, "Two-Dimensional Concentration Dependent Diffusion," *Bell Systems Technical J.*, Vol. 59, 1 (1980).

2. R. Tielert, "Two-Dimensional Numerical Simulation of Impurity Redistribution in VLSI Processes," *IEEE Trans. Electron Dev.*, Vol. ED-27, 1479 (1980).

3. K. Taniguchi, M. Kashiwagi, and H. Iwai, "Two-Dimensional Computer Simulation Models for MOSLSI Fabrication Processes," *IEEE Trans. Electron Dev.*, Vol. ED-28, 574 (1981).

4. D. J. Chin, M. R. Kump, H. G. Lee, and R. W. Dutton, "Process Design Using Coupled 2D Process and Device Simulators," *IEEE Int. Electron Devices Meet.*, Tech. Dig., p. 223 (Dec. 1980).

5. E. M. Buturla, P. E. Cottrell, B. M. Grossman, and K. A. Salsburg, "Finite-Element Analysis of Semiconductor Devices: The FIELDAY Program," *IBM J. Res. Develop.*, Vol. 25, 218 (1981).

6. F. Morehead, private communication.

7. K. Inoue, T. Hirao, Y. Yaegashi and S. Takayanagi, "Asymmetrical Profiles of Ion Implanted Phosphorus in Silicon," *Japanese J. Applied Phys.*, Vol. 18, a367 (1979).

8. H. Okabayashi and D. Shinoda, "Lateral Spread of P and B Ions Implanted in Silicon" *J. Appl. Phys.*, Vol. 44, No. 9, (September 1973), pp. 4220-4221.

9. D. A. Antoniadis and R. W. Dutton, "Models for Computer Simulation of Complete IC Fabrication Processes," *IEEE Trans. on Electron Devices*, Vol. ED-26, 490 (1979).

10. F. Jahnel, H. Ryssel, G. Prinke, K. Hoffmann, K. Muller, J. Biersack and R. Henkelmann, "Description of Arsenic and Boron Profiles Implanted in SiO2, Si3N4 and Si Using Pearson Distributions With Four Moments," *Nuc. Inst. Meth.*, Vol. 182, 183, 223 (1981).

11. M. Y. Tsai, F. F. Morehead, J. E. E. Baglin, and A. E. Michel, "Shallow Junctions by High-dose As Implants in Si: Experiments and Modeling," *J. Appl. Phys.*, Vol. 51, No. 6, (June 1980), pp. 3230-3235.

12. F. J. Morin and J. P. Maita, *Phys. Rev.*, "Electrical Properties of Silicon Containng Arsenic and Boron," Vol. 96, 28 (1954).

13. B. L. Crowder, J. F. Ziegler, F. F. Morehead, and G. W. Cole, "The Application of Ion Implantation to the Study of Diffusion of Boron in Silicon," in *Ion Implantation in Semiconductors and Other Materials*, edited by B. L. Crowder, Plenum Press (1973).

14.   R. B. Fair and J. C. C. Tsai, "A Quantitative Model for the Diffusion of Phosphorous in Silicon and the Emitter Dip Effect," *J. Electrochemical Soc.*, 124, 1107 (1977).

15.   O. C. Zienkiewicz, *The Finite Element Method in Engineering Science*, McGraw-Hill (1977).

16.   A. George and J. W. H. Liu, *Computer Solution of Large Sparse Positive Definite Systems*, Prentice-Hall, Inc. (1981).

17.   F. G. Gustavson, "Some Basic Techniques for Solving Sparse Systems of Equations", in *Sparse Matrices and Their Applications*, edited by D. J. Rose and R. A. Willoughby, Plenum Press, New York (1972), pp. 41-52.

18.   G. W. Stewart, *Introductions to Matrix Computations*, Academic Press, New York (1973).

19.   A. George and J. Liu, *Users Guide for SPARSPAK: Waterloo Sparse Linear Equations Package*, Department of Computer Science, University of Waterloo, Ontario, Canada (1979).

20.   B. E. Deal and A. S. Grove, "General Relationship for the Thermal Oxidation of Silicon," *Journal of Applied Physics*, Volume 36, Number 12, (December 1965).

OPTICAL AND DEEP UV LITHOGRAPHY[*]

W. G. Oldham
Department of Electrical Engineering and Computer Science
and Electronics Research Laboratory
University of California
Berkeley California, 94720

Abstract

The principal models for optical lithography are reviewed: Image formation in near diffraction-limited systems; positive resist exposure (bleaching) ; post-exposure processing; and resist development. The algorithms which implement these models are reviewed, and complete examples of typical process sequences discussed. The limits of optical lithography are examined, considering the variety of machines available, or soon to be available.

1. Review of Optical Lithography

Modern optical lithography may be conveniently divided into three processes: (1) image projection, (2) exposure/bleaching, and (3) resist development. An optical tool is used to project a high-contrast image of the desired pattern onto a thin film of photosensitive material. The image information is recorded in the photosensitive medium in the form of broken chemical bonds (positive photoresist) or crosslinked polymeric bonds (negative photoresist). A selective solvent is used to transform the latent chemical image into a physical image by selective dissolution of exposed resist (positive tone) or unexposed resist (negative tone).

1.1 Optical Systems for Image Projection

Image projection, as opposed to contact printing, is a requirement for VLSI, primarily on economic grounds. A very low defect density and a very high placement accuracy are required. Inexpensive copies of a master mask are inadequate. The cost of mask manufacture and repair is quite high, and must be recovered by achieving a long useful life. The damage introduced by even a "soft" contact is unacceptable. The lithographer has only proximity printing and projection printing as

* Research supported by National Science Foundation under Grant ECS-8106243

alternatives. However the resolution of simple proximity printing is inadequate for features in the $1\mu m$ range, and projection printing is the preferred technique [1].

A variety of optical tools are presently available for high-resolution projection lithography. While physically quite distinct, the ultimate resolution performance may be compared simply on the basis of the parameter λ / NA, where NA is the numerical aperture of the imaging optics, and λ is the wavelength (or effective wavelength in multi-wavelength systems) [2]. For example with ideal (diffraction-limited) optics using an illuminator operating at a filling factor of 0.7, an image contrast of 90% or greater is achieved for linewidth of $0.58\lambda / NA$ or greater (e.g., resolution of $0.58\mu m$ lines and spaces at $\lambda / NA = 1$). The actual contrast requirement depends on the linewidth control needed, but 90% is typical for imaging on wafers [2].

Recent measurements on several optical tools suggest that present-day projection systems can come reasonably close to diffraction limited performance [3]. An approximate rule of thumb can be stated that the equivalent residual aberrations plus tool focus error can be reduced to equivalent defocus of about 0.8 Rayleigh Units. Again assuming a 90% contrast requirement, the limiting resolution is increased from $0.58\lambda / NA$ to about λ / NA [3].

Presently available machines are both of the scanning type and stepping type. The former operate typically at an NA of 0.17, with 1 X demagnification, and can utilize wavelengths to as short as $0.24\mu m$. The well-corrected image field in such systems is a narrow slit of the full wafer height. To expose the entire wafer, mask and wafer are simultaneously scanned. Optical overlay is necessarily achieved on a full-wafer basis. Consequently wafer, mask, and optical distortion requirements are very stringent.

Projection steppers typically expose a field area of about $1 cm^2$ at each step. Presently available refractive optics are primarily 10:1 demagnifying systems operating at NA of 0.28 and λ of $0.436\mu m$. Recently systems with NA of 0.35 covering the same field have become available. Another type of projection stepper uses a two-wavelength (0.405, $0.436\mu m$) catadioptric lens operating 1:1 with NA = 0.31.

1.2 Resist systems for Fine-Line Lithography

Lithography with feature sizes below about $2\mu m$ relies almost exclusively on positive photoresist. The essential resist properties have been discussed by Dill and co-workers [4]. The typical resist system consists of a diazo type sensitizer in a novalak base resin. In normal operation photo-initiated destruction of the photo active compound [PAC] results in an enhanced solubility rate in basic developers. The exposure process is accompanied by a optical bleaching owing to PAC destruction, and the exact state of exposure may be determined from the absorption coefficient [4].

The development of positive photoresist is generally performed by batch immersion in a basic aqueous solution. Spray development is also used. The development rate is a highly nonlinear function of the exposure condition, typically varying between a few tenths of a nm per second in exposed resist up to hundreds of nm per second in fully exposed resist. It is the resist nonlinearity which is responsible for the very high resolution patterns at the very limits of optical imaging [5]. The nonlinearity may be quantified as "resist contrast," defined as the slope of the resist thickness remaining vs the logarithm of the exposure dose [5]. Typically positive photoresists operate with effective contrast in the range of 2.5 to 5.

2. Models

The simulator SAMPLE has three basic subprograms which model the processes of optical imaging, resist exposure, and resist development. The details of the models have been elaborated in other publications, and will only be briefly reviewed here [6-8].

2.1 Optical Model

The optical model is based on the theory of imaging with partially coherent light. In the basic algorithm only 1-dimensional periodic objects are considered, and a circular aperture is assumed. Periodic patterns of lines and spaces or isolated lines or spaces may be effectively simulated. The method is based on the Hopkins theory of partially coherent imaging, and calculates the image intensity pattern by means of the transmission cross-coefficient weighting the Fourier transform of the object transmittance. The cross-coefficient is calculated from the pupil function, given the degree of partial coherence, σ, and the defocus μ. In the SAMPLE implementation, a combination of analytical and numerical integration is used to evaluate the convolution integrals [7]. The user specifies the image pattern (linewidth and spacewidth), wavelength, numerical aperture and coherence, the focus error, and the imaging window. Typical execution times are a few seconds using a VAX 11/780 operating under UNIX and using the f77 compiler.

2.2 Resist Exposure

The resist model proposed by Dill et al. [4] is used to simulate the exposure process. It is extended to multiple wavelengths in a manner similar to the model of Narasimham and Dill [9]. The vertical standing wave pattern is computed at each instant of time for each wavelength present. The resist is bleached at each vertical position according to the weighted intensity and sensitivity summed over the wavelengths used. Time is advanced, and the vertical intensity profiles recomputed. At each time interval the state of the resist is described by the relative inhibitor concentration "M." At appropriate intervals the vertical profile M(z) is saved, and the exposure is continued until some total energy dose E_{max} is achieved. The result of the exposure simulator is a two-dimensional array describing M(Z,E) as a function of depth and energy.

In a separate operation the horizontal intensity profile and the specified exposure (dose) are combined with the M(Z,E) array to produce a new array M(X,Z) which describes the state of the resist at each point across the image (X) and at each depth (Z). If a variety of total exposure doses are used, it is straightforward to compute a new M(X,Z) without recomputing either the horizontal intensity profile or M(Z,E). Of course, this method of separating the exposure into independent vertical and horizontal processes necessarily assumes that all rays enter the resist normal to the surface. For lenses of very large numerical aperture and thick resists, this assumption is somewhat aggressive. For example, at NA = 0.28, the cone of rays in the resist includes rays up to about 10° from the normal. For a $1\mu m$ resist thickness, the lateral spread would be about $1700\mathring{A}$

The effect of post-exposure predevelopment processing on the resist is modeled by means of modification of the M(X,Z) array. For example, post-exposure baking is known to smooth the standing wave pattern in the vertical resist profile [10], [11]. This effect has been modeled by a simple diffusion in which the standard deviation is specified.

2.3 Resist Development

The resist is assumed to dissolve in the developer solution at a rate which is only a function of the local inhibitor concentration M(X,Z). In effect, a surface reaction-rate limited process is assumed. The rate function R(M) is an empirical fit to measured data. The fitting function originally proposed by Dill [4] is convenient, and is specified by three parameters E_1, E_2, and E_3.

The etching of any layer in which the kinetics are surface-rate limited may be simulated by means of the string development algorithm of Jewett et al. [8]. The boundary between the etched and unetched region is approximated by a series of points joined by straight line segments (a string). Each point advances along the angle bisector of the two adjoining segments according to the local value of etch rate. A typical string, composed of 40-100 line segments, is started on the surface and, as time proceeds, advances through the layer being etched. During the simulation. the segments are kept roughly equal in length by adding points in regions of expansion of the etch front, and deleting points in regions of contraction. Further details may be found in [8]. For the development of positive resist, the M(X,Z) array is combined with the etch rate versus M curve to generate the etch-rate function.

The basic etching algorithm can be adapted to more general situations. For example, multiple layers of resist are treated by including a separate etch rate versus M curve for each layer. Plasma ashing is simulated by adjusting the etch rate to the appropriate constant value.

3. Simulation of Optical Lithography

The organization of Sample has been described in detail in reference [6]. The FORTRAN source code is divided into block of subroutines or "machines" which simulate various process steps such as imaging the mask on the wafer, exposing the resist and developing the resist. Instead of changing the FORTRAN source code the user simply interacts with a controller and a data bank through an interpreter. In effect the user speaks in a high level language whose vocabulary is processing jargon. This allows the program to be used without a knowledge of the programming language. In this way SAMPLE is similar to many CAD tools, and can be used with no computer or programming experience.

An example of the key words and commands used in SAMPLE for projection printing is shown in Table 1 below. The first column indicates various process steps and is followed by typical process specifications and then their translation into SAMPLE input. The program assumes a nominal or standard process with all process parameters already specified. Thus the parameters shown here will be inserted in place of the default values initially stored in memory. Only the parameters that are changed from their nominal values need be inputed. For example the layers card specifies the silicon substrate refractive index and the refractive index and thickness of the oxide layer. The resist card gives the wavelength, exposure parameters (A,B,C's) and refractive index and thickness of the resist layer. The next three cards specify the use of a projection printing tool with a numerical aperture of 0.28 at a wavelength of of $0.436\mu m$ and a dose of $110mJ/cm^2$. The etch rate vs chemical inhibitor concentration (M) after bleaching is described by the three constants on the rate card. The development time is then specified as 60 seconds. Finally the run card instructs the controller to perform the entire imaging, bleaching and developing simulation using the updated parameters.

TABLE 1. A TYPICAL PROCESS SEGMENT AND THE SIMULATOR INPUT		
Process Step	Typical Specification	Simulator Specification
Start Oxidize	Si (100) 10Ωcm wafers Grow dry oxide 0.068μm	layers (4.75 -.138) (1.47 -0.0 0.068)
Spin Prebake	Spin positive resist 0.94μm Bake 80°C 30min.	resmodel 0.436 (0.551 0.058 0.10) (1.68 0.0 0.94)
Mask	Pattern: 2μm line, 6μm spaces	linespace 2 6
Expose	Expose: 110mJ/cm^2	proj 0.28 lambda 0.436 dose 110
Develop	Develop 60 sec	etchrate analytic 5.63 7.43 -12.6 devtime 60
Run wafer		run

Numerous examples of the results of process simulation have been given in references [2,3,and 5-7]. Additional examples include the study of the effects of chromatic aberration on projection printing [12], and the effects of process control on limiting the ultimate resolution of optical lithography [13], to be described in the next section.

4. Limits of Optical Lithography

Modern optical systems for projection lithography are capable of resolving submicron features under ideal conditions. However in the production of integrated circuits the practical limit is in the neighborhood of 1.5μm features. The factor of two loss in useful resolution stems primarily from the problem of process control. Unavoidable variations in process parameters produce either unacceptable variations in resist linewidth, or result in 'unresolved' features - lines missing or spaces not cleared.

TABLE 2. PROCESS VARIABLES IN OPTICAL LITHOGRAPHY	
Variable	Source of Variations
resist thickness	steps and spin striations
reflectivity	thin film thickness
source uniformity	optical non-idealities
system contrast	defocus and flare
development rate	temperature and aging
resist properties	batch and process differences
development time	operator errors
exposure time	photometry errors

Table 2 lists, in approximate order of importance, the variables which dominate linewidth control in real circuit fabrication. Chief among these is the resist thickness variable, which dominates lithography today. Even if no large steps are present, small thickness striations, normally present in positive photo-resist films, can produce 50% changes in apparent dose by modulating the reflectivity. The remaining variables of Table 1 are responsible for much smaller variations in linewidth, and can be approximately treated by means of a sensitivity analysis. Source intensity variations are equivalent to exposure time variations; similarly development time can be traded approximately for development rate. Variations in resist properties result in changes in sensitivity (equivalent to dose changes) or changes in development rate. Thus a much more limited set of variables may be used to explore the control of linewidth in resist exposure and development.

In this section simulation is used to perform quantitative studies on the sensitivity of critical dimensions of positive resist profiles with respect to variations in resist thickness T, illumination dose and development time t_{dev}. A modified version of the simulation program SAMPLE was used.

4.1 Linewidth Sensitivity on Flat Substrates

In this study, critical dimension CD_{min} is defined as the minimum reproduced linewidth in resist for a mask pattern of either isolated lines or equal lines and spaces. The total linewidth loss from the original mask dimension is labeled δCD. The objective of the first study reported is to evaluate the sensitivity of CD to dose and development time. We explore the combination of these two variables which produces the minimum sensitivity of CD_{min}; i.e., we attempt to find an optimum operating point. In the simulations we have used two different linewidths of both isolated (resist) lines and patterns of equal lines and spaces. The simulated positive resist is AZ1350J, and the parameters used are those reported in [3]. Both conventional development, and development with a post-exposure bake are investigated. The simulated optics use a projection printing system with NA = 0.28, a partial coherence factor of 0.7, and a defocus of 1.4μm (corresponding to one half a Rayleigh unit). The single wavelength used is 436 nm. The geometrical parameters are: resist thickness of 0.9$l\,\mu m$ spun on a 75 nm oxide film on a silicon substrate. With such thicknesses we obtain an intensity maximum at the resist-oxide and air-resist interfaces.

The results of the investigation are reported in Fig. 1, for a 1.5μm line. The development time (t_{dev}) is given as a function of dose, for different values of linewidth loss δCD, for both the standard and the post-exposure-bake (PEB) case. The effect of PEB is to smooth the large vertical inhibitor fluctuations caused by standing waves in the resist. A mild reduction in resist contrast and a slight surface inhibition of development are observed and accounted for in the simulation.

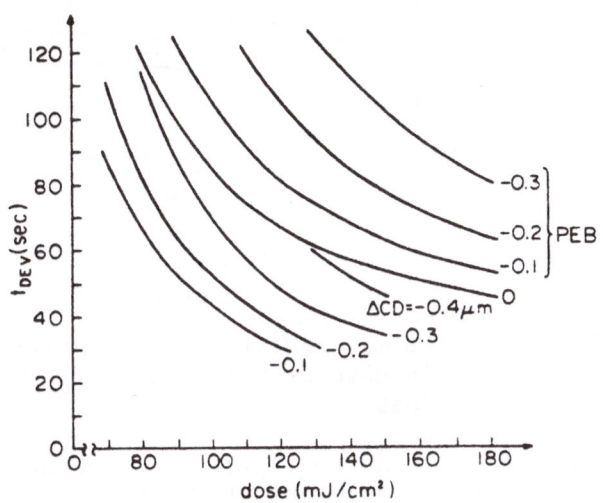

Fig. 1

Development time-dose tradeoffs
as function of linewidth loss.

From these data we obtain the sensitivity S of normalized linewidth change to dose variations. In Fig. 2 we plot the sensitivity S versus t_{dev}, for different values of δCD. Linewidth loss can be traded for decreased linewidth sensitivity. The usefulness of post-exposure bake is obvious from the sensitivity plot. Not only is the sensitivity reduced to less than 0.5, but no linewidth loss is necessary. Figure 2 also illustrates the general result that over a reasonable range, dose may be traded for development time with virtually no change in process sensitivity. Thus development time may be safely determined by other considerations, such as resist thickness loss and adhesion.

4.2 Linewidth Control Over Steps

Photoresist linewidth variations are particularly difficult to control in the vicinity of steps in the substrate, where abrupt resist thickness changes occur. In the second study reported, simulation is used to quantitatively explore linewidth control in the vicinity of steps; the goal of this investigation is to determine the minimum obtainable value of the linewidth discontinuity at the step as a function of dose and t_{dev}, for a given step thickness.

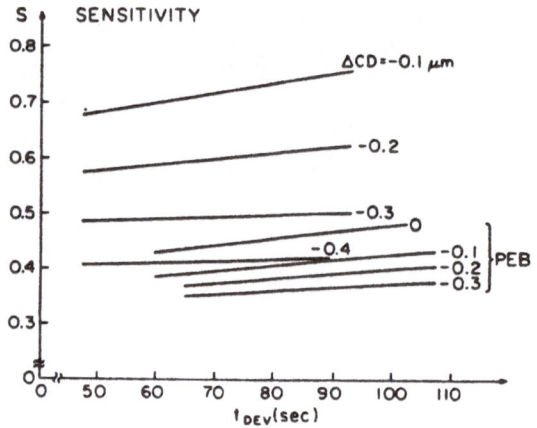

Fig. 2

Normalized linewidth sensitivity
to dose as function of linewidth
loss.

Figure 3 illustrates the typical behavior observed in fine lines crossing steps. The linewidth undergoes a major jump at the step and is modulated periodically on both sides of the step. The latter effect arises from
the periodic minima and maxima in reflectance as the resist thickness
tapers from the extremes at the step. To explore this effect we have
defined some relevant quantities in Fig. 3. The most interesting parameter is d, the linewidth discontinuity at the step, while m_1 and m_2 are the
modulation parameters.

In the simulation we have used a $1.1\mu m$ step in silicon and a maximum resist thickness $T_{max}=1.42\mu m$ (11 half waves). For this thickness
we have determined a suitable dose to safely develop the resist within the
chosen time   such a procedure has been repeated for three different
times, 90, 120 and 150 sec. For each time and corresponding dose we
have obtained the resist profiles for the critical resist thicknesses in
proximity of the step, i.e., $0.325\mu m$ (T_{min}), $0.39\mu m$ and $1.355\mu m$. Such
thickness values represent a possible worst case; in fact T_{max} is such to
produce a minimum in the intensity distribution coupled inside the resist,
while T_{min} produces a maximum. In this case we have used a $1.5\mu m$ line-
space mask and an Al substrate.

The exposure is chosen at a value 20% higher than needed to just clear the lines at the
point of maximum resist thickness.

The results, reported according to the parameters of Fig. 3 are summarized in Fig. 4; with and without the PEB. From this figure we see that, just as in the previous sensitivity analysis, dose can be traded for develop time with very little process control impact. The high reflectivity of aluminum results in a deterioration of the linewidth control under normal operating conditions, a fact that is well-known experimentally. The standing waves are very pronounced and the intensity minima, combined with resist nonlinearities, result in very long develop times and/or large doses. Consequently considerable lateral development occurs. These deep minima, in the case of aluminum, also affect simulation, in that the grid point occurs at each of the minima and maxima. In this case the improvement obtained with the PEB technique is quite dramatic, especially with regard to the parameter CD min and to the discontinuity parameter d. Without PEB, the $1.5\mu m$ line is reduced to an unacceptable $0.2\mu m$ at the neck, whereas PEB increases this value to about $0.8\mu m$.

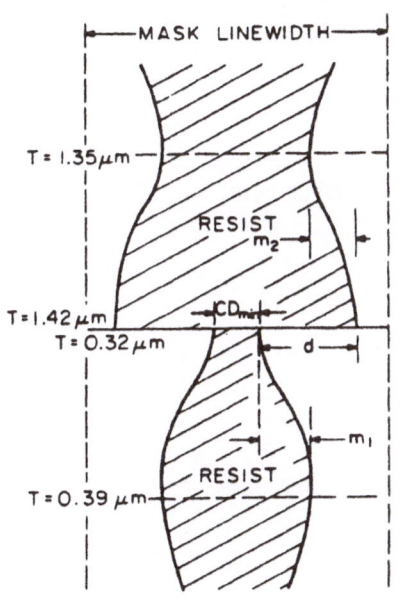

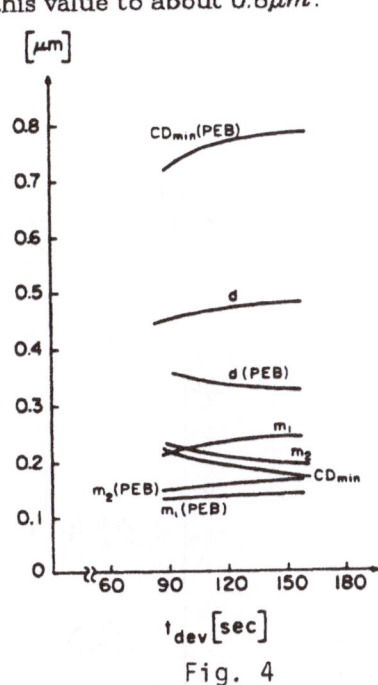

Fig. 4

Step modulation parameters: lithography on aluminum.

Fig. 3

Resist line modulation in crossing a step.

As a final example, we show in Fig. 5a a sketch (to scale) based on simulations, of the resist line crossing a $1.1\mu m$ step on aluminum with and without post-exposure bake; this sketch clearly indicates the advantage of using the PEB technique. Figure 5b shows the corresponding experimental results for resist lines crossing a $1.5\mu m$ aluminum step [14].

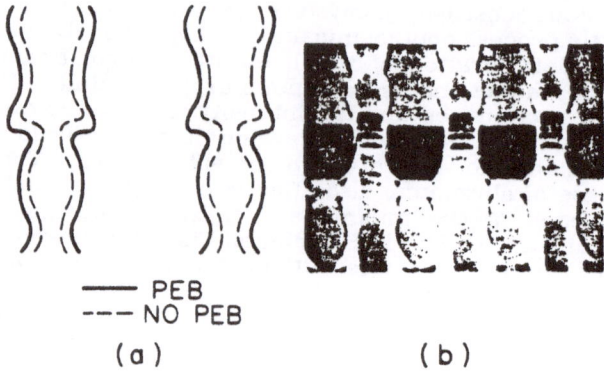

Fig. 5

Simulated and observed linewidth
modulation for resist on aluminum.

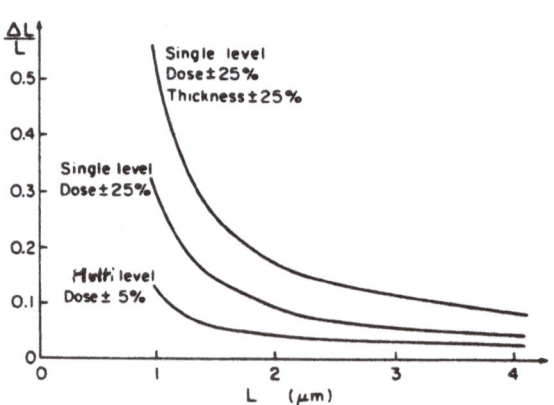

Fig. 6

Linewidth control versus linewidth
for different resist technologies.

4.3 Modifications in Resist Technology

A planarization technique can be used in order to virtually eliminate the steps in the substrate; this planarization is obtained by spinning a passive underlayer. Also an absorbing medium in the planar layer eliminates variations in reflectance, thus reducing the effect of dose variations on linewidth sensitivity. The goal of the third study reported is to estimate the potential improvements in linewidth control for such multilevel resist systems. The very significant gain in control with such systems leads to higher practical resolution.

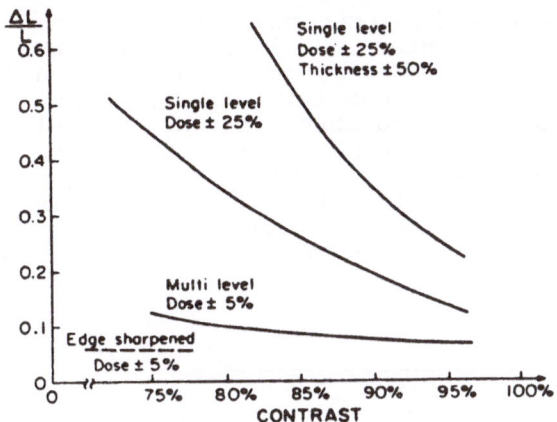

Fig. 7

Linewidth control versus optical contrast for different resist technologies.

The optics and resist parameters used in the simulation are equal to those reported in the previous examples. In the single-level resist system we have used a quarter wave of oxide (750 A) on silicon, covered by $1\mu m$ of resist. We have chosen the value of exposure at D_0 such that large areas of resist clear in a 30 sec. development time, one half the actual nominal t_{dev} of 60 sec. We have then found the value of CD_{min} for nominal conditions with patterns of equal lines and spaces. We have then considered a $\pm25\%$ variation of the value of dose and found the values of CD_{min} for the same masks. Moreover we have simulated a thickness change $\pm25\%$ thickness change added to dose variation and again we have found the values of CDmin.

From these three sets of values of CDmin we have obtained the percentage variations of linewidth control for the changes in dose and resist thickness. The results for the single-level case are reported in Fig. 6 as a function of the different mask linewidths. Figure 6 also shows the results for a multi-level resist system. In this case we have a silicon substrate which is covered by a planarizing layer of $2\mu m$ in which there is an absorbing medium. On top of this layer there is a $0.5\mu m$ film of resist. The only variable in multi-level resist was an assumed $\pm5\%$ dose variation.

Finally we have considered a fixed mask pattern of $1.5\mu m$ line-space and we have repeated the whole previous procedure for both single-level and multi-level resist systems; in this case we have changed the values of defocus to get different values of image contrast. The results are shown in Fig. 7.

Also shown in Fig. 7 is the simulated linewidth control for a resist system with "edge sharpening." This effect, reported for certain inorganic resists [15], enhances the edge slope of the final photosensitizer distribution. Consequently the linewidth is improved and the useful contrast range is extended.

4.4 Impact on Future System Resolution

Figure 8 is a universal plot of optical system contrast versus normalized linewidth, with focus error as a parameter. As a practical matter, we have found that well-tuned optical systems perform at a minimum effective focus error of about 0.8 Rayleigh units; curve 2 in Fig. 8. If we assume, based on the simulations reported above, that the contrast requirements are in the vicinity of 95% for conventional resists, 90% on flat substrates, 80% for multilevel technology, and 70% for "edge-sharpened" resists we can read off the useful resolution. For example with a state-of-the-art lens with NA - 0.35 at 436 nm, we obtain 1.5, 1.1, 0.75, and $0.6\mu m$ respectively for the four cases considered.

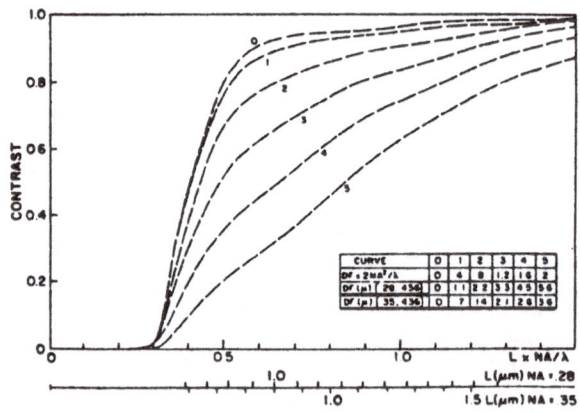

Fig. 8

Diffraction-limited contrast of optical systems.

References

1. B. J. Lin "Optical Methods for Fine-Line Lithography", *Fine-Line Lithography*, R. Newman Ed., North Holland, 1980.

2. W. G. Oldham, S. Subramanian, and A. R. Neureuther, "Optical Requirements for Projection Lithography", *Solid State Electronics*, *Vol 24*, pp 975-980, 1981.

3. W. G. Oldham, P. Jain, A. R. Neureuther, C. Ting, and H. Binder, "Contrast in High-Performance Projection Optics", *Kodak Microelectronics Seminar, Interface 81*, Dallas, Oct 1981.

4. F. H. Dill, W. P. Hornberger, P. S. Hauge, and J. M. Shaw, "Characterization of Positive Photoresist", *IEEE Trans. Electron Devices, ED-22*, pp 445-452, 1975.

5. W. G. Oldham, and A. R. Neureuther, "Projection Lithography with High Numerical Aperture Optics", *Solid State Technology*, pp106-111, May 1981.

6. W. G. Oldham, S. N. Nandgaonkar, A. R. Neureuther, and M. M. O'Toole, "A General Simulator for VLSI Lithography and Etching Processes: Part I -- Application to Projection Lithography" *IEEE Trans. Electron Devices, ED-26*, pp 717-722, 1979.

7. S. Subramanian, "Rapid Calculation of Defocused Partially Coherent Images", *Applied Optics, Vol 20*, pp1854-1857, 1981.

8. R. E. Jewett, P. I. Hagouel, A. R. Neureuther, and T. Van Duzer, "Line-Profile Resist Development Simulation Techniques", *Polymer Eng. Sci, Vol 17*, pp381-384, 1977.

9. M. A. Narasimham and F. H. Dill, "Projection Printed Photolithographic Images under Polychromatic Exposure", presented at *SPIE 30th Ann. Conf.*, No. Hollywood, CA., May 1977.

10. E. J. Walker, "Reduction of Photoresist Standing Waves Effects by Post-Exposure Bake", *IEEE Trans. Electron Devices, ED-22*, pp464-466, 1975.

11. F. H. Dill and J. M. Shaw, "Thermal Effects on the Photoresist AZ1350J", *IBM J. Res. Devel., Vol 21*, pp210-218, 1977.

P. K. Jain, A. R. Neureuther, and W. G. Oldham, "Influence of Axial Chromatic Aberration in Projection Printing", *IEEE Trans. Electron Devices, ED-28*, pp 1410-1416, 1981.

13. W. G. Oldham, P. Antognetti, and C. Fasce, "Toward Finer Lines: Process Control in Optical Lithography", *Semicon 82*, San Mateo Ca, May 1982.

14. H. Binder and M. Lacombat, "Step-and-repeat Projection Printing for VLSI Circuit Fabrication," *IEEE Trans. Elec. Dev.*, vol. Ed-26, pp. 698-704, 1979.

15. K. L. Tai, R. G. Vadimsky, C. T. Kemmerer, J. S. Wagner, V. E. Lamberti, A. G. Timko, "Submicron Optical Lithography Using an Inorganic Resist/Polymer Bilevel Scheme," *J. Vac. Sci. Technol.*, vol. 17, p. 1169, Sept./Oct. 1980.

WAFER TOPOGRAPHY SIMULATION

A. R. Neureuther

Department of Electrical Engineering
and Computer Sciences
and Electronics Research Laboratory
University of California, Berkeley, California 94720.

ABSTRACT
As dimensions of the "classical" planar process are reduced the device features in the third dimension pose major fabrication and performance problems. Establishing techniques to characterize and optimize these non-planar device features is a major goal of research on modeling and simulation. Lithography , etching and deposition models have been established which agree well with experiment. In many cases the dominant physical mechanism is a surface reaction process which can be simulated by a surface advancing algorithm such as the cell, string or ray approach. Prime examples of the usefulness of simulation are in understanding projection printing, step coverage in deposition, wafer planarization and linewidth bias and control for composite process sequences. Simulations for individual IC fabrication processes are being combined in a user oriented program for Simulation And Modeling of Profiles in Lithography and Etching (SAMPLE).

1 INTRODUCTION
As IC fabrication technology pushes toward finer linewidths the topographical features of devices are becoming an increasing concern. Not only is greater linewidth control required for high resolution but it must be accomplished in a context where features in the third dimension are as large as the linewidth itself. SEM cross sections of wafers are frequently used to examine problems related to topography. While these device cross sections or line edge profiles give true process feedback they are time consuming to perform and require unusually accurate process control to systematically explore a matrix of experimental parameters. If process models can be established, algorithms can often be used to explore pro-

cess effects by simulating the time evolution of the line edge profile. Besides the advantages of controllability and observability, simulation can be done much more rapidly and economically.

This paper gives an overview of simulation and modeling of IC processes. It focuses primarily on the areas of lithography, etching and deposition and their particular embodiment in the user oriented program SAMPLE[1-3]. The thermal related process of ion implantation, diffusion and oxidation and their incorporation in to programs SUPREM and ICECREM have recently been described elsewhere[4,5] and are recommended for additional reading. Section 2 gives an overview of the status and uses of simulation. The computer requirements and algorithms for carrying out process simulation are described in section 3. A survey of simulation results and comparisons with experiment are given in section 4. Finally, the future trends and issues are discussed in section 5.

2 STATUS OF SIMULATION

It is potentially possible to find both physical models and efficient algorithms to simulate most IC fabrication processes. Figure 1 lists a variety of examples. They are divided into four major areas: lithography, etching, deposition and thermal processes. Simulation work has been carried out for most of these processes by a number of authors. The simulation of the time-evolution of line edge profiles began about ten years ago although the models themselves in some cases are related to basic work extending back even further.

IC PROCESS APPLICATION AREAS

LITHOGRAPY
OPTICAL
E-BEAM
X-RAY
ION-BEAM

ETCHING
.WET
ION MILLING
PLASMA ETCHING
ION ASSISTED ETCHING

THERMAL PROCESSING
OXIDATION
DOPING
ION IMPLANTATION
ANNEALING

DEPOSITION
EVAPORATION
SPUTTERING
CHEMICAL VAPOR DEPOSITION
ELECTROPLATING
EPITAXIAL GROWTH

Fig. 1 IC processing areas for which modeling and simulation are potentially applicable.

Simulation has been used to study key issues in all types of lithography. Optical lithography simulation was made possible by the establishment of an exposure-bleaching development-etching model for positive photoresist [6]. This permitted the modeling of projection printing [7-11] and has been useful in the characterization of modern optical lithography tools [12-19] and resist materials [14,20-26]. Electron beam lithography simulation has been based on Monte Carlo simulation of the latent energy deposition profile [27,28] and a solubility versus dose model for PMMA resist [29,30]. Profile simulation has been useful in characterizing effects of machine, wafer and resist parameters on lithographic performance [31-45]. X-ray and ion beam lithography have also been simulated using energy deposition models with the PMMA resist model[46-49].

A variety of etching and deposition models have also been developed. Basic etching concepts have evolved from crystal growth models and have been applied to plasma etching, reactive-ion-etching and ion milling [50-57]. Sputter deposition and evaporation have also been simulated [62-67]. The conditions necessary for metal "lift-off" [66,67] and step coverage have been explored [65].

Simulation and modeling of processing have been utilized by process and device designers for a number of purposes. A very common use is for education and understanding of new processes or equipment. The ability to investigate individual physical phenomena and interactions of of different phenomena through the huge number of process parameters is especially useful. This leads frequently into more quantitative use of simulation in the determination of optimum processing conditions for equipment such as projection printers. A third type of use is in exploration studies of new processes and new process sequences. A good example here is the study of the use of isotropic deposition combined with anisotropic etching to planarize the wafer surface. Another important use is the assessment of the impact of hypothetical and future technology advances such as new lithography exposure tools or new resist technologies. It is also possible to use simulation in diagnostic and discovery modes in which the physical models are modified to create new effects or establish plausible explanations of experimental observations.

3 ALGORITHMS

Simulation is typically carried out on large computer systems with the aid of graphics terminals. Most source codes are in FORTRAN and require about 5 to 60s of execution time ($10^{**}7$-8 machine instructions). If more extensive calculations are required such as in the case of Monte Carlo techniques, a reusable library of data is first generated and then accessed by the program for rapid execution. A file handling system and editor are convenient for

414

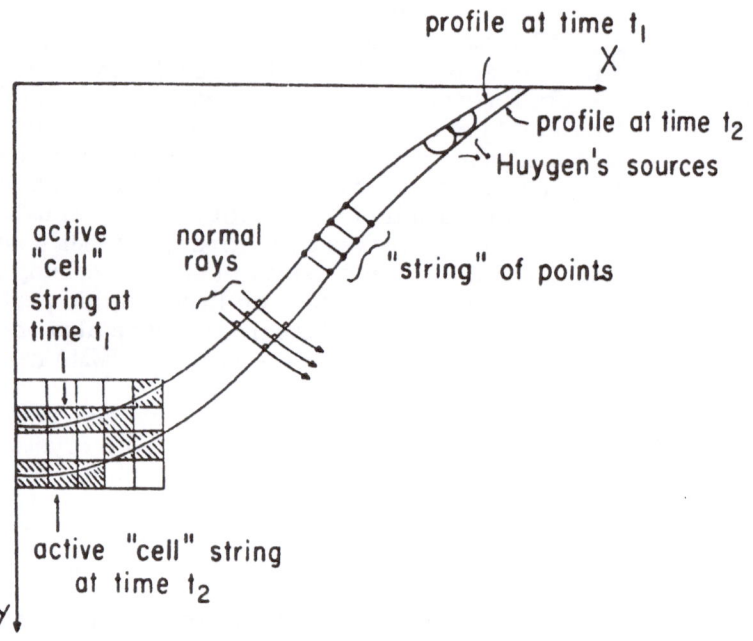

Fig. 2 Three algorithms for implementing surface etching and the analogy to Huygen's source problem.

creating and selectively running data sets. Terminals and support packages for interactive operation and graphics are also desirable.

Many of the processes can to first order be considered as surface reaction rate limited processes. This is true for positive resist development, dry etching and even deposition (negative etching). Thus the simulation of many of the IC processes can be based on a generalized surface etching algorithm. Figure 2 graphically depicts the surface etching problem for lithography along with several algorithmic approaches. Each point on the initial surface acts like an infinitesimal Huygen's source and the advancing contour is the locus of tangents to all the Huygen's spheres of influence. Computationally this process has been implemented by cell removal, by analogy to optical ray tracing, and by advancing a string of line segments. The string approach has been commonly used for two dimensional problems while for three dimensional problems the ray approach may have inherent advantages.

To make simulation accurate and efficient the physical mechanisms and boundary conditions of the process being simulated must be incorporated into the algorithms. For this reason the string advance algorithm is evolving into a variety of algorithms for special purposes. One example is the segment motion algorithm shown in Figure 3 which is suitable for plasma etching and

Segment Motion Algorithm

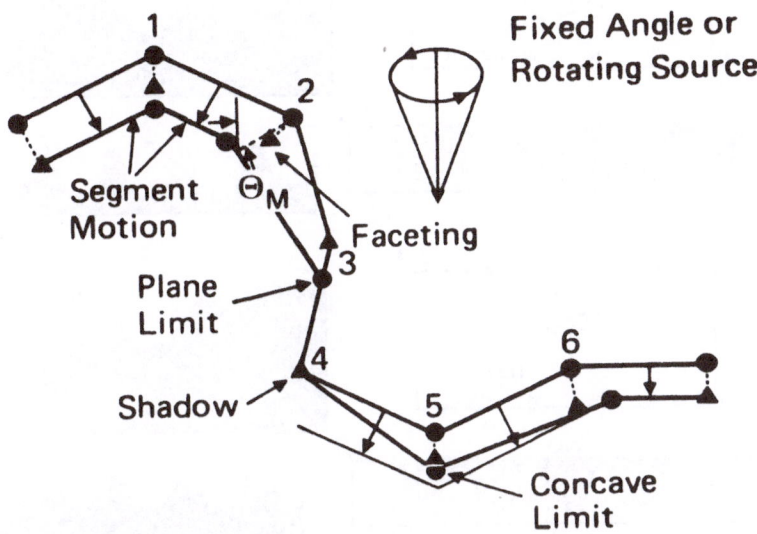

Fig. 3 Improved segment motion algorithm for reactive ion etching and ion milling which incorporates more information about the physical mechanisms and boundary conditions.

ion milling [54] For this algorithm the segments advance to form new segment intersections. Faceting at exterior corners and the preservation of angles at interior corners have also been included. The addition of new segments and deletion of small segments must be made in accordance with these conditions as well. Another example of the need for modifications of the algorithms occurs in deposition. Here a negative etch rate can be used to make the etch front move backward. However, the mechanisms of motion and boundary conditions for evaporation and sputtering differ. Care must also be taken to insure that the etching algorithm which is stable in the forward direction is also stable when advanced in reverse.

4 SIMULATION RESULTS

The impact which simulation is capable of making on device processing is best illustrated by surveying a few examples from the lithography, etching and deposition areas. The most extensive use of simulation has been in optical lithography and this application is covered in the companion article by W.G. Oldham. A comparison of experimental and simulated resist profiles for electron beam lithography is shown in Figure 4 [36]. Here the ninth or next to last

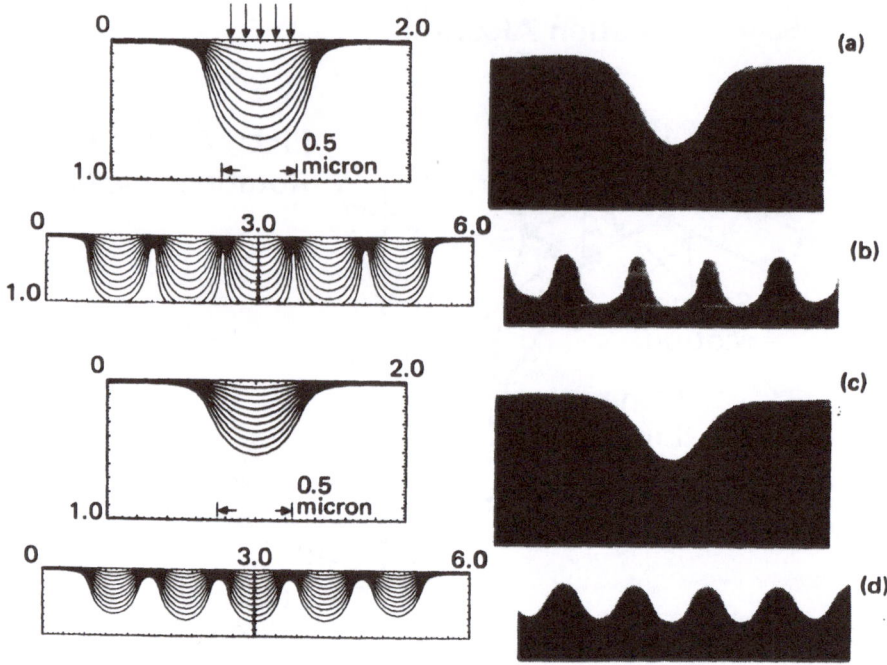

Fig. 4 Comparison of simulated and experimental electron beam exposed PMMA resist profiles.

profile corresponds to the 90 s experimental development time. The profile depth and size agrees well with experiment as a function of dose and proximity of other features. The experimental profile is also slightly more vertical indicating that an additional physical mechanism may need to be modeled.

Numerous simulation studies have been carried out in electron beam lithography. The magnitude of intra-line and inter-line proximity effects on various substrates has been studied [31-45]. Three dimensional development effects have been explored [45]. The role of scattering by high energy secondary electrons has been shown to be relatively small [28]. The optimum bias is predictable from a pluralistic scattering model [40]. The advantages of multilayer resists [39] and the relative advantages of electron beam compared to optical lithography [43,44] have been explored. Profile description parameters have been used to show that bias correction for proximity effects gives better linewidth control and profile quality although it may be more difficult to implement in machine design. These profile description parameters [41] are shown in Figure 5 and are applicable to quantitative profile comparisons in other lithography and etching processes as well.

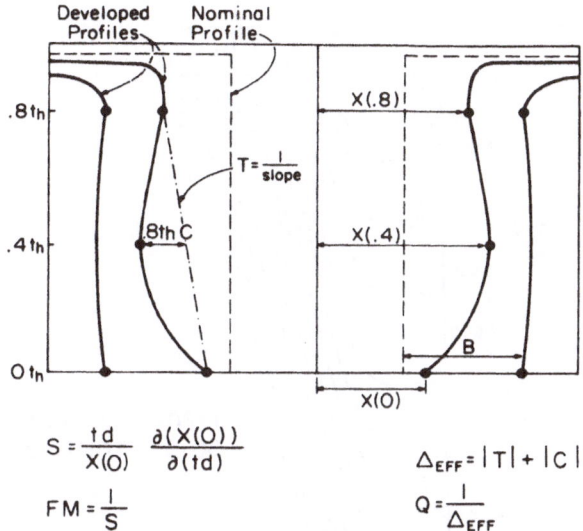

Fig. 5 Line edge profile description parameters convenient for quantitative experimental and simulation studies.

For x-ray and ion beam lithography the exposure mechanisms for depositing energy into the resist are well understood and can readily be simulated. Generally it is assumed that the etch rate for development is the same function of deposited energy as used for electron beam lithography. A example of simulation of an x-ray lithography resist profile in multilayer resist from a mask with a tapered edge is shown in Figure 6 [46]. Photoelectrons can be generated in the substrate which may undercut positive resists [47,48]. To simulate ion beam lithography profiles Monte Carlo exposure simulation must include nuclear as well as electronic scattering [49].

An example of simulating plasma etching is shown in Figure 7. Here isotropic and anisotropic etching components determined from published etch rates have been used to simulate the profile [53]. The simulated profile generally tracks the experimental profiles well as a function of power, pressure and voltage. The task of determining the etch rate components directly from the physical, electrical and chemical tool parameters is a much more formidable task [55,58-59]. Ion milling is fairly well characterized by etch rate versus angle of incidence curves for various materials [60]. Ion reflection effects such as trenching and redeposition of materials with low etch rates can greatly complicate the basic process. A comparison of an ion milled v-groove in silicon is shown in Figure 8 [54]. It is interesting to note that the angle at the bottom of the groove tends to be preserved.

418

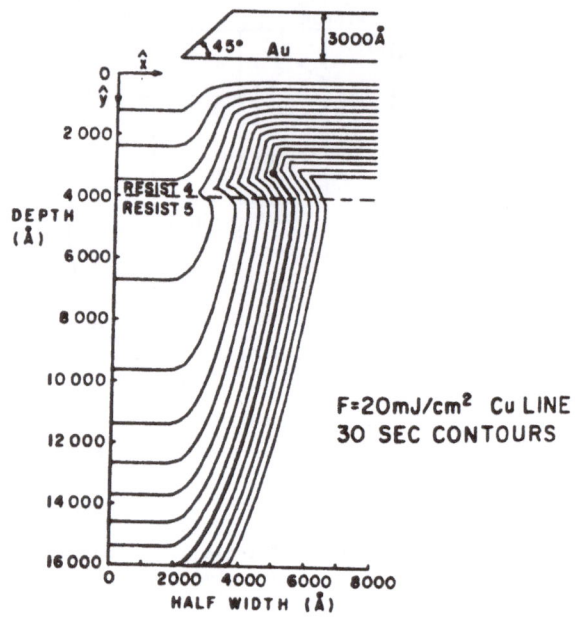

Fig. 6 Multilayered X-ray resist profile after exposure through a mask with a tapered edge.

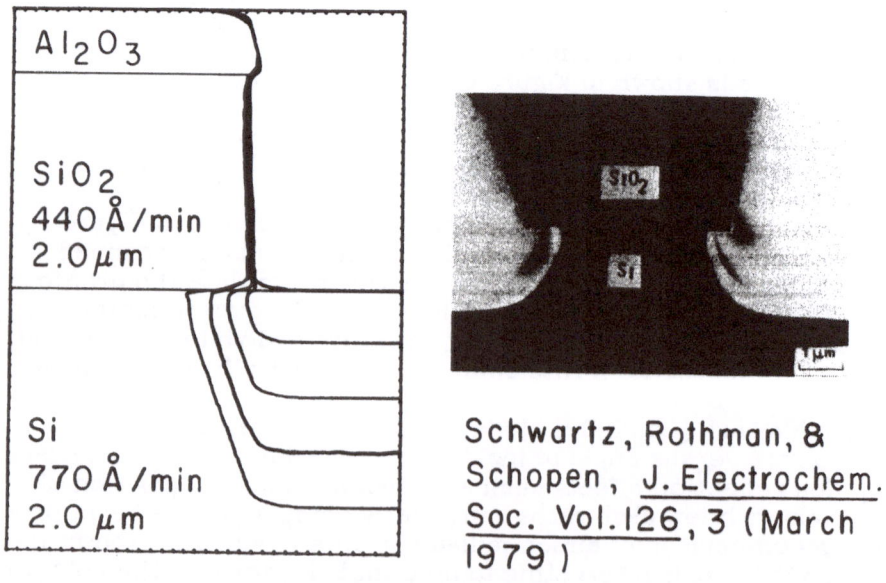

Schwartz, Rothman, & Schopen, J. Electrochem. Soc. Vol.126, 3 (March 1979)

Fig. 7 Comparison of simulated and experimental profiles for reactive ion etching.

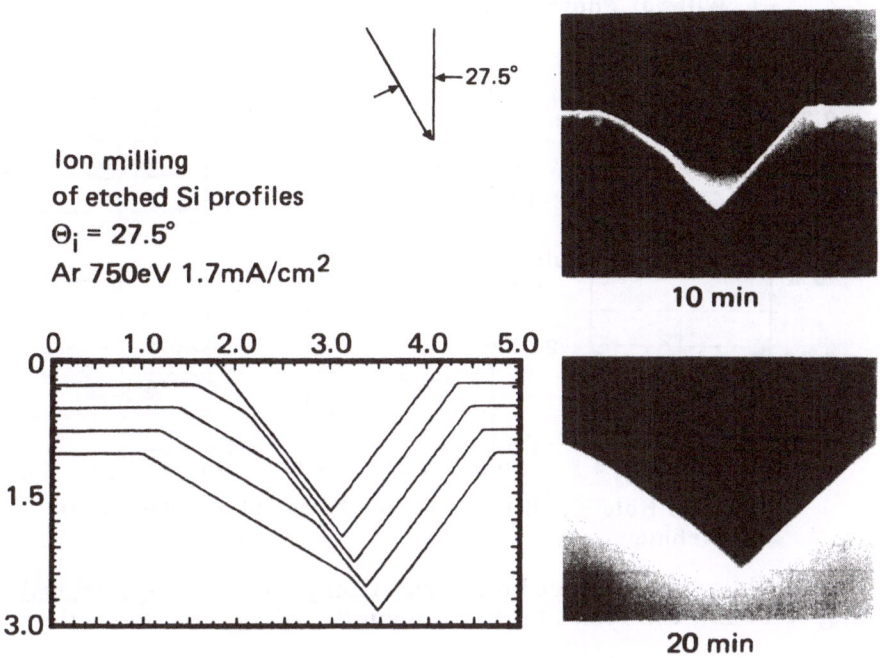

Fig. 8 Comparison of simulated and experimental profiles for ion milling of a v-groove in silicon.

The simulation of plasma etching can be used to optimize tradeoffs between process steps and competing effects as a function of operating conditions. Figure 9 shows a contact deposition process [57] where the resist has been reflowed to produce a tapered oxide step for improved coating. A example of simulating the residue problem in dry etching [57] is shown in Figure 10. Here a CVD polysilicon layer (isotropically grown) is being patterned by RIE (anisotropic etching). The residue that results can be removed by over etching at the expense of the underlying oxide thickness due to poor selectivity. A more isotropic removal with good selectivity results in overetching the polysilicon gate. The tradeoff between anisotropy and selectivity in etching is shown in Figure 11 [61]. Simulation was used to generate the design graph in Fig. 12 which shows the tradeoff in oxide etching and gate undercut as a function of the anisotropy of the etching.

Many geometrical effects occur in deposition and etching which can be investigated with simulation. Figure 13 shows an experimental-simulation comparison for a test structure in a planetary evaporator system [65]. This structure illustrates the tradeoff in symmetry which must be made in configuring the tooling for reasonable throughput. The coating process becomes asym-

420

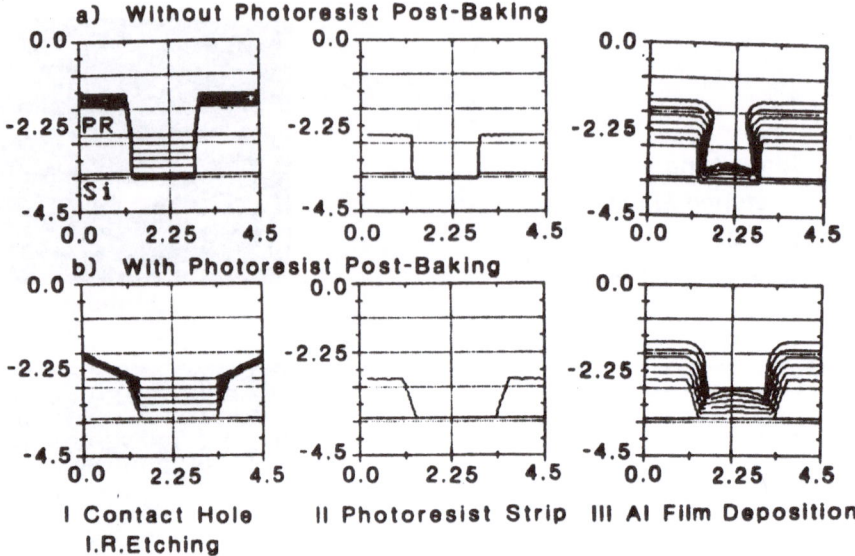

a) Without Photoresist Post-Baking

b) With Photoresist Post-Baking

I Contact Hole   II Photoresist Strip   III Al Film Deposition
I.R.Etching

Fig. 9 Simulation of contact hole formation process using infrared baking to reflow the resist prior to plasma etching and deposition.

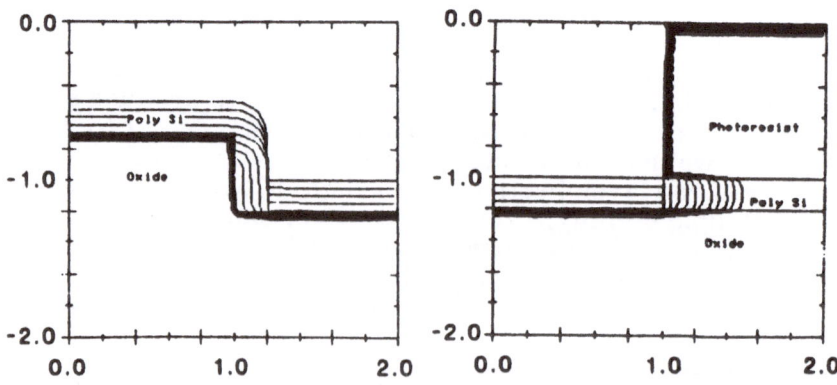

Fig. 10 Conflicting problems of residue removal at steps and under etching of the gate structure. Note that over etching attacks the underlying oxide.

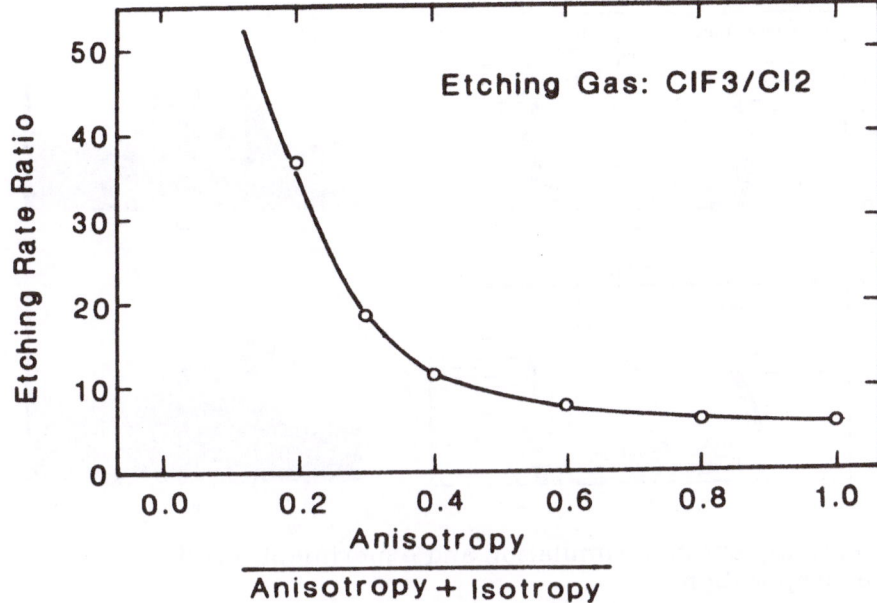

Fig. 11 Tradeoff in etch rate ratio selectivity with anisotropy [61].

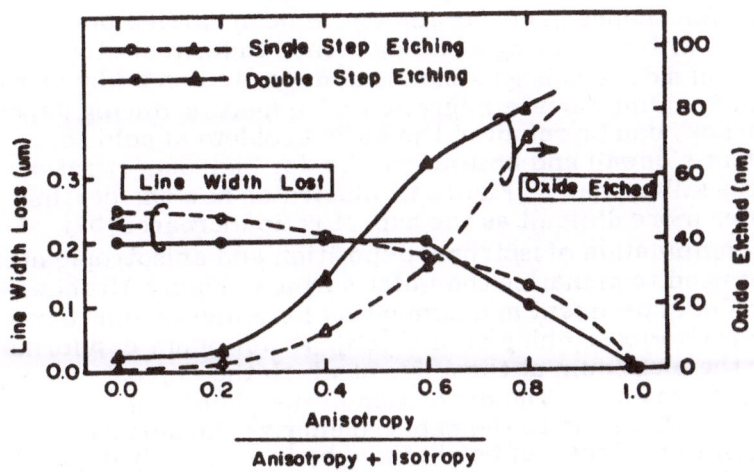

Fig. 12 Design graph for loss in gate linewidth versus etching of underlying oxide.

422

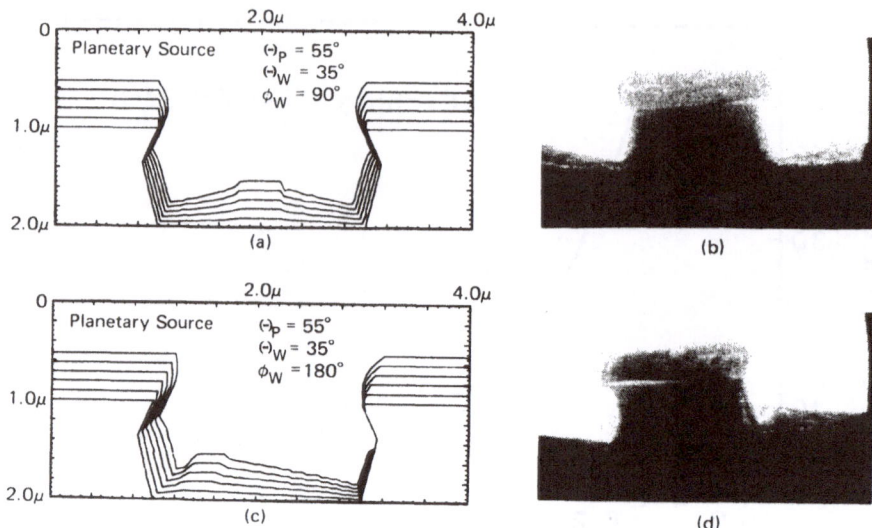

(a)

(b)

(c)

(d)

Fig. 13 Comparison of simulation and experiment for planetary metal evaporation.

metrical for lines oriented perpendicular to the planet radius on the outboard wafers. This is because the source is located below the intersection of the planet axis with the system axis to achieve good step coverage and the wafer is not free to rotate independent of the planet rotation. Studies of the effects of orientation angles and profile shapes in evaporation have been carried out [62-67].

Even with sputtering a "crack" tends to form at steps and the coating of sidewalls is less than that deposited on a planar surface. Although adding surface migration with heating during deposition the "crack" can be removed the basic problem af achieving sufficient sidewall and bottom coating for high aspect ratio features still remains. Figure 14 illustrates how the coating becomes more difficult as the aspect ratio increases [57].

A combination of isotropic deposition and anisotropic etching can be used to planarize the wafer surface. Figure 15 shows a comparison of experiment and simulation for simultaneous deposition and reactive-ion-etching of SiO2 [65]. A stationary cap forms on top of the aluminum at the angle at which the deposition and etching rates are equal. The deposition in the planar regions continues and thus catches up to the cap to planarize the surface. A similar planarization effect can be obtained by isotropically depositing a film sufficiently thick to fill laterally across small openings. The film is then anisotropically etched so that the residue fills in the small openings as shown in Figure 16 [57].

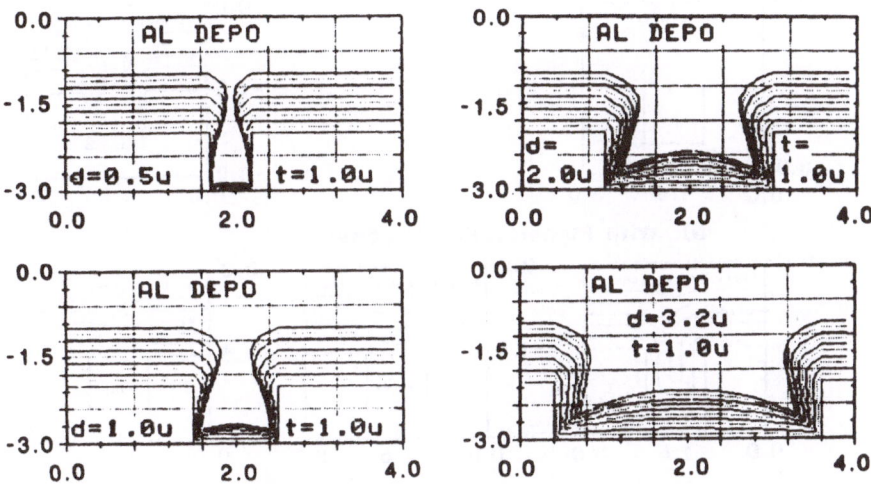

Fig. 14 Illustration of how the diameter d of a contact hole influences the amount of material which can be sputtered on the bottom and side walls.

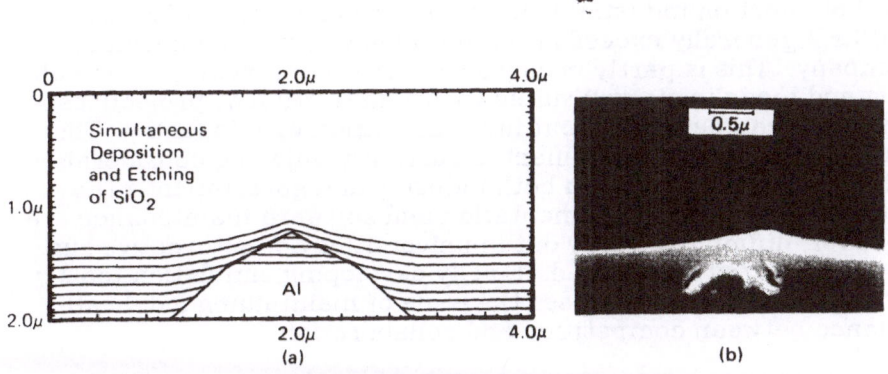

Fig. 15 Simultaneous deposition and etching of SiO2 to planarize aluminum lines.

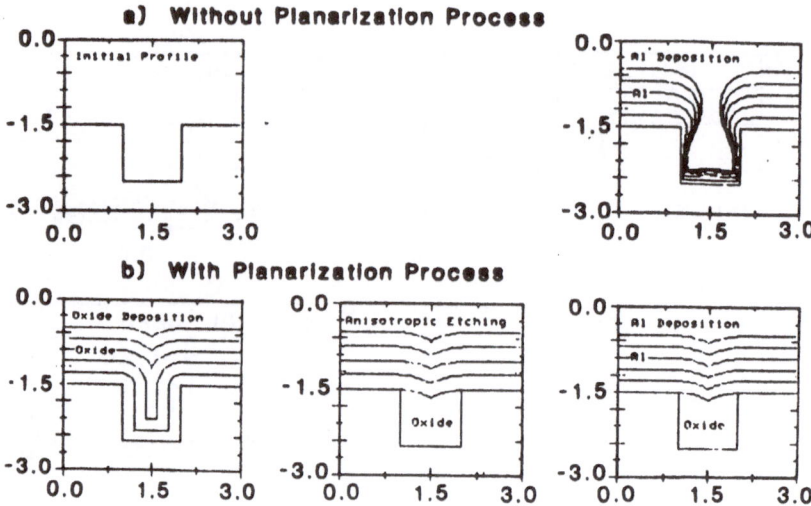

Fig. 16 Simulation of oxide deposition and anisotropic etching to planarize prior to metalization with aluminum.

5 FUTURE PROGNOSIS

Simulation and modeling of IC fabrication processes have reached an important point where a critical mass of tools can be invoked to address technology problems. The progress and current level of effort on industry wide programs such as SAMPLE and SUPREM generally exceeds that available within any individual company. This is partly because the nature of research on simulation and modeling is well suited to graduate student projects especially if it can be carried out in collaboration with industry. This mode of program development is currently enjoying considerable interest and support from both industry and government. The problems of careful documentation and software maintenance are however, difficult to carry out in universities and alternative support structures are needed. Finally developing simulation tools on an industry wide basis raises the issue of maintaining a delicate balance between competition and collaboration.

6 SUMMARY

IC process modeling and simulation will likely be a leading contributor to future VLSI design. It is well accepted as a means of characterizing individual lithography, etching and deposition processes. Today process simulation is being applied to the study of the complex tradeoffs between conflicting physical mechanisms in the context of complete multi-step process sequences. The success of modeling and simulation has created a demand for better process parameter data, the extensions of the capability to new manufacturing tools, the incorporation of new physical process into models and more flexible and robust algorithms. In the future topography emulation, simulation and electrical parameter extraction will likely be driven directly from the layout system to provide the designers feedback on the consequences of their designs. These developments will likely be made on an industry wide basis at universities through collaborative support with industry.

ACKNOWLEDGEMENT

The author would like to thank his colleagues, students and industrial collaborators whose assistance over the last several years have made possible the results presented here. A special thanks to Yoshio Sakai for his interest in many of the new applications of the SAMPLE program. This work was supported in part by a National Sciences Foundation grant ECS-8106234 and in part by industrial grant in aid.

REFERENCES

[1] W. G. Oldham, S. Nandgaonkar, A. R. Neureuther, and M. M. O'Toole, "A General Simulator for VLSI Lithography and Etching Processes: Part I - Application to Projection Lithography," *IEEE Trans. on Electron Devices*, Vol. ED-26, No. 4, pp. 717-722 April 1979.

[2] W. G. Oldham, A. R. Neureuther, C. Sung, J. L. Reynolds and S. N. Nandgaonkar, "A General Simulator for VLSI Lithography and Etching Processes: Part II-Application to Deposition and Etching," *IEEE Trans. on Electron Devices*, Vol. ED-27, No. 8, pp. 1455-1459, August 1980.

[3] A. R. Neureuther, "Simulating VLSI Wafer Topography," *1980 IEDM Technical Digest*, pp 214-218, (December).

[4] R.W. Dutton and S.E. Hansen, "Process Modeling of Integrated Circuit Device Technology," *Proceedings of the IEEE*, Vol. 69, No. 10, pp. 1305-1320, October 1981.

[5] H. Ryssel, K.Haberger, K. Hoffmann, G. Prinke, R. Dumcke and A. Sachs, "Simulation of Doping Processes," *IEEE Trans on Electron Devices*, Vol. ED-27, No. 8, pp. 1484-1492, August 1980.

[6] F.H. Dill, "Optical Lithography," *IEEE Trans. Elec. Dev.*, ED-22, No. 7, pp. 440-444, July 1975.

[7] F.H. Dill, A. R. Neureuther, T.A. Tuttle, and E.J. Walker, "Modeling Projection Printing of Positive Photoresist," *IEEE Trans. Elec. Dev.*, Vol. ED-22, No. 7, pp. 456-464, July 1975.

[8] A. Brochet, G.M. Dubrceucq, and M. Lacombat, "Modelisation des process d'exposition et de development d'une resine photosensible positive. Application ou masquage par projection," *Revue Technique Thompson-CSF*, Vol. 9, No.2, pp. 287-335, Juin 1977.

[9] S. Fujimori, "Computer Simulation of Exposure and Development of a Positive Resist," *J. Appl. Phys.*, Vol. 50, No. 2, pp. 615-623, February 1979.

[10] J. Bauer and H. Haferkorn, "Berechnung zur Polychromatischen Projektionsfotolithografie," *Feingeratetchnik*, 28. Jg, Heft 4, pp. 166-171, 1979.

[11] M. M. O'Toole and A. R. Neureuther, "The Influence of Partial Coherence on Projection Printing," *SPIE Vol. 135*, pp.22-27, 1979.

[12] M.A. Narasimham and J.H. Carter, Jr., "Effects of Defocus on Photolithographic Images Made with Projection Printing Systems," *SPIE Proceedings*, Vol. 135, pp. 2-9, April 1978.

[13] C.N. Ahlquist, P. Schoen and W.G. Oldham, "A Study of a High-Performance Stepper Lens," *Kodack Microelectronics Seminar Proceedings*, 1979, and Proceedings of "Microcircuit Engineering" Aachen Germany, 25-27 September, 1979.

[14] D.C. Hofer, C.G. Willson, A. R. Neureuther and M. Hakey, "Characterization of the 'Induction Effect' at Mid UV Exposure: Application to AZ2400 at 313 nm," SPIE Vol. 334, Optical Microlithography, pp. 196-205, March 1982.

[15] W. G. Oldham and A. R. Neureuther, "Projection Lithography with High Numerical Aperture Optics," *Solid State Technology*, Vol. 24, NO. 5, pp 106-111, 140, May 1981.

[16] W. G. Oldham, S. Subramanian and A. R. Neureuther, "Optical Requirements for Projection Lithography," *Solid State Electronics*, Vol. 24, No. 10, pp. 975-980, 1981.

[17] P. K. Jain, A. R. Neureuther and W. G. Oldham, "Influence of Axial Chromatic Aberration in Projection Printing," *IEEE Trans. on Electron Devices*, Vol.ED-28, No.11, pp. 1410-1416, (November),1981.

[18] A. R. Neureuther, P. K. Jain and W. G. Oldham, "Factors Affecting Linewidth Control Including Multiple Wavelength Exposure and Chromatic Aberration," *SPIE*, Vol. 275, pp. 110-116, 1981.

[19] P.D. Robertson, F.W. Wise, A.N. Nasr, A. R. Neureuther and C.H. Ting, "Proximity Effects and Influences of Nonuniform Illumination in Projection Lithography," SPIE Vol. 334, Optical Microlithography, pp. 37-43, March, 1982.

[20] F.H. Dill, W.P. Hornberger, P.S. Hauge, and J.M. Shaw, "Characterization of Positive Photoresist," *IEEE Trans. on Electron Devices*, Vol. ED-22, No. 7, pp. 445-452, July 1975.

[21] F.H. Dill and J.M. Shaw, "Thermal Effects on the Photoresist AZ1350J," *IBM J. Res. Dev.*, Vol. 21, pp. 210-218, May 1977.

[22] J.M. Shaw and M. Hatzakis, "Performance Characteristics of Diazo-Type Photoresists Under e-Beam and Optical Exposure," Trans. of IEEE on Electron Devices, Vol. ED-25, No. 4, pp. 425-430, April 1978.

[23] J.M. Shaw and M. Hatzakis, "Developer Temperature Effects on Positive Photoresists," Proceedings of the Kodak Microelectronics Seminar, San Diego, October 1978.

[24] M.M. O'Toole and W.J. Grande, "Characterization of Positive Resist Development," *IEEE Trans. Electron Devices*, Vol. EDL-2, No. 12, December 1981.

[25] T. Matsuzawa, A. Kishimoto, and H. Tomioka, "Profile Simulation of Negative Resist MRS Using the SAMPLE Photolithography Simulator," *IEEE Trans. Electron Device Letters*, Vol. EDL-3, No. 3, pp. 58-60, March 1982.

[26] M. Exterkamp, W. Wong, H. Damar, A. R. Neureuther and W.G. Oldham, "Resist Characterization: Procedures, Parameters, and Profiles," SPIE Vol. 334, Optical Microlithography, pp. 182-187, March, 1982.

[27] R.J. Hawryluk, "Exposure and Development Models Used in Electron Beam Lithography," *J. Vac. Sci. Technol.*, Vol. 19, pp. 1-17, May/June 1981.

[28] K. Murata, D.F. Kyser and C.H. Ting, "Monte Carlo Simulation of Fast Secondary Electron Production in Electron Beam Resists," *Journal of Applied Physics*, Vol. 52, No. 7, pp. 4396-4405, July 1981.

428

[29] M. Hatzakis, C.H. Ting, and N.S. Viswanathan, "Fundamental Aspects of Polymeric Resist Systems," *Electron and Ion Beam Science and Technology, Sixth International Conference Proceedings*, R. Bakish, Ed., ECS, pp. 542, 1974.

[30] J.S. Greeneich, "Developer Characteristics of PMMA Electron Resist," *J. ECS*, Vol. 122, pp. 970, 1975.

[31] J.S. Greeneich, "Time Evolution of Developed Contours in PMMA Electron Resist," *J. Appl. Phys.*, Vol. 45, pp.5264-5268, 1974.

[32] N.S. Viswanathan, R. Pyle and D. Kyser, "Simulation of Lithographic Images in Electron-Beam Technology," *Electron and Ion Beam Science and Technology, Seventh International Conference Proceedings*, R. Bakish, Ed., ECS, pp. 218-232, 1976.

[33] J.S. Greeneich, "Impact of Electron Scattering on Linewidth Control in Electron-Beam Lithography," *J. Vac. Sci. Technol.*, Vol. 16, No. 6, pp.1749-1753, Nov./Dec 1979

[34] J.C.H. Phang and Ahmed, "Line Profiles in Thick Electron Resist Layers and Proximity Effect Correction," *J. Vac. Sci. Technol.*, Vol. 16, No. 6, pp. 1759-1763, Nov/Dec 1979.

[35] K. Murata, E. Nomura and K. Nogami, "Experimental and Theoretical Study of Cross-Sectional Profiles of Resist Patterns in Electron-Beam Lithography," *J. Vac. Sci. Technol.*, Vol. 16, No. 6, pp. 1734-1736, Nov/Dec 1979.

[36] A. R. Neureuther, D. F. Kyser, and C. H. Ting, "Electron Beam Resist Edge profile Simulation," *IEEE Trans. on Electron Devices*, Vol. ED-26, No. 4, pp. 686-692, April 1979.

[37] D.F. Kyser and R. Pyle, "Computer Simulation of Electron Beam Resist Profiles," *IBM J. Res. and Dev.*, Vol. 24. No. 4, pp. 426-437, July 1980.

[38] M.P.C. Watts, P. Rissman and J. Kahn, "Solubility Ratio, Sensitivity and Line Profile Control in Positive e-Beam Resists," *Electron Ion Beam Science and Technology, Ninth International Conference Proceedings*, R. Bakish Ed., ECS, pp. 375-381, 1980.

[39] J.S. Greeneich, *Electron Ion Beam Science and Technology, Ninth International Conference Proceedings*, R. Bakish Ed., ECS, pp. 282-303, 1980.

[40] J.S. Greeneich, "Electron Beam Processes," Chapter 2 of *Electron-Beam Technology in Fabrication*, Academic Press, New York 1980.

[41] M. G. Rosenfield, A. R. Neureuther and C. H. Ting, "The Use of Bias in Electron-Beam Lithography for Improved Linewidth Control," *J. Vac. Sci. and Technol.*, Vol. 19, No. 4, pp. 1242-1247, Nov/Dec 1981.

[42] M. G. Rosenfield and A. R. Neureuther, "Exploration of Electron-Beam Writing Strategies and Resist Development Effects," *IEEE Trans. on Electron Devices*, Vol.ED-28, No.11, pp 1289-1294, (November),1981.

[43] T.S. Chang, D.F. Kyser and C.H. Ting, "Comparison of Electron Beam and Optical Projection Lithography in the Region of One Micrometer," *SPIE Proceedings*, Vol. 275, pp. 117-121, 1981.

[44] T.S. Chang, D.F. Kyser and C.H. Ting, "Exploration of Electron-Beam Writing Strategies and Resist Development Effects," *IEEE Trans. Electron Devices*, Vol. ED-28, No. 11, pp. 1295-1300, November 1981.

[45] F. Jones and J. Paraszczak, "RD3D (Computer Simulation of Resist Development in Three Dimensions," *IEEE Trans. Electron Devices*, Vol. ED-28, No. 12, pp. 1544-1552, December 1981.

[46] A. R. Neureuther, "Simulation of X-ray Resist Line Edge Profiles," *J. Vac. Sci. Technol.*, Vol. 15, No. 3, pp 1004-1008, May/June 1978.

[47] P.Tischer and E. Hundt, "Profiles of Structures in PMMA by x-ray lithography," *Electron and Ion Beam Science and Technology, Eighth International Conference Proceedings*, R. Bakish, Ed., ECS, pp. 444-457, 1978.

[48] K. Heinrich, H. Betz, A. Heuberger, S. Pongraz, "Computer Simulations of Resist Profiles in x-Ray Lithography," *J. Vac. Sci. Technol.*, Vol. 19, No. 4, pp. 1254-1258, Nov./Dec. 1981.

[49] L. Karapiperis, I. Adesida, C.A. Lee and E.D. Wolf, "Ion Beam Exposure Profiles in PMMA - Computer Simulation," *J. Vac. Sci. Technol.* Vol. 19, No. 4, pp. 1259-1263, Nov/Dec 1981.

[50] J.P. Ducommun, M. Cantagrel and M. Moulin, "Evolution of Well-Defined Surface Contour Shapes Submitted to Ion Bombardment: Computer Simulation and Experimental Investigation," *J. Mater. Sci.*, Vol. 10, pp.52-62, 1975.

[51] H.W. Lehmann, L. Krausbauer and R. Widmer, "Redeposition - A serious Problem in rf Sputter Etching of Structures with Micrometer Dimensions," *J. Vac. Sci. Technol.*, Vol. 14, No. 1, pp. 281-284, Jan/Feb 1977.

[52] N.S. Viswanathan, "Simulation of Plasma Etched Lithographic Structures," *J. Vac. Sci. Technol.*, Vol. 16. No. 2, pp.388-390, Mar/Apr 1979.

[53] J. L. Reynolds, A. R. Neureuther, "Simulation of Dry Etched Line Etched Profiles," *J. Vac. Sci Technol.*, Vol 16, No 6, pp. 1772-1775, Nov/Dec 1979.

[54] A. R. Neureuther, C. Y. Liu and C. H. Ting, "Modeling Ion Milling," J. Vac. Sci. and Technol., pp. 1167-1171, 1979.

[55] L. Mei, S. Chen and R.W. Dutton, "A Surface Kinetics Model for Plasma Etching," *1980 IEDM Technical Digest*, pp. 831-832, December 1980.

[56] R.W. Dutton, L. Mei, D. Chin and M. Kump, "Two Dimensional Process Modeling for High Density (LOCOS) Technology," *1981 Symposium on VLSI Technology, Technical Digest*, pp. 90-91.August 1981.

[57] Y. Sakai, J.L. Reynolds and A. R. Neureuther, "Topography Simulation for Dry Etching Process," ECS Spring Meeting, Abstract No. 166, Montreal, Canada, May 9-14, 1982.

[58] M.J. Kushner, "A Kinetic Study of the Plasma-Etching Process. II. A Model for the Etching of Si and SiO2 in CnFm/O2 Plasmas," Journal of Applied Physics, Vol. 53., No. 4, pp. 2923-2938, April 1982.

[59] M.J. Kushner, "A Kinetic Study of the Plasma-Etching Process. II. Probe Measurements of Electron Properties in an rf Plasma-Etching Reactor," Journal of Applied Physics, Vol. 53., No. 4, pp. 2939-2946, April 1982.

[60] A.B. Jones and G.S. Plonski, "Ion Milling of Thin Films for Magnetic Bubble Circuits," *ECS Extended Abstracts*, San Francisco, 1974, and in L.D. Bollinger, "Ion Milling for Semiconductor Production Process," *Solid State Technol.*, pp. 66-70, Nov. 1977.

[61] D.L. Flamm, D.N.K. Wang and D. Maydan, 1981 ECS Fall Meeting.

[62] I.A. Blech, "Evaporated Film Profiles Over Steps in Substrates," *Thin. Solid Films*, Vol. 6, pp. 113-118, 1970.

[63] I.A. Blech, D.B. Fraser and S.E. Hasyko, "Optimization of Al Step Coverage Through Computer Simulation and Scanning Electron Microscopy," *J. Vac. Sci. Technol.*, Vol. 15, No. 1, pp. 13−19, 1978.

[64] *Process and Device Modeling for Intergrated Circuit Design*, Ed by F. Van de Weile, W.L. Engle and P.G. Jespers, Noordhoff International Publishing, Leyden, The Netherlands/Reading, Mass., 1978.

[65] A. R. Neureuther, C. H. Ting and C. Y. Liu, "Application of Line-Edge Profile Simulation to Thin-Film Deposition Process," *IEEE Trans. on Electron Devices*, Vol. ED-27, No, 8, pp. 1449-1455, August 1980.

[66] T. Batchelder, "Simple Metal Lift-Off Process for 1 Micron AL/5% Cu Lines," *SPIE Proceedings Vol. 275*, pp. 143-149, 1981. also *Solid State Technology*, Vol. 25, No. 2, pp. 111-114, February, 1982.

[67] Y. Homma,A. Yajima and S. Harada, "Feature Size Limit Analysis of Lift-Off Metalization Technology," *1981 IEDM Technical Digest*, pp. 570-573, December 1981.

ANALYSIS OF NONPLANAR DEVICES

James A. Greenfield, Craig H. Price and Robert W. Dutton

Integrated Circuits Laboratory
Stanford University
Stanford, California 94305

1 INTRODUCTION

As stated in the companion chapters, the objective of process modeling is to assist in the design of integrated devices. The focus of this chapter involves device analysis tools and particularly two-dimensional simulation, with emphasis on technology-oriented device design.

Over the past two decades integrated circuit technology has powered an electronics revolution as potent as the industrial revolution. Shrinking device dimensions allow the increased packing density of gates—presently more than a million devices can be created on a single chip. Yet scaled-down device dimensions create new problems concerning reliable operation of circuits. For example isolation between neighboring devices becomes more difficult—field oxides are thinner and packing density requirements seek to minimize drain-to-source spacing. The evolution towards CMOS technology adds potential bipolar effects to the list of parasitics and concern about latchup is critical in defining the next generation of scaling. Both the field isolation and latchup problems reflect current problems in device scaling.

Moreover, they provide an effective platform from which device analysis can be viewed with proper perspective.

Previous work in device analysis is extensive and recent conferences provide an excellent review of progress in the field [1]. In this work a specific approach is taken to emphasize a central theme—technology-oriented device design. In the sections which follow, the discussion will evolve toward a demonstration of technology-based device effects, beginning with the details of analysis techniques specifically suited to nonplanar devices. It is our belief that this work provides a synergistic coupling to that of others presented in this volume. Specifically, nonplanar technology effects are inevitable in silicon technology—particularly near the isolation regions of both bipolar and MOS devices. While other papers tend to focus on the planar aspects of intrinsic devices, this work considers nonplanar extrinsic as well as intrinsic devices.

2 ANALYSIS TECHNIQUES

As stated above, the emphasis of this paper is on the analysis of non-planar devices. Figure 1a shows the cross-section of an NMOS transistor created using a locally oxidized isolation technology. The boxed region indicates the intrinsic portion of the device, while the p^+ regions indicate the adjoining "channel stop" diffusions with locally oxidized regions and characteristic "birds-beaks". Opening the technology "view-port" further, Fig. 1b shows adjacent intrinsic devices with a number of the parasitic effects indicated. For example, a region enclosed by dashes shows the coupling of adjacent source and drain regions via the field isolation device. Moreover, the parasitic device effects of sidewall capacitance and narrow width effects of the p^+ diffusion are indicated. These cross-sections clearly indicate that technology-oriented device modeling requires consideration of nonplanar device effects.

Turning to the analysis of nonplanar devices, one must consider carefully the choice of grid and analysis methods used. In this section two

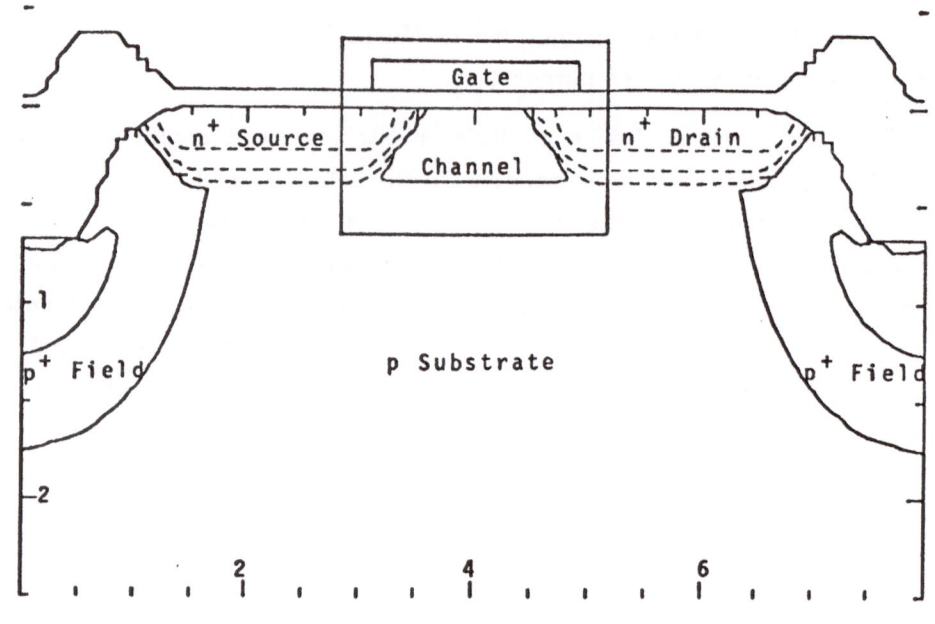

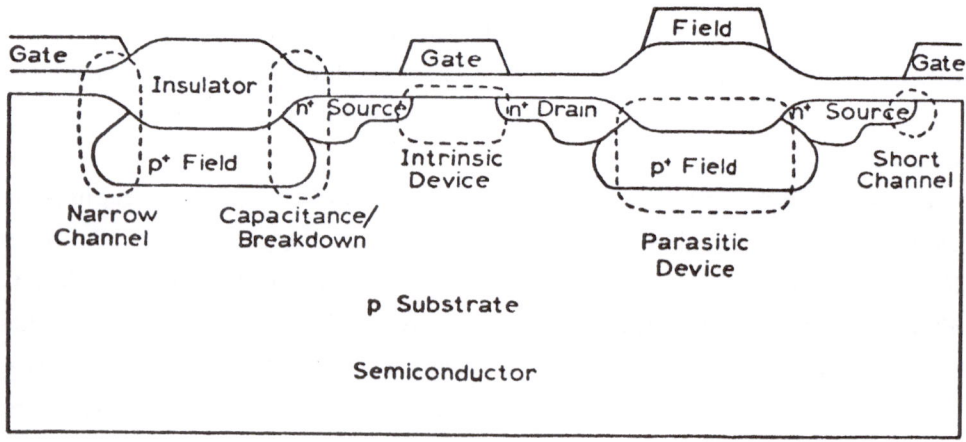

Fig. 1. Two views of NMOS device technology: (a) the cross-section of the intrinsic device including isolation regions (b) the cross-section of both intrinsic and extrinsic devices.

distinct aspects of nonplanar device analysis are considered. First, Poisson analysis on a nonplanar grid is discussed. Here the emphasis is on maximum efficiency—both in terms of storage and computation. As can be seen from Fig. 1b, the region required to analyze parasitic device effects extends over substantial distances and hence necessitates the use of cost-effective numerical solutions. The second portion of this section presents single-carrier, steady-state solutions for nonplanar devices. Although efficiency is of some concern, the present state-of-the-art reveals that robust, fail-safe algorithms are yet to be developed. Hence, the discussion seeks to point out some of the important considerations of nonplanar structures using iterative methods.

2.1 Poisson's Equation

Consider the formulation of Poisson's equation in a rectangular coordinate system (x, y) with the potential assumed to remain invariant in the third spatial dimension. The second order nonlinear elliptic partial differential equation defining the potential is of the form

$$\vec{\nabla} \cdot \left[\epsilon(x, y) \vec{\nabla} \psi(x, y) \right] = \begin{cases} -\rho_I(x, y), & \text{insulator} \\ -\rho_V(x, y, \psi) - \rho_I(x, y), & \text{semiconductor} \end{cases} \quad (1a)$$

$$\rho_V(x, y, \psi) = q[p(x, y, \psi) - n(x, y, \psi) + N_D^+(x, y, \psi)$$

$$-N_A^-(x, y, \psi)] \quad (1b)$$

$$\rho_I(x, y) = qN_I(x, y). \quad (1c)$$

where the subscripts V and I refer to the volume and interface contributions of charge, respectively. The potential is defined in terms of the conduction band edge, while the hole and electron concentrations are given by the half-order Fermi-Dirac integrals as discussed elsewhere [2]. Fermi-Dirac statistics are also incorporated to express the ionized donor and acceptor distributions using the appropriate degeneracy factors.

The elimination of the current-continuity equations from the solution, as described in this section, requires that ϕ_{Fn} and ϕ_{Fp} be chosen indepen- dently. The accurate solution of Poisson's equation under quasi-equilibrium conditions can be achieved by taking ϕ_{Fn} and ϕ_{Fp} to be locally constant within various regions of a device structure. For example, in an n-channel MOSFET, constant values of ϕ_{Fn} consistent with the source and drain biases are used within the source and drain regions, respectively. A constant value of ϕ_{Fp} consistent with the substrate bias is used within the substrate region. The use of locally constant values for ϕ_{Fn} and ϕ_{Fp} within a region is equiv- alent to the assumption that no carrier transport occurs in the region. Any discontinuities in the values of ϕ_{Fn} and ϕ_{Fp} must occur only in regions where the values of n and p, respectively, are negligible.

The solution of Poisson's equation must satisfy continuity conditions internally and boundary conditions along the outside boundary of the spa- tial region in which the solution is obtained. The continuity of the poten- tial ψ must be maintained everywhere to achieve the physically realistic condition of finite stored electrostatic field energy. In addition, the applica- tion of the integral form of Gauss's law indicates that, in the absence of interface charge, the normal component of the electric displacement $\epsilon \vec{\nabla} \psi$ must remain continuous across interfaces between materials with different dielectric permittivities. The standard choices for boundary conditions are Dirichlet, Neumann, or periodic. A Dirichlet condition specifies the value of ψ along the boundary and is used where an electrode with high conduc- tivity contacts the device surface. A Neumann condition specifies the value of the normal gradient of potential $\hat{n} \cdot \vec{\nabla} \psi$ along the boundary, where $\hat{n}$ is a unit vector normal to the boundary and directed out of the enclosed region. This condition is used along boundaries where the potential has reflection symmetry ($\hat{n} \cdot \vec{\nabla} \psi = 0$) or where a known surface charge concentration Q_s (coulomb/cm^2) is present ($\hat{n} \cdot \vec{\nabla} \psi = - Q_s/\epsilon$). Periodic boundary conditions are used at the sides of structures such as charge-coupled devices, when the device repeats periodically in the direction parallel to the device surface.

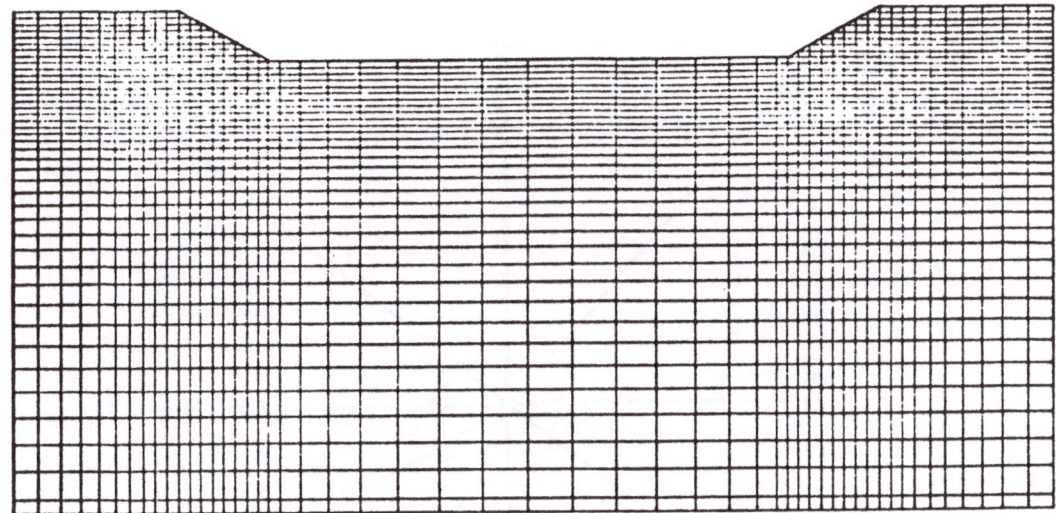

Fig. 2. A typical nonuniform rectangular node structure used to discretize Poisson's equation. The nodes occur at the intersections of horizontal and vertical lines.

2.1.1 Equation Discretization

The potential which solves Poisson's equation as defined by Eq. (1) is a continuous function of the spatial coordinates. The numerical solution of Poisson's equation requires that the potential and charge concentrations be represented by discrete values at a network of nodes within the solution region. This representation is achieved by discretizing Eq. (1) to yield a system of nonlinear equations. One equation arises for each node except those where boundary conditions fix the potential.

The node structure chosen for use in this work is rectangular with nonuniform spacing in the horizontal and vertical directions. A typical example of this structure is illustrated in Fig. 2, where the nodes occur at the intersections of horizontal and vertical lines. The nonplanar surface is represented by horizontal and oblique line segments joining adjacent nodes.

Boundaries such as surfaces and material interfaces must be accounted

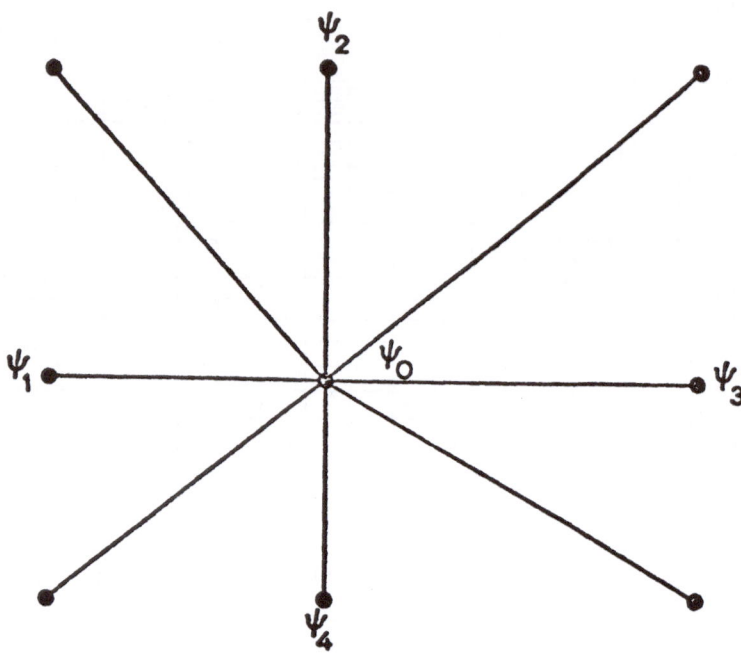

Fig. 3. The eight line segments joining a node to its nearest and next nearest neighbors. Surfaces and material interfaces are composed of sequences of these segments.

for during the discretization of Poisson's equation. The line segments used in this work to represent boundaries in the vicinity of one node are illustrated in Fig. 3. The eight line segments shown join the center node to its horizontally, vertically, and diagonally adjacent neighbor nodes. To represent various boundaries, each of the triangular regions between adjacent line segments in Fig. 3 can be eliminated from the solution region or composed of various materials, such as semiconductor or insulator. When the center node in Fig. 3 lies along the boundary of the solution region, triangular regions are eliminated from the solution region to represent a portion of the boundary. When the center node lies along a material interface, triangular regions on either side of the interface can be composed of different materials to represent the interface.

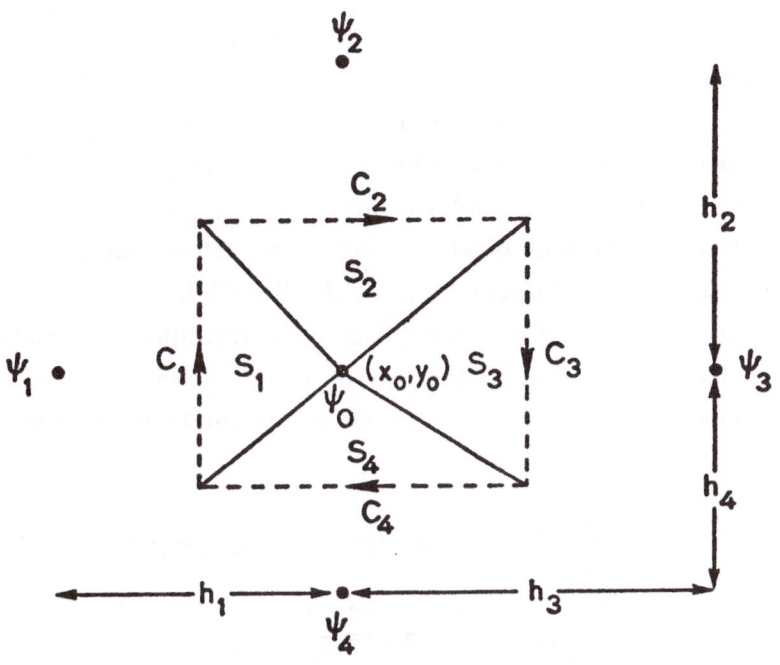

Fig. 4. A node and its four horizontally and vertically adjacent neighbors. The node locations and potentials are indicated. The C_m are perpendicular bisectors of the lines joining the center node to the outside nodes and define a rectangle S with triangular subregions S_m.

The discrete approximation to the continuous Poisson equation relates the potential value at a node to the values at neighboring nodes. The simplest approach, called a five-point finite difference approximation, includes the four neighbor nodes which are horizontally and vertically adjacent to the center node. The resulting equation associated with the center node depends on the potential values at this node and its four neighbors. This approach is used in this work because of its simplicity and the ability to adequately treat nonplanar structures.

The discretization of Poisson's equation is accomplished by integrating Eq. (1) over the local area surrounding each node in the solution region.

These integrals are then approximated to yield a discrete finite difference equation associated with each node. The local node structure used to derive one of these equations is illustrated in Fig. 4 for a node that does not lie along a boundary of the solution region or along a material interface. The potential at the center node, located at (x_0, y_0), is defined as ψ_0 and the potentials at the four neighbor nodes are defined as ψ_1, ψ_2, ψ_3, and ψ_4. The perpendicular bisectors C_m of the lines joining the center node to the outside nodes define a rectangle S with triangular subregions S_m. The finite difference approximation at the center node is derived by integrating Poisson's equation over the rectangle S and the resulting difference equation is

$$\sum_{m=1}^{4} B_m(\psi_0 - \psi_m) = \rho_V(x_0, y_0, \psi_0) + \rho_{IS} \tag{2a}$$

$$B_m = \frac{L_m}{h_m A} \tag{2b}$$

$$L_m = \int_{C_m} \epsilon \, dl \tag{2c}$$

$$A = \sum_{m=1}^{4} A_m \tag{2d}$$

$$\rho_{IS} = \frac{\displaystyle\int\int_S \rho_I(x, y) \, dx \, dy}{A} \tag{2e}$$

where ρ_{IS} is the total surface or interface charge concentration (coulomb/cm^3) within the rectangule S and A_m is the area of the triangle S_m. The volume charge concentration ρ_V is taken to vanish within insulator regions and is taken to have the constant value $\rho_V(x_0, y_0, \psi_0)$ within the semiconductor portion of the rectangle S. An equation of the form of Eq. (2a) can be written for each node in the solution region where the potential is not fixed by a Dirichlet boundary condition. No equation is written for nodes where a Dirichlet boundary condition is specified because the node potential is already determined by the boundary condition.

The system of nonlinear equations which results when Eq. (2) is applied at each node within the solution region can be written in a standard form for all nodes. The quantities ψ_0, ψ_m, B_m, ρ_V, and ρ_{IS} associated with each node are represented as elements of two-dimensional arrays and are labelled by the horizontal (i) and vertical (j) node indices. For a solution region containing N_x vertical lines of nodes with the vertical indices in line i ranging between $j_s(i)$ and N_y, the complete discrete approximation to Poisson's equation is written as

$$B_0(i,j)\psi(i,j) + B_1(i,j)\psi(i-1,j) + B_2(i,j)\psi(i,j-1)$$
$$+B_3(i,j)\psi(i+1,j) + B_4(i,j)\psi(i,j+1) = Q(i,j)$$
$$i = 1,\ldots,N_x, \qquad j = j_s(i),\ldots,N_y. \tag{3}$$

For a single node (i,j), the following correspondence exists between the quantities in Eqs. (2) and (3)

$$B_m(i,j) = -B_m, \qquad m = 1,2,3,4 \tag{4a}$$

$$B_0(i,j) = \sum_{m=1}^{4} B_m \tag{4b}$$

$$\psi(i,j) = \psi_0 \tag{4c}$$

$$\psi(i-1,j) = \psi_1 \tag{4d}$$

$$\psi(i,j-1) = \psi_2 \tag{4e}$$

$$\psi(i+1,j) = \psi_3 \tag{4f}$$

$$\psi(i,j+1) = \psi_4 \tag{4g}$$

$$Q(i,j) = \rho_V(x_0,y_0,\psi_0) + \rho_{IS}. \tag{4h}$$

A node lying along a boundary of the solution region or a material interface results in the modification of the definitions of L_m and A_m in Eq. (2). The values for A_m depend on whether any portion of the rectangle S contains semiconductor material where ρ_V could be nonzero. Details of the choice of these modified coefficients is given elsewhere [3].

2.1.2 Iterative Solution

The discrete form of Poisson's equation is the system of nonlinear equations represented by Eq. (3). The nonlinearity of this system of equations requires that its solution be obtained by using an iterative technique. Even the size of the system, typically involving from one thousand to ten thousand simultaneous equations, necessitates an iterative solution.

The development of an iterative solution technique is facilitated by representing the system of equations as a single matrix equation. Matrix techniques can then be used to develop various solution techniques and analyze their properties. Equation (3) can be written in matrix form as

$$B\vec{\psi} = \vec{Q} - \vec{F} \tag{5}$$

where the vector elements $\vec{\psi}(k)$ and $\vec{Q}(k)$ consist of the elements $\psi(i,j)$ and $Q(i,j)$ for each node in the solution region where $\psi(i,j)$ is unknown. Equation (5) represents a system of N simultaneous equations, where N is the total number of elements in $\vec{\psi}$. The vector $\vec{F}$ accounts for boundary conditions and contains elements consisting of those products of $B_m(i,j)$ and $\psi(i,j)$ in Eq. (3) where the value of $\psi(i,j)$ is specified directly by the boundary conditions.

The form of the matrix B in Eq. (5) depends on the method of ordering the nodes within the solution region and the equations associated with these nodes. The nodes are sequenced in the order of increasing i and j, with i held constant at each of its values while j takes on all values in its range. This is illustrated in Fig. 5 for a simple node structure where the numbers define the node ordering. The elements $\psi(i,j)$ and $Q(i,j)$ are placed in $\vec{\psi}$ and $\vec{Q}$ using the same order as the node sequence.

Each row of B contains at most five nonzero entries given by $B_m(i,j)$, $m=0,1,2,3,4$ for one pair of indices (i,j). For the node ordering described above, three of the five nonzero values in each row occur along the principal diagonal and its two adjacent diagonals. For the equation

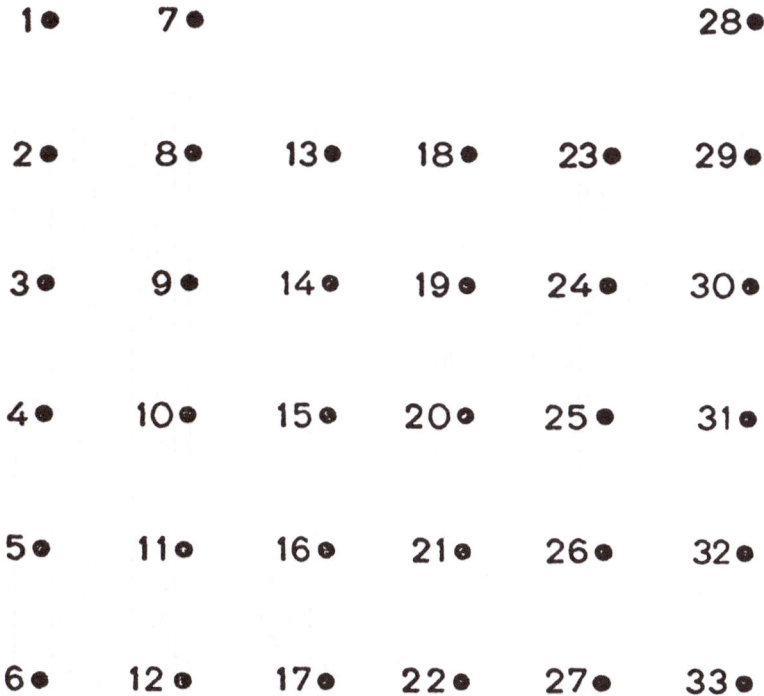

Fig. 5. Sequential node ordering for a simple node structure.

associated with node (i, j), the remaining two nonzero values occur along diagonals situated $N_y - j_s(i) + 1$ locations to the left and $N_y - j_s(i+1) + 1$ locations to the right of the principal diagonal. For the node structure of Fig. 5 with Dirichlet boundary conditions along the top and bottom boundaries and Neumann boundary conditions along the side boundaries, the resulting form of Eq. (5) is illustrated in Fig. 6, where omitted elements have zero value. The use of the same coefficients B_m, $m=0, 1, 2, 3, 4$ in each row of B is illustrative only and seldom occurs in practice. In the general case, each nonzero element of B may have a unique value. Examination of Fig. 6 indicates that for a nonplanar device structure, the nonzero elements of B farthest from the principal diagonal do not lie along single diagonals. This can introduce complications in some matrix solution techniques.

An iterative technique can be used to obtain an approximate solution

444

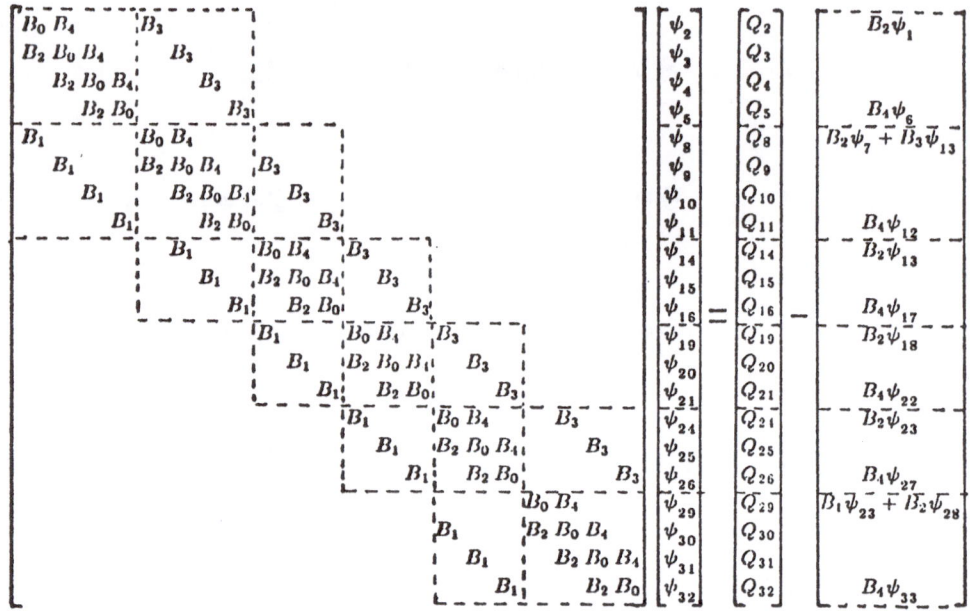

Fig. 6. The form of Eq. (5) for the node structure of Fig. 5 with Dirichlet boundary conditions along the top and bottom boundaries and Neumann boundary conditions along the side boundaries.

$\vec{\psi}$ which is acceptably close to the correct solution of Eq. (5). The most common nonlinear iterative solution techniques use Newton's method, which is based on a first order Taylor series expansion of the nonlinear terms in the equation about the most recent approximation to $\vec{\psi}$. The iteration is started by choosing an initial approximation to $\vec{\psi}$. For each iteration, Eq. (5) is linearized by a Taylor series expansion and a linear solution technique is used to solve the resulting linear matrix equation for the new approximation to $\vec{\psi}$. The iteration is halted when adequate convergence is achieved.

The nonlinear iteration requires the choice of an initial approximation $\vec{\psi}^0$ to the solution $\vec{\psi}$ of Eq. (5). In this work the vector elements $\vec{\psi}^0(k)$ are chosen such that $\vec{Q}(k)=0$, yielding a condition of charge neutrality at each node within the solution region. This approach is simple to implement and guarantees that the iteration is initially numerically stable.

The linearized form of Eq. (5) at the beginning of iteration $n+1$ is

$$B\vec{\psi}^{n+1} = \vec{Q}^n + J^n\left(\vec{\psi}^{n+1} - \vec{\psi}^n\right) - \vec{F} \tag{6}$$

where J^n is the Jacobian matrix with elements defined by

$$J^n(k,l) = \left.\frac{\partial \vec{Q}(k)}{\partial \vec{\psi}(l)}\right|_{\vec{\psi}^n}, \qquad k,l = 1, \ldots, N. \tag{7}$$

The superscript n in $\vec{\psi}^n$ indicates that this vector was obtained as a result of the nth nonlinear iteration. The superscript n in $\vec{Q}^n$ and J^n indicates that the elements of these vectors are evaluated using $\vec{\psi}^n$. The matrix J^n contains nonzero elements only along the principal diagonal because $\vec{Q}^n(k)$ depends only on $\vec{\psi}^n(k)$.

The rearrangement of Eq. (6) yields the following linear matrix equation for the unknown $\vec{\psi}^{n+1}$

$$(B - J^n)\vec{\psi}^{n+1} = \vec{Q}^n - J^n\vec{\psi}^n - \vec{F}. \tag{8}$$

The new approximation to $\vec{\psi}$ at the end of iteration $n+1$ is based on the complete or partial solution of Eq. (8) for $\vec{\psi}^{n+1}$. This iteration procedure is repeated until the approximations to $\vec{\psi}$ resulting from successive iterations satisfy the condition

$$\left|\frac{\vec{\psi}^{n+1}(k) - \vec{\psi}^n(k)}{\vec{\psi}^{n+1}(k)}\right| < \delta_1, \qquad k = 1, \ldots, N \tag{9}$$

where δ_1 is a specified error bound. Satisfaction of Eq. (9) generally indicates that the iteration has converged to yield a self-consistent solution $\vec{\psi}$ of Eq. (5).

A number of matrix solution techniques were investigated in the course of this work. Standard techniques such as Gaussian elimination and point successive overrelaxation (SOR) were considered, as well as block SOR and alternating direction implicit (ADI). All techniques solve the basic linear matrix equation

$$A\vec{x} = \vec{b} \qquad (10)$$

where the elements of Eqs. (8) and (10) are related by

$$A = B - J^n \qquad (11a)$$

$$\vec{x} = \vec{\psi}^{n+1} \qquad (11b)$$

$$\vec{b} = \vec{Q}^n - J^n\vec{\psi}^n - \vec{F}. \qquad (11c)$$

The elements of A, $\vec{x}$, and $\vec{b}$ are represented as $a_{i,j}$, x_i, and b_i, respectively, for $i, j = 1, \ldots, N$.

The iterative techniques use the notation $\vec{x}^m$ to represent an approximation to $\vec{x}$, where the superscript m indicates that $\vec{x}^m$ is obtained as a result of the mth linear iteration used to solve Eq. (10). The linear iteration procedure is started with $\vec{x}^0 = \vec{\psi}^n$ and is repeated until the approximations to $\vec{x}$ resulting from successive iterations satisfy the condition

$$\left| \frac{\vec{x}_i^{m+1} - \vec{x}_i^m}{\vec{x}_i^{m+1}} \right| < \delta_2, \qquad i = 1, \ldots, N \qquad (12)$$

where δ_2 is a specified error bound. Satisfaction of Eq. (12) generally indicates that the iteration has converged to yield a self-consistent solution $\vec{x}$ of Eq. (10).

The Block Iterative Techniques— Block iterative linear solution techniques are based on a sequential update of the solution according to blocks of nodes. For these techniques, $\vec{x}$ is divided into N_x blocks, corresponding to the vertical lines of nodes in the solution region. The elements of A, $\vec{x}$, and $\vec{b}$ then become $A_{i,j}$, $\vec{x}_i$, and $\vec{b}_i$, respectively, for $i, j = 1, \ldots, N_x$. The matrices $A_{i,j}$ are tridiagonal for $j = i$, zero for $j < i-1$ and $j > i+1$, and have a single nonzero diagonal for $j = i-1$ and $j = i+1$. Treatment of the nodes by blocks improves the computational efficiency of the block iterative techniques over their point iterative counterparts. This is demonstrated below by means of numerical comparisons.

The block successive overrelaxation (SOR) technique is defined by

$$\vec{x}_i^{m+1} = (1 - \omega)\vec{x}_i^m + \omega A_{i,i}^{-1}\left(\vec{b}_i - A_{i,i-1}\vec{x}_{i-1}^{m+1} - A_{i,i+1}\vec{x}_{i+1}^m\right),$$

$$i = 1, \ldots, N_x \quad (13)$$

where ω is the relaxation parameter in the range $0 < \omega < 2$. This parameter can be chosen to achieve a maximum convergence rate and minimum computation [2],[4]. The primary storage for this technique is required for the vector $\vec{b}$ and the principal diagonal of A, which are calculated as part of the Newton iteration prior to the linear solution, and the vector $\vec{x}^m$, which contains the approximation to $\vec{x}$ during the linear solution. This requires storage for $3N$ values. The block SOR technique requires $18N$ operations for each iteration.

The Alternating Direction Implicit Technique— The alternating direction implicit (ADI) technique differs substantially from the previous technique. The matrix A is first separated into the two matrices H and V which account for the horizontal and vertical spatial derivatives, respectively, in the partial differential equation on which A is based. H and V can be made tridiagonal by ordering the nodes by successive horizontal and vertical lines, respectively. This characteristic allows the use of Gaussian elimination to efficiently solve the two linear matrix equations which arise during each iteration.

One variation of the ADI technique is defined by

$$\vec{x}^{m+\frac{1}{2}} = (H + r_{m+1}D)^{-1}\left(\vec{b} - V\vec{x}^m + r_{m+1}D\vec{x}^m\right) \quad (14a)$$

$$\vec{x}^{m+1} = (V + r_{m+1}D)^{-1}\left(\vec{b} - H\vec{x}^{m+\frac{1}{2}} + r_{m+1}D\vec{x}^{m+\frac{1}{2}}\right) \quad (14b)$$

where D is a diagonal matrix containing the principal diagonal elements of A and r_{m+1} is one of a sequence of iteration parameters in the range

$0 < r_{m+1} \leq 1$. In this work, a set of six geometrically spaced iteration parameters is used cyclically, with the largest value of unity being used first in each cycle and the minimum value being chosen to achieve minimum computation [2].

The ADI iterative technique requires storage for the vector $\vec{b}$ and the principal diagonals of D, H, and V, which are calculated as part of the Newton iteration prior to the linear solution, and the vector $\vec{x^m}$, which contains the approximation to $\vec{x}$ during the linear solution. In addition, two temporary storage areas, each capable of saving N values, are needed during the solution of the tridiagonal matrix equations in Eq. (14). Thus, storage is required for a total of $7N$ values. This technique requires $36N$ operations for each iteration.

The One-Step Block SOR Technique— The linear iterative matrix solution techniques discussed above are used to obtain an accurate solution $\vec{x}$ to Eq. (10), within the error bound allowed by the convergence criteria defined by Eq. (12). It is possible to incompletely solve Eq. (10) by terminating the linear iteration after a predetermined number of iterations without satisfying Eq. (12). The result is an inaccurate solution $\vec{\psi}^{n+1}$ of Eq. (8). The disadvantage of this approach is that more nonlinear iterations will be required to satisfy the convergence criteria of Eq. (9), increasing the computational requirement associated with the nonlinear iterations. However, the incomplete solution of Eq. (10) has the advantage of eliminating the computation associated with the accurate solution of Eq. (8) when $\vec{\psi}^n$ is far from $\vec{\psi}$ and the accuracy of $\vec{\psi}^{n+1}$ is less important.

The approach taken in this work is called a one-step block SOR technique, which is implemented by applying Eq. (13) for a single iteration to obtain $\vec{\psi}^{n+1} = \vec{x}^1$ from $\vec{x}^0 = \vec{\psi}^n$. This technique has the additional advantage of requiring less storage than techniques which use more than one linear iteration. The primary storage for this technique is needed for those elements of the vector $\vec{b}$ and the principal diagonal of A which are associated with a single block $\vec{x}_i$ of $\vec{x}$. These elements are calculated during the implementation of Eq. (13), <u>not</u> prior to the linear solution, as for the techniques discussed previously. Consequently, this linear solution technique requires storage for approximately $2N^{\frac{1}{2}}$ values. The technique requires $18N$ operations for each linear iteration, as is the case for the standard block SOR technique.

Comparision of the Linear Matrix Solution Techniques— A realistic comparison of the efficiency of linear solution techniques can be performed by determining the total storage and computation required to obtain a complete solution of Poisson's equation for a practical example. The total storage required by several common techniques, including storage for the nonlinear Newton iteration, are compared in the second column of Table 1. Although the convergence rate of a solution technique can generally be analyzed theoretically for model problems, the problems encountered in practice seldom exhibit the features which facilitate such an analysis. Thus, the computational requirements of the linear solution techniques considered in this work are compared by using the techniques to solve Eq. (8) during the solution of Poisson's equation for a typical n-channel, enhancement-mode MOS device structure.

Each technique is used to solve the example problem for values of N_x varying from 20 to 60. For those solution techniques requiring the choice of an iteration parameter, such as ω in the block SOR technique and r_{m+1} in the ADI technique, a value yielding the minimum computation is determined and used for the comparison.

The discretization nodes within the solution region are placed on a rectangular grid and spaced nonuniformly in the horizontal and vertical directions. The same number of nodes is used in both the horizontal and vertical directions. For each change in N_x, the node spacings are all changed by approximately the same multiplicative factor to maintain similar node distributions for all node densities.

Table 1

Storage and Computation Requirements of Solution Techniques

Solution Technique	Storage	Computation	Operations ($\times 10^6$) ($N=2500$)
Gaussian elimination	$N^{\frac{3}{2}}$	$20N^2$	125.7
point Jacobi	$11N$	$25N^2$	154.4
point SOR	$10N$	$11N^2$	67.6
block SOR	$10N$	$171N^{\frac{3}{2}}$	22.1
ADI	$14N$	$54N^{\frac{3}{2}}$	7.5
one-step block SOR	$7N$	$52N^{\frac{3}{2}}$	6.5

N = number of discretization nodes

For each linear solution technique, the example problem is solved for each node density with error bounds of $\delta_1 = 0.001$ and $\delta_2 = 0.001$. Equation (9) is used to test the convergence of the Newton iteration and Eq. (12) is used to test the convergence of the linear iteration. For all solution techniques except the one-step block SOR technique, ten Newton steps are normally required to obtain solution convergence. A consistent comparison of the various solution techniques is provided by using the total number of linear iteration steps needed for the first ten Newton iterations as a measure of the computational requirements of a technique. For the one-step block SOR technique, the total number of Newton steps required to achieve convergence is used to measure the computational requirement. Note that for this technique, the number of linear iteration steps equals the number of Newton steps.

The total number of linear iterations necessary to obtain a solution of Poisson's equation for each solution technique and for the series of node densities described above has been determined for the linear solution techniques shown in Table 1. The most important information which can be extracted from this data is the variation of the total number of iterations with N. For each solution technique, the function $I=\alpha N^{\beta}+\gamma$ is least-square fitted to the data. I is the number of iterations and α, β, and γ are unknown parameters determined by the fit. In all cases, β is very close to either 0.5 or 1.0, exhibiting the type of behavior expected based on theoretical calculations for model problems. To provide a basis for comparing the results for different solution techniques, a second fit is performed using $\beta=0.5$ or $\beta=1.0$, as appropriate, to determine a new value for α. Combining this new expression for the number of iterations with the known number of computations required for each iteration yields the results shown in column three of Table 1 for the dominant component αN^{β}. These results make it possible to determine how quickly the computational requirements for a solution increase with increasing node density and, equivalently, with solution accuracy.

The storage and computational requirements summarized in Table 1 indicate that the one-step block SOR technique exhibits better efficiency than the other techniques considered. Only the ADI technique achieves comparable computational efficiency. However, the ADI technique requires twice as much storage as the one-step block SOR technique and is much more difficult to implement in a computer program. Thus, the one-step block SOR technique is chosen to provide efficient solutions of Poisson's equation. The numerical techniques described in this section have been implemented in the GEMINI computer program [3]. Numerical results obtained using GEMINI are presented in later sections to demonstrate the accuracy and efficiency of device analysis based on solutions of Poisson's equation.

2.2 Single-Carrier Analysis

Discussions of the Poisson solutions for nonplanar devices have been presented in the last section. It will be shown in the next section that for many applications the Poisson solutions, when coupled with suitable auxiliary calculations, can give results which accurately estimate current transport in devices under certain conditions. From the standpoint of engineering design this is of extreme value since Poisson analysis is most cost effective. However, there are applications and regions of device operation which require the solution of carrier transport as well as the Poisson results. This section describes single-carrier analysis for nonplanar devices.

The grid chosen for Poisson analysis facilitates consideration of non-planar structures. Figure 7a shows again a portion of the grid used for a typical analysis of a nonplanar junction device. The nonuniform rectangular grid can be adjusted to increase density near features such as corners; however, the total grid allocation is somewhat inefficient owing to "wasted" grid in regions away from features of interest. By contrast, the simplicity and computational efficiency of the one-step SOR analysis technique offers major advantages as discussed earlier. Figure 7b shows the same basic device structure with a distorted rectangular grid. The key advantage of this second grid is its efficiency in allocating grid primarily in regions of interest by means of "distorting" grid away from less important regions. The single-carrier analysis discussed next uses this second grid and treats analysis techniques suitable for nonplanar structures modeled in this way.

2.2.1 Equation Discretization

The solution of carrier flow in a semiconductor requires additional equations to that of the Poisson relationships. The complete set of equations which must be solved consists of Eq. (1) and

$$\vec{\nabla} \cdot \vec{J}_n = qU_n \tag{15a}$$

$$\vec{\nabla} \cdot \vec{J}_p = -qU_p \tag{15b}$$

$$\vec{J}_n = q\mu_n\left(n\vec{E} + V_T\vec{\nabla}n\right) \tag{16a}$$

$$\vec{J}_p = q\mu_p\left(p\vec{E} - V_T\vec{\nabla}p\right) \tag{16b}$$

$$n = n_{Ie}^{\left(\psi - \phi_{Fn}\right)/V_T} \tag{17a}$$

$$p = n_{Ie}^{\left(\phi_{Fp} - \psi\right)/V_T} \tag{17b}$$

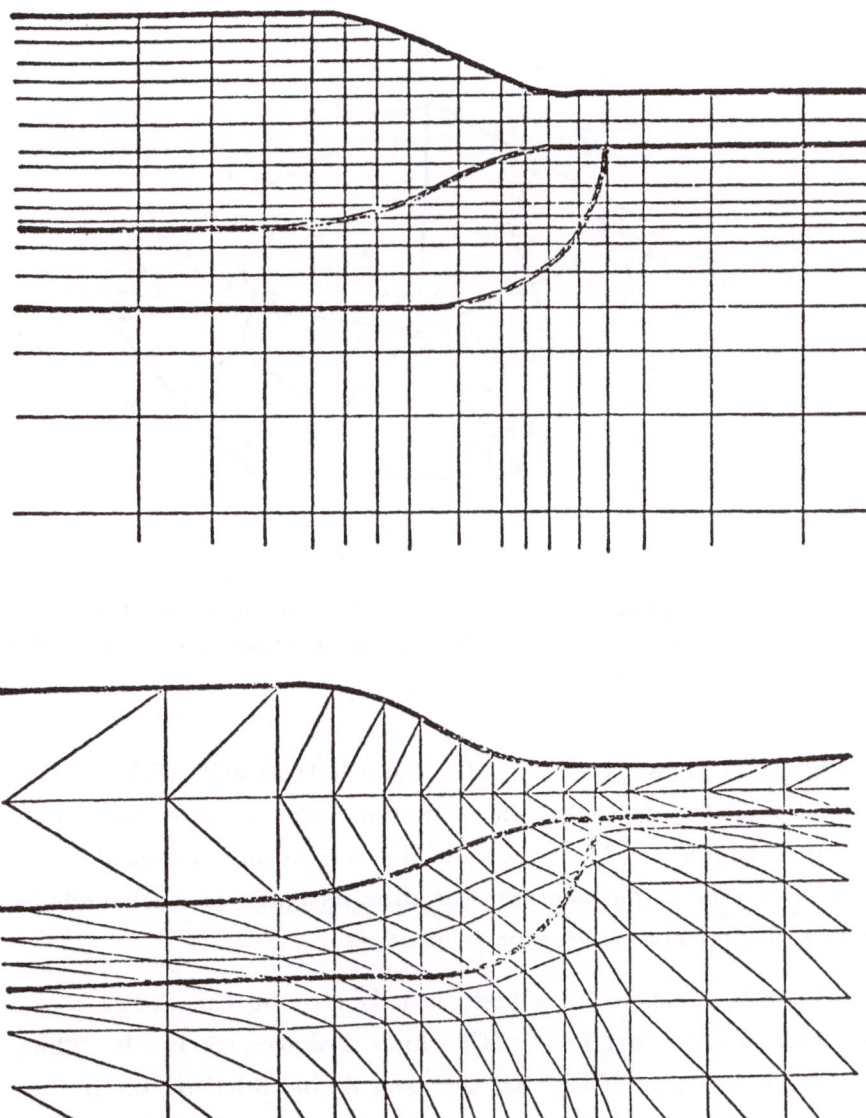

Fig. 7. Two grid structures for nonplanar device analysis: (a) rectangular grid used for SOR Poisson analysis (b) distorted rectangular grid used for single-carrier analysis.

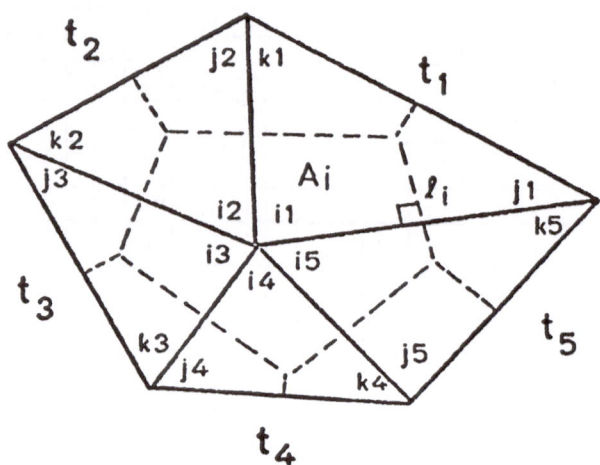

Fig. 8a. Sample grid with five triangles. The triangles are labeled t_1–t_5, A_i is the area associated with the central node, and l_i is the boundary of that area.

Figure 8a shows a section of a hypothetical grid with five triangular sections labeled t_1–t_5 having one common node variously labeled i_1–i_5 and referred to as node i. The process of discretization involves the determination of two sets of parameters: the area assigned to each node, and the coupling coefficients between pairs of nodes.

The area assigned to a node is taken to be the area closer to that node than to any other node with which it shares a triangle. Thus, in the five triangle example of Fig. 8a, the area A_i bounded by the dashed line l_i represents the boundary of the area assigned to node i. This boundary is conveniently formed by the perpendicular bisector of each edge common to node i.

Poisson's Equation— The coupling coefficients are obtained from the discretization of Poisson's equation. This is achieved by applying Gauss' law to Eq. (1) and converting integrals into summations. Gauss's law is applied

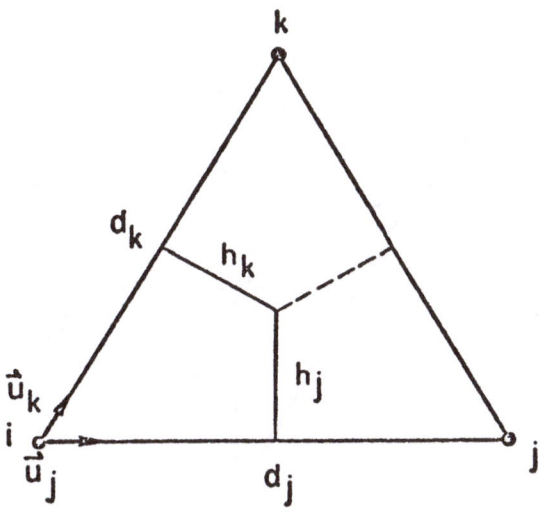

Fig. 8b. Labeling convention for a triangle. The distance between nodes is d, the length (height) of the boundary segments is h, and $\vec{u}$ is a unit vector.

over line segments labeled l_{im} and encompassing an area A_{im} for each triangle t_m.

Making the assumption that material boundaries lie only along triangle edges, and that the electric field is constant allows the line integrals to be replaced by dot products. Furthermore, since the net charge assigned to a node is considered to be evenly distributed throughout its area, ρ is a spatial constant and may be taken out of the integral.

Figure 8b shows the labeling convention for a sample triangle. The vectors $\vec{u}_j$ and $\vec{u}_k$ are the unit vectors in the ij and ik directions respectively. The subscript denoting the triangle number has been dropped in the figure and the boundary segments l_{ij} and l_{ik} have been relabeled with their length h_j and h_k. Writing the electric field in terms of potential and the distance d_j between nodes i and j, the discretized form of Poisson's equation becomes

$$\sum_{1 \leq m \leq M} \epsilon_m \left((\psi_i - \psi_{jm}) \frac{h_{jm}}{d_{jm}} + (\psi_i - \psi_{km}) \frac{h_{km}}{d_{km}} \right) = \rho_i A_i. \tag{18}$$

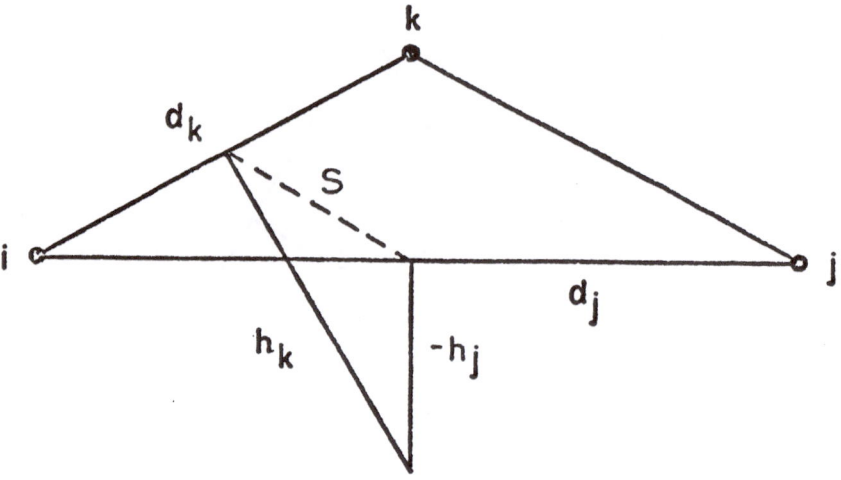

Fig. 9. Coupling coefficients for Poisson's equation on an obtuse triangle.

where the problem has been generalized to include any number M of triangles containing the common node i.

Obtuse Triangles— The above derivations work quite nicely for acute triangles; however, when the triangles become obtuse, adjustments need to be made [5]. The need arises from the fact that the intersection of the perpendicular bisectors of the sides of an obtuse triangle occurs outside of the triangle as illustrated in Fig. 9. In this example, the coupling coefficients are exactly as derived earlier except that the coupling coefficient between nodes i and j becomes negative, $-\frac{\epsilon h_j}{d_j}$ since the segment h_j has reversed direction. It can be shown that this choice of coupling coefficients still accurately accounts for the flux passing through the line segment s. Intuitively, it may be reasoned that the vector sum of the directed line segment h_k minus the directed line segment h_j is the directed line segment s, so the vector sum of their normals also equate. Note that the coupling coefficient goes from positive to zero to negative as the triangle goes from acute to right to obtuse, so there is no discontinuity in the transition from acute to obtuse.

The total coupling coefficient for nodes i and j includes a contribution from both triangles which share the side joining nodes i and j. Thus, although one component of the coupling coefficient may be negative, the total coupling coefficient may still be positive if the contribution from the adjacent triangle is sufficiently large and positive. It can be shown that the total coupling coefficient will be positive if the sum of the opposite angles is less than 180 degrees. This condition can always be satisfied for triangles which are not on the solution region boundary or on a material interface by reconnecting the triangles.

The occurrence of obtuse angles also complicates the area allocation method. Looking again at Fig. 9, the boundary for the area allocated to node i is no longer defined by the line segments h_j and h_k as they were in Fig. 8b. One possible choice of boundary is that portion of line segment h_k which lies within the triangle. Unfortunately, this choice does not meet the prerequisite that the boundary include the midpoints of the adjacent sides, thus there would not be conservation of flux within the triangle using this boundary.

A better choice for the boundary of the area allocated to node i is the line segment s in Fig. 9, the segment joining the midpoints of the adjacent sides. This choice satisfies the condition for conservation of flux and is somewhat better than the centroid method in area weighting in that it allocates one-fourth of the triangle area to node i, one-fourth to node j, and one-half to node k. The acute and obtuse area allocation schemes are identical for an angle of 90 degrees so there is a smooth transition from one to the other.

Continuity Equation— The discretization of the continuity equation will be described only for electrons since the discretization for holes is analogous with the exception of signs. The discretization process parallels that of the Poisson equation except that quantities on the left hand side of the equation are taken to be constant only along an edge of the triangle instead of over the entire triangle. The discretization leads to a form which is similar to

458

that of Eq. (18)

$$\sum_{1 \le m \le M} \left(\vec{J}_{njm} \cdot (h_{jm} \vec{u}_{jm}) + \vec{J}_{nkm} \cdot (h_{km} \vec{u}_{km}) \right) = qU_n A_i \qquad (19)$$

where $\vec{J}_{njm}$ is the electron current density along edge ij in triangle m and $\vec{J}_{nkm}$ is similarly defined. Note that since $\vec{J}_n$ cannot be assumed constant over the triangle, $\vec{J}_{njm} \ne \vec{J}_{nkm}$. Since $\vec{J}_{njm}$ and $\vec{J}_{nkm}$ are one-dimensional current density vectors in the $\vec{u}_{jm}$ and $\vec{u}_{km}$ directions respectively, the dot products become multiplies and Eq. 19 becomes

$$\sum_{1 \le m \le M} (J_{njm} h_{jm} + J_{nkm} h_{km}) = qU_n A_i \qquad (20)$$

where the current density is taken to be positive in the direction away from node i. Thus the electron current out of the area of node i into the area of node j in triangle m is the current density, J_{njm}, times the length of the perpendicular boundary through which the current flows, h_{jm}. This same current density also flows through a perpendicular segment of the node i boundary in the adjacent triangle. The total electron current out of the node i area is the summation of the current crossing the area boundary into each adjacent node, and this total must equal the recombination rate $qU_n A_i$.

The fully discretized electron continuity equation is

$$\sum_{1 \le m \le M} \left(\frac{\mu_{njm} h_{jm}}{d_{jm}} (\psi_i - \psi_{jm}) \left(\frac{n_{jm}}{1 - e^{-(\psi_i - \psi_{jm})/V_T}} + \frac{n_i}{1 - e^{(\psi_i - \psi_{jm})/V_T}} \right) \right.$$
$$\left. + \frac{\mu_{nkm} h_{km}}{d_{km}} (\psi_i - \psi_{km}) \left(\frac{n_{km}}{1 - e^{-(\psi_i - \psi_{km})/V_T}} + \frac{n_i}{1 - e^{(\psi_i - \psi_{km})/V_T}} \right) \right)$$
$$= U_n A_i. \qquad (21)$$

where the terms with exponentials in the denominator represent the Scharfetter-Gummel form [6] for discretization of the electron transport equation in terms of adjacent node potentials. The unknowns are the electron concentrations n_i, n_{jm}, and n_{km} while the mobilities, potentials, and electron recombination rate are assumed to be known; although, each is actually a function of the electron concentration. The potentials are a strong function of the electron concentration through Poisson's equation, thus alternating solutions of the Poisson and continuity equations are required until convergence is obtained. The mobilities and electron recombination rates may also be functions of the electron concentrations, but the functional relationships are generally very weak so that merely updating the mobility and recombination rate after new electron concentrations are computed is sufficient.

When Eq. (21) is assembled for every node in the grid, one again has N equations in N unknowns as in the Poisson case; however, the continuity equation does not have to be solved in the insulator regions of the device so that one may limit the solution to only those nodes lying in the semiconductor region. The coefficient matrix does not have the desirable iteration properties of the Poisson coefficient matrix, but since iteration is not required to solve the continuity equation, this point is not critical. Also, the coefficients look rather formidable at first, but a careful look will show that the coefficients are well behaved and, in fact, no difficulties are observed in obtaining a solution. The well behaved nature of the coefficients is demonstrated in the fact that

$$\frac{\psi_1 - \psi_2}{1 - e^{(\psi_1 - \psi_2)/V_T}} \approx \begin{cases} 0, & \text{for } \dfrac{\psi_1 - \psi_2}{V_T} \gg 0; \\ -V_T, & \text{for } \dfrac{\psi_1 - \psi_2}{V_T} \approx 0; \\ \psi_1 - \psi_2, & \text{for } \dfrac{\psi_1 - \psi_2}{V_T} \ll 0. \end{cases} \tag{22}$$

Finally, note that the coupling terms $\frac{\mu_n h}{d}$ in Eq. (21) are analogous to the terms $\frac{\epsilon h}{d}$ in Poisson's equation, Eq. (18).

Obtuse Triangles— As in the case of Poisson's equation, adjustments need to be made to the continuity equation discretization in obtuse triangles.

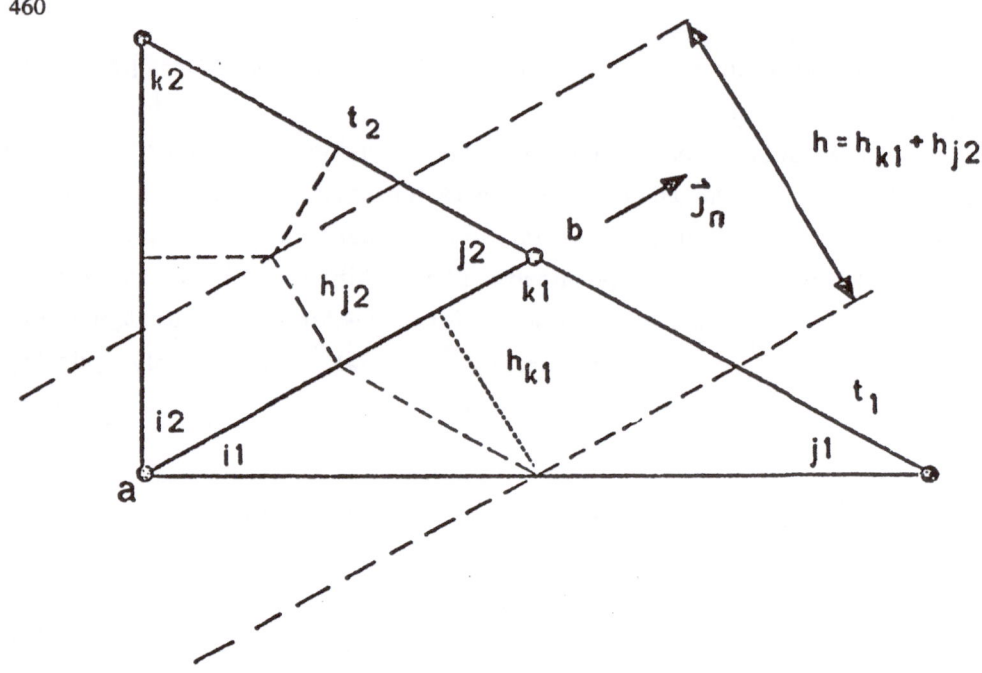

Fig. 10. Coupling coefficient derivation for the continuity equation on acute and obtuse triangles.

In the Poisson deiscretization, the node area boundary was pinned at the midpoints of the sides as the triangle became obtuse while the coupling coefficients, $\frac{\epsilon h}{d}$, were allowed to become negative in a rather natural manner. The continuity equation necessarily uses the same boundary as Poisson's equation but cannot allow its $\frac{\mu_n h}{d}$ terms to become negative as the trangle becomes obtuse. This discrepancy stems from the quasi-two-dimensonal nature of the continuity equation discretization.

Figure 10 graphically depicts the reasoning behind the choice of coupling coefficients for the continuity equation in acute and obtuse triangles. The current density flowing from node a node b is shown as $\vec{J}_n$. In triangle t_2 this current density passes from the node i_2 area into the node j_2 area through a cross-sectional window of width h_{j_2} which is the length of the boundary segment. (Note that node a is equivalent to both i_1 in t_1 and i_2 in t_2). However, in the obtuse triangle, t_1, the current density passes through a cross-sectional width of h_{k1} which is the projected length of the boundary segment onto the perpendicular to the current density vector. The value of h used in the total coupling coefficient between nodes a and b is the sum of these two terms, *i.e.* the total cross-sectional width of the common boundary between the two nodes.

Note in obtuse triangle t_1, that there is no common boundary between nodes i_1 and j_1 so $h_{j1} = 0$ and no current is allowed to flow between them. Current may be allowed along this same edge in an adjacent triangle, however, if the opposite angle is acute. This condition can be assured, as was shown previously, as long as the triangle edge is not a solution region boundary or a material interface. Even these cases can be handled by sub-division of the triangles, so it is never necessary for the continuity equation coupling coefficient between two adjacent semiconductor nodes to be zero (except in the degenerate case of a perfectly rectangular grid structure where the triangle edges representing the diagonals of the rectangles have coefficients of zero). Note also that as in the Poisson discretization, there is no discontinuity in the choice of coupling coefficients as the triangle passes from acute to obtuse.

2.2.2 Iterative Solution

Discretization of Poisson's equation and the continuity equation results in two systems of equations which must be solved in order to determine the potentials and carrier concentrations. There are two aspects to the solution of the equations—one is the solution of the matrix equation $M\vec{x}=\vec{y}$ representing either the discretized Poisson or continuity equation by itself, the other is the consistent solution of the two coupled equations together. In this section the problems of solving coupled equations will be considered.

In contrast to Poisson's equation, the continuity equation does not yield a symmetric or positive definite matrix. As a result, some of the matrix solution algorithms discussed in conjunction with the solution of Poisson's equation cannot be applied to the solution of the continuity equation. Since no strong statement may be made concerning the eigenvalues of the continuity equation coefficient matrix, the SOR and SBOR methods may converge very slowly or even diverge for some conditions. *LU* decomposition, on the other hand, is suitable for use on the continuity equation and no instabilities have been observed in using this method.

In this work the matrix equation solution package VEGES (Vectorized General Sparsity) algorithms [7] was used which puts all vectorizable steps in the *LU* decomposition process into vector operation form. On true vector architecture computers, significant time savings can be realized by exploiting the vectorizable operations; however, even on non-vector machines (such as the HP-1000F) some savings can be achieved with microcoded vector instructions. For example, the vector version shows a factor of three improvement in solution time over the scalar version. Depending on the computer, operations on short vectors can take longer than equivalent scalar operations due to the overhead associated with vector operation startup. The point of diminishing returns for vector versus scalar operation depends solely on the computer being used.

Figure 11 shows flow charts of the two principal approaches to solving the coupled equations—the simultaneous method and the alternating (or Gummels) method. The comparison between these two methods is somewhat analogous to the comparison of direct versus iterative matrix solution methods. The alternating method takes less work per pass but may require many passes and thus more work for a consistent solution than the simultaneous method.

The simultaneous method involves appending the two matrix equations together to form a single matrix equation which is twice as large. In addition, the partial derivatives of all combinations of potential and carrier concentration for adjacent nodes must be included. The result is a matrix equation with nearly four times the number of non-zero entries as either coefficient matrix alone, and more than four times as many operations per

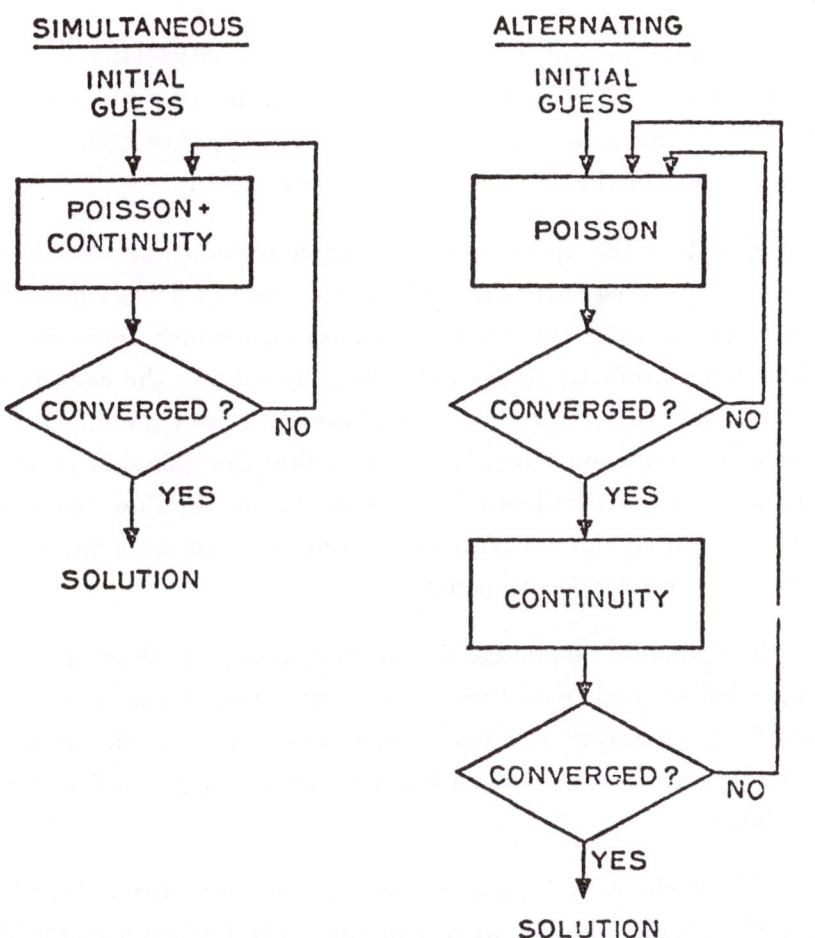

Fig. 11. Algorithm flow of simultaneous and alternating methods for solution of the coupled equations.

solution. This rapid multiplication of effort is prohibitive for grids of any practical size.

The alternating method of Gummel [8] is the most commonly used method for obtaining consistent solutions to the Poisson and continuity equations. As illustrated in the flow chart of Fig. 11, beginning with an initial guess of potentials and carrier concentrations, the Poisson equation is solved for node potentials. The carrier concentrations are updated based on

464

the new potentials and appropriate carrier statistics (Boltzmann or Fermi-Dirac) and the Poisson equation is solved again. This procedure is repeated until the potential change is below some convergence criteria limit. This is the inner loop of the flow chart of Fig. 11b.

Implicit in the updating of carrier concentrations based on potential changes and carrier statistics is the assumption of a fixed quasi-Fermi level. In fact, the quasi-Fermi level is an unknown which must be determined. This is done implicitly in the outer loop by solving the continuity equation and updating the carrier concentrations without changing the potentials. These new carrier concentrations require that the Poisson equation be solved again so the algorithm loops back to the top of the flow chart. Eventually, potentials and carrier concentrations converge and are consistent with both the Poisson and continuity equations.

The principal advantage of the alternating method is that less work is expended on each pass through the outer loop than in the simultaneous method. A significant disadvantage, however, is that the convergence rate of the method is dependent on the device operating conditions and may be very slow.

Table 2 shows a typical convergence pattern for a MOSFET biased below threshold. The left hand column is the iteration count of the outer loop of the alternating method, the center column is the error measure for each iteration of the inner loop (Newton iteration on Poisson's equation), and the right hand column is the error measure for each iteration of the outer loop. The inner loop error measure is the average of the absolute values (the *one norm* of numerical analysis) of the incremental potentials resulting from the Poisson solution. The outer loop error measure is the one norm of the net change in the node potentials from one pass through the outer loop to the next.

Analyzing the convergence of Table 2, on the first pass through the outer loop, Poisson's equation converges in two iterations. The continuity

Table 2
Subthreshold Convergence

Outer Loop Count	Inner Loop Count	Outer Loop Error
	3.7×10^{-4}	
	5.8×10^{-12}	
1		3.2×10^{-4}
	1.5×10^{-1}	
	1.4×10^{-2}	
	1.7×10^{-3}	
	3.4×10^{-4}	
	1.3×10^{-5}	
	3.3×10^{-8}	
2		1.4×10^{-2}
	1.2×10^{-12}	
3		1.2×10^{-12}
Iterations	9	3

equation solution, however, alters the carrier concentrations and results in six Poisson iterations on the second pass through the outer loop. On the third pass, only one Poisson solution is required and the change in potential is extremely small indicating that the algorithm has converged. Twelve matrix solutions are required for this simulation, nine for Poisson and three for continuity. This rapid convergence is due to the very weak coupling between the Poisson and continuity equations. When a MOSFET is biased below threshold, the dominant charge (besides boundary charge) is the space charge of ionized impurities in the depletion regions. Since this charge is immobile, the continuity equation has little effect and a Poisson solution is essentially all that is required.

Another point worth noting in Table 2 is the convergence rate of the inner loop. Newton's method has quadratic convergence which means that when the solution estimate is in the vicinity of the correct solution, the error in the solution estimate will be squared with each iteration. This type of behavior is evident in the inner loop error measure shown in the table.

A different convergence pattern is seen in the MOSFET linear region simulation. In this case, there is an inversion layer of free carriers at the semiconductor surface resulting in stiff coupling between the Poisson and continuity equations. Each continuity solution alters the carrier concentrations enough to negate the previous Poisson solution and cause several more iterations of the inner loop. A total of 151 inner loop and 46 outer loop iterations are required. It is this behavior which undermines the utility of the alternating method in device simulation. In the search for ways to reduce the simulation time required for devices above threshold, four techniques were derived which may be used to reduce this time by a factor of three to four. These techniques are outlined below.

The projection method involves extrapolating the initial guess for the new bias condition based on the solutions at two previous bias conditions. Only one contact bias may be varying between the two previous biases and the new bias. The assumption is that the potentials and quasi-Fermi levels for each node will vary linearly with the bias. The projection method provides the greatest convergence acceleration when stepping the drain bias of devices in saturation, typically reducing the number of iterations by a factor of two. It is nearly as effective in the linear region but tends to lose some of its effectiveness in projecting initial guesses through the transition from linear to saturation. For subthreshold bias conditions the method does not significantly reduce the number of iterations; however it does allow larger bias steps to be taken without loss of stable convergence.

The second convergence acceleration technique is the use of a *single Poisson* solution per outer loop iteration. Each continuity equation solution severely alters the previous Poisson solution as indicated by the large inner

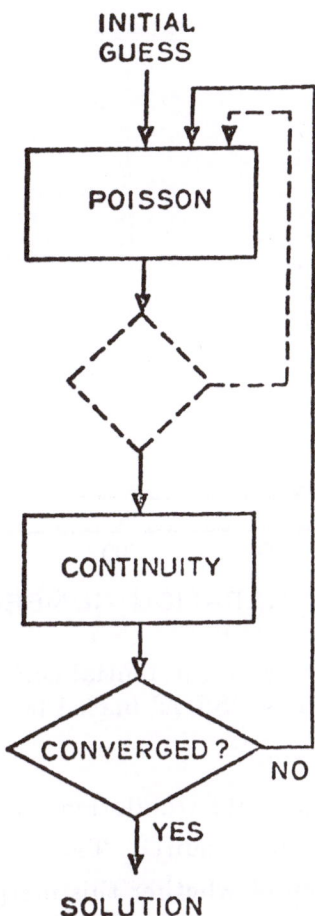

Fig. 12. Algorithm flow of the single Poisson acceleration scheme showing elimination of the inner loop.

loop error seen after each outer loop iteration for strong inversion. In short, the accurate Poisson solution achieved through several iterations of the inner loop is unnecessary. By performing only one inner loop iteration on each pass through the outer loop, the number of outer loop iterations increases by roughly 20% but the total number of matrix solutions decreases by roughly 40%. Figure 12 shows the revised flow chart with the elimination of the inner loop. This technique applies only to the linear and saturation bias

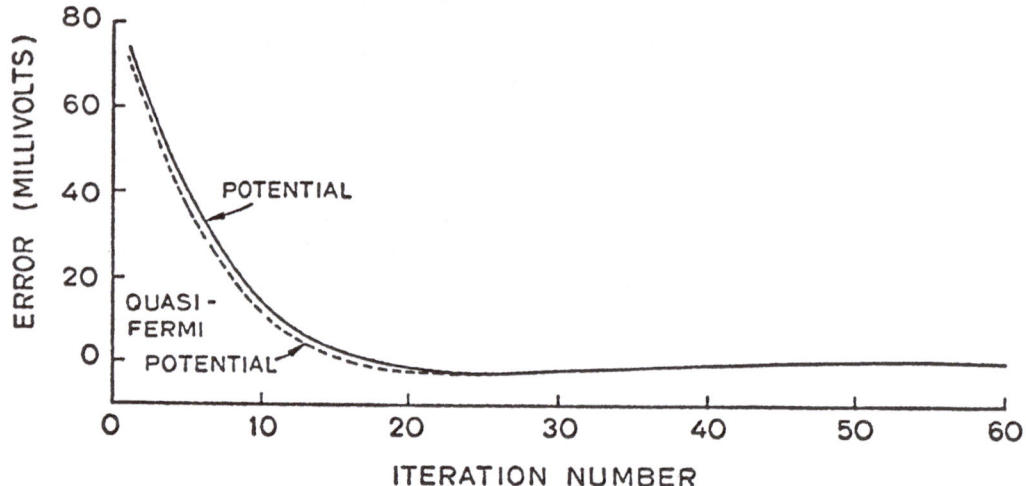

Fig. 13. Potential and quasi-Fermi potential convergence of a node in the channel region of a MOSFET biased in saturation.

conditions. Its use in subthreshold simulations will generally increase the total number of matrix solutions required. The success and stability of this method provokes the question of whether this merging of the two iterative loops could be taken a step further when iterative methods are used for the matrix equation solutions. This postulate is consistent with the results discussed earlier for one-step block SOR.

The third convergence acceleration technique is the use of a form of *overrelaxation*. This technique must be used only with the single Poisson iteration described above. The method was developed by observing the details of the convergence of simulations using projection and a single Poisson solution. Figure 13 shows the potential and quasi-Fermi potential convergence of a node in the channel region of a MOSFET biased in saturation. The error is plotted versus the outer loop iteration count, thus each

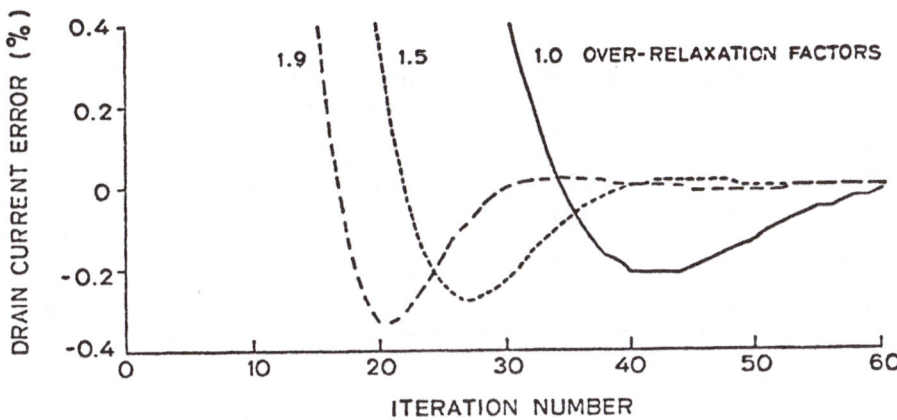

Fig. 14. Drain current convergence acceleration using overrelaxation.

iteration represents two matrix solutions. Since the potential increments are nearly constant for every iteration, it appears that faster convergence can be obtained by merely increasing the size of the potential increments. This is somewhat analogous to overrelaxation in iterative matrix solution methods. When the vector of potential increments out of the Poisson solution is multiplied by a factor greater than one before being added to the previous potentials, faster convergence does indeed result.

Figure 14 shows the improvement in drain current convergence obtained using overrelaxation. A factor of 1.0 means no overrelaxation and 1.5 or 1.9 mean that the Poisson solution vector is multiplied by 1.5 or 1.9 each time before updating the potentials. Factors of two or larger cause immediate instability in the iterations, suggesting a similarity between this method and the successive overrelaxation matrix iterative technique. A variable overrelaxation factor which changes every three iterations, typically starting at a large value and then decreasing. This factor is computed using an algorithm published by Carré [4]; however, fixed factors between 1.5 and

Table 3
Number of Matrix Solutions to Convergence

	Acceleration Method				
Drain Bias	None	Projection	1-Poisson	1-Poisson Projection	Overrelaxation 1-Poisson Projection
1.5	189	98	120	74	68
2.0	188	110	86	84	68
2.5	183	103	88	84	48
3.0	183	78	88	70	58
3.5	184	82	90	52	48
Total	927	471	472	364	290
Ratio	1.00	0.51	0.51	0.39	0.31

1.9 generally work equally as well. The overrelaxation technique can be used only when using the single Poisson method also, thus it is applicable only to the linear and saturation regions of operation.

The convergence acceleration obtainable from combinations of these acceleration methods on a MOSFET operating in the saturation mode is exhibited in Table 3. The left hand column shows the five drain biases at which the simulations were run. The second column shows the number of matrix solutions (Poisson plus continuity) required for convergence at each drain bias with no acceleration. The total number required for all five biases is shown at the bottom along with a ratio of the total number for each column to the total of the unaccelerated column. The third column shows the acceleration achieved using projection alone. The fourth and fifth columns show the acceleration achieved using the single Poisson iterations alone and with projection. The final column shows the acceleration achieved by using all three: projection, single Poisson, and overrelaxation.

3 EXAMPLES

In the previous section both Poisson analysis and single-carrier solution techniques for nonplanar devices have been discussed. In this section the application of these tools to IC technology are considered. The simplicity and efficiency of Poisson analysis offers a useful and powerful design aid. In the first subsection the necessary analytic means to fully exploit Poisson analysis to extract current-voltage behavior is discussed. Equations are derived which estimate current flow for the subthreshold and linear regions based only on potential distributions computed from Poisson analysis. The results are verified using the CADDET computer program [9] which solves for single-carrier behavior. Following this benchmark comparison, the results of Poisson-based calculations are compared with experimental data. In a final example, results for the single-carrier nonplanar program are compared with CADDET and results based on analytical calculations using only solutions of Poisson's equation. While the examples presented here are restricted to planar topologies in order to provide CADDET cross-comparisons, the ultimate potential of the nonplanar analysis tools reflects the coupling to real technology cross-sections—especially involving locally oxidized structures— as discussed in the preceeding paper on process analysis.

3.1 Analytic Calculations of Current Using Poisson Analysis

One very .important use for solutions of Poisson's equation is in the calculation of mobile carrier transport current. Poisson's equation can be efficiently solved under quasi-equilibrium conditions without simultaneously solving the current-continuity equations. The resulting potential distribution can be analyzed to determine the transport current under a limited set of operating conditions. Although this approach is not applicable to some high current conditions, it is possible to calculate the current in two very important regions of MOS .device operation. In the first case, current can be calculated under weak to strong inversion conditions when the bias applied across the conducting channel is small and does not significantly affect the potential and mobile carrier distributions within the device. In the second case, current can be calculated under low current conditions when the potential distribution is not significantly affected by the presence of mobile carriers in the conducting channel. The analysis required to perform the above calculations is presented in the remainder of this section.

It is a straightforward exercise [2] to show that Eq. (16a) can be written as

$$\vec{J}_n(r, \theta, w) = -q\mu_n(r, \theta, w)n(r, \theta, w)\vec{\nabla}\phi_{Fn}(r, \theta, w) \tag{23}$$

where the three coordinate axes are defined such that r lies along the center of the channel, θ is orthogonal to r and directed into the semiconductor substrate, and w is orthogonal to r and θ and directed along the width of the device.

For the cases considered in this work, the current density is oriented primarily along the direction of the channel, allowing the θ and w components of $\vec{J}_n$ to be neglected in comparison to the r component. This approximation allows the reformulation of Eq. (16a) to

$$\vec{J}_n(r, \theta, w) \simeq \left[-q\mu_n(r, \theta, w)n(r, \theta, w)\frac{\partial \psi(r, \theta, w)}{\partial r} + qD_n(r, \theta, w)\frac{\partial n(r, \theta, w)}{\partial r}\right]\hat{r} \tag{24}$$

where $\hat{r}$ is a unit vector in the r direction. Applying the same approximation to Eq. (23) yields

$$\vec{J}_n(r, \theta, w) \simeq -q\mu_n(r, \theta, w)n(r, \theta, w)\frac{d\phi_{Fn}(r)}{dr}\hat{r} \tag{25}$$

where elimination of the θ and w components of $\vec{J}_n$ implies that ϕ_{Fn} depends only on r.

The drain-source current I_{DS} is independent of r and can be obtained by calculating the total current density passing through any θ-w plane. After integrating the component of $\vec{J}_n$ normal to the 0-w plane, I_{DS} is given by

$$I_{DS} = -\int_{w_1}^{w_2} \int_{\theta_1}^{\theta_2} \vec{J}_n(r, 0, w) \cdot \hat{r} \, d0 \, dw \qquad (26)$$

where the channel is confined within the region of semiconductor bounded by $\theta_1 < \theta < \theta_2$ and $w_1 < w < w_2$. The approximations of small channel bias or small mobile carrier concentration in the channel can now be applied to Eqs. (24)–(26) to obtain the mobile carrier transport current from solutions of Poisson's equation.

The Region of Low Channel Bias V_{DS}— This section develops approximations which allow the calculation of current from solutions of Poisson's equation for small values of bias applied across the conducting channel. These approximations are useful for the analysis of the subthreshold and linear regions of MOSFET operation. They are generally valid for channel biases V_{DS} which satisfy $|V_{DS}| < 0.1$ V. Under this restriction, $\vec{J}_n$ results mainly from mobile carrier drift in the presence of a potential gradient, so that the diffusion component of $\vec{J}_n$ can be neglected. This approximation reduces the current density defined by Eq. (24) to

$$\vec{J}_n(r, 0, w) \simeq -q\mu_n(r, 0, w)n(r, 0, w)\frac{\partial\psi(r, \theta, w)}{\partial r}\hat{r}. \qquad (27)$$

The θ and w dependencies of $\frac{\partial\psi}{\partial r}$ in Eq. (27) can be neglected without seriously affecting the accuracy of $\vec{J}_n$. Defining $V(r)$ as the potential along the center of the conducting channel and introducing the approximation $\frac{\partial\psi}{\partial r} \simeq \frac{dV}{dr}$, Eq. (27) is reduced to

$$\vec{J}_n(r, 0, w) \simeq -q\mu_n(r, 0, w)n(r, 0, w)\frac{dV(r)}{dr}\hat{r}. \qquad (28)$$

The drain-source current I_{DS} is obtained by substituting Eq. (28) into Eq. (26) to yield

$$I_{DS} = q\frac{dV(r)}{dr}\int_{w_1}^{w_2} \int_{\theta_1}^{\theta_2} \mu_n(r, 0, w)n(r, 0, w) \, d0 \, dw. \qquad (29)$$

Equation (29) can be rearranged as

$$I_{DS}\left(\int_{w_1}^{w_2}\int_{\theta_1}^{\theta_2}\mu_n(r,\theta,w)n(r,\theta,w)\,d\theta\,dw\right)^{-1}dr = q\,dV \tag{30}$$

and integrated from the source ($r=0, V=0$) to the drain ($r=L, V=V_{DS}$) to obtain

$$I_{DS} = q\left[\int_0^L\left(\int_{w_1}^{w_2}\int_{\theta_1}^{\theta_2}\mu_n(r,\theta,w)n(r,\theta,w)\,d\theta\,dw\right)^{-1}dr\right]^{-1}V_{DS}. \tag{31}$$

The calculation of I_{DS} through Eq. (31) requires knowledge of the electron concentration $n(r,\theta,w)$ in a three-dimensional coordinate space. Although Poisson's equation could be solved in three dimensions to obtain n, the storage and computational requirements of such a solution are prohibitively large. One alternative is the application of approximations to Eq. (31) which enable I_{DS} to be determined from solutions of Poisson's equation in two dimensions. These solutions can be efficiently performed with the bias condition $V_{DS}=0$ to obtain an accurate electron distribution. The assumption is then made that for small values of V_{DS} the electron distribution is independent of V_{DS}. The electron distribution determined for $V_{DS}=0$ is used in approximate forms of Eq. (31) to calculate I_{DS} for small values of V_{DS}.

The first approximation to Eq. (31) consists of taking n and μ_n to be independent of r and solving Poisson's equation to obtain n in the two-dimensional θ-w plane at $r=\frac{L}{2}$. Elimination of the r dependence in Eq. (31) yields

$$I_{DS} = \frac{q}{L}\left(\int_{w_1}^{w_2}\int_{\theta_1}^{\theta_2}\mu_n(\theta,w)n(\theta,w)\,d\theta\,dw\right)V_{DS} \tag{32}$$

where L is the channel length. This expression is used to evaluate the effect of narrow channel widths on MOSFET operation.

A second approximation to Eq. (31) consists of taking n and μ_n to be independent of w and solving Poisson's equation to obtain n in the two-dimensional r-θ plane at $w{=}0$. Elimination of the w dependence in Eq. (31) yields

$$I_{DS} = qW \left[\int_0^L \left(\int_{\theta_1}^{\theta_2} \mu_n(r,0)n(r,0)\,d\theta \right)^{-1} dr \right]^{-1} V_{DS} \qquad (33)$$

where W is the channel width. This expression is used to evaluate the effect of short channel lengths on MOSFET operation. The validity of Eq. (33) will be demonstrated by comparing simulated subthreshold characteristics with measurements for a short-channel MOSFET.

The Low Current Region— Two modes of MOSFET current transport which can be distinguished are surface current, with the conducting channel lying along the insulator-semiconductor interface, and bulk current, with the channel lying in the bulk of the semiconductor substrate. The operation of a MOSFET under particular bias conditions may depend on the presence of one or both of these modes of current transport. Each type of current is controlled by a negative potential barrier to electron transport which exists along the channel between the source and drain. The height of this potential barrier can be controlled through the biases applied to the device, resulting in control of the drain-source current I_{DS}.

A point P can be located at the center of the potential barrier region which controls the device current. The origin of the coordinate system (r,θ,w) is taken to occur at P for the purpose of the analysis performed in the remainder of this section. Figure 15 illustrates the location of P in the r-θ plane at $w{=}0$ for a case in which P occurs in the bulk of the substrate. The potential distribution near P is such that the potential at P corresponds to a minimum value along the r direction and a maximum value along the θ and w directions. This type of distribution is normally described as a *saddle* potential when P occurs in the bulk of the substrate. The potential in the

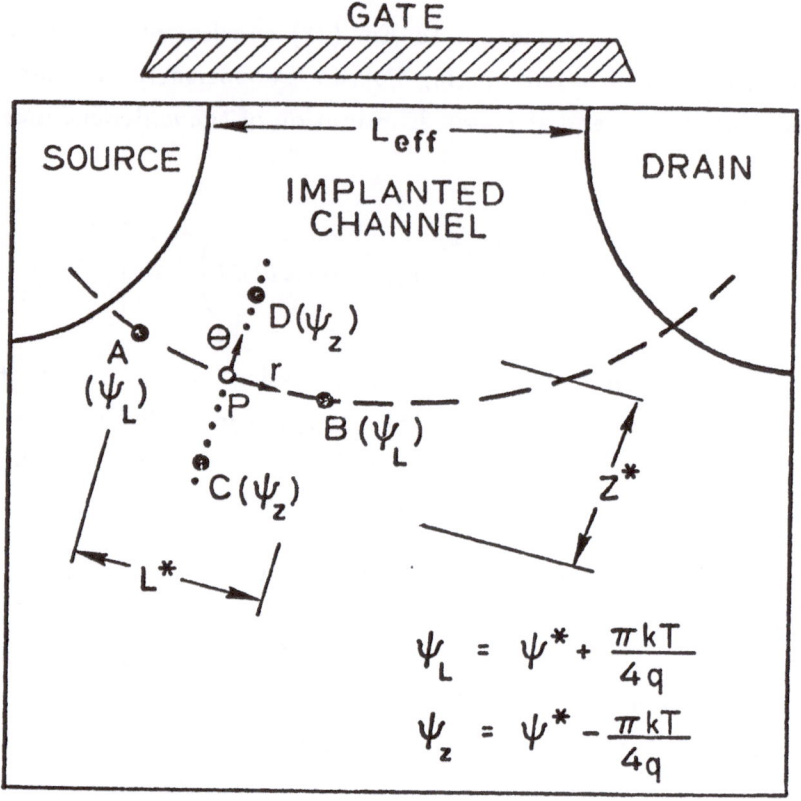

Fig. 15. Cross section in the r-θ plane at $w=0$ for a MOSFET with a conducting channel occurring in the bulk substrate. The point P, the r and θ coordinate axes, and the points defining the approximate effective length L^* and effective depth Z^* of the channel are shown.

vicinity of P can be represented as

$$\psi(r, \theta, w) = \psi^* + h(r, \theta, w) \tag{34}$$

where ψ^* is the potential at P and h defines the variation of ψ away from ψ^*.

It can be shown [2] that the drain-source current can be obtained from Eq. (25) and, by using the expansion in Eq. (34), represented as

$$I_{DS} = qD_n \frac{WZ^*}{L^*} \frac{n_i^2}{N_B} \exp\left(\frac{q}{kT}[\psi^* - V_S]\right)\left[1 - \exp\left(-\frac{qV_{DS}}{kT}\right)\right] \quad (35)$$

where

$$\frac{WZ^*}{L^*} = \left(\int_{rs}^{r_D} \left\{\int_{w_1}^{w_2} \int_{\theta_1}^{\theta_2} \exp\left[\frac{qh(r, 0, w)}{kT}\right] d\theta\, dw\right\}^{-1} dr\right)^{-1} \quad (36)$$

The current is expressed in terms of the channel width W and the effective length L^* and depth Z^* of the region of electron concentration near P which controls the current transport.

The calculation of I_{DS} through Eqs. (35) and (36) requires knowledge of the function $h(r, \theta, w)$, which depends on the potential distribution $\psi(r, \theta, w)$, in a three-dimensional coordinate space. As for the region of low channel bias, an approximation can be applied to Eq. (36) which enables I_{DS} to be determined from solutions of Poisson's equation in two dimensions. This consists of taking ψ to be independent of w and solving Poisson's equation to obtain h in the two-dimensional r-θ plane at $w=0$. After eliminating the w dependence in Eq. (36), Z^*/L^* is given by

$$\frac{Z^*}{L^*} = \left(\int_{rs}^{r_D} \left\{\int_{\theta_1}^{\theta_2} \exp\left[\frac{qh(r, \theta)}{kT}\right] d\theta\right\}^{-1} dr\right)^{-1} \quad (37)$$

The solution of Poisson's equation in two dimensions can be efficiently performed by neglecting the electron concentration in the conducting channel. The assumption is made that the actual electron concentration in the channel is small enough that neglecting its presence does not significantly affect the potential distribution near P. This is a reasonable assumption for devices where these calculations are applicable because the potential distribution is determined primarily by the device geometry, the bulk impurity concentration, and the applied biases. The electron concentration in the conducting channel has only a minor effect on the potential distribution when the electron concentration is small compared to the ionized impurity concentration or when these concentrations are comparable and the electrons are localized near the center of the channel. The validity of Eqs. (35) and (37) is demonstrated by comparing results based on the solution of only Poisson's equation with simultaneous solutions of Poisson's equation and a single current-continuity equation. Comparisons are also performed which show simulated surface punchthrough characteristics and measurements for a short-channel MOSFET.

The techniques discussed above for calculating current from solutions of Poisson's equation yield a very efficient approach to device analysis. These techniques are implemented in the GEMINI computer program [3]. The following two sections present comparisons which illustrate the accuracy and efficiency of this approach to device analysis.

3.2 Accuracy of the Low Current Calculations

The technique just described for calculating current under low current conditions is only approximately correct. The analysis uses Eqs. (35) and (37) to determine the current from quasi-equilibrium solutions of Poisson's equation. The removal of the simultaneous solution of the current-continuity equations requires that the electron concentration in the conducting channel be neglected during the Poisson solutions. Although this technique is an approximation, it provides a very efficient approach for calculating current in the subthreshold and punchthrough regions of MOSFET operation. Much less computation is required than for a simultaneous solution of Poisson's equation and the electron current-continuity equation.

The accuracy of the values of Poisson-based currents calculated using Eqs. (35) and (37) is evaluated by comparing them with results obtained from the CADDET computer program [9]. CADDET simultaneously solves Poisson's equation and the electron current-continuity equation using finite difference approximations and a nonuniform rectangular node structure. The current-continuity equation is treated with the stream function approach [10]. The numerical solution of the discrete form of Poisson's equation uses Stone's strongly implicit technique [11], while the solution

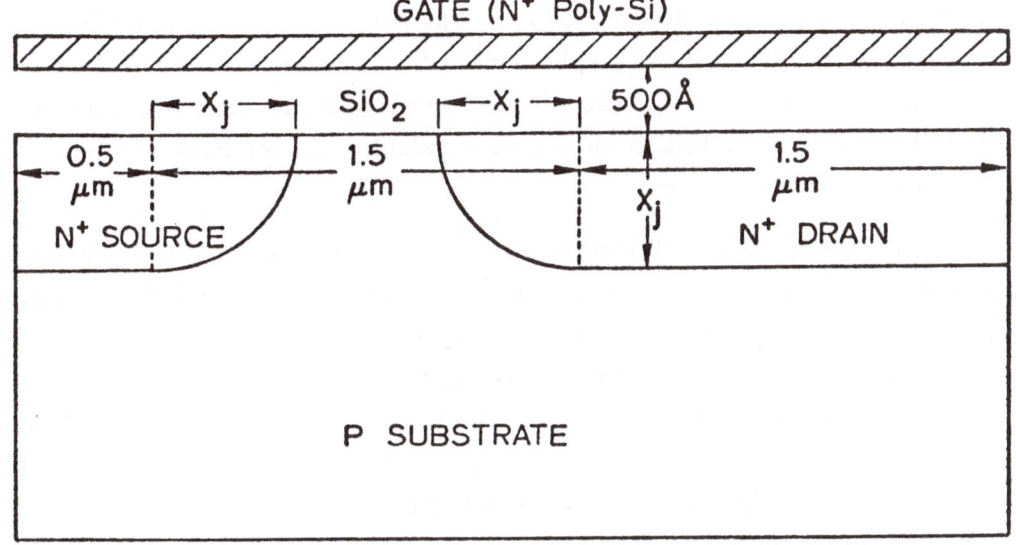

Fig. 16. Cross section of the n-channel MOSFET for which a comparison is made of the calculation of current using CADDET and the solution of Poisson's equation. The device width is 10 μm and the substrate impurity distribution is given by Eq. (38).

of the discrete form of the current-continuity equation uses the successive overrelaxation technique.

The evaluation of the accuracy of the low current calculations is based on the determination of surface and bulk punchthrough currents for the n-channel, enhancement-mode MOS device structure illustrated in Fig. 16. This figure shows a vertical cross section of the device along the channel direction. The device has a masked channel length of 1.5 μm and a width of 10 μm in the direction orthogonal to the plane illustrated in the figure. The impurity distribution in the implanted substrate is given by

$$N_A(y) = N_{A0} + 8.0 \times 10^{15} \exp\left[-\left(\frac{y - 0.18}{0.24}\right)^2\right] \text{cm}^{-3} \tag{38}$$

where N_{A0} is the bulk impurity concentration and y is the vertical coordinate in micrometers. The positive y axis is directed into the substrate with the origin at the silicon surface. The source and drain impurity distributions are Gaussian with cylindrical edges and have a peak concentration of 4.0×10^{19} cm^{-3} at the silicon surface. The device parameters are chosen to provide a test of the current calculations for both surface and bulk conducting channels.

The use of CADDET results as the basis for evaluating the accuracy of the Poisson-based current calculations imposes several restrictions on the geometry of the test device. It is necessary to use a planar silicon surface, a uniform silicon dioxide thickness, simple Gaussian source and drain impurity distributions, and a gate electrode which extends over the entire device surface. These simplifications are not generally appropriate for the analysis of VLSI devices, but they do not affect the validity of the calculation of current from solutions of Poisson's equation.

The accuracy comparisons are performed by calculating I_{DS} versus V_{DS} using both CADDET and the application of Eqs. (35) and (37) to solutions of Poisson's equation. The gate-source bias V_{GS} is varied during these comparisons to test a variety of bias conditions. The variations also illustrate the application of these calculations to the analysis of the sensitivity of device operation. The calculations use a gate electrode flatband voltage of -0.841 V and an electron mobility of 600 cm^2/V-s. Although this value of mobility is appropriate mainly for surface conduction, it is used for both the surface and bulk currents to facilitate the comparisons. Identical finite difference node structures and boundary conditions are used in CADDET and for the solutions of Poisson's equation.

The comparison is illustrated in Fig. 17, which shows the effect on I_{DS} of varying the gate-source bias V_{GS} from 0.0 to -1.0 V. The other structural and bias parameters are fixed at the nominal values $N_{A0}=2.0 \times 10^{15}$ cm^{-3}, $X_j=0.40$ μm, and $V_{BS}=0.0$ V. For $V_{GS}=0.0$ V, a conducting channel occurs only at the silicon surface. More negative values of V_{GS} decrease I_{DS} and induce the formation of a bulk conducting channel for sufficiently large values of V_{DS}. For $V_{GS}= -0.5$ V, the current in the bulk channel becomes

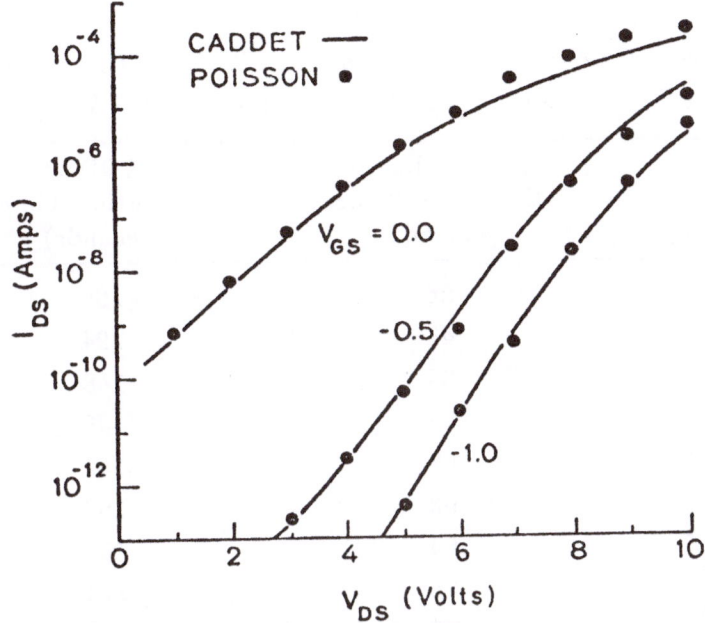

Fig. 17. Comparison of I_{DS} versus V_{DS} calculated using CADDET and the application of Eqs. (35) and (37) to solutions of Poisson's equation. V_{GS} is varied from 0.0 to -1.0 V, $N_{A0}{=}2.0 \times 10^{15}$ cm^{-3}, $X_j{=}0.40\,\mu$m, and $V_{BS}{=}0.0$ V.

the dominant component of I_{DS} for $V_{DS}>6.0$ V. For $V_{GS}={-}1.0$ V, the surface channel disappears and I_{DS} is composed entirely of bulk channel current for all values of V_{DS}.

The efficiency of calculating current by applying Eqs. (35) and (37) to solutions of Poisson's equation is demonstrated by comparing the amounts of computation required by CADDET and the Poisson-based approach for the calculation of one curve in Fig. 17. The calculations are performed on a Hewlett-Packard 1000-F minicomputer using six-digit single precision arithmetic. Table 4 summarizes the execution times required to calculate I_{DS} for $N_{A0}{=}2.0 \times 10^{15}$ cm^{-3}, $X_j{=}0.40\,\mu$m, $V_{GS}={-}0.5$ V, $V_{BS}{=}0.0$ V, and V_{DS} ranging from 1.0 to 10.0 V. The calculations use 1776 nodes and iteration convergence criteria designed to achieve comparable accuracy with

Table 4

Execution Time for Poisson-based and CADDET I_{DS} Calculations

$(N_{A0}=2.0 \times 10^{15} \text{ cm}^{-3}; \; X_j=0.40 \; \mu\text{m}; \; V_{GS}=-0.5 \text{ V}; \; V_{BS}=0.0 \text{ V})$

V_{DS}	Poisson-based Execution Time (seconds)	CADDET Execution Time (seconds)
1	24	165
2	41	124
3	55	148
4	55	149
5	57	162
6	63	198
7	64	158
8	75	191
9	77	229
10	66	144

both calculation methods. It is evident from Table 4 that the elimination of the simultaneous solution of the electron current-continuity equation substantially reduces the execution time required to calculate I_{DS}, with typical reductions by a factor of from two to three.

3.3 Comparison of Simulations and Measurements

The application of the Poisson-based current calculations to the prediction of electrical device characteristics is demonstrated by comparing the results of simulations with experimental measurements for short-channel MOSFET's [12]. Devices were fabricated on a <100> 6 Ω-cm p-type silicon substrate using a normal n-channel polysilicon-gate process. The devices have a gate oxide thickness of 500 Å, a substrate impurity concentration of 2.8×10^{15} cm^{-3}, a channel width of 10 μm, and masked channel lengths of

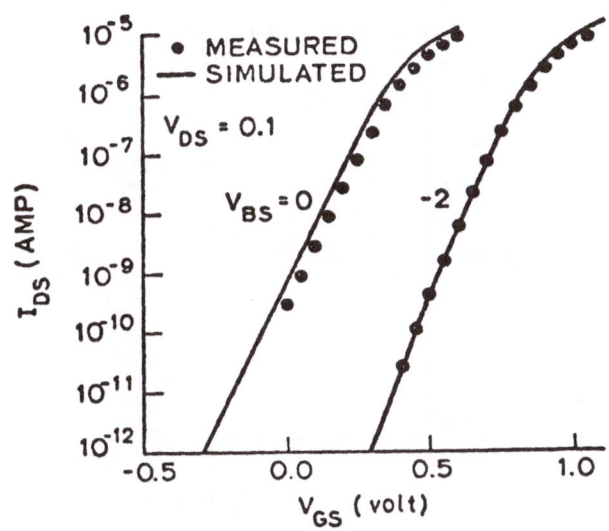

Fig. 18. Measured and simulated subthreshold I_{DS} versus V_{GS} for $V_{DS}=0.1$ V and $V_{BS}=0.0, -2.0$ V for a device with an effective channel length of 0.89 μm. The current is calculated using Eq. (33).

10.2, 5.1, 3.8, 2.6, and 1.3 μm. The effective channel lengths, oxide thickness, and surface mobilities of the devices are determined experimentally from electrical measurements.

Figure 18 compares the simulated and measured variation of I_{DS} versus V_{GS} in the subthreshold region for $V_{DS}=0.1$ V and $V_{BS}=0.0, -2.0$ V. The current is calculated using Eq. (33) for a device with an effective channel length of 0.89 μm. The simulations agree well with measured data over six decades of current. The slopes of the simulated and measured data in the figure agree within several percent. The low current portion of Fig. 18 could also be calculated with Eqs. (35) and (37), although the results would be somewhat less accurate because the electron distribution is only approximated with this approach.

Punchthrough is induced by increasing the reverse bias on the drain-substrate junction. The impurity concentration in the channel is sufficiently

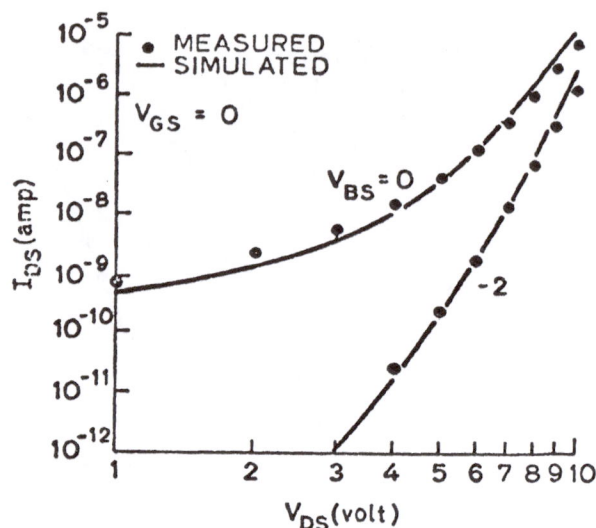

Fig. 19. Measured and simulated surface punchthrough I_{DS} versus V_{DS} for V_{GS}=0.0 V and V_{BS}=0.0, −2.0 V for a device with an effective channel length of 0.89 μm. The current is calculated using Eqs. (35) and (37).

small that current in a surface conducting channel dominates I_{DS}. Figure 19 compares the simulated and measured variation of I_{DS} versus V_{DS} for V_{GS}=0.0 V and V_{BS}=0.0, −2.0 V. The current is calculated using Eqs. (35) and (37) for a surface channel.

The variation of threshold voltage versus the effective channel length L_{eff} is illustrated in Fig. 20 for V_{DS}=0.1 V and V_{BS}=0.0 V. Measured data is shown for 140 devices on a single silicon wafer and having five different masked channel lengths ranging from 10.2 to 1.3 μm. Variations in the fabrication process across the wafer result in an uncertainty in L_{eff} of approximately ±0.25 μm. As shown in Fig. 20, this uncertainty is responsible for changes in the threshold voltage. For the four largest masked channel lengths the threshold voltage variation is limited to ±0.025 V, while for the 1.3 μm masked channel length the threshold voltage varies by as much as ±0.15 V. It is evident that the ability to accurately predict this type of

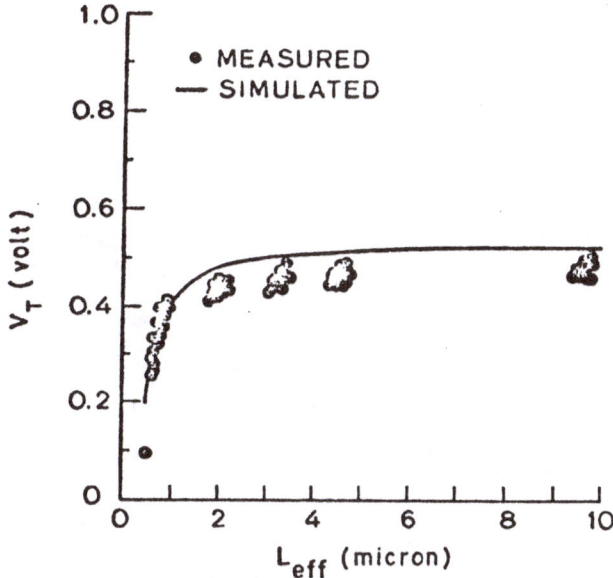

Fig. 20. Measured and simulated threshold voltage sensitivity to effective
channel length for V_{DS}=0.1 V and V_{BS}=0.0 V. Measured data
is shown for five masked channel lengths.

behavior is important to the successful design of a device fabrication process.

3.4 A Benchmark of the Nonplanar Single-Carrier Analysis

The previous subsection has considered the extraction of low bias
and subthreshold current-voltage characteristics from Poisson analysis. As
stated in the discussion, this approach is invalid for some bias conditions.
For example, for increasing drain-source bias including saturated behavior
the carrier transport must be included. The intent of this subsection is
not to explore the limitations of Poisson-based versus single-carrier analysis.
Rather, in a final example a benchmark is run to compare the nonplanar
analysis tools discussed here with the CADDET program. Again, a planar
example is presented owing to the restrictions of the version of CADDET
used.

The analysis of punchthrough operation of a 0.7 μm channel length

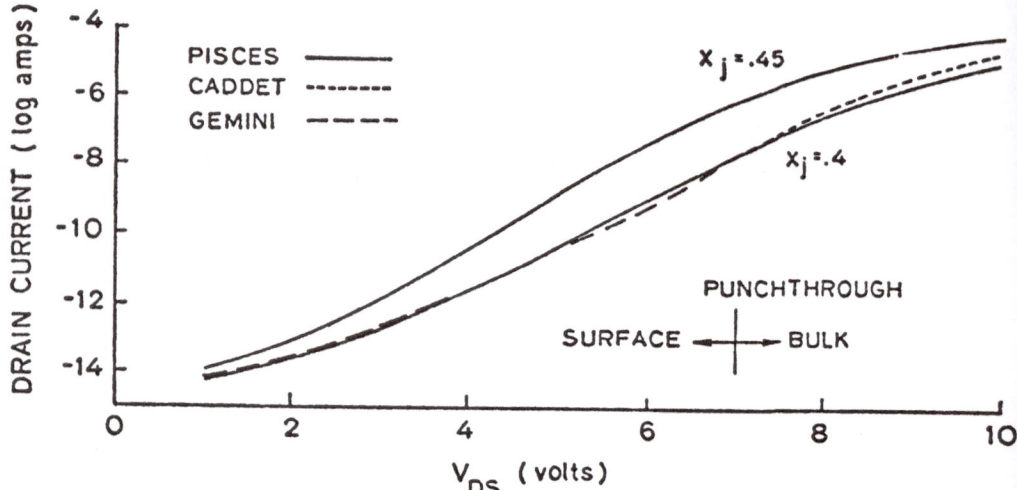

Fig. 21. MOSFET punchthrough characteristics using single-carrier analysis, GEMINI, and CADDET. The current path in the 0.4μm junction device is at the semiconductor surface for biases below 7V and in the bulk for higher biases. The 0.45μm junction device results are from the single-carrier analysis only.

device was performed using CADDET and the nonplanar single-carrier program as well as the Poisson analysis technique. The device structure was planar with an oxide thickness of 500Å. The background doping was $2 \times 10^{15}/cm^2$ with a channel implant of $3.4 \times 10^{11}/cm^2$ to a depth of 0.18μm. The results are shown for source-drain junction depths of 0.4 and 0.45μm.

Figure 21 shows the results of the punchthrough simulations using single-carrier analysis, GEMINI, and CADDET. The drain bias was varied in one volt steps from one to ten volts while holding the gate at −0.5V. An analysis of equipotential contour plots would reveal that the punchthrough current path is at the surface for $V_{DS} < 7V$ and in the bulk for $V_{DS} > 7V$. The 0.4μm source and drain regions are modeled as Gaussian implants with the concentration peak exactly at the semiconductor surface. The 0.45μm junction depth device source and drain have the same Gaussian shape but the peak is shifted 0.05μm below the semiconductor surface. Only single-carrier was used to simulate the 0.45μm junction device. The punchthrough current for this device exceeds that of the 0.4μm device by as much as a factor of 50. This fifty-fold increase in punchthrough current with a 12% increase in junction depth is a rather startling result and illustrates the utility of numerical simulation in device design. This extreme sensitivity and similar sensitivities to junction curvature and impurity gradients make anlytical modeling of these effects extremely difficult.

The three simulation programs used to analyze the 0.4μm junction depth device agree quite well throughout the range of drain biases. The reason for the discrepancy in the GEMINI solution at $V_{DS} = 6V$ is un-known but may be related to the shifting of the dominant punchthrough current path from the surface to the bulk. The CADDET program shows a variation from the other two at drain voltages above $8V$. The largest varia-tion between the three programs is approximately a factor of two. This is larger than one would like but certainly sufficient for the accuracies typi-cally achieved for subthreshold device characteristics. Similar comparisons between the single-carrier method discussed here and CADDET for devices operating above threshold show excellent agreement with variations typi-cally less than 5%. Note that GEMINI is applicable only in the subthreshold and low-drain-bias linear regions of operation.

4 DISCUSSION AND CONCLUSIONS

This paper has presented a discussion of device analysis techniques suitable for nonplanar devices. For Poisson analysis a one-step block SOR technique is discussed and shown to be exceptionally efficient. Moreover, analytical approximations to estimate current flow based solely on Poisson analysis results are discussed and compared with both experiments and single-carrier device analysis. A discussion of nonplanar analysis for single-carrier solutions demonstrated several important results. Obtuse triangles pose substantial problems when using nonrectangular grid—a problem which must also be solved when using finite-element approaches [13]. Several acceleration techniques demonstrated for iterative solutions resulted in the reduction of simulation times by as much as a factor of three-to-four. Benchmark results for the single-carrier analysis agree well with those produced using CADDET.

It should be emphasized that both the Poisson and single-carrier approaches discussed here are intended for nonplanar device analysis. Hence, one may speculate that they will find greatest value in studying technologies with locally oxidized surfaces. Moreover, these analysis tools provide an excellent compliment to planar analysis tools intended for efficient analysis of intrinsic devices [14].

5 REFERENCES

1. B. T. Browne and J. J. H. Miller, *Numerical Analysis of Semiconductor Devices and Integrated Circuits,* , Proceedings of NASECODE I (June 1979) and II (June 1981) Conferences, Dublin: Boole Press Ltd.

2. J. A. Greenfield, *Analysis of Intrinsic MOS Devices and Parasitic Effects Using Solutions of Poisson's Equation,* Stanford University Ph.D. Dissertation, Aug. 1982.

3. J. A. Greenfield and R. W. Dutton, "Nonplanar VLSI Device Analysis Using the Solution of Poisson's Equation," *IEEE Trans. Electron Devices,* vol. ED-27, pp. 1520–1532, Aug. 1980.

4. B. A. Carré, "The Determination of the Optimum Accelerating Factor for Successive Over-relaxation," *Computer J.,* vol. 4, pp. 73–78, 1961.

5. C. H. Price, *Two-Dimensional Numerical Simulation of Semiconductor Devices,* Stanford University Ph.D. Dissertation, May 1982.

6. D. L. Scharfetter and H. K. Gummel, "Large-Signal Analysis of a Silicon Read Diode Oscillator," *IEEE Trans. Electron Devices*, vol. **ED-16**, pp. 64–77, Jan. 1969.

7. D. A. Calahan and P. G. Buning, *Vectorized General Sparsity Algorithms with Backing Store, SEL Report 96*, Systems Engineering Laboratory, Ann Arbor, MI, 1977.

8. H. K. Gummel, "A Self-Consistent Iterative Scheme for One-Dimensional Steady State Transistor Calculations," *IEEE Trans. Electron Devices*, vol. **ED-11**, pp. 455–465, Oct. 1964.

9. T. Toyabe and S. Asai, "Analytical Models of Threshold Voltage and Breakdown Voltage of Short-Channel MOSFET's Derived from Two-Dimensional Analysis," *IEEE Trans. Electron Devices*, vol. **ED-26**, pp. 453–461, Apr. 1979.

10. M. S. Mock, "A Two-Dimensional Mathematical Model of the Insulated-Gate Field-Effect Transistor," *Solid-State Electronics*, vol. **16**, pp. 601–609, 1973.

11. H. L. Stone, "Iterative Solution of Implicit Approximations of Multi-dimensional Partial Differential Equations," *SIAM J. Numer. Anal.*, vol. **5**, pp. 530–558, Sep. 1968.

12. E. Demoulin, J. A. Greenfield, R. W. Dutton, P. K. Chatterjee, and A. F. Tasch, Jr., "Process Statistics of Submicron MOSFET's," *IEEE Int. Electron Devices Meet., Dig. Tech. Papers*, 1979.

13. E. M. Buturla, P. E. Cottrell, B. M. Grossman, and K. A. Salsburg, "Finite-Element Analysis of Semiconductor Devices: The FIELDAY Program," *IBM J. Res. Develop.*, vol. **25**, pp. 218–231, July 1981.

14. S. Selberherr, A. Schütz, and H. W. Pötzl, "MINIMOS—A Two-Dimensional MOS Transistor Analyzer," *IEEE Trans. Electron Devices*, vol. **ED-27**, pp. 1540–1550, Aug. 1980.

TWO DIMENSIONAL MOS-TRANSISTOR MODELING

S. SELBERHERR, A. SCHÜTZ and H. PÖTZL

Institut für Allgemeine Elektrotechnik und Elektronik
Abteilung für Physikalische Elektronik
Technische Universität Wien
Gußhausstraße 27
A-1040 Wien, AUSTRIA
    and
Ludwig Boltzmann Institut für Festkörperphysik

ABSTRACT - The advent of Very Large Scale Integration has been an incentive to concentrate persistently on device modeling. The fundamental properties which represent the basis for all device modeling activities are summarized. The sensible use of physical and technological parameters is discussed and the most important physical phenomena which are required to be taken into account are scrutinized. The assumptions necessary for finding a reasonable trade-off between efficiency and effort for a model synthesis are recollected. Methods to bypass limitations induced by these assumptions are pin-pointed. Simple and easy to use formulae for the physical parameters of major importance are presented. The necessity of a careful parameter-selection, based on physical information, is shown. Some glimpses on the numerical solution of the semiconductor equations are given. The discretisation of the partial differential equations with finite differences is outlined. Linearisation methods and algorithms for the solution of large sparse linear systems are sketched. Results of our two dimensional MOSFET model - MINIMOS - are discussed with typical applications. Much emphasis is laid on the didactic potential of such a complex high order model. In addition to its academic importance, the role of modeling as a tool to optimize transistor performance is stressed.

1. INTRODUCTION

The first integrated circuits which just contained a few devices became commercially available in the early 1960's. Since that time an evolution has taken place so that today the manufacture of integrated circuits with over 400.000 transistors per single chip is possible. This advent of the so-called Very-Large-Scale-Integration (VLSI) certainly revealed the need of a better understanding of the basic device physics. The miniaturization of the single transistor, which is one of the inseparable preconditions of VLSI, brought about a collapse of the classical device models, because totally new phenomena emerged and even dominated the device behaviour. One consequence of this evidence led to an unimaginable number of suggestions of how to modify the classical models to incorporate various of the new phenomena. Additionally new activities have been initiated to explore the physical principles which make a device operationable. The number of scientific publications which utilize the terms "device analysis", "device simulation" and "device modeling" (c.f./4/, /53/, /83/) grew in an incredible manner.

At first it seems necessary to clarify these frequently used terms to facilitate the intelligibility of the subsequent chapters. Consulting a dictionary one will find among many more the following interpretations:

Analysis
- separation of a whole into its component parts, possibly with comment and judgement
- examination of a complex, its elements, and their relations in order to learn about

Simulation
- imitative representation of the functioning of one system or process by means of the functioning of another
- examination of a problem not subject to experimentation

Modeling
- to produce a representation or simulation of a problem or process
- to make a description or analogy used to help visualize something that cannot be directly observed

Therefore, analysis is at least intended to mean "Exact Analysis" and simulation must inferentially mean "Approximate Simulation" using only to some extent physically motivated models. Modeling is necessary for analysis and simulation, but with a different objective. However, any model should at least reflect the underlying physics.

The characteristic feature of early modeling was the separation of the interiour of the device into different regions, the treatment of which could be simplified by various assumptions like special doping profiles, complete depletion and quasineutrality. These separately treated regions were simply connected to produce the overall solution. If analytic results are intended, any other approach is prohibitive. Fully numerical modeling based on partial differential equations /172/ which describe all different regions of semiconductor devices in one unified manner was first suggested by Gummel /69/ for the one dimensional bipolar transistor. This approach was further developed and applied to pn-junction theory by De Mari /39/, /40/ and to IMPATT diodes by Scharfetter and Gummel /132/.

A two dimensional numerical analysis of a semiconductor device was carried out the first time by Kennedy and O'Brien /85/ who investigated the junction field effect transistor. Since then two dimensional modeling has been applied to fairly all important semiconductor devices. There are so many papers of excellent repute that it would be unfair to cite only a few. Recently also the first results on three dimensional device modeling have been published. The time dependence has been investigated by e.g. /96/, /107/ and models for three space dimensions have been announced by e.g. /25/, /185/, /186/.

In spite of all these important and successful activities, the need for economic and highly user oriented computer programs became more and more apparent in the field of device modeling. Especially for MOS devices which have evolved since their invention by Kahng and Atalla /82/ to an incredible standard, modeling has become inherently important because current flow controlled by a perpendicular field is an intrinsically two dimensional problem. One such program which has been applied successfully in many laboratories is called CADDET /165/. We have also tried to bridge that gap and developed MINIMOS /140/ for the two dimensional static analysis of planar MOS transistors.

In the next chapter the fundamental properties which are the basis for all device models are summarized. Much effort is laid on the documentation of various physical effects which possibly have to be taken into account when synthezising a device model for some special application. The assumptions which are usually made to ease modeling are presented and their validity is, at least qualitatively, discussed. Simple and easy to use formulae are presented which allow phenomenological simulation of the most important physical parameters with which the modelist has to deal.

In the third chapter the numerical solution of the basic semiconductor equations is discussed. The two main methods for the solution of differential equations (i.e. finite differences and finite elements) are briefly compared. The discretisation of the general quasiharmonic equation is explained, because Poisson's equation as well as the continuity equations can be classified as quasiharmonic equations. A few linearisation schemes are presented and judged for adaequacy in terms of effort and efficiency. Classical algorithms for the solution of the sparse algebraic systems which are obtained by linearisation of the discrete semiconductor equations are explained.

The fourth chapter entirely deals with applications of MINIMOS. A didactic example which should make transparent the applicability of a high order model like MINIMOS for the analysis of device behaviour is given first. This example has been chosen to demonstrate the influence of ion implantation in the channel region of a very small scale MOSFET on threshold voltage and punch through. Secondly, an example of a process sensitivity analysis is presented which is an application ideally suited for numerical investigations. A more sophisticated application is shown next to demonstrate that it is possible to comprehend some complex interaction of different physical phenomena with device modeling. This is accomplished by explaining the reasons for the snap-back effect in the characteristics of a miniaturized MOSFET. The final sections deal with the analysis of coupled devices. For that purpose an n-Mos Inverter with a depletion load transistor and a CMOS inverter are investigated. These few examples are intended to stress the power and versatility of device modeling and its informative potential for the semiconductor industry.

This paper cannot intend to be any more than a summary with only moderate impact on details. The influence of process modeling, for instance, on the accuracy of the final result, important as it is, could not be dealt with. The reader with specific interest in any single subject will find the references useful for further information.

Throughout this paper all constants and quantities are given in the following units, if not specified otherwise: lengths in cm, times in s, temperature in K, voltages in V, currents in A. The units are often omitted to gain transparency in the formulae.

2. SOME FUNDAMENTAL PROPERTIES

To accurately analyze an arbitrary semiconductor structure which is intended as a self-contained device under various operating conditions, a mathematical model has to be given. The equations which form this mathematical model are often called the fundamental semiconductor equations; these will be discussed in the first section of this chapter.

The second section will deal with assumptions which have to be made for special applications additionally to those which have already been used in the derivation of the equations and which are beyond the scope of this presentation. Furthermore, all quantities which are involved in the basic equations will be outlined more or less qualitatively.

It will become apparent that the fundamental equations employ a set of physical and technological parameters. An in-depth analysis of all these parameters is far from being finished at the moment - or the results of such an analysis are of overwhelming complexity - because of inherent methodical difficulties.

The third section will deal with additional assumptions which can be made to ease and speed up models for MOS-devices.

The topic of the fourth section of this chapter is the description of some suggestions for a heuristic simulation of the most important parameters based, as it were, on physical principles.

2.1 The Fundamental Semiconductor Equations

The most familiar model of carrier transport in a semiconductor device has been proposed by Van Roosbroeck /172/. It consists of Poisson's equation (2.1-1), the current continuity equations for electrons (2.1-2) and holes (2.1-3) and the current relations for electrons (2.1-4) and holes (2.1-5)

$$\text{div } \mathcal{E} \text{ grad } \Psi = -q \; (\; p - n + C \;) \tag{2.1-1}$$

$$\text{div } \vec{J}_n = -q \; (\; G - R \;) \tag{2.1-2}$$

$$\text{div } \vec{J}_p = \; q \; (\; G - R \;) \tag{2.1-3}$$

$$\vec{J}_n = -q \; (\; \mu_n \; n \text{ grad } \Psi - D_n \text{ grad } n \;) \tag{2.1-4}$$

$$\vec{J}_p = -q \; (\; \mu_p \; p \text{ grad } \Psi + D_p \text{ grad } p \;) \tag{2.1-5}$$

These relations form a system of coupled partial differential equations. Poisson's equation, which is one of Maxwell's laws, describes the charge distribution in the interior of a semiconductor device. The balance of sinks and sources for electron- and hole currents is characterized by the continuity equations. The current relations describe the absolute value, direction and orientation of electron- and hole currents. The continuity equations and the current relations can be derived from Boltzmann's equation by not at all trivial means. The ideas behind these considerations cannot be presented here due to limited space. The interested reader should refer to /172/ and its secondary literature or text books on semiconductor physics e.g. /18/, /78/, /136/, /148/.

However, it is of prime importance to note that the equations (2.1-4) and (2.1-5) do not characterize effects which are caused by degenerate semiconductors (e.g. heavy doping). /97/, /171/, /174/ discuss some modifications of the current relations, which partially take into account the consequences introduced by degenerate semiconductors (e.g. invalidity of Boltzmann's statistics, bandgap narrowing). These modifications are not at all simple and lead to problems especially for the formulation of boundary conditions /116/, /173/. In case of modeling MOS devices, degeneracy is, owing to the relatively low doping in the channel region, practically irrelevant. For modern bipolar devices, though, bearing in mind shallow and extraordinarily heavily doped emitters, it is an absolute necessity to account for local degeneracy of the semiconductor.

Furthermore, (2.1-4) and (2.1-5) do not describe velocity overshoot phenomena which become apparent at feature lengths of $0.1\mu m$ for silicon and $1\mu m$ for gallium-arsenide /60/; and certainly no effects which are due to ballistic transport, the existence of which is still questionable /77/, are included. The latter start to become important for feature sizes below $0.01\mu m$ for silicon and $0.1\mu m$ for gallium-arsenide /61/. Considering the state of the art of device miniaturization, neither effect has to bother the modelists of silicon devices. For gallium-arsenide devices new ideas are mandatory for the near future /60/, /110/, /111/.

2.2 Assumptions and Discussion of Parameters

It is imperative to discuss the parameters of the semiconductor equations to get some insight into the complexity of that mathematical model and the difficulty of a more or less rigorous solution.

The permittivity ε in Poisson's equation in the most general case is a rank two tensor. Because all common semiconductor materials grow in cubic crystal structure and because silicondioxide is amorphous no anisotropy exists and the permittivity can be treated as a scalar quantity. Furthermore, one can savely assume that the permittivity is homogenous with sufficient accuracy for even degenerate semiconductors.

The electrically active net doping concentration C in Poisson's equation is the most important technological parameter. To obtain this quantity by mathematical analysis /51/ is at least as cumbersome as to accurately analyze some semiconductor device, because the physics of the technological processes which determine the doping concentration still lacks basic understanding. The need of modeling in this area is drastically increasing in view of VLSI devices. One-dimensional process modeling is fairly well established nowadays, but two-dimensional simulation is just appearing /51/, /164/. Some glimpses of modeling doping profiles with handy analytical expressions will be given in section 2.4.1. One assumption which is usually made with fairly satisfactory success is the total ionization of all dopants (2.2-1).

$$C = N_D - N_A = N_D^+ - N_A^- \qquad (2.2-1)$$

As long as the Fermi level is separated several thermal voltages from the impurity level, this assumption holds quite nicely. For modern bipolar transistors, however, it certainly becomes questionable for the emitter region (degenerate material).

The electron density n and the hole density p are commonly assumed to obey Boltzmann's statistics (2.2-2).

$$n = n_i \cdot e^{(\psi - \varphi_n)/U_T} \qquad p = n_i \cdot e^{(\varphi_p - \psi)/U_T} \qquad (2.2-2)$$

This assumption principally neglects degeneracy; but moderate degeneracy can be included /55/ by introducing an effective, doping dependent intrinsic number (2.2-3).

$$n_i = n_i(T,N) \qquad (2.2-3)$$

$$n_i(T,N) = n_i(T) \; e^{52.7(\ln(N/10^{17})+\sqrt{(\ln(N/10^{17}))^2 +0.5})/T}$$

$$n_i(T) = 3.88 \cdot 10^{16} \cdot T^{1.5} \cdot e^{-7000/T}$$

$$N = N_D + N_A$$

The temperature dependence of the intrinsic number is based on the influence of the effective carrier masses and the bandgap. More elaborate formulae for these effects which might be imperative for low temperature applications can be found in /62/. The formula for bandgap narrowing in (2.2-3) was first suggested by Slotboom /146/. For a doping concentration of $1.3 \cdot 10^{17}$ cm^{-3} the intrinsic number has already increased by twenty percent.

The mobility of electrons μn and holes μp is in principle a rank two tensor function of many arguments. One ends up with a so called "mobility" after averaging and combining various physical mechanisms which are still not analyzed thouroghly enough to be modeled satisfactorily /79/. Some formulae for a mobility model for silicon will be summarized in section 2.4.2.

Another assumption which is unfortunately not at all free of doubts is the validity of the Einstein-Nernst relations (2.2-4).

$$D_n = \mu_n \cdot U_T \qquad D_p = \mu_p \cdot U_T \qquad\qquad (2.2-4)$$

Some guidelines on how to extend these relations for degenerate material are given in e.g. /8/. It is important to remember that the current relations (2.1-4) and (2.1-5) do not differentiate between lattice temperature and electron temperature. Therefore, if one has to deal with hot electrons in a precise manner, the current relations have to be updated; in particular the mathematical structure of the diffusion current term has to be refined.

The last parameter which remains to be dealt with for a qualitative characterization is the net generation/recombination rate (G-R) in (2.1-2) and (2.1-3). This quantity has to describe a number of physical processes which are responsible for generation/recombination of electron-hole pairs. These processes and their interactions are also not analyzed to a satisfactory level so that one has to use heuristic expressions for a model which is at least plausible in the underlying physics. Some suggestions for these formulae will be given in section 2.4.3.

2.3 Additional Assumptions for MOS-Models

The fundamental semiconductor equations describe the internal behavior of any semiconductor device. However, for certain devices these equations may be simplified without significant loss of accuracy. As the MOSFET is a minority carrier device, the current is given mainly by the continuity equation of one carrier type. If avalanche is neglected, only little carrier generation occurs in the MOSFET.

Therefore, the eqs. (2.1-2)-(2.1-3) may be rewritten as

$$\mathrm{div}\ \vec{J}_n = 0 \qquad\qquad (2.3-1)$$

$$\vec{J}_p = 0 \qquad\qquad (2.3-2)$$

for the n-channel device and

$$\mathrm{div}\ \vec{J}_p = 0 \qquad\qquad (2.3-3)$$

$$\vec{J}_n = 0 \qquad\qquad (2.3-4)$$

for the p-channel device. However, it should be kept in mind that these assumptions are valid only if the avalanche effect is neglected.

The channel width of a MOSFET is usually much larger than the depletion widths. As a consequence the partial derivatives in that direction can be neglected and the semiconductor equations reduce to two dimensions. The neglection of the derivative of the potential in source-drain direction is a proper assumption only for long-channel devices. The so called "gradual-channel approximation" was the basis of a lot of one-dimensional models. These models fail to predict accurately the behavior of modern miniaturized devices.

If the avalanche effect should be included, the assumptions (2.3-1)-(2.3-4) are no longer valid and both continuity equations have to be solved with inhomogeneity terms. As a consequence, the ionization-generated majority carriers (holes for an n-channel MOSFET) flow to the substrate as they are repelled from the source and drain junctions. There are several options to account for the voltage drop which is induced by the substrate current: (a) a truly three-dimensional analysis; (b) extension of the simulation over the entire bulk area; (c) extension of the two-dimensional simulation over the depletion region and using an (effective) bulk resistor (Fig. 2.3-1). If one wants to avoid excessive computing time associated with (a), option (c) is to be preferred because it allows inclusion of current spread into the third dimension and, also, consumes less computing time than (b). In that way the voltage drop across the parasitic bulk resistor simulates a more positive bulk bias and, if large enough, is able to forward-bias the parasitic bipolar npn transistor (according to source, bulk, and drain). This causes a larger drain current and facilitates the breakdown which then will occur at smaller drain voltages /133/.

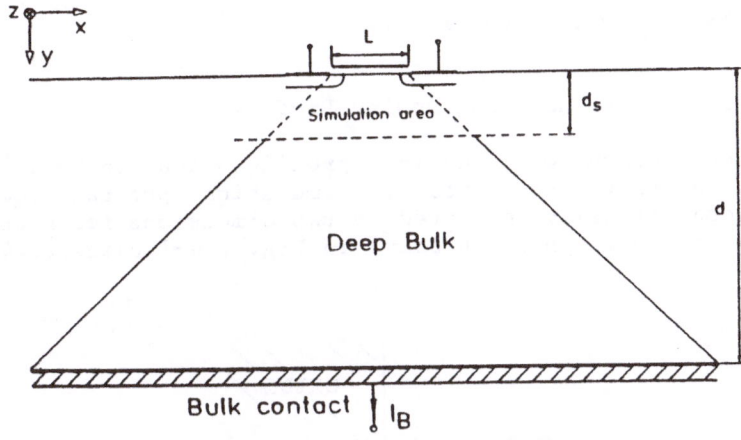

Fig. 2.3-1: Current flow in deep bulk

In the following we should like to suggest an easy method to estimate the value of the bulk resistor. It is assumed that the current spreads at an angle of 45 degrees /15/ into both directions perpendicular to its flow (x- and z- direction in Fig. 2.3-1). This assumption is arbitrary but not implausible, and, furthermore, if we neglect any diffusion current, we obtain the following expression for the electric field in the deep substrate.

$$\frac{d\Psi}{dy} = \frac{I_B}{KA} = \frac{I_B}{K(L+2y)(W+2y)} \qquad (2.3-5)$$

with K standing for the conductivity of the substrate and A the area of the current flow. L and W are channel length and channel width, respectively. Integrating this equation along y from the end of the simulation area d_s to the bulk contact we obtain

$$R_{Bulk} = \frac{\displaystyle\int_{d_s}^{d} \frac{d\Psi}{dy}\,dy}{I_B} = \frac{1}{2K(W-L)}\left(\ln(\frac{L+2d}{L+2d_s}) - \ln(\frac{W+2d}{W+2d_s}) \right) . \qquad (2.3-6)$$

For L=W this equation simplifies to

$$R_{Bulk} = \frac{d-d_s}{K(L+2d)(L+2d_s)} \qquad (2.3-7)$$

This calculation is fairly crude compared to the elaborate solution of the basic equations. However, any more precise calculation would be very complicated and the present method is sufficient to investigate the influence of the parasitic bulk resistance at least qualitatively.

2.4 Models of Physical Parameters

2.4.1 Formulae for Modeling Doping Profiles

A one dimensional doping profile which can be calculated fairly accurately with a process simulation program (e.g. /6/) may be heuristically converted to two dimensions for a structure with an ideal oxide mask as shown in Fig. 2.4-1 using (2.4-1).

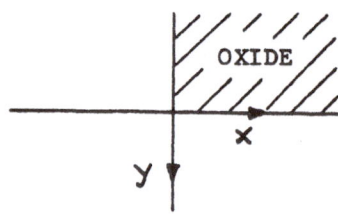

Fig. 2.4-1 Coordinate nomenclature for an ideal oxide mask

$$C(x,y) = C(\sqrt{y^2 + \max(x/f,0)^2})$$

(2.4-1)

This formula is extraordinarily simple to use and needs only one fitting parameter: f which controls the amount of lateral diffusion. For most applications f lies in the range of 0.5 to 0.9. An elliptic rotation at x=0 (c.f. Fig. 2.4-1) of the one-dimensional profile is performed to obtain the doping concentration below the oxide mask. Out-diffusion effects which occur near the mask edge are not at all taken into account.

Lee /89/, /90/ recently published expressions which are still fairly simple to use, but which are based on more physical reasoning. (2.4-2) can be used for the simulation of a predeposition step. Ld denotes the diffusion length; D: diffusion constant, t: diffusion time, N_s: desired surface concentration.

$$Ld = 2 \cdot \sqrt{D \cdot t}$$

(2.4-2)

$$C_p(x,y) = 0.5 \cdot N_s \cdot e^{-(y/Ld)^2} \cdot \mathrm{erfc}(x/Ld)$$

The distribution of implanted ions under mask edges has also)een investigated extensively e.g. /128/. The formulae (2.4-3) vhich have been taken from /89/, /90/ allow simulation of diffusion with an initial ion-implantation. Rp denotes the projected range, ΔRp: projected standard deviation, Dose: implantation dose.

$$a = (2 + (Ld/\Delta Rp)^2)^{-1/2} \qquad\qquad (2.4-3)$$

$$K(y) = e^{-(a \cdot (Rp-y)/\Delta Rp)^2} \cdot \mathrm{erfc}(-a \cdot ((Rp/\Delta Rp) + \sqrt{2 \cdot y/Ld}))$$

$$C_i(x,y) = (a/(4 \cdot \Delta Rp \cdot \sqrt{\pi})) \cdot \mathrm{Dose} \cdot (K(y)+K(-y)) \cdot \mathrm{erfc}(x/Ld)$$

In the derivation of (2.4-2) and (2.4-3) it is assumed that the diffusion "constant" is really constant. This limits the application to relatively low peak values of the implanted profile. For high peak values one might fit the diffusion lengths Ld to obtain a desired junction depth.

The diffusion constant D can be estimated, again for fairly low concentrations, with the classical exponential law (2.4-4).

$$D = D_0 \cdot e^{T_a/T} \qquad\qquad (2.4-4)$$

Element	$D_0/(\mathrm{cm}^2 \mathrm{s}^{-1})$	$T_a/(K)$
B	0.5554	$-3.975 \cdot 10^4$
P	3.85	$-4.247 \cdot 10^4$
Sb	12.9	$-4.619 \cdot 10^4$
As	24.	$-4.735 \cdot 10^4$

The projected range parameters Rp and ΔRp which are nonlinear functions of the implantation energy can be looked up in standard tables /64/. These tables are principally tedious to implement in computer programs, so that one might prefer some polynomial fit (2.4-5); x denotes here the implantation energy.

$$Rp = \sum_{i=1}^{n} a_i \cdot x^i \quad (\mu m) \qquad\qquad (2.4-5)$$

$$\Delta Rp = \sum_{i=1}^{n} b_i \cdot x^i \quad (\mu m)$$

The coefficients for such polynomials are given in Fig. 2.4-2 for Rp in silicon, in Fig. 2.4-3 for $\triangle$Rp in silicon and in Fig. 2.4-4 for Rp in silicon-dioxide.

Element	B	P	Sb	As
a_1	$3.338 \cdot 10^{-3}$	$1.259 \cdot 10^{-3}$	$8.887 \cdot 10^{-4}$	$9.818 \cdot 10^{-4}$
a_2	$-3.308 \cdot 10^{-6}$	$-2.743 \cdot 10^{-7}$	$-1.013 \cdot 10^{-5}$	$-1.022 \cdot 10^{-5}$
a_3		$1.290 \cdot 10^{-9}$	$8.372 \cdot 10^{-8}$	$9.067 \cdot 10^{-8}$
a_4			$-3.056 \cdot 10^{-10}$	$-3.442 \cdot 10^{-10}$
a_5			$4.028 \cdot 10^{-13}$	$4.608 \cdot 10^{-13}$

Fig. 2.4-2 Coefficients for Rp in silicon

Element	B	P	Sb	As
b_1	$1.781 \cdot 10^{-3}$	$6.542 \cdot 10^{-4}$	$2.674 \cdot 10^{-4}$	$3.652 \cdot 10^{-4}$
b_2	$-2.086 \cdot 10^{-5}$	$-3.161 \cdot 10^{-6}$	$-2.885 \cdot 10^{-6}$	$-3.820 \cdot 10^{-6}$
b_3	$1.403 \cdot 10^{-7}$	$1.371 \cdot 10^{-8}$	$2.311 \cdot 10^{-8}$	$3.235 \cdot 10^{-8}$
b_4	$-4.545 \cdot 10^{-10}$	$-2.252 \cdot 10^{-11}$	$-8.310 \cdot 11^{-10}$	$-1.202 \cdot 10^{-10}$
b_5	$5.525 \cdot 10^{-13}$		$1.084 \cdot 10^{-13}$	$1.601 \cdot 10^{-13}$

Fig. 2.4-3 Coefficients for $\triangle$Rp in silicon:

Element	B	P	Sb	As
a_1	$3.258 \cdot 10^{-3}$	$9.842 \cdot 10^{-4}$	$7.200 \cdot 10^{-4}$	$7.806 \cdot 10^{-4}$
a_2	$-2.113 \cdot 10^{-6}$	$-2.240 \cdot 10^{-7}$	$-8.054 \cdot 10^{-6}$	$-7.899 \cdot 10^{-6}$
a_3			$6.641 \cdot 10^{-8}$	$7.029 \cdot 10^{-8}$
a_4			$-2.422 \cdot 10^{-10}$	$-2.653 \cdot 10^{-10}$
a_5			$3.191 \cdot 10^{-13}$	$3.573 \cdot 10^{-13}$

Fig. 2.4-4 Coefficients for Rp in silicon-dioxide

The maximum error of the projected range parameters culated with these coefficients and (2.2-5) is in the energy ge of 5keV to 300keV only a few percent compared to /64/. e data are given in /141/.

If an implantation is performed through an oxide, the projected range in the semiconductor has to be reduced /129/ e.g. with (2.4-6).

$$Rp = Rp_{se} \cdot (1 - T_{iox}/Rp_{ox}) \qquad (2.4-6)$$

T_{iox} denotes the thickness of the oxide, Rp_{se}/Rp_{ox}: projected range in semiconductor/oxide.

2.4.2 Formulae for Mobility Modeling

The mobility of carriers is, as already mentioned, an eminently complex quantity. Additionally it is an important parameter, because all errors in the mobility lead to a proportional error of the current through the multiplicative dependence. This is certainly one of the primary results any model should yield reliably. The formulae which will be given below describe phenomenologically the mobility in silicon; the subscripts n and p denote electrons and holes, respectively.

To model mobility at least plausibly, several scattering mechanisms have to be taken into account, the basis of which is lattice scattering. This effect can be described by a simple power law /79/, /136/ in dependence of temperature (2.4-7).

$$\mu_L(T) = A \cdot T^{-g} \qquad (cm^2/Vs) \qquad (2.4-7)$$

$$A_n = 7.12 \cdot 10^8 \qquad\qquad A_p = 1.35 \cdot 10^8$$
$$g_n = 2.3 \qquad\qquad g_p = 2.2$$

The pure lattice mobility is reduced through the scattering processes at ionized impurities. (2.4-8) is a well established formula which models temperature dependent ionized impurity scattering /24/ and electron-hole scattering /55/. The latter is extremely important in low doped regions where high injection takes place.

$$\mu_{LI}(N,T) = \mu_L(T) \cdot a + \mu_{min} \cdot (1 - a) \qquad (cm^2/Vs) \qquad (2.4-8)$$

$$a = \frac{1}{1 + (T/300)^b \cdot (N/N_0)^c}$$

$$N = 0.67 \cdot (N_D^+ + N_A^-) + 0.33 \cdot (n + p)$$

$$\mu_{minn} = 55.24 \qquad\qquad \mu_{minp} = 49.7$$
$$b_n = -3.8 \qquad\qquad b_p = -3.7$$
$$c_n = 0.73 \qquad\qquad c_p = 0.7$$
$$N_{0n} = 1.072 \cdot 10^{17} \qquad\qquad N_{0p} = 1.606 \cdot 10^{17}$$

Similar expressions which have been partly deduced from measurement and/or theory have been presented in /7/, /41/, /47/, /92/, /132/.

To properly simulate the mobility in MOS transistors, one has to deal with surface roughness and field dependent surface scattering. /30/, /130/, /153/ presented interesting measured results on inversion layer mobility; /162/, /163/ gave some excellent ideas on how to treat theoretically these and other scattering mechanisms; /182/ suggested a heuristic formula for field dependent surface scattering which is applicable for two-dimensional simulations, but whose adequacy is questioned in /162/. However, we have developed (2.4-9) which models phenomenologically with best fit to measurement surface roughness as well as field dependent surface scattering /143/.

$$\mu_{LIS}(y,E_p,E_t,N,T) = \mu_{LI}(N,T)\cdot\frac{y+y_r}{y+b\cdot y_r} \quad (cm^2/Vs) \tag{2.4-9}$$

$$y_r = y_0/(1+E_p/E_{p0})$$
$$b = 2+E_t/E_{t0}$$
$$E_p = max(0,(E_x\cdot J_x+E_y\cdot J_y)/(J_x^2+J_y^2)^{1/2})$$
$$E_t = max(0,(E_x\cdot J_y-E_y\cdot J_x)\cdot J_x/(J_x^2+J_y^2))$$

$$y_{0n} = 5\cdot10^{-7} \qquad\qquad y_{0p} = 4\cdot10^{-7}$$
$$E_{p0n} = 10^4 \qquad\qquad E_{p0p} = 8\cdot10^3$$
$$E_{t0n} = 1.8\cdot10^5 \qquad\qquad E_{t0p} = 3.8\cdot10^5$$

In regions with a high electric field component parallel to current flow, drift velocity saturation has to be taken into account. (2.4-10) combines, also phenomenologically, this physical effect and the lattice-impurity-surface mobility using a Mathiessen-type rule with a weakly temperature dependent saturation velocity /23/, /79/, /80/.

$$\mu_{tot}(y,E_p,E_t,N,T) = (\mu_{LIS}(...)^{\beta}+(v_s/E_p)^{\beta})^{1/\beta} \tag{2.4-10}$$

$$v_{sn} = 1.53\cdot10^9\cdot T^{-0.87} \qquad\qquad v_{sp} = 1.62\cdot10^8\cdot T^{-0.52}$$
$$\beta_n = -2 \qquad\qquad\qquad\qquad \beta_p = -1$$

2.4.3 Formulae for Modeling Generation/Recombination

To simulate satisfactorily transfer phenomena of majority carrier current and minority carrier current in just a simple diode, it is an absolute necessity to model carrier recombination and generation as carefully as possible. (2.4-11) represents the well known Shockley-Read-Hall term for modeling thermal generation/recombination. The carrier lifetimes can be simulated as being doping dependent /35/, /103/.

$$(G - R)_{th} = \frac{n_i^2 - p \cdot n}{\tau_n(p+p_1)+\tau_p(n+n_1)} \qquad (1/cm^3 s) \qquad (2.4-11)$$

$$\tau_n = 3.95 \cdot 10^{-5}/(1+N/7.1 \cdot 10^{15}) \qquad \tau_p = 3.52 \cdot 10^{-5}/(1+N/7.1 \cdot 10^{15})$$

Surface generation/recombination /74/ can be treated in a fairly similar manner by (2.4-12).

$$(G - R)_s = \frac{n_i^2 - p \cdot n}{(p+p_1)/s_n+(n+n_1)/s_p} \cdot \delta(y) \qquad (1/cm^3 s) \qquad (2.4-12)$$

$\delta(y)$: Dirac-Delta function, y=0 denotes an interface

$$s_n = 100 \qquad\qquad s_p = 100$$

Impact ionization can be modeled by an exponentially field dependent generation term /27/, /28/. The constants in (2.4-13) are essentially taken from /170/.

$$G_a = \frac{|\vec{J}_n|}{q} A_n \exp \left(- \frac{B_n |\vec{J}_n|}{\vec{E} \cdot \vec{J}_n} \right) +$$

$$+ \frac{|\vec{J}_p|}{q} A_p \exp \left(- \frac{B_p |\vec{J}_p|}{\vec{E} \cdot \vec{J}_p} \right) \qquad (1/cm^3 s) \qquad (2.4-13)$$

$$A_n = 7 \cdot 10^5 \qquad\qquad A_p = 1.588 \cdot 10^6$$
$$B_n = 1.23 \cdot 10^6 \qquad\qquad B_p = 2.036 \cdot 10^6$$

It should be noted that this form of simulating avalanche is relatively crude compared to more exact considerations, but the underlying physical principles are so complex that a trade-off in accuracy and complexity leads to that type of formula. The ionization probabilities $\alpha_{n,p}$ for silicon as a function of the electric field have been measured by various authors: Mc Kay /101/, /102/, Miller /105/, /106/, Chynoweth /27/, /28/, Lee /88/, Moll /112/, /113/, Ogawa /118/, Van Overstraeten /170/, Grant /65/, Dalal /36/. Their results are summarized in Fig. 2.4-5 for electrons and in Fig. 2.4-6 for holes. Additionally, the measured results are compared to theoretical results of Baraff /10/ (material constants from Sze /157/, /158/). Also drawn in Fig. 2.4-5 and Fig. 2.4-6 are theoretical limits published by Okuto /122/, /123/, which imply that all the energy the carriers can obtain from the electric field is used to generate additional carriers. Furthermore, the energy loss per single ionization has been taken to be 1.6eV for electrons and 1.8eV for holes (see also /75/). A more concise treatment of the ionization probabilities has been undertaken theoretically by /5/, /26/, /91/, /145/, /160/, /161/, /162/, /169/, /181/ and experimentally by /95/, /131/, /149/.

To analyze high injection conditions, Auger recombination has to be included as an antagonism to avalanche generation. Already the use of a simple formula like (2.4-14) in general gives satisfactory results /31/, /35/, /52/, /55/.

$$(G - R)_{Aug} = (n_i^2 - p \cdot n) (C_n \cdot n + C_p \cdot p) \qquad (1/cm^3 s) \quad (2.4-14)$$

$$C_n = 2.8 \cdot 10^{-31} \qquad\qquad C_p = 9.9 \cdot 10^{-32}$$

Finally, all generation/recombination phenomena have to be combined to one total quantity. The usual way to do so is to simply sum up all terms (2.4-15). However, that means that no interaction of the different phenomena does exist.

$$(G-R)_{tot} = (G-R)_{th} + (G-R)_s + (G-R)_{Aug} + G_a \qquad (2.4-15)$$

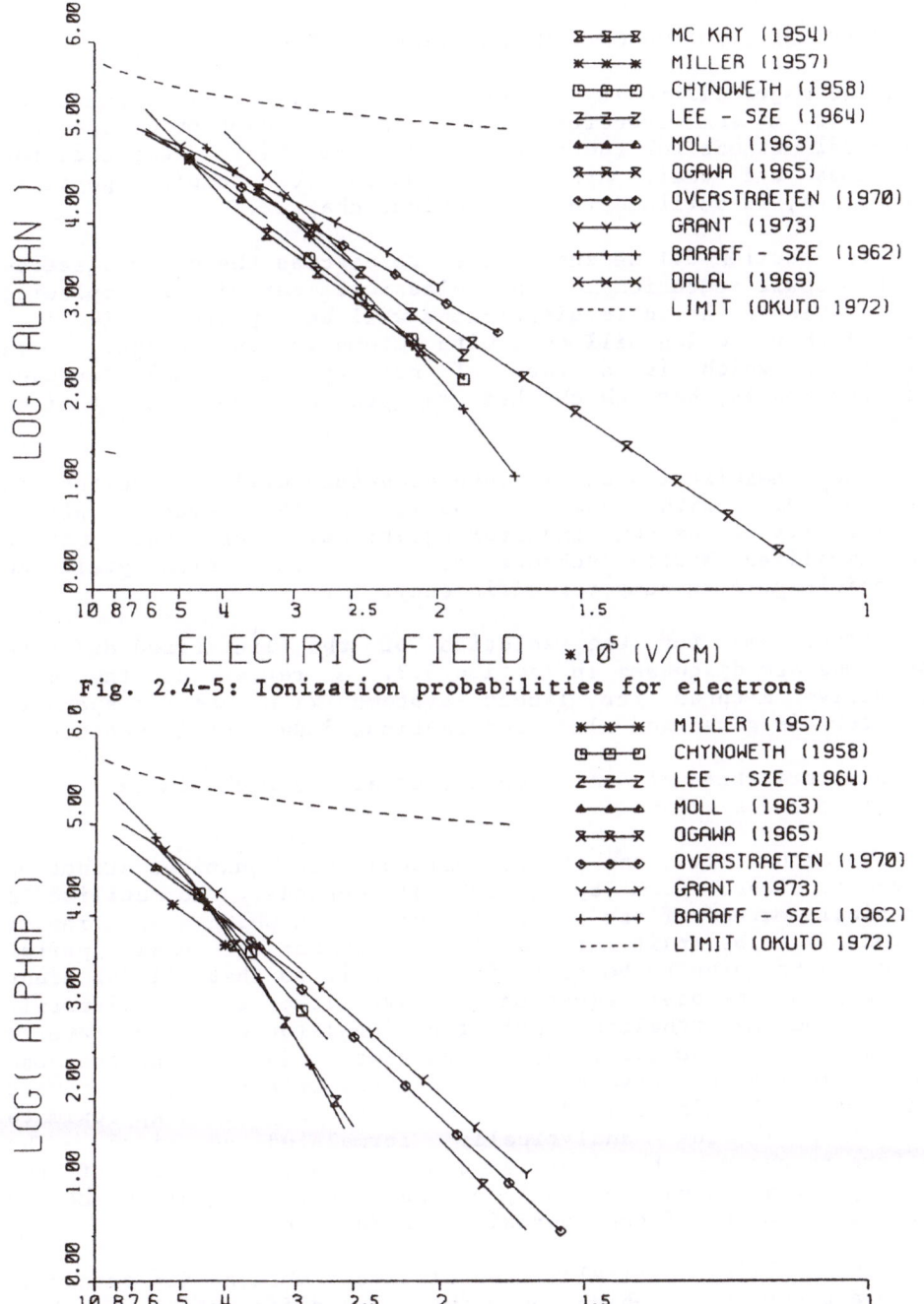

Fig. 2.4-5: Ionization probabilities for electrons

Fig. 2.4-6: Ionization probabilities for holes

3. NUMERICAL SOLUTION OF SEMICONDUCTOR EQUATIONS

The major difficulty in designing a high order numerical model of a semiconductor device is the adaption of adequate numerical methods for the solution of the basic semiconductor equations and their associated, often very complex, physical parameters, as outlined in the previous chapter.

In section 3.1 we should like to discuss the discretisation of the basic equations. The classical method of replacing derivatives with finite differences will be explained. The last part of that section will deal with automatic and adaptiv mesh generation which is a task of primary importance for user oriented models, but which has as yet not been scrutinized thoroughly.

The linearization of discrete equations will be treated in section 3.2 with some emphasis on the severely strong nonlinearity of the semiconductor equations. For that purpose some modified Newton schemes are presented which yield an incredible gain in computer efficiency.

Algorithms for the solution of the linearized discrete equations are discussed in section 3.3. A review of the most attractive methods for linear systems with special sparsity structure is given and also some cautious judgement is ventured.

3.1 Discretisation of Semiconductor Equations with Finite Differences

Unfortunately, the basic semiconductor equations cannot be solved in closed form by analytical methods. To utilize a numerical method, first of all the domain in which a solution is wanted has to be split into a finite number of small parts. These parts have to be sufficiently small so that all dependent variables of the basic equations behave like some arbitrarily chosen, but nevertheless simple functions; the equations have to be discretised. However, one should always bear in mind that one can, following the above sketched outline, obtain only an exact solution of the discretised problem, which is just an approximate solution of the analytically formulated equations. The difference between the discrete solution and the solution of the real problem depends obviously on the partitioning of the domain and the selection of the approximating functions.

There exist basically two classical methods for obtaining algebraic equations, which approximate the differential equation and which can be solved numerically, namely: the Finite Difference method and the Finite Element method. The fundamental difference between these two methods can be summarized, at least qualitatively, as follows. By applying the finite difference method, all derivatives in the differential equation are replaced by finite differences between discrete points in the interior of the domain and the residual of the resulting difference equation is set to zero on every discrete point. The finite element method in its residual formulation demands that the weighted residuum integrated over the whole domain be zero. This can be achieved algebraically by setting all residual integrals for every finite element for which the solution is assumed to obey some simple functional relation to zero. From our point of view it is impossible to favourize one method distinctly; both methods have their advantages and bottlenecks. Following the literature many renowned authors have concentrated their work on finite elements e.g. /1/, /13/, /20/, /21/, /22/, /25/, /32/, /70/, /71/, /119/, or finite differences e.g. /56/, /66/, /67/, /76/, /84/, /85/, /93/, /94/, /107/, /108/, /109/, /121/, /154/. We have also concentrated our activities on finite differences, because the mathematical background required to produce a running program seems to be somewhat smaller for the finite difference method than for the finite element method. Some interesting extensions of the finite difference method have been recently proposed by Adler /2/, /3/. When fully utilizing these ideas, one advantage of the finite element method, high flexibility at the partitioning task, should also be reached with the finite difference method.

We should like to explain the discretisation with five-point-star differences, which is probably the best known approach of the finite difference method for two dimensional partial differential equations (PDEs). The domain in which the solution of a PDE is desired is first partitioned into small areas by grid lines parallel to some arbitrary coordinate system. For the sake of simplicity a rectangular domain and a cartesian coordinate system will be assumed. By laying NX vertical grid lines (parallel to y-axis) and NY horizontal grid lines (parallel to x-axis) one gets NX·NY intersections. On these intersections one wants to obtain an approximate solution of the PDE of sufficient accuracy. For that purpose the PDE is replaced on every inner point (i,j) (see Fig. 3.1-1) by a difference equation which uses the inner point (i,j) and its four nearest neighbours $(i+1,j)$, $(i-1,j)$, $(i,j+1)$ and $(i,j-1)$. The major assumption for the derivation of the difference equation is that the solution can be approximated with a piecewise linear function along the verteces between the inner point (i,j) and its neighbours. Thus one gets $(NX-2)\cdot(NY-2)$ difference equations because that is exactly the number of inner points. At the boundary of the domain the solution of the PDE has to fulfill some boundary conditions from which one can obtain equations for the boundary

points in a similar manner; there exist 2·(NX+NY-2) boundary
points. The total number of equations equals, therefore, the
total number of points and a unique solution can be found.

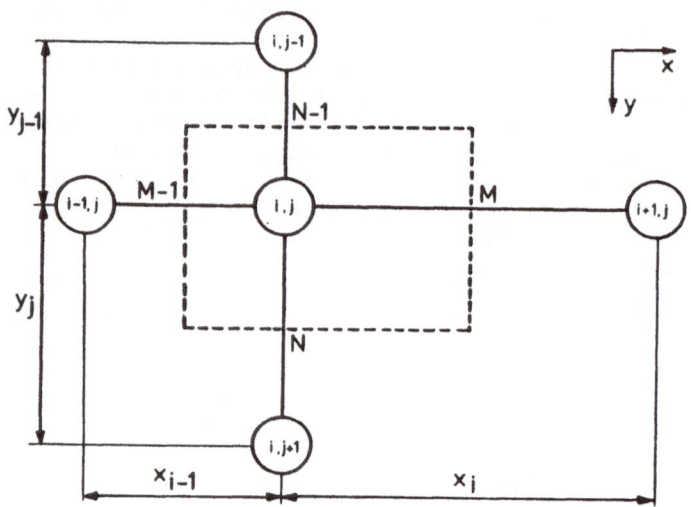

Fig. 3.1-1: The index convention used

In the next section the discretisation of the quasiharmonic
equation in a precise manner will be dealt with, because the
linearized forms of Poisson's equation as well as continuity
equations belong to this important category of PDEs and many
publications have been written on that subject e.g. /58/, /59/,
/98/, /147/.

3.1.1 The Quasiharmonic Equation

Let G be a finite domain in the (x,y) plane bounded by R
which is piecewise continuously differentiable. Furthermore, let
the functions P(x,y), S(x,y) and F(x,y) be piecewise continuous
in G. P(x,y) be positive and not vanishing anywhere; S(x,y) be
positive or zero. Then (3.1-1) represents the quasiharmonic
equation with solution u(x,y).

$$\text{div}(P(x,y) \cdot \text{grad}(u(x,y))) - S(x,y) \cdot u(x,y) = F(x,y) \qquad (3.1-1)$$

subject to the boundary conditions:

$$A(x,y) \cdot u(x,y) + B(x,y) \cdot u(x,y)_n = C(x,y) \qquad (3.1-2)$$

where $A(x,y)$, $B(x,y)$ and $C(x,y)$ are defined in R, piecewise continuous and positive or zero, and $A(x,y)+B(x,y)$ is not vanishing anywhere. $u(x,y)_n$ denotes the derivate of $u(x,y)$ perpendicular to the boundary.

For a solution of this problem the differential equation has to be integrated in every area g_{ij}, obtained by partitioning as outlined above, around the inner point (i,j). The area g_{ij} is drawn with dashed lines in Fig. 3.1-1; it is represented by the rectangle around point (i,j).

$$\iint_{g_{ij}} \text{div}(P \cdot \text{grad}(u)) \cdot dx \cdot dy - \iint_{g_{ij}} S \cdot u \cdot dx \cdot dy = \iint_{g_{ij}} F \cdot dx \cdot dy \qquad (3.1\text{-}3)$$

Using Green's theorem, the area integral can be transformed into a closed boundary integral around g_{ij}.

$$\iint_{g_{ij}} \text{div}(P \cdot \text{grad}(u)) \cdot dx \cdot dy = \int_{r_{ij}} (P \cdot (\partial u/\partial x) \cdot dy - P \cdot (\partial u/\partial y) \cdot dx) \qquad (3.1\text{-}4)$$

Let x_i be the geometrical distance between the i.th and i+1.st vertical grid line and y_j the distance between the j.th and j+1.st horizontal grid line (cf. Fig. 3.1-1). Let P_M be the value of function $P(x,y)$ at point M which is placed exactly between points (i,j) and $(i+1,j)$; and assume the analogous relations for P_{M-1}, P_N and P_{N-1}, which can be easily made clear with Fig. 3.1-1. Then the following holds:

$$\int_{r_{ij}} (P \cdot (\partial u/\partial x) \cdot dy - P \cdot (\partial u/\partial y) \cdot dx) =$$

$$= 0.5 \cdot (y_j + y_{j-1}) \cdot (P_M \cdot (u_{i+1,j} - u_{i,j})/x_i +$$

$$+ \qquad P_{M-1} \cdot (u_{i-1,j} - u_{i,j})/x_{i-1}) +$$

$$+ 0.5 \cdot (x_i + x_{i-1}) \cdot (P_N \cdot (u_{i,j+1} - u_{i,j})/y_j +$$

$$+ \qquad P_{N-1} \cdot (u_{i,j-1} - u_{i,j})/y_{j-1}) +$$

$$+ o(x_{i-1} + x_i) + o(y_{j-1} + y_j) \qquad (3.1\text{-}5)$$

The second and third integral of Eq. 3.1-3 can be approximated straightforwardly under the assumptions that the functions $S(x,y)$ and $F(x,y)$ and the solution $u(x,y)$ are sufficiently smooth in the area g_{ij}.

$$\iint_{g_{ij}} S \cdot u \cdot dx \cdot dy \doteq 0.25 \cdot S_{i,j} \cdot u_{ij} \cdot (x_i + x_{i-1}) \cdot (y_j + y_{j-1}) \quad (3.1-6)$$

$$\iint_{g_{ij}} F \cdot dx \cdot dy \doteq 0.25 \cdot F_{ij} \cdot (x_i + x_{i-1}) \cdot (y_j + y_{j-1}) \quad (3.1-7)$$

After combining (3.1-5), (3.1-6) and (3.1-7) and separating the unknowns, one obtains for each inner point (i,j) a linear equation of the following form:

$$\begin{aligned}
u_{i,j} \cdot &((y_j + y_{j-1}) \cdot (P_M / x_i + P_{M-1} / x_{i-1}) + \\
&(x_i + x_{i-1}) \cdot (P_N / y_j + P_{N-1} / y_{j-1}) + \\
&0.5 \cdot S_{i,j} \cdot (x_i + x_{i-1}) \cdot (y_j + y_{j-1})) = \\
= u_{i+1,j} \cdot &((y_j + y_{j-1}) \cdot P_M / x_i) + \\
+ u_{i-1,j} \cdot &((y_j + y_{j-1}) \cdot P_{M-1} / x_{i-1}) + \\
+ u_{i,j+1} \cdot &((x_i + x_{i-1}) \cdot P_N / y_j) + \\
+ u_{i,j-1} \cdot &((x_i + x_{i-1}) \cdot P_{N-1} / y_{j-1}) - \\
- 0.5 \cdot F_{i,j} \cdot &(x_i + x_{i-1}) \cdot (y_j + y_{j-1})
\end{aligned} \quad (3.1-8)$$

In Eq. 3.1-8 no estimate of the discretisation error is given. For a non-aequidistant mesh $(x_i \neq x_{i-1}, y_j \neq y_{j-1})$ the discretisation error decreases approximately linearly with the mesh spacings. Some ideas on proper mesh selection to sufficiently bound that error will be given in section 3.1.4. However, more exact consideration should be looked up in the classical mathematical literature e.g. /58/, /147/.

The discretisation of the boundary conditions is basically no problem. It is treated quite carefully in many lecture books on numerical mathematics e.g. /47/ so that we can refrain from an explanation.

All equations obtained by the discretisation procedure can be combined to a sparse linear operator B (3.1-9) applied to a vector u of unknowns which represent the solution at the mesh points.

$$B(u) = 0 \qquad (3.1\text{-}9)$$

Therefore, the rank of the operator B equals the total number of meshpoints which is usually rather large. However, B is also very sparse; there exist at most five elements per row. The treatment of these specially structured equations will be outlined in the following sections.

3.1.2 Poisson's Equation

Poisson's equation (3.1-10) is an exponentially nonlinear elliptic equation.

$$\text{div } \varepsilon \text{ grad } \psi = -q \cdot (n_i \cdot e^{(\varphi_p - \psi)/Ut} - n_i \cdot e^{(\psi - \varphi_n)/Ut} + C) \qquad (3.1\text{-}10)$$

The geometry for the simulation of MOS transistors which we and many others use is shown in Fig. 3.1-2. Poisson's equation (also the continuity equations) has to be solved for the rectangular area A-F-G-H which represents the silicon region. In the area C-D-E-B which represents the gate oxide, only the Laplacian equation has to be solved because no space charge exists there. The boundary conditions are usually treated as follows: The contacts (A-B: source, E-F: drain, G-H: bulk) are assumed to be ideally ohmic. The potential is kept constant at the sum of the applied bias plus the built-in potential which is caused by the doping. At the vertical boundaries (A-H, F-G) the derivative of the potential perpendicular to the boundary (i.e. the lateral electrical field component) has to be zero. Certainly, this condition is only valid from the physical point of view if the source contact A-H and the drain contact E-F are sufficiently long. At the silicon to silicon dioxide interface the potential must obey Gauß's law (3.1-11). The existence of fixed surface states can be treated directly with Gauß's law, confer to /155/; however, we think it is more economic and sufficiently accurate to account for fixed surface states with the flatband voltage, because fixed surface states should be kept small anyway and, should, therefore, not effect the solution very much.

$$\varepsilon_{ox} \cdot (\partial \psi / \partial y)_{ox} = \varepsilon_{si} \cdot (\partial \psi / \partial y)_{si} \qquad (3.1\text{-}11)$$

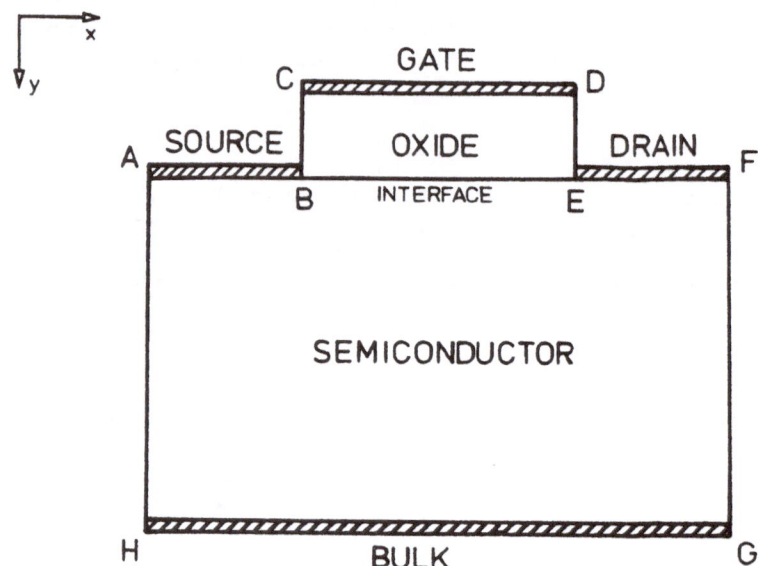

Fig. 3.1-2: The simulation geometry for planar MOSFETs

The Laplacian equation in the oxide is coupled with Poisson's equation via (3.1-11). At the gate contact (C-D) the potential is kept constant at the applied bias minus the flatband voltage; at the vertical boundaries of the oxide (C-B, D-E) the lateral electric field has to vanish.

It is interesting to note that many authors suggest a one dimensional voltage drop in the oxide e.g. /166/. In that manner one can obtain a mixed boundary condition (c.f. Eq. 3.1-2) for the potential at the interface. However, we feel that this is too crude an assumption for miniaturized MOS transistors.

3.1.3 Current Continuity Equations

Only the discretisation of the continuity equation for electrons will be treated in the following because the continuity equations for holes can be handled in an analogous manner. The major difficulty of the discretisation of the continuity equations is to find a proper, numerically stable formulation of the divergency of the current by just using information of physical quantities at meshpoints. A naive discretisation of the current relation (c.f. (3.1-12) for electrons) as only published in very early papers has been proved by various authors to be unstable.

$$\text{div } \vec{J}_n = -q \cdot \text{div } (\mu_n \cdot n \cdot \text{grad } \psi - D_n \cdot \text{grad } n) \qquad (3.1-12)$$

However, Scharfetter and Gummel suggested already in 1969 a stable discretisation which has been physically motivated /132/. Their work can certainly be interpreted mathematically which helps in understanding various numerical phenomena associated with that discretisation. By assuming the validity of Boltzmann's statistics one gets with the substitution:

$$s = e^{-\psi_n/Ut} \qquad (3.1-13)$$

the following expression for the divergence of the electron current:

$$\text{div } \vec{J}_n = q \cdot \text{div}(D_n \cdot n_i \cdot e^{\psi/Ut} \cdot \text{grad } s) \qquad (3.1-14)$$

This substitution is, as a matter of fact, essential because we have now a self-adjoint elliptic operator in s for the divergence of the current. For that type of operators the mathematical analysis is relatively easy and well investigated.

Recalling now the elliptic operator of the quasiharmonic equation (3.1-1), an analogy becomes evident. With:

$$P(x,y) = D_n \cdot n_i \cdot e^{\psi/Ut} \qquad (3.1-15)$$

we can use the results of section 3.1.1 for the discretisation. The fundamental problem, however, is to find a proper interpolation of (3.1-15) to obtain the mid-vertex values P_M etc. A naive linear interpolation of the exponential of the electric potential is definitely not appropriate. The very best one can do from the mathematical point of view is to use an exponential interpolation. (3.1-16) is an example for this type of interpolation between points (i,j) and (i+1,j). We should like to refrain from a proof of this relation as it is fairly lengthy. The interested reader should consult e.g. /124/.

$$e^{\psi_M} = e^{\psi_{i,j}} \cdot (\psi_{i,j} - \psi_{i+1,j})/(e^{\psi_{i,j} - \psi_{i+1,j}} - 1)$$

$$= e^{\psi_{i,j}} \cdot \text{ber}(\psi_{i,j} - \psi_{i+1,j}) \qquad (3.1-16)$$

with: $\text{ber}(x) = x/(e^x - 1)$ (Bernoulli function) $\qquad (3.1-17)$

The actual programming of the Bernoulli functions (3.1-17) has to be undertaken with care to avoid underflow and overflow traps /73/. Furthermore, it should be noted that only differences of potential values occur in the Bernoulli functions.

Numerical stability is, therefore, greatly increased. The leading explicit exponential of Ψ in (3.1-16) vanishes if it is combined with the s values (3.1-13) to electron densities which are numerically in an acceptable range compared to the exponentials of Ψ and φ_n.

The mystery of the stability of the discretisation which has just been outlined lies in the fact that it represents a so-called "windward" difference approximation. For a large potential drop between neighbouring meshpoints the windward scheme degenerates in a forward- or backward difference scheme depending on the sign of the potential difference (i.e. electric field). Therefore the propagation of local errors is very small.

The boundary conditions are very simple for the continuity equations. At the contacts (A-B, E-F, G-H in Fig. 3.1-2) the carrier densities are set constant to their equilibrium value. At the remaining boundaries (B-E, F-G, A-H) no current component perpendicular to the boundary must exist.

A very interesting and successfully applied alternative to the outlined discretisation has been proposed by Mock /108/ for the MOS transistor. Through the introduction of so-called "stream-functions" one also obtains a self-adjoint operator for the divergence of the current with similar problems in the interpolation of exponentials of the electric potential. The treatment of inhomogeneities i.e. recombination/generation, however, is more complicated with stream-functions. Therefore we favour the other discretisation.

3.1.4 Grid Generation

To keep computer time as well as memory requirements reasonably small, it is necessary to limit the number of mesh points. A suitable tradeoff between accuracy and computing costs can be found once the discretisation errors are estimated. In critical regions with large discretisation errors grid spacing has to be kept small whereas it may be large in regions in which only small errors occur. Such considerations make it evident that an equidistant mesh is not suitable because in that case grid spacing has to be adapted to the critical regions and the number of mesh points would be very large.

As the discretisation errors depend on the distribution of the quantities Ψ,n,p a suitable mesh cannot be estimated a priori, that is without knowledge of the solution. Therefore, grid generation is performed adaptively, i.e. a priliminary solution is calculated on the basis of an initial mesh, then the mesh is adapted to this solution and again the basic equations are solved. Regeneration of the mesh can be done repeatedly if necessary.

Let f(x) be a four times continuously differentiable function; then one can savely write;

$$f_{i+1} \doteq f_i + f_i' h_i + f_i'' \frac{h_i^2}{2} + f_i''' \frac{h_i^3}{6} + f^{IV}(\xi)\frac{h_i^4}{24} \qquad (3.1\text{-}18)$$

$$f_{i-1} \doteq f_i - f_i' h_{i-1} + f_i'' \frac{h_{i-1}^2}{2} - f_i''' \frac{h_{i-1}^3}{6} + f^{IV}(\xi)\frac{h_{i-1}^4}{24} \qquad (3.1\text{-}19)$$

and we get for the second order differential quotient:

$$f''(x_i) \doteq 2 \cdot \frac{(f_{i+1}-f_i)/h_i + (f_{i-1}-f_i)/h_{i-1}}{h_i + h_{i-1}} +$$

$$+ f_i''' \frac{h_i - h_{i-1}}{3} + f^{IV}(\xi) \cdot \frac{h_i^2 - h_i h_{i-1} + h_{i-1}^2}{12} \qquad (3.1\text{-}20)$$

The first term on the right hand side of eq. (3.1-20) is the finite difference approximation. The other two terms represent the discretisation error. For an aequidistant mesh ($h_i = h_{i-1}$) the second term on the right hand side vanishes and only differentials of at least fourth order cause discretisation errors. Principially eq. (3.1-20) can be used to fix the mesh spacing h_i for given largest acceptable error, knowledge of the fourth differential provided, and also to bound the maximum mesh progression γ_i when knowing the third differential

$$\gamma_i = \max \left(\frac{h_i}{h_{i-1}}, \frac{h_{i-1}}{h_i} \right). \qquad (3.1\text{-}21)$$

The errors which are induced by the finite difference approximation of the inhomogeneity terms are going to be considered in the following. The inhomogeneities around each mesh point up to the next midpoint are approximated by

$$\int_{x_i - \frac{h_{i-1}}{2}}^{x_i + \frac{h_i}{2}} \int_{y_j - \frac{k_{j-1}}{2}}^{y_j + \frac{k_j}{2}} F(x,y) \cdot dx \cdot dy \doteq 0.25 \cdot F_{i,j}(h_i + h_{i-1})(k_j + k_{j-1}) \qquad (3.1\text{-}22)$$

with F(x,y) being the inhomogeneity term. For a one dimensional error estimation a series expansion of F(x) yields:

$$\int_{x_i-\frac{h_i}{2}}^{x_i+\frac{h_i}{2}} F(x) \cdot dx \doteq \frac{1}{2}F_i(h_i+h_{i-1})+F_i'\frac{h_i^2-h_{i-1}^2}{8}+F''(\xi)\frac{h_i^3+h_{i-1}^3}{24} \qquad (3.1-23)$$

The first term in (3.1-23) is the finite difference approximation as given in (3.1-22) and the two other terms describe the local error. The global error is extremely difficult to estimate. However, it is often of by one reduced order.

If we consider Poisson's equation once again, the second differential of the space charge limits the mesh spacing and the first differential limits mesh progression.

$$\gamma_i - 1 = 4 \frac{\varepsilon_{max}}{F_i' h_{i-1}} \qquad (3.1-24)$$

$$h_i^2 = 12 \frac{\varepsilon_{max}}{F''(\xi)} \qquad (3.1-25)$$

As already discussed earlier ionisation rates are very sensitive to the electric field. Therefore, the generation rate exhibits an abrupt peak in the pinch-off region which can only be kept under control with a very fine discretisation. The integral of the generation rate over the total area gives the substrate current, and the discretisation error is, therefore, proportional to the error of the substrate current. If we consider only the first derivative of the electric field

$$E(x) \doteq E(x_i) + (x-x_i) \cdot E'(x_i)$$

we get for α'' on the basis of Chynoweth's law:

$$\alpha(x) \doteq A \cdot \exp\left[\frac{-B}{E_i} \cdot \left(1 - \frac{(x-x_i)E_i'}{E_i}\right)\right]$$

$$\alpha'(x_i) \doteq \frac{B}{E_i^2} \cdot E_i' \cdot \alpha(x_i) \qquad (3.1-26)$$

$$\alpha''(x_i) \doteq \left[\frac{B}{E_i^2} \cdot E_i'\right]^2 \cdot \alpha(x_i) \qquad (3.1-27)$$

By substituting these expressions in (3.1-24), (3.1-25) we get some rules for the mesh generation. If one likes to limit the maximum relative error in the substrate current, it is useful to divide (3.1-26), (3.1-27) by the maximum ionization rate which occurs in the device.

3.2 Linearization of the Coupled System

In this section we should like to discuss some properties of one-step stationary iterative methods of the form (3.2-2) for the solution of systems of nonlinear equations (3.2-1).

$$F(x) = 0 \qquad\qquad\qquad (3.2-1)$$

$$x^{k+1} = G \cdot x^k \qquad\qquad\qquad (3.2-2)$$

The problem (3.2-1) be properly defined /124/. Suppose the operator G has a fixed point x^*. Then (3.2-2) will converge to x^* if G is F-differentiable /124/ at x^* and if the spectral radius of it's Jacobian $G'(x^*)$ satisfies Eq. (3.2-3).

$$\rho(G'(x^*)) < 1 \qquad\qquad\qquad (3.2-3)$$

This very important theorem (Ostrowski theorem) is the basis of all investigations on the convergence of one-step iterations for the solution of nonlinear equations. In the following we should like to write the discretised semiconductor equations in a more abstract form to simplify the formulae. Let (3.2-4) be Poisson's equation and (3.2-5), (3.2-6) the continuity equations for electrons and holes in the unknowns ψ, φ_n and φ_p which represent vectors of the meshpoint values of the electrostatic potential, the quasifermilevel of electrons and holes, respectively.

$$F1 = F1(\psi, \varphi_n, \varphi_p) = 0 \qquad\qquad\qquad (3.2-4)$$

$$F2 = F2(\psi, \varphi_n, \varphi_p) = 0 \qquad\qquad\qquad (3.2-5)$$

$$F3 = F3(\psi, \varphi_n, \varphi_p) = 0 \qquad\qquad\qquad (3.2-6)$$

Then F1, F2 and F3 represent a nonlinear system of equations with the rank $3 \cdot (NX \cdot NY)$ which is three times the total number of meshpoints, usually a large number. We have now to find some operators G which are relatively easy to calculate and for which condition (3.2-3) holds. The best known method is certainly the classical Newton-Raphson method.

3.2.1 Newton's Method and Modified Newton Methods

For Newton's method the iteration is defined as follows:

$$
\begin{pmatrix} \psi \\ \varphi_n \\ \varphi_p \end{pmatrix}^{k+1}
=
\begin{pmatrix} \psi \\ \varphi_n \\ \varphi_p \end{pmatrix}^{k}
-
\begin{pmatrix}
\partial F1/\partial\psi & \partial F1/\partial\varphi_n & \partial F1/\partial\varphi_p \\
\partial F2/\partial\psi & \partial F2/\partial\varphi_n & \partial F2/\partial\varphi_p \\
\partial F3/\partial\psi & \partial F3/\partial\varphi_n & \partial F3/\partial\varphi_p
\end{pmatrix}_k^{-1}
\cdot
\begin{pmatrix} F1 \\ F2 \\ F3 \end{pmatrix}^{k}
\qquad (3.2-7)
$$

The operator G is, therefore, defined as:

$$G = (I - F'^{-1} \cdot F) \qquad (3.2-8)$$

It can be proved that the Jacobian G' of this operator has only zero eigenvalues at x^*, $F(x^*)=0$ and fulfills trivially condition (3.2-3). As all eigenvalues are zero, even quadratic convergence is anticipated as the solution is approached /124/, /126/. Although Newton's method is very attractive from the mathematical point of view, there are practical difficulties.

The main implementation problem is the evaluation of the derivative terms in equation (3.2-7), since the total equation (including modeled physical parameters) must be differentiated accurately with respect to the variables ψ, φ_n and φ_p. This can be done analytically, in principle, but one loses much flexibility in changing the models of the physical parameters. Therefore, a numerical algorithm is necessary to automatically calculate the required derivatives. The best algorithm known at the moment has been published by Curtis and Reid /34/. Some very interesting comments on numerical differentiation have also been given in /81/.

Another fairly difficult problem when considering Newton's method is overshoot. The iteration process does not neccesarily converge monotonously to the solution. Especially, if one starts the iteration with a bad initial guess - which is the usual situation - monotonic convergence or convergence at all cannot be guaranteed. Therefore, one has to introduce a mechanism which dampens the increments resulting from the iteration process so that convergence is monotonic. The naive algorithms simply limit the increments to some maximum value e.g. /108/ or they use some function to continuously limit large increments e.g. /19/. Deuflhardt suggested a more elaborate method /44/, /45/. Roughly explained, he calculates a parameter ν in the range $]0,1]$ so that condition (3.2-9) holds.

$$\| F'^{-1}(x^k) \cdot F(x^k + \nu \cdot (x^{k+1} - x^k)) \| < \| F'^{-1}(x^k) \cdot F(x^k) \| \qquad (3.2-9)$$

After having found ν, the solution x^{k+1} is calculated with Eq. (3.2-10).

$$x^{k+1} = x^k + \nu \cdot (x^{k+1} - x^k) \qquad (3.2-10)$$

This procedure guarantees monotonic convergence; for $\nu=1$ it is the classical Newton method.

Another excellent method was proposed by Meyer /104/. He suggested to modify the iteration operator G by introducing a positive parameter λ as given in Eq. (3.2-11).

$$G = (I - (\lambda \cdot I + F')^{-1} \cdot F) \qquad\qquad (3.2\text{-}11)$$

λ has to be chosen as small as possible so that the norm of $F(x^{k+1})$ is smaller than the norm of $F(x^k)$. Some practical guidelines on how to find this parameter λ with reasonable effort have been recently presented by Bank and Rose /9/.

It is very time and also memory consuming to solve the large linear system which, nevertheless, has to be done for every Newton step. Therefore, many authors use another iterative algorithm, suggested by Gummel /69/, for the linearization of the semiconductor equations.

3.2.2 Block-Nonlinear Iteration

Gummel's idea, in essence, was to solve the semiconductor equations by independently linearizing each equation consecutively with respect to its dominant variable. The first step is to solve Poisson's equation with Newton's method assuming that the quasifermilevels are known functions of position; i.e. the best guess of the quasifermilevels is assumed to be correct. In the second step one of the continuity equations is solved assuming that the electric potential and the quasifermilevel of the other continuity equation are correct. In the third step the second continuity equation is solved under similar assumptions. These three steps are performed repeatedly until a consistent solution is found. The effort for one cycle of this block-nonlinear iteration is obviously less than for one Newton step because the rank of one decoupled equation system is only a third of the rank of the total system.

A complete theoretical proof of the convergence of the block-nonlinear iteration algorithm has as yet not been published. However, strong theoretical indications have been given by Mock /109/. One should also not underestimate the "practical" indications; many authors have used Gummel's iterative scheme with excellent success.

It should also be noted that a few authors have published modifications of the original Gummel method e.g. /14/, /133/, /156/. These modifications are basically motivated on intuition. They are, therefore, as arbitrary as the original approach itself. However, for special purpose applications some improvement in efficiency could be gained.

3.3 Solution of Large Sparse Linear Systems

For any of the linearization procedures which have been outlined in section 3.2 a large sparse linear equation system (3.3-1) has to be solved repeatedly.

$$A \cdot x = b \qquad (3.3-1)$$

We assume that A has been derived by linearizing five-point-star discretized PDEs. Hence matrix A has at most five nonzero entries per row; A is very sparse. For a full Newton scheme these entries are 3x3 matrices; for Gummel's scheme they are scalars. For the solution of these special types of linear equation systems two classes of methods, can, in principle, be used: direct methods which are based on elimination and iterative methods. An excellent survey on that subject has been published recently by Duff /48/.

3.3.1 Direct Methods

Classical Gaussian elimination is not feasible for our systems of equations because the rank of A in (3.3-1) is very large and A has many coefficients which are zero. Therefore, some modifications of the classical Gaussian elimination algorithm have to be introduced to account for the zero entries. There exist quite a few activities on that subject (c.f. /49/) and powerful algorithms which treat the nonzero coefficients only are available. Another serious drawback of direct methods lies in the fact that the upper triangular matrix which is created by the elimination process has to be stored for back substitution. This matrix has usually more nonzero entries than the matrix A. Therefore, memory requirement of direct methods is substantial.

In spite of all drawbacks of direct methods, their major advantage is high accuracy of the solution. However, we feel that for the semiconductor problems iterative algorithms are to be slightly favoured.

3.3.2 Relaxation Methods

The fundamental idea of relaxation methods is the splitting of the coefficient matrix A (3.3-1) into three matrices D, E, F (3.3-2).

$$A = D - E - F \qquad (3.3-2)$$

D denotes the diagonal entries of A; -E denotes a lower triangular matrix which is formed from all sub-diagonal entries of A; and -F denotes an upper triangular matrix which is formed from all super-diagonal entries of A.

With an arbitrary non-singular matrix B which has the same rank as A the linear system (3.3-1) can be rewritten to (3.3-3).

$$B \cdot x + (A-B) \cdot x = b \qquad (3.3-3)$$

One obtains an iterative schema by setting:

$$B \cdot x^{k+1} = b - (A-B) \cdot x^{k} \qquad (3.3-4)$$

(3.3-4) can be solved for x^{k+1}:

$$x^{k+1} = (I-B^{-1} \cdot A) \cdot x^{k} + B^{-1} \cdot b \qquad (3.3-5)$$

The iterative scheme (3.3-5) will converge if condition (3.3-6) holds.

$$\rho(I-B^{-1} \cdot A) < 1 \qquad (3.3-6)$$

(3.3-6) is a necessary and sufficient condition. The various relaxation methods can be won by differently setting matrix B with the matrices obtained by the splitting of A (3.3-2).

The simplest scheme, the point-Jacobi method, uses D for B. Matrix D is a diagonal matrix and is, therefore, easily invertible.

The Gauss-Seidel method uses D-E for B. The matrix D-E is a lower triangular matrix. Therefore one has only to perform a forward substitution process for its inversion.

The successive overrelaxation method (SOR) makes use of a parameter ω within the range $]0,2[$. The iteration matrix B is defined:

$$B = D/\omega - E \qquad (3.3-7)$$

As B is again a lower triangular matrix, its inversion is instantly reduced to a substitution.

The major advantage of these iterative methods lies in their simplicity. They are very easy to program and demand only low memory requirement. As already noted, they converge if condition (3.3-6) holds. However, it is difficult to prove that condition. A sufficient condition for convergence is that A is positive definite (3.3.-8) which is the regular case for five-point-star discretized PDEs.

$$x^{T} \cdot A \cdot x > 0 \text{ for all } x \neq 0 \qquad (3.3-8)$$

These point-iterative schemes can by accelerated quite remarkably with the conjugate gradient method or the Chebyshev method. An excellent survey on these topics can be found in /68/.

3.3.3 Strongly Implicit Iterative Methods

The convergence rate of relaxation methods is relatively poor. Therefore, various activities can be observed for the development of more powerful algorithms with the advantages of iterative schemes.

One of the best known algorithms which has been established in semiconductor device analysis is perhaps Stone's strongly implicit procedure /151/. Stone's idea was to modify the original coefficient matrix A by the addition of a small matrix N so that a factorization of (A+N) involves much less computational effort than the standard decomposition of A and the norm of N is much smaller than the norm of A. Assuming this has been done, the development of an iterative procedure is then fairly straightforward because the equation can be written as:

$$(A+N) \cdot x = (A+N) \cdot x + (b - A \cdot x) \tag{3.3-9}$$

which suggests the iterative procedure:

$$(A+N) \cdot x^{k+1} = (A+N) \cdot x^k + (b - A \cdot x^k) \tag{3.3-10}$$

When the right hand side is known and (A+N) can be factorized easily, (3.3-10) gives an efficient method for directly solving for x^{k+1}. Furthermore, one would intuitively expect a rapid rate of convergence if N is sufficiently small compared to A. We will refrain from explaining in detail Stone's suggestion of how to choose the perturbation matrix N because this has been done thoroughly in many publications e.g. /59/, /147/, /151/.

There exist a few algorithms which are similiar in terms of underlying ideas compared to Stone's method. The most attractive are perhaps the method of Dupont et al. /50/, the "alternating direction implicit" methods e.g. /16/, /59/, /178/ and the Fourier methods /150/, /177/.

One disadvantage of all strongly implicit methods and also the direct methods is that they cannot be implemented efficiently on a computer with a pipe-line architecture (vector processor). Some comments on that subject have been given in /48/.

+. TYPICAL APPLICATIONS OF MINIMOS

+.1 A Didactic Example

It is rather difficult to provide an interesting example for the experienced reader, which is also impressive and easy to understand for readers with general interest in modeling but without specific knowledge of device physics. We have chosen the effects of ion implantation on short channel MOS transistors for the purpose of demonstrating the use of two dimensional simulation. Three devices are calculated whose properties become apparent from the original simulation input decks presented in Fig. 4.1-1. The following discussion of Fig. 4.1-1 shall also demonstrate the ease of using MINIMOS /138/, /139/, /140/, /142/, our simulation program.

The first line is a title line, which is used only to identify the output of the program. The input syntax is totally based on a master key, key and value structure. The next input line which is the "DEVICE" statement, characterizes the device. Specified is an n-channel device (CHANNEL=N) with an n-doped polysilicon gate (GATE=NPOLY), an oxide thickness of 35 nanometers (TOX=350.E-8), a channel width of 10 micrometers (W=10.E-4) and a channel length of one micrometer (L=1.E-4). The "BIAS" statement specifies the operating point. A drain voltage of 3 volts (UD=3.) and a gate voltage of zero volts (UG=0.) has been chosen. The substrate voltage is assumed to be zero by MINIMOS, if not specified otherwise.

The "PROFILE" statement is used to specify the substrate doping and the source/drain diffusion. In the examples presented here we used the simplest way of defining a doping profile, that is the direct calculation by MINIMOS. Another possibility would be to make use of a technology simulation program like SUPREM, the Stanford University PRocess Engineering Models program /6/, for the more accurate calculation of vertical profile shapes which are fitted in the lateral direction. For our simulation a substrate doping of 10^{15} cm^{-3} (NB=1.E15) and a source/drain implantation with phosphorus (ELEM=PH), an implantation dose of 10^{15} cm^{-2} (DOSE=1.E15) and an implantation energy of 40keV (AKEV=40) is specified. The implantation is performed through an isolation oxide of 35 nanometers (TOX=350.E-8) and followed by an annealing step at 1000 centigrades (TEMP=1000) for 1200 seconds (TIME=1200).

```
ONE-MICRON ANALYSIS (DEVICE 1)
DEVICE      CHANNEL=N  GATE=NPOLY  TOX=350.E-8  W=10.E-4  L=1.E-4
BIAS        UD=3.  UG=0.
PROFILE     NB=1.E15  ELEM=PH  DOSE=1.E15  AKEV=40  TOX=350.E-8
+           TEMP=1000  TIME=1200
END


ONE-MICRON ANALYSIS (DEVICE 2)
DEVICE      CHANNEL=N  GATE=NPOLY  TOX=350.E-8  W=10.E-4  L=1.E-4
BIAS        UD=3.  UG=0.
PROFILE     NB=1.E15  ELEM=PH  DOSE=1.E15  AKEV=40  TOX=350.E-8
+           TEMP=1000  TIME=1200
IMPLANT     ELEM=B  DOSE=3.5E11  AKEV=25  TEMP=925  TIME=1800
END


ONE-MICRON ANALYSIS (DEVICE 3)
DEVICE      CHANNEL=N  GATE=NPOLY  TOX=350.E-8  W=10.E-4  L=1.E-4
BIAS        UD=3.  UG=0.
PROFILE     NB=1.E15  ELEM=PH  DOSE=1.E15  AKEV=40  TOX=350.E-8
+           TEMP=1000  TIME=1200
IMPLANT     ELEM=B  DOSE=3.5E11  AKEV=25  TEMP=925  TIME=1200
IMPLANT     ELEM=B  DOSE=1.5E11  AKEV=100
END
```

Fig. 4.1-1: Some typical input decks for MINIMOS

The second input deck further includes an "IMPLANT" statement which defines a channel implantation with boron (ELEM=B), a dose of $3.5 \cdot 10^{11} cm^{-2}$ (DOSE=3.5E11), an energy of 25keV (AKEV=25), annealed at 925 centigrades (TEMP=925) for 1800 seconds (TIME=1800). The third input deck has an additional "IMPLANT" statement specifying a second, deeper channel implantation with boron (ELEM=B), a dose of $1.5 \cdot 10^{11} cm^{-2}$ (DOSE=1.5E11) and an energy of 100keV (AKEV=100). It is assumed that both channel implantation steps are annealed at the same time. It is fairly well known that the first of these three devices is, owing to the short channel effect, "normally-on" and that the shallow implantation of device 2 effects a threshold shift to obtain a "normally-off" device. Furthermore, the deep implantation of device 3 is necessary to avoid punch through. These effects will now be demonstrated by birds-eye-view- and contour-plots of physically relevant quantities in the interior of the three model devices.

The calculated doping density distributions for our devices are shown in Figs. 4.1-2, 4.1-3, 4.1-4.

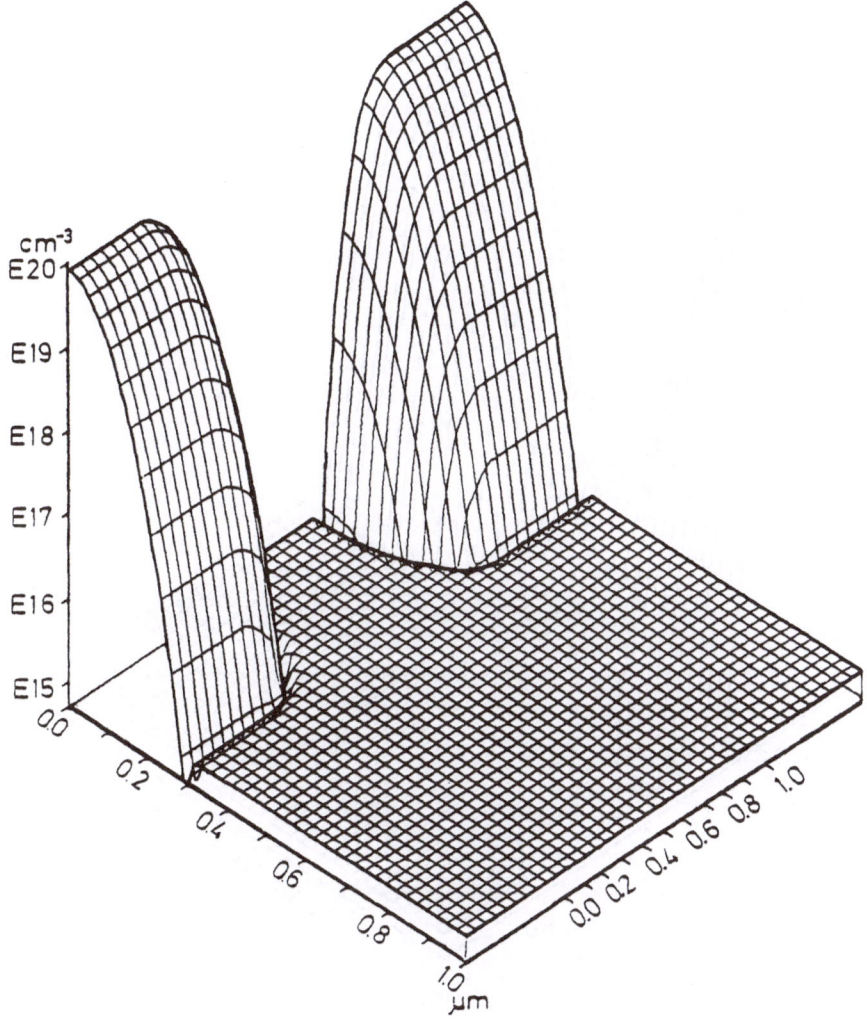

Fig. 4.1-2: Doping profile for device 1

From these figures one can read off the depth of the pn-junctions under source and drain being approximately 300 nanometers. The surface concentration of the source and drain regions is about $10^{20} cm^{-3}$. The effective channel length is reduced by the lateral subdiffusion to about 0.6 micrometers. The shallow channel implantation for threshold tailoring can be seen in Figs. 4.1-3, 4.1-4. Additionally, Fig. 4.1-4 shows the deep implantation for punch through suppression. The threshold voltage is only marginally affected by the deep implantation.

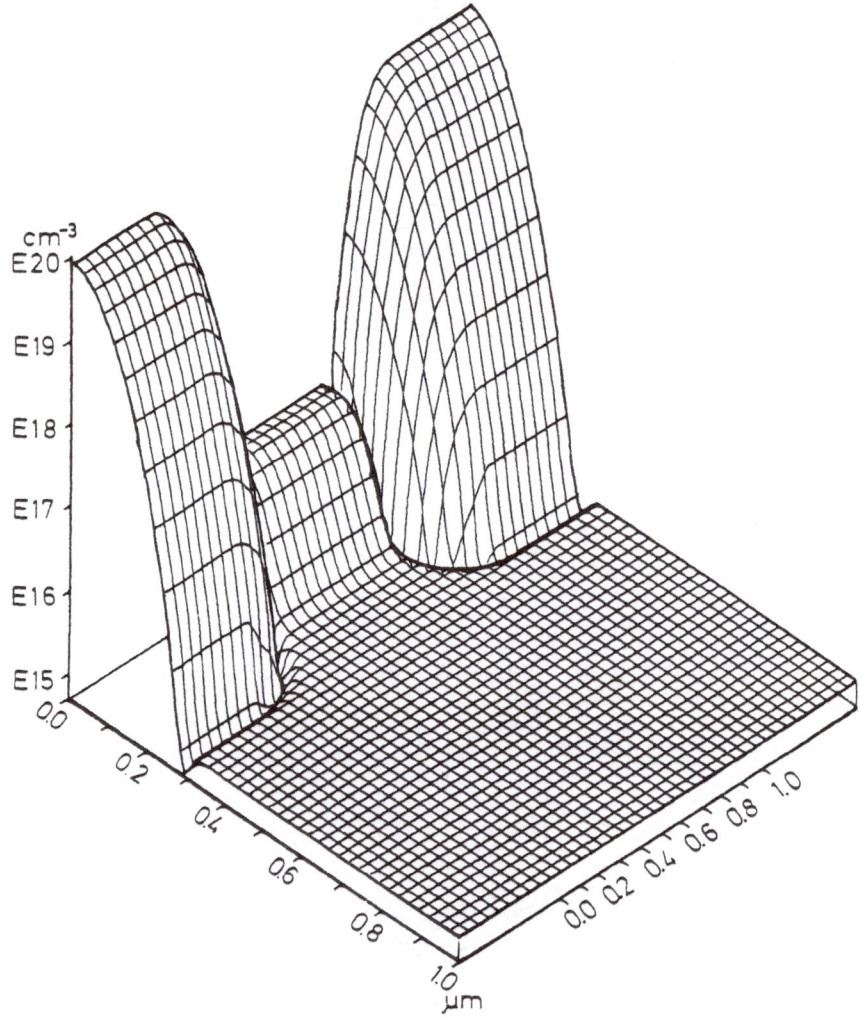

Fig. 4.1-3: Doping profile for device 2

Fig. 4.1-5 shows the distribution of the electric potential for the first device. The drain contact is on the right. In the depletion region of the reverse biased drain-bulk diode the potential decreases monotonously and it is more or less constant in the highly doped source and drain regions. The barrier at the source channel diode is relatively small /168/. Fig. 4.1-6 shows the potential distribution in the second device. The birds-eye-view plot looks very similar to the plot in Fig. 4.1-5.

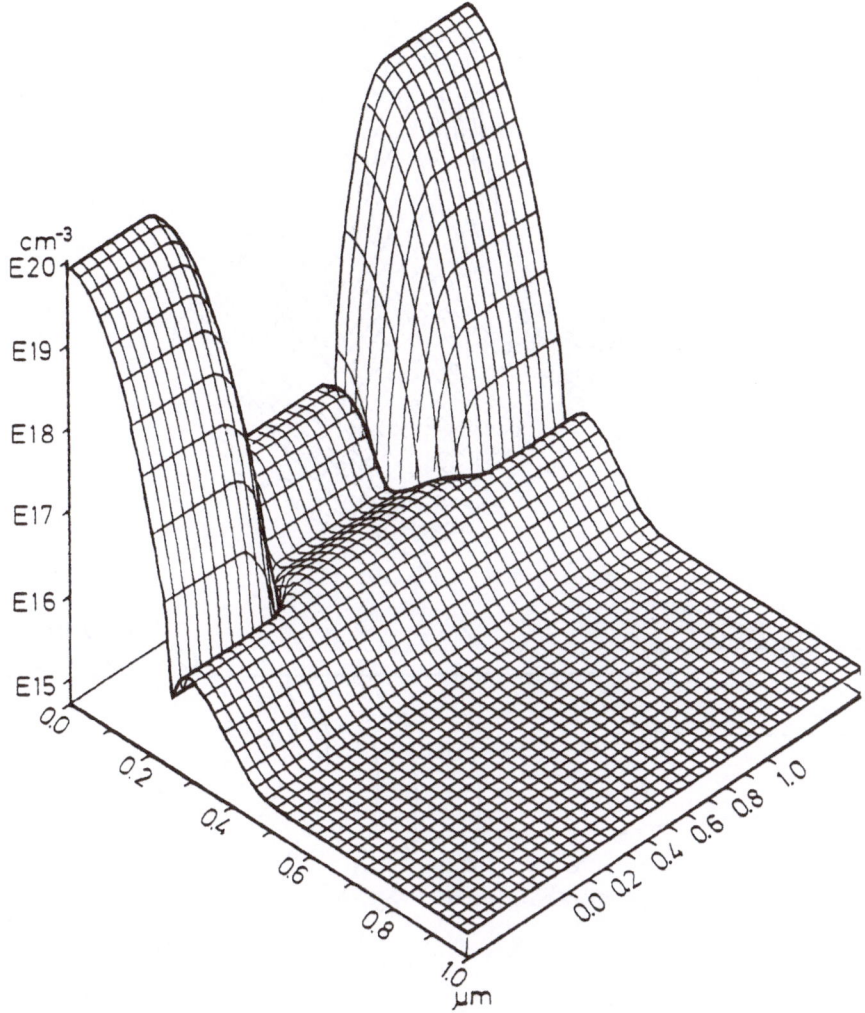

Fig. 4.1-4: Doping profile for device 3

    The contour-plot, however, shows quite a pronounced potential basin directly below the interface. Of even greater importance than this basin itself is the saddlepoint below the basin. At this saddlepoint the electric field vanishes and current only can flow by carrier diffusion. This sort of saddle-point is, following the proposition of many authors (e.g. /12/, /87/), a typical indication of the punch-through effect. The electric field which is induced by the gate is unable to separate the depletion regions of source and drain.

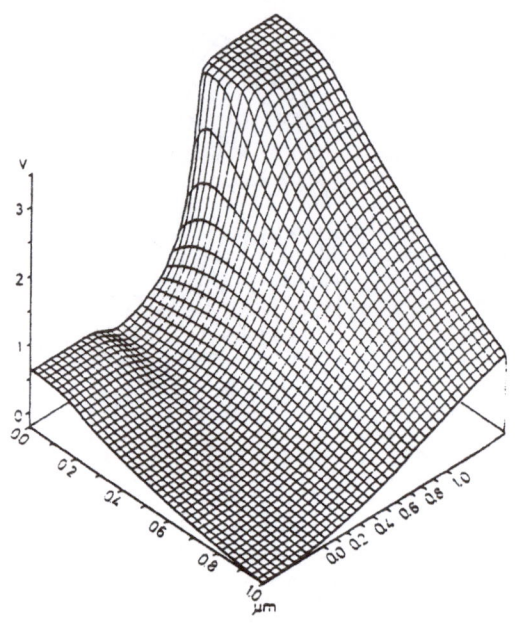

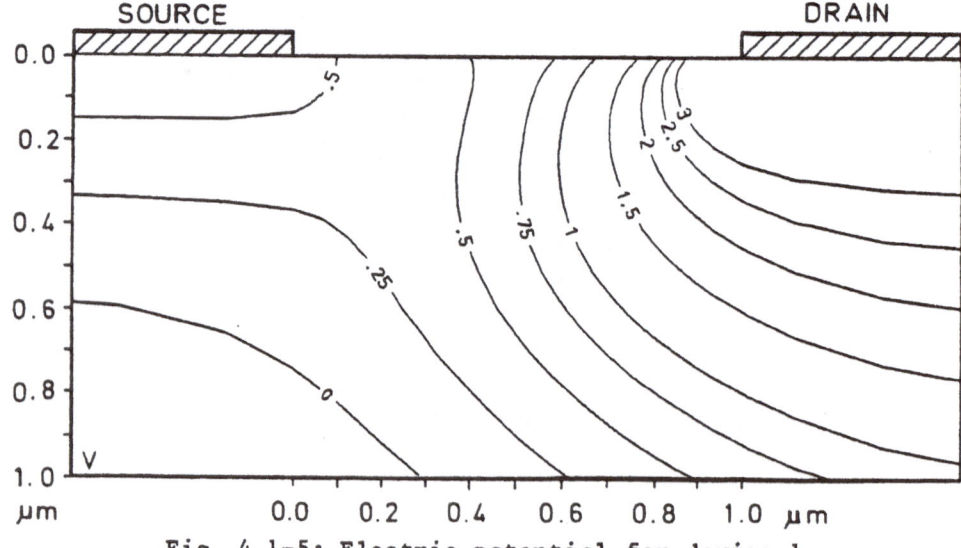

Fig. 4.1-5: Electric potential for device 1

These depletion regions are in contact below the region of control by the gate. As it will become apparent later , the saddlepoint is a reliable indication of the punch-through effect, but it need not exist.

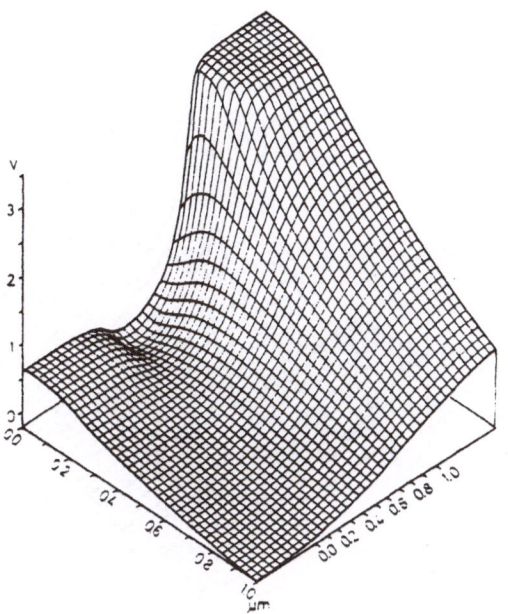

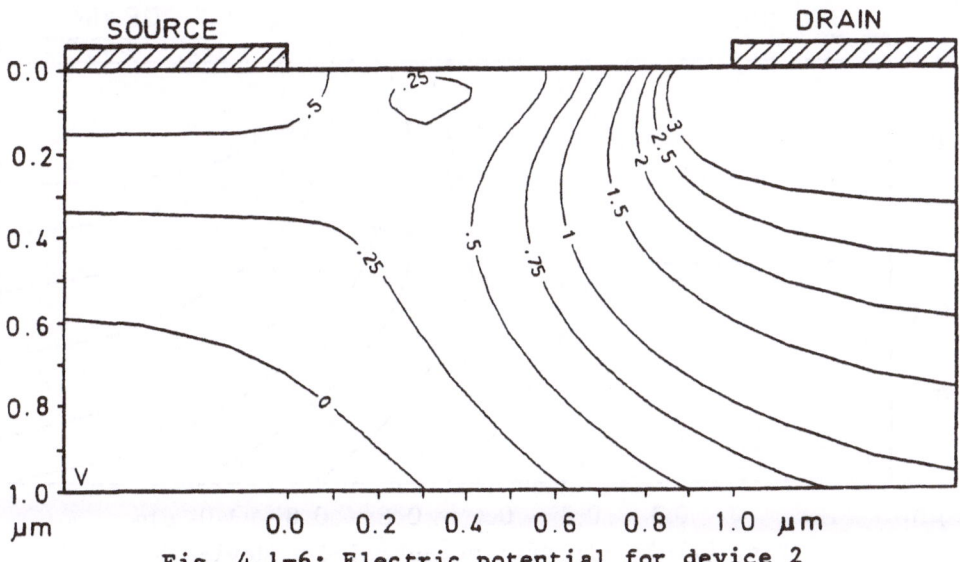

Fig. 4.1-6: Electric potential for device 2

Fig. 4.1-7 shows the potential distribution in the third device. The birds-eye-view plot differs just marginally from the plot in Fig. 4.1-6. But from the contour plot one can see a well pronounced barrier between source and channel which guarantees the "normally-off" behaviour.

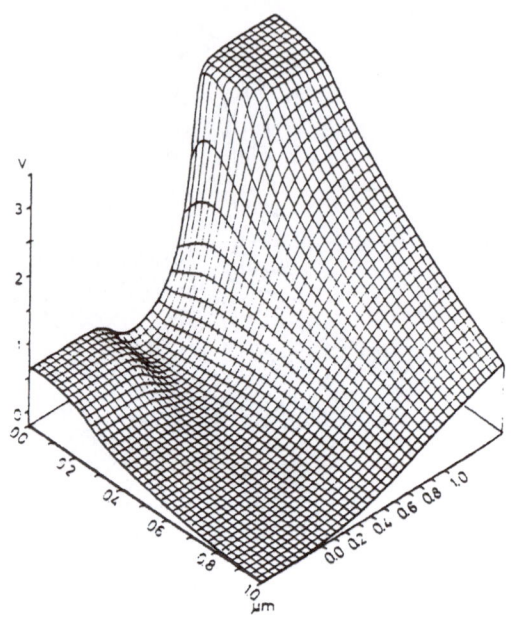

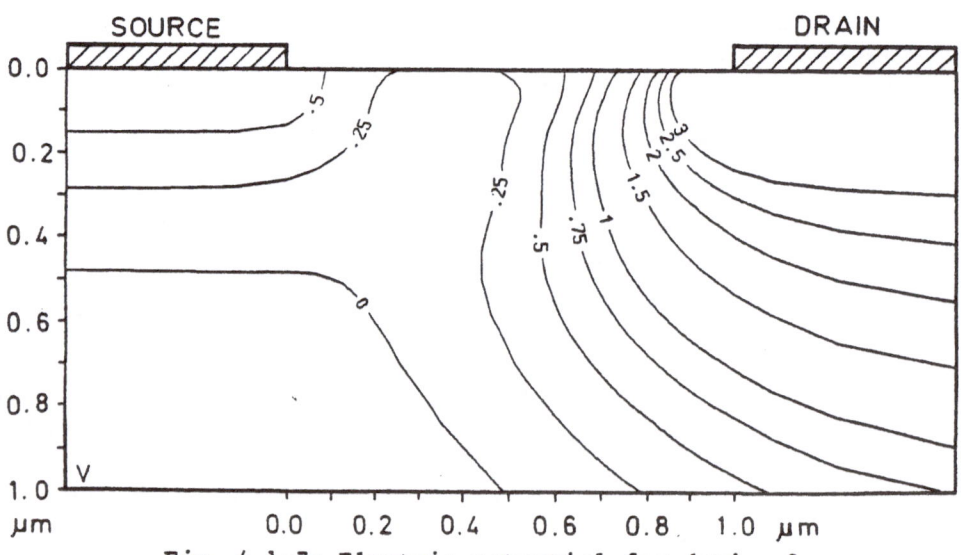

Fig. 4.1-7: Electric potential for device 3

Fig. 4.1-8 shows the lateral current density distribution in the first device. For better visibility, the plot on the right shows the mirror image to give better insight into the channel region. In the channel near the source the current is forced to flow at the surface by the transversal component of the field.

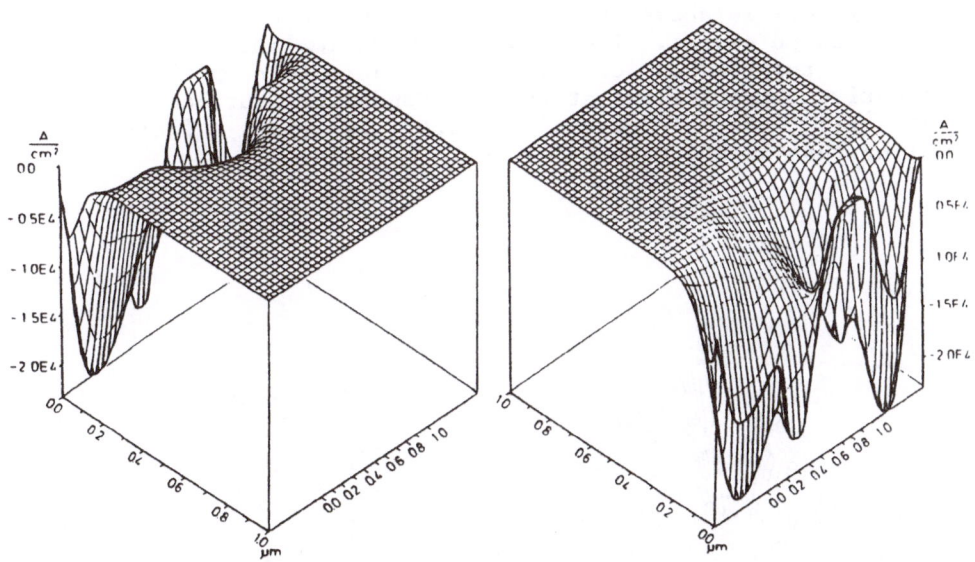

Fig. 4.1-8: Lateral current density for device 1

But already in the middle of the channel one can watch current spreading caused by the drain influence, a typical short channel effect.

It also should be noted that the current channel is fairly wide. The reason for this phenomenon is to be found in a superposition of an inversion channel and a punch through channel. The maximum of the lateral current density surprisingly lies below the contacts. This fact becomes clear when we consider current continuity. Current can only pass through the contact in transversal direction. Current flow in the semiconductor, however, takes place globally in the lateral direction from source to drain. As current flow is continuous, the lateral current component has to be large below the contacts, because the flux in the channel, which is relatively wide, as mentioned, is large too.

The lateral current distribution for the second device is shown in Fig. 4.1-9. As one can see, this device is operating in the punch through mode. The current flow takes place in a wide channel in the bulk. Surface current does effectively not exist. Furthermore, the maximum of the current density has decreased more than an order of magnitude compared to the first device.

Fig. 4.1-10 shows the lateral current density distribution for the third device. The second channel implantation results in a total suppression of punch through in this operating point. The entire current flows at the semiconductor surface, but the peak value of the current density is about a factor of 200 smaller than in the second device.

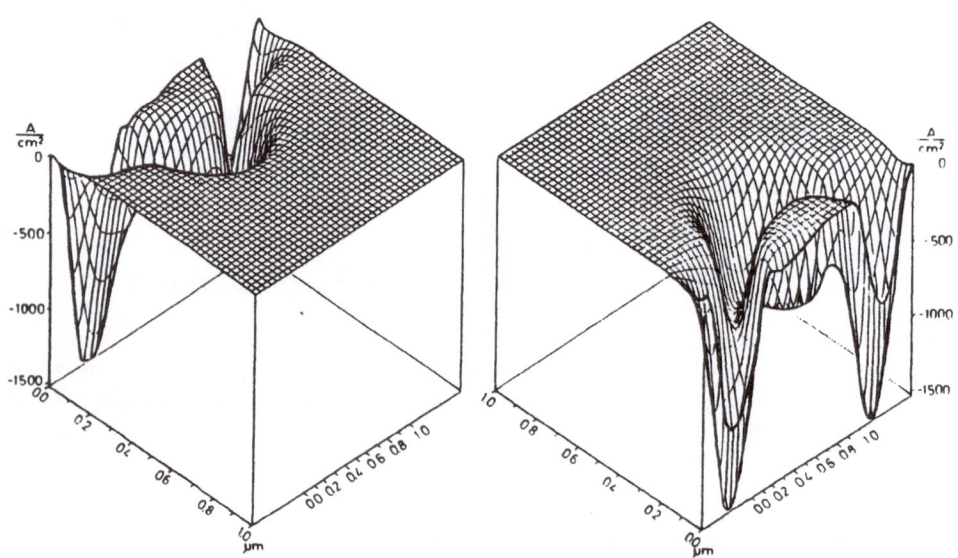

Fig. 4.1-9: Lateral current density for device 2

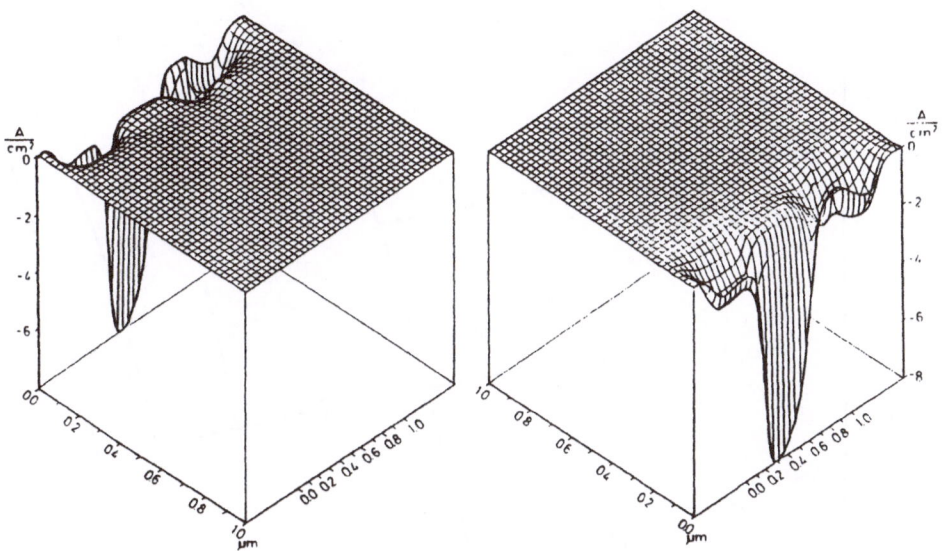

Fig. 4.1-10: Lateral current density for device 3

Current density distributions of this shape are typical for regularly operating transistors in subthreshold and can be used as criterion for valuation.

Fig. 4.1-11 shows the subthreshold characteristics for two different drain voltages. The solid lines denote 100mV, the dashed lines 3V drain bias. The slope is the same for all three devices at a drain voltage of 100mV. It is dramatically decreased at 3V drain bias for devices 1 and 2 by the punch through current. The shift of the characteristics for different drain voltages, which is caused by the short channel effect, is a minimum for the third transistor thus verifying the success of the channel implantation steps.

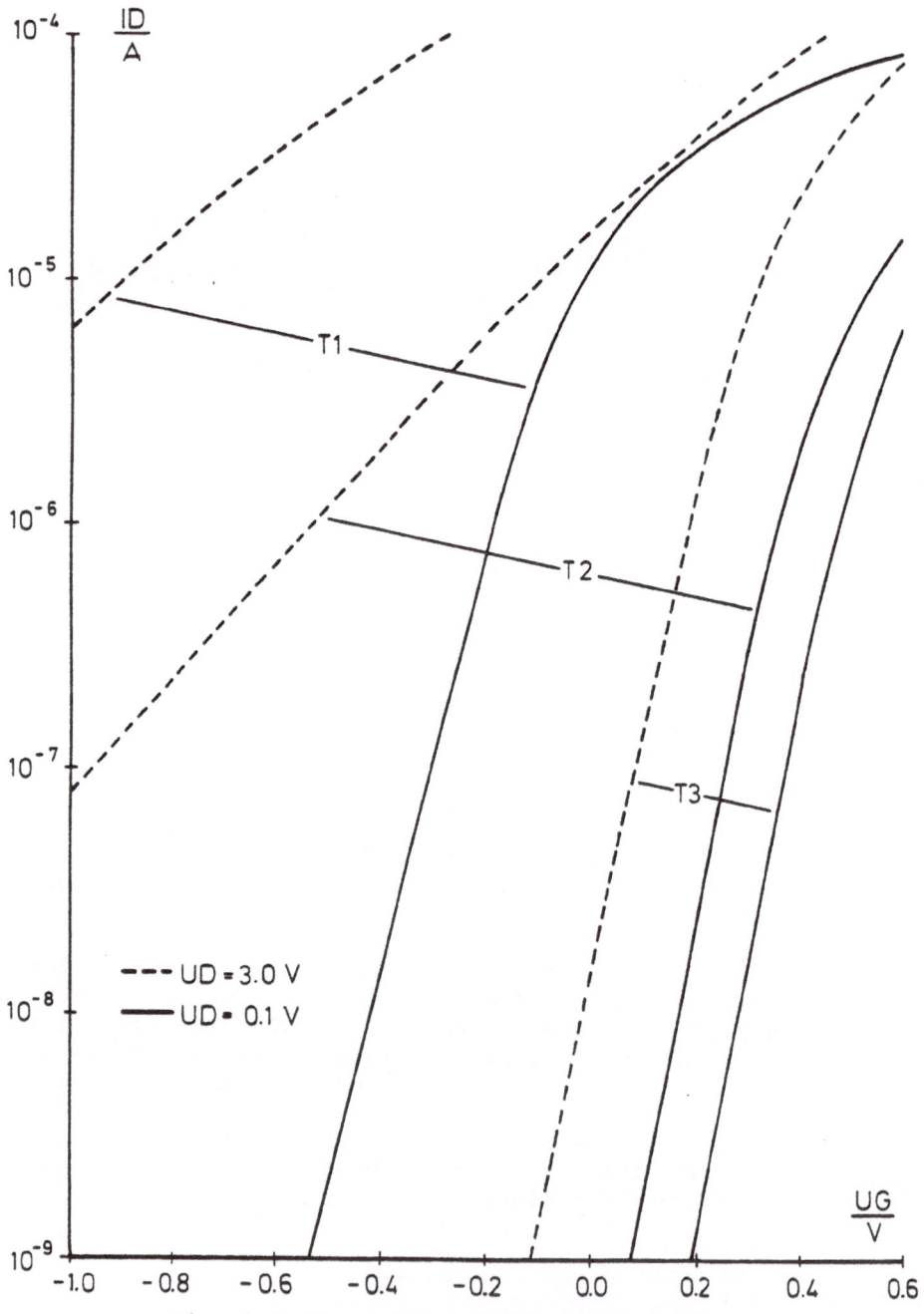

Fig. 4.1-11: Subthreshold characteristics

4.2 Process Sensitivity

VLSI is evidently connected to the miniaturization of the single transistor. Merely shrinking the physical device dimensions usually poses serious problems concerning device behaviour. Instead, all device parameters have to be scaled (e.g. /43/, /99/) together with the device geometry according to certain rules. In general, lower voltages, heavier doping, shallower junctions and thinner oxides help to maintain applicable device characteristics as channel length is reduced.

Down to about two microns channel length the device behaviour can be controlled excellently by the relevant technological steps (implantation, diffusion, oxidation, photolithography). However, as often observed in experimental investigations, this controlability is no longer ensured for devices with further reduced channel length. Reproducability tends to become worse with decreasing size, posing increasingly severe problems on tracking down the parameters of adjacent transistors, which should behave identically for certain kinds of circuits (e.g. latches).

To verify the increased process sensitivity of scaled devices, we have performed an analysis of certain device parameters[1] with MINIMOS /137/. In this chapter the sensitivity of the threshold voltage, which is usually the most important device property[2] for the designer, will be outlined for a well established short channel MOS process to determine the practical limit of miniaturization for a given technology. However, the analysis of threshold sensitivity is just an example for a strategy which is applicable to examine the sensitivity of any device property.

An n-channel silicon gate process with arsenic source/drain doping and a double channel implantation for threshold tailoring and punch-through suppression has been chosen. Fig. 4.2-1 shows the doping distribution logarithmically drawn in a quasi-three-dimensional plot for a one micron transistor. The channel implantation is performed with boron as the dopant, a dose of $3.10^{11} cm^{-2}$ and an energy of 35keV for the shallow layer and a dose of $10^{11} cm^{-2}$ and an energy of 160keV for the deeper layer.

---

1) <u>parameter</u> = a variable which one can choose arbitrarily, e.g. L, W, T_{ox} .
2) <u>property</u> = a physical attribute which is influenced by choice of parameters, e.g. UT, breakdown voltage, transconductance.

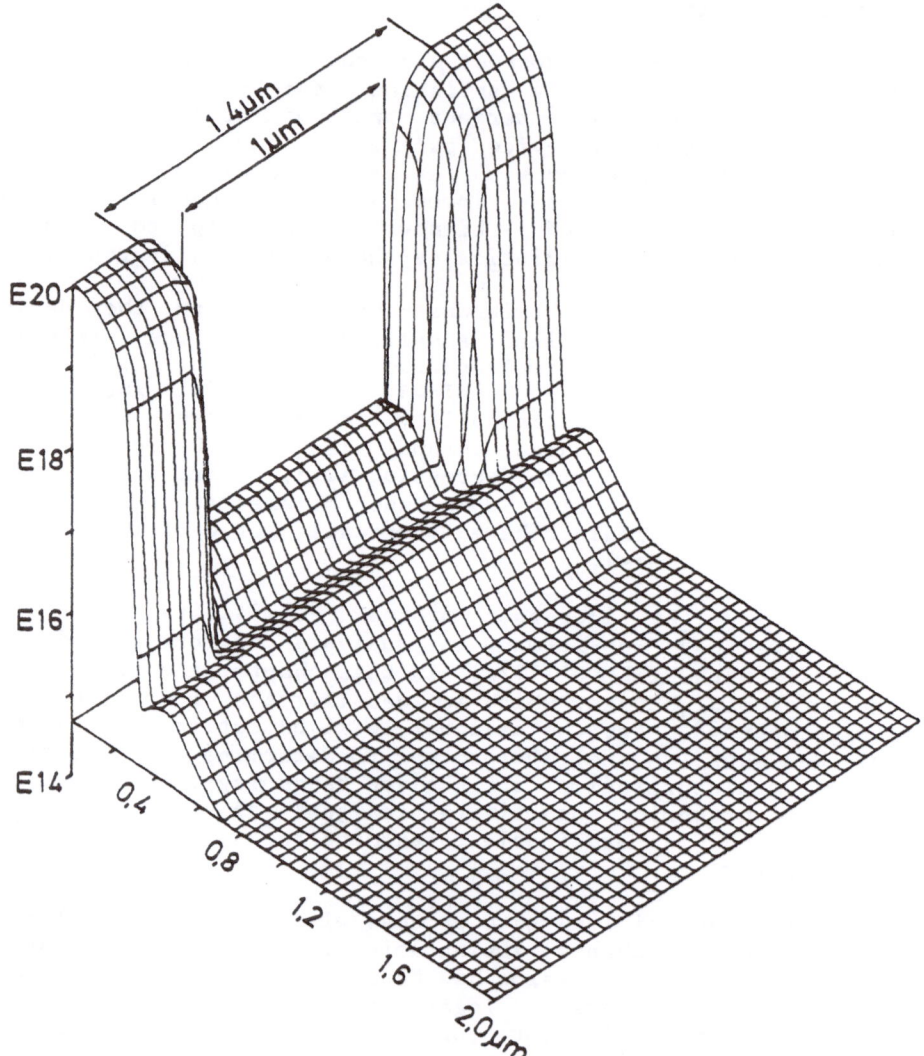

Fig. 4.2-1: Doping profile of the analyzed devices

A junction depth of 320nm and a lateral diffusion of about 200nm is obtained by this process. The extremely steep gradient at the junctions is typical for arsenic. The oxide thickness — the oxide is not drawn in these figures — is about 50nm for these devices. The whole process was designed for two micron lateral dimension.

For an analysis of the behaviour of the threshold voltage one first has to formulate an adequate definition of the threshold voltage. The most common definitions are based on the extrapolation of an output characteristic. However, one drawback of extrapolation methods lies in their inaccuracy and in the experimental effort. Mainly owing to these mentioned reasons we define threshold voltage in the following simple and definite way: It is that applied gate voltage, at which the device sinks 0.1 microamps times the channel width per channel length. The channel length is defined as the distance between the metallurgical junctions. With this definition it is ensured that no threshold voltage shift versus channel length for long devices occurs and we can, therefore, directly obtain a quantitative measure for the influence of the short channel effects. It is probably necessary at this point to mention that drain bias and bulk bias are not explicit parameters in our definition of the threshold voltage. The dependence on those parameters has to be obtained by certain characteristics, namely: threshold voltage versus drain bias, threshold voltage versus bulk bias. Our definition is naturally arbitrary - as arbitrary as any definition - so one might have to argue about the quantitative value of the used 'constant (0.1µA). For devices with a steep subthreshold characteristic, and only such devices are of practical relevance, we think that the constant we use is quite suitable. For devices with a degraded subthreshold characteristic any definition of a subthreshold characteristic becomes meaningless.

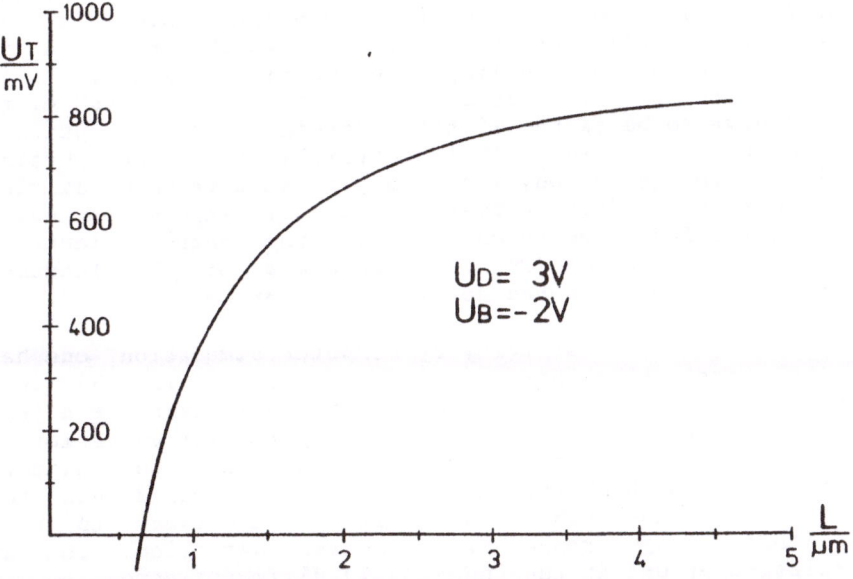

Fig. 4.2-2:   Threshold voltage versus channel length

Fig. 4.2-2 shows the threshold voltage versus channel length for our devices. An operating point of 3V drain bias and -2V bulk bias has been chosen for the comparison of different channel lengths. To avoid confusion, all the following figures will also refer to this operating point. Fig. 4.2-2 reflects the well known decrease of the threshold voltage with shrinking device length, which becomes dramatic at a length of below one micron.

Usually in papers on short channel MOS transistors a comparison between theoretical curves and selected experimental results is given. Some of them report on statistical measurements (e.g. /42/), but only one paper /184/, to our knowledge, deals explicitly with the sensitivity of an electrical property, namely the threshold voltage. However, with respect to the inherent dependence of most properties on the dispersion of geometry and technology, it seems to be a real necessity to analyse and present these dependences directly. Therefore we carried out numerical investigations to extract the most important sensitivities. A two dimensional simulation program like MINIMOS is excellently suited for numerical investigation of the sensitivity of device properties to dispersion of design and process parameters. First, the physical model parameters of the computer program have to be matched to those corresponding to measured characteristics, i.e. the program has to be "calibrated". This procedure has to be done with non critical transistors with relatively long channels because the measured characteristics should deviate only minimally with inaccuracies in geometry and in technology. This "calibration" procedure should certainly be done for every technology which is to be analyzed numerically as the formulae which are used in a simulation program for modeling the physical parameters (e.g. mobility) are partly heuristic. A few "constants" of those formulae have to be fitted if total agreement of simulation and measurement is desired. It is certainly absurd, and physically invalid, to change the physical model parameters when simulating transistors with just different channel lenghts (for example) because all effects due to changes in the channel length are principally included in the structure of the fundamental semiconductor equations and not in their parameters.

To obtain a sensitivity by computer simulation, one has to vary the interesting parameter (e.g. channel length) in the vicinity of its nominal value and then differentiate after the results (e.g. threshold voltage). This parameter variation must certainly be done within a small range because the validity of linearization on which the whole strategy is based has to be ensured. On the other hand, it is necessary to have a sufficiently large range of parameter variation to avoid cancellation errors at the (numerical) differentiation.

This parameter variation within a small range cannot, in general, be performed experimentally. A minute change of a process parameter which is reproducable piles up tremendous fabrication problems or inherent costs. However, with a fast modeling program the partial derivative of any electrical property with respect to any technological or geometrical parameter can be calculated easily with the outlined strategy. Thus numerical investigations are ideally suited for the performance of sensitivity analysis.

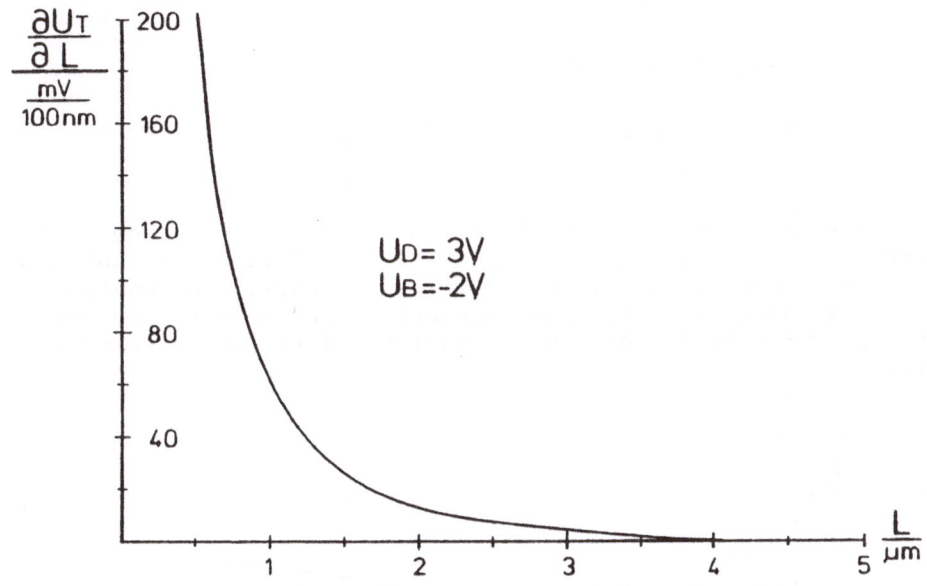

Fig. 4.2-3: Sensitivity on channel length tolerances

Fig. 4.2-3 shows the partial derivative of the threshold voltage with regard to the channel length versus channel length for our devices; that is, the sensitivity of the threshold voltage on tolerances of the channel length. Assume a transistor with an effective channel length of one micron accurate to ten percent. With this figure one can read an uncertainty of the threshold voltage of +/-60mV.

Fig. 4.2-4 shows the sensitivity of the threshold voltage to the deviation of the oxide thickness. As one probably has not expected at first glance this sensitivity decreases for devices with short channels. This is due to the decreasing influence of the bulk charge with shrinking channel lengths. Note that this figure is qualitatively very similar to the figure showing the threshold voltage versus channel length (Fig. 4.2-2). This fact can be understood analytically by recalling the simple formula for the threshold voltage:

$$UT \doteq \Phi_{MS} + 2 \cdot \Phi_F - (Q_{fs} + Q_b)/C_{ox}$$

(without short-channel effect)

$$C_{ox} = \mathcal{E}_{ox}/T_{ox}$$

$$\partial UT/\partial T_{ox} \doteq -(Q_{fs} + Q_b)/\mathcal{E}_{ox}$$

$$UT \doteq (\partial UT/\partial T_{ox}) \cdot T_{ox} + \text{const.}$$

With an uncertainty of 5% of the oxide thickness, one has an uncertainty of about +/-40mV for a 5 micron device and not even half this value for a 1 micron device. However, one should not revel in this fact. The decrease of the sensitivity results from the decrease of the controllability of the transistor by the gate.

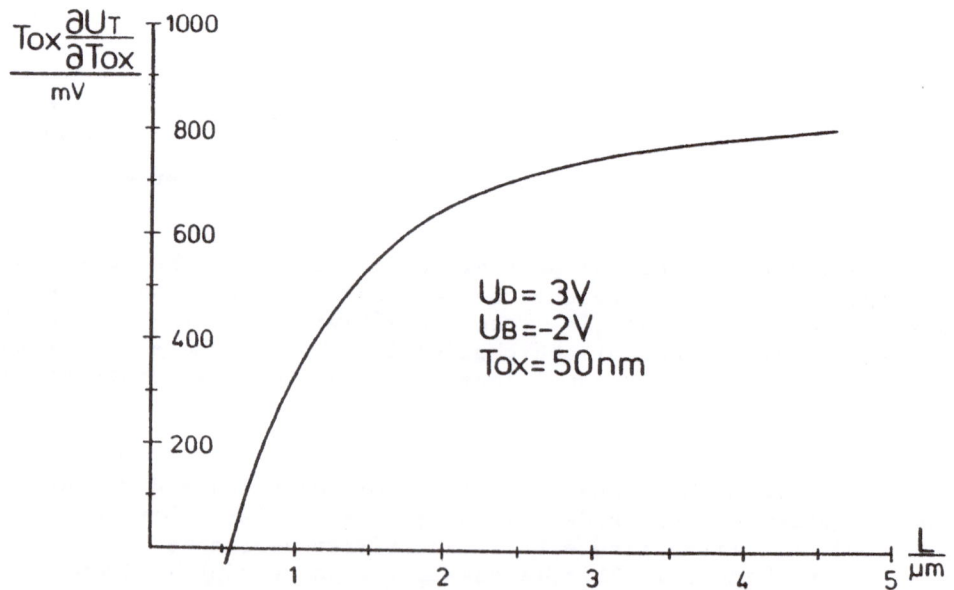

Fig. 4.2-4: Sensitivity on oxide thickness tolerances

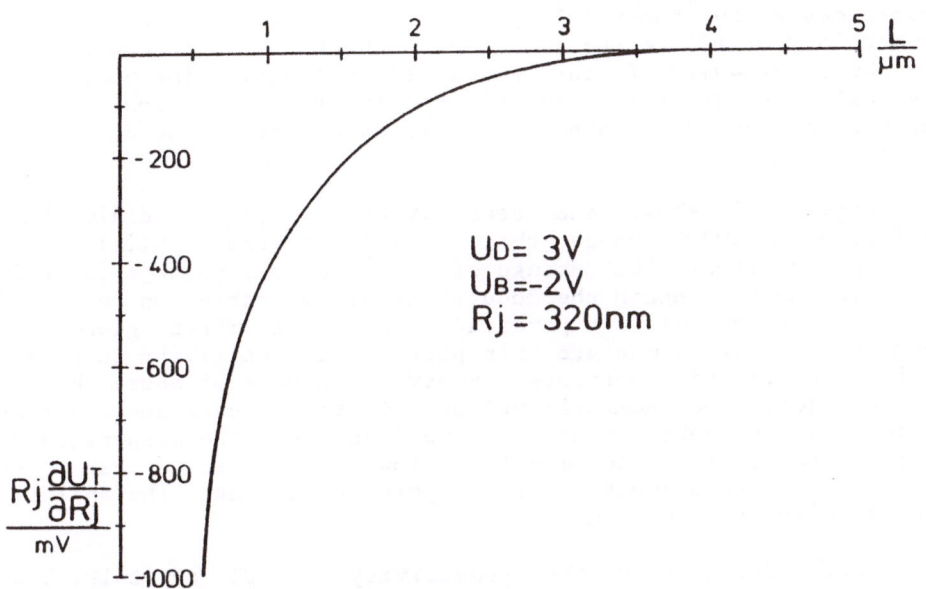

Fig. 4.2-5: Sensitivity on junction depth tolerances

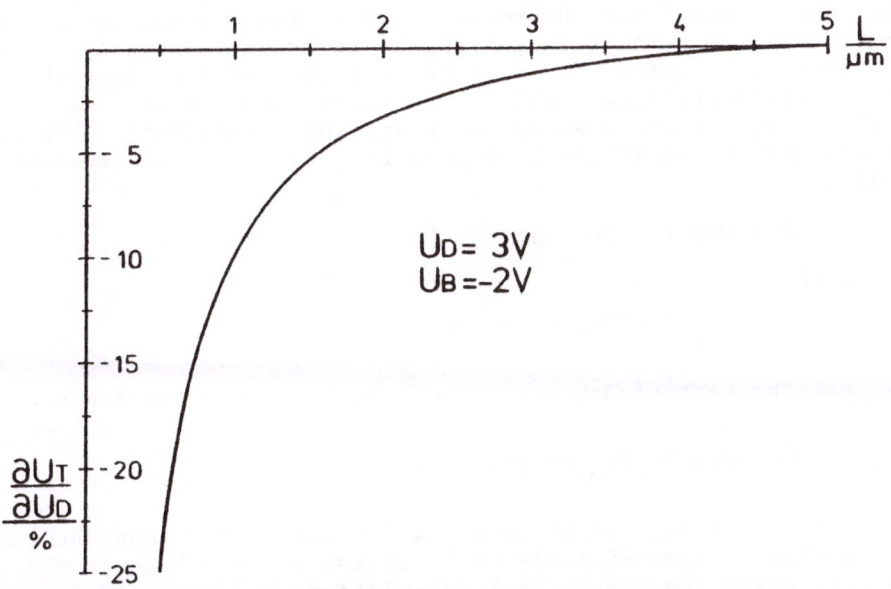

Fig. 4.2-6: Sensitivity on drain bias variation

Fig. 4.2-5 shows the sensitivity of UT on junction depth tolerances versus channel length. A one micron device with an uncertainty of 10% in the junction depth, thus has an uncertainty of about -/+40mV of the threshold voltage. The underlying physical cause of this sensitivity is the reduction of the channel charge by the depletion regions of source and drain (cf. /168/).

Fig. 4.2-6 shows the sensitivity of UT on drain bias variation. A 300mV change, that is 10% of the applied bias, results in about 30mV change of the threshold voltage for this operating point. Again the modulation of the depletion region of the drain is the relevant physical effect. At first glance it seems to be easy to measure this particular sensitivity in even a minimally equipped laboratory. However, in case of short channel devices just the nominal values of the process and geometry parameters are known for an individual device. The dispersion of these parameters would merely allow to extract bars by statistical measurements which again will make the analysis expensive and time consuming.

Fig. 4.2-7 shows the sensitivity of UT on bulk bias variation. A 200mV change, that is again 10%, results in a threshold shift of about 11mV, which is usually not dramatic. (For the practical problem, however, one has to deal with a sum of all uncertainties. Therefore this influence may also become important.) An interesting detail of this figure is the fact that the sensitivity decreases first with shrinking channel length and at a certain length begins to increase rapidly. This behaviour is caused by a superposition of the short channel effect, which decreases this particular sensitivity, and the punch-through effect, which increases the sensitivity. For long channel devices it is fairly simple to estimate this sensitivity analytically:

$$\partial UT/\partial UB \doteq -1/C_{ox} \cdot \partial Q_b/\partial UB$$

with:

$$Q_b \doteq q \cdot (N_b \cdot y_c + Dose)$$

For the partial derivative only y_c has to be considered:

$$\partial UT/\partial UB \doteq -(T_{ox}/y_c) \cdot (\varepsilon_{si}/\varepsilon_{ox})$$

With a value of about two micrometer for y_c one obtains a sensitivity of approximately -7.5 percent which is confirmed by the more exact two-dimensional calculations.

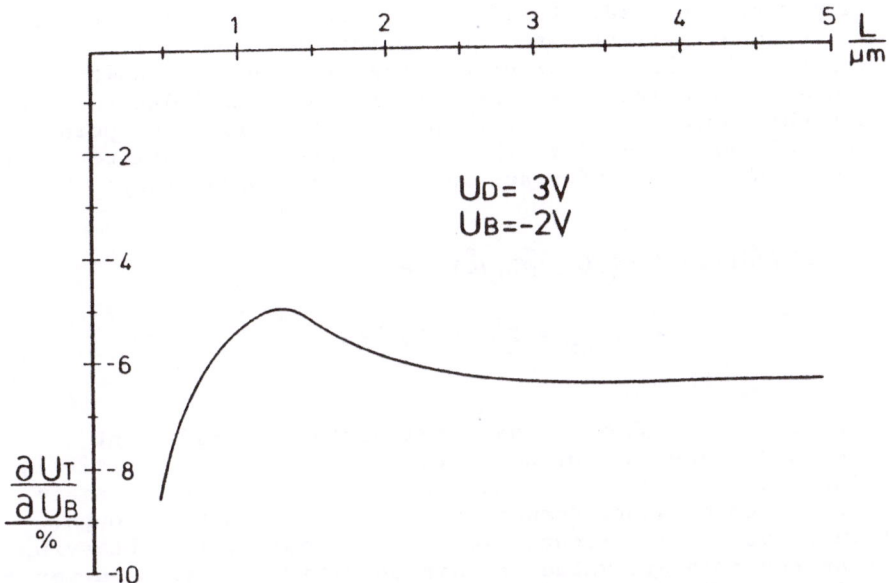

Fig. 4.2-7:   Sensitivity on bulk bias variation

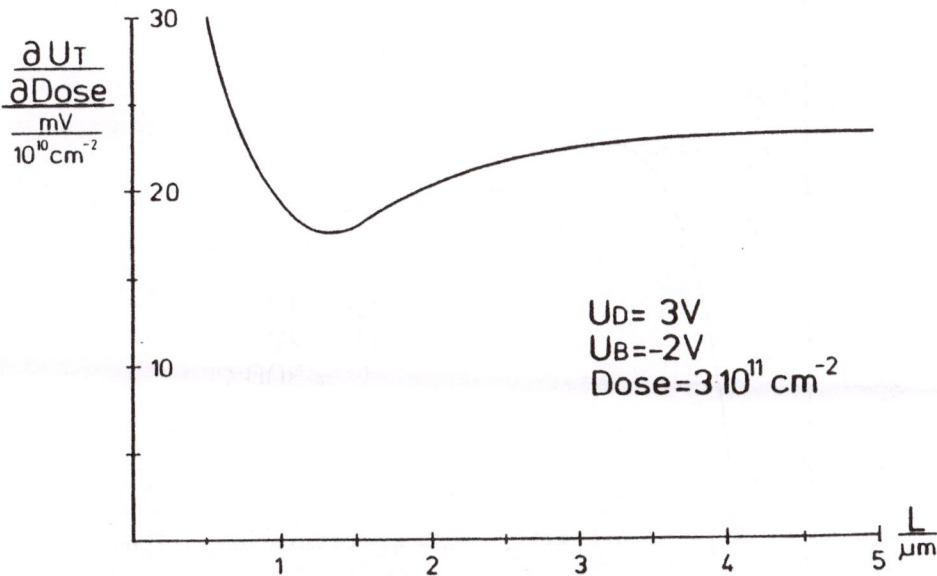

Fig. 4.2-8:   Sensitivity on implantation dose tolerances

Fig. 4.2-8 shows the sensitivity of UT on uncertainties of the implantation dose. Qualitatively the superposition of the short channel effect and the punch-through effect is again apparent. The absolute value of this particular sensitivity is low due to the fact that the depletion region below the channel covers the whole implanted region at this operating point. An analytical estimate for the long channel transistor can be obtained in a straightforward way for this sensitivity:

$$\partial UT / \partial Dose \doteq -1/C_{ox} \cdot \partial Q_b / \partial Dose$$

$$= q/C_{ox} = 23 \ mV/10^{10} cm^{-2}$$

Fig. 4.2-9 shows the temperature coefficient of the threshold voltage for our devices. We have, qualitatively, a similar behaviour to that already discussed, namely the superposition of short channel effect and punch through. The absolute value is around -1mV/K. The qualitative behaviour as well as the absolute value of this sensitivity have been verified by fairly complicated experiments /159/.

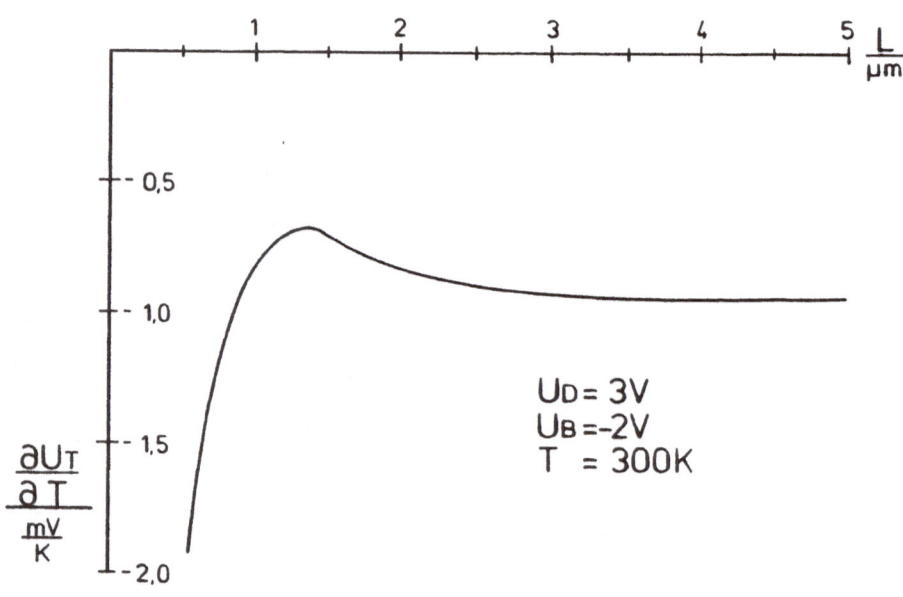

Fig. 4.2-9: Sensitivity on temperature variation

The partial derivatives denote isolated sensitivities on a certain set of parameters. These values show which parameters are the most critical ones. However, in addition, a global sensitivity number indicating the cumulative effect of the isolated sensitivities is useful. The global sensitivity is related to a certain technology and its expected application. It should indicate the limit of channel length reduction. To obtain such a global number typical ranges of deviation of design parameters have to be specified. The table in Fig. 4.2-10 is an example for such a specification. In this example a rather small value of the absolute uncertainty of the channel length (100nm) has been chosen. For long devices this value is unrealistic, but in consideration of a one micrometer technology 100nm absolute uncertainty represents 10 percent relative dispersion, which is relatively large. The tolerances of the remaining parameters in Fig. 4.2-10 , however, represent a good laboratory standard.

Parameter	X	DX	%
L		100nm	
TOX	50nm	2.5nm	5
RJ	320nm	32nm	10
UD	3V	150mV	5
UB	−2V	100mV	5
AKEV	35keV	0.7keV	2
DOSE	$3.10^{11}cm^{-2}$	$6.10^{9}cm^{-2}$	2

Fig. 4.2-10: Desired process and operating tolerances

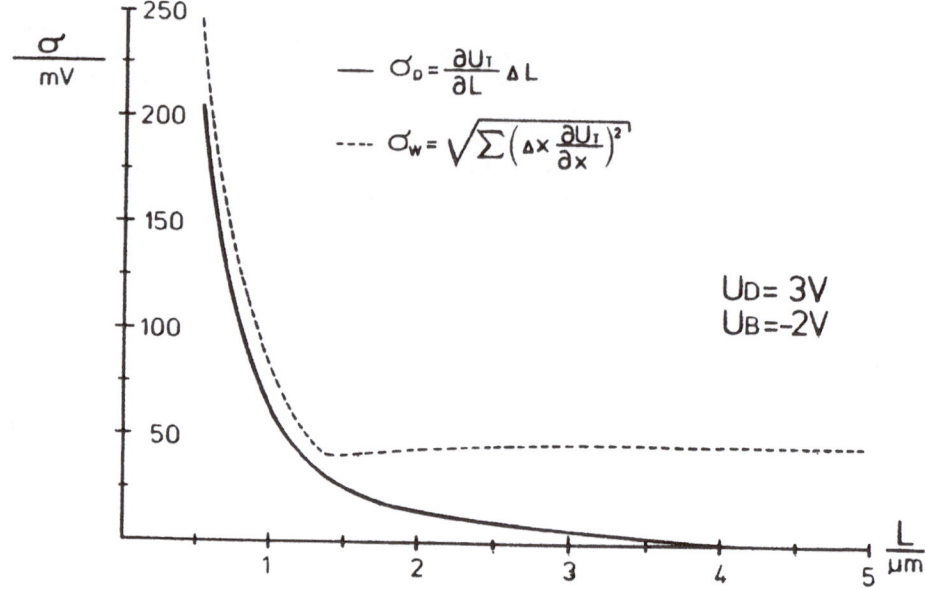

Fig. 4.2-11: Reproducability of the analyzed devices

Fig. 4.2-11 shows the global threshold voltage sensitivity based on the specifications of Fig. 4.2-10. σ_D denotes the uncertainty of the threshold voltage for identical devices on the same chip. D stands for device. This sensitivity is given by just the length influence, as the other parameters are commonly very homogeneous across one chip. σ_W, W stands for wafer, denotes the uncertainty for identical devices on wafers, which have been fabricated with different charges. Here one has to use a Euclidian norm over all deviations. Note that this value is highly constant down to a certain channel length, but then increases dramatically. The channel length at which the excellently pronounced knee is located, 1.4 microns for our devices, can, therefore, be interpreted as the practical limit of channel length reduction due to threshold uncertainty. Nevertheless should it be noted that the data in Fig. 4.2-10 are to be understood as an example which is mainly of importance for our technology.

4.3 Breakdown Phenomena

To increase the number of functional units per chip, it is necessary to decrease the size of the devices (e.g. channel length and channel width for MOS transistors). As the performance of a device depends strongly on its geometry, any reduction requires certain design rules (c.f. /43/ for MOS devices). However, in recent years devices have been miniaturized without reduction of supply voltage, mainly to stay compatible with existing circuits and to maintain an acceptable signal to noise ratio. Unfortunately problems with punch through and avalanche breakdown arise from the reduction of size. Punch through can be controlled relatively well by technological steps as already outlined in the last chapter for an MOS transistor. But there exists an increasing demand for a transparent description of the physical processes which lead to avalanche breakdown.

Avalanche problems have so far been treated /86/, /165/ in the following manner: First Poisson's equation is solved to obtain a solution for the electrical potential distribution and then the ionization integral is evaluated by integrating the strongly field dependent ionization coefficients over the high field region. As result multiplication factors are obtained which describe the increase of current due to avalanche. Since the carrier densities need not be calculated, this method seems to be very efficient in calculating breakdown voltages. However, any feedback of the increase in carrier densities on the electrical field is, therefore, neglected. A more serious treatment requires the solution of both carrier continuity equations with proper modeling of the generation term /133/, /135/.

In this section calculations for a one micron gate length n-channel MOS transistor are presented. The lateral subdiffusion and the junction depth of the source and drain regions are 0.2 and 0.3 μm, respectively. A deep channel implantation with fairly high dose was supposed to have been performed to suppress punch-through.

Fig. 4.3-1 shows calculated drain and bulk currents versus drain voltage for that transistor. For U_{GS}=1V breakdown is reached at U_{DS}=5.6V whereas 8.4 Volts are necessary to lead the device into breakdown if no gate voltage is applied. On first glance that seems to be paradox, if one considers that U_{GS}=0V certainly causes larger peak values of the electric field. The explanation of this phenomenon lies in the low current level.

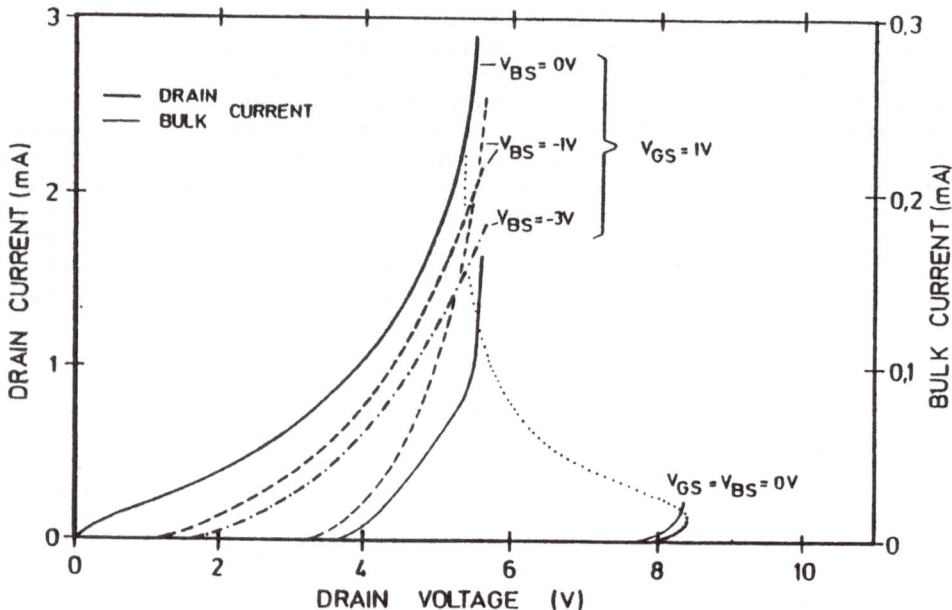

Fig. 4.3-1 Drain and bulk current characteristics

Although the probability of ionization is larger for U_{GS}=0V than for U_{GS}=1V, the generation rate still remains small as there is little current flow causing ionization. With increasing drain voltage the drain current and consequently avalanche generation as well as hole density increase. This additional space charge even lowers the potential barrier between source and bulk. Now an internal feedback mechanism exists which acts as follows: Because of the lower potential barrier the electron current injected by source, and consequently, the avalanche generation increase. Thus the hole density rises even more and, in turn, further lowers the potential barrier. Once the feedback gain becomes unity, the node currents rise unlimited unless controlled by external resistors in the current paths. Furthermore, owing to the higher current level, the situation now becomes more and more similar to the situation at larger gate voltages. The I-V characteristic, therefore, has to move towards the U_{GS}=1V characteristic and the drain voltage decreases with increasing drain current. This effect implies negative resistance and is usually called "snap-back". The voltage drop of the hole current at the parasitic resistor of the deep bulk also lowers the potential barrier and thus enhances the feedback gain.

Applying a negative bulk voltage renders breakdown more difficult although it increases the bulk current level. The reason for this lies in the hole density which is decreased by applying a more negative bulk bias which attracts the holes.

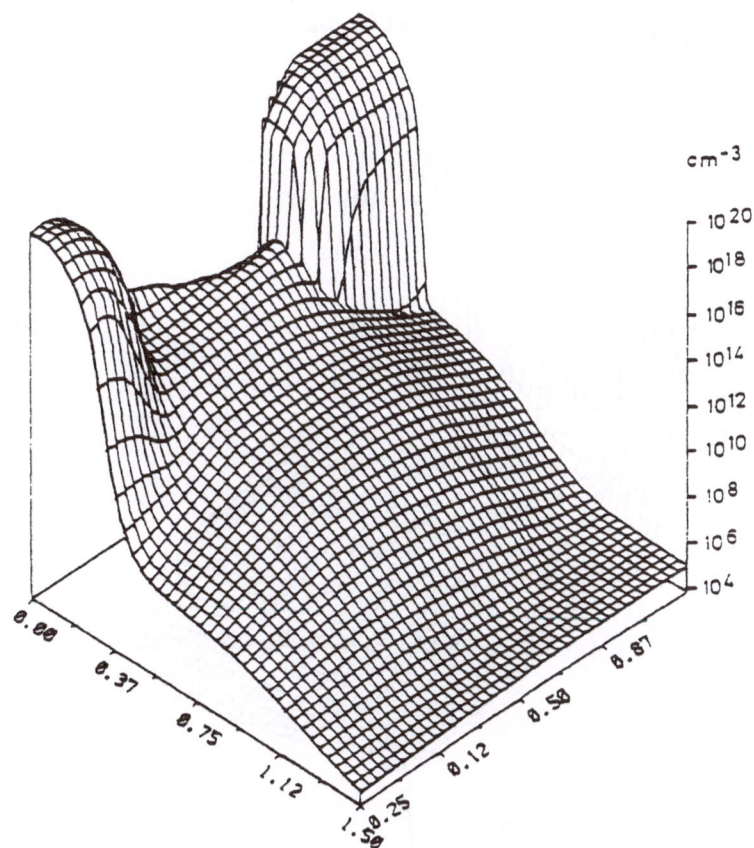

Fig. 4.3-2 Concentration of electrons (U_{GS}=0V, U_{DS}=8V)

There exists an additional feedback mechanism apart from the one just mentioned: The carriers generated by ionization cause again ionization. This effect leads to an "avalanche-like" increase of both carrier densities, and determines the breakdown voltage of a p-n junction. The feedback depends on the ionization ability of both carrier types and is of little

552

importance in our case. For MOS transistors the mechanism described above is much stronger and it is active with even vanishing ionization ability of holes.

In the following we should like to discuss internal physical quantities at U_{GS}=0V, U_{DS}=8V, and U_{GS}=2V U_{DS}=5.6V. These operating points have been chosen to explain clearly the physical phenomena which eventually lead to the snap back effect. The computed drain currents are about 20µA and 15mA, respectively. Since the U_{GS}=2V characteristic was out of locus bounds, it is not drawn in Fig. 4.3-1.

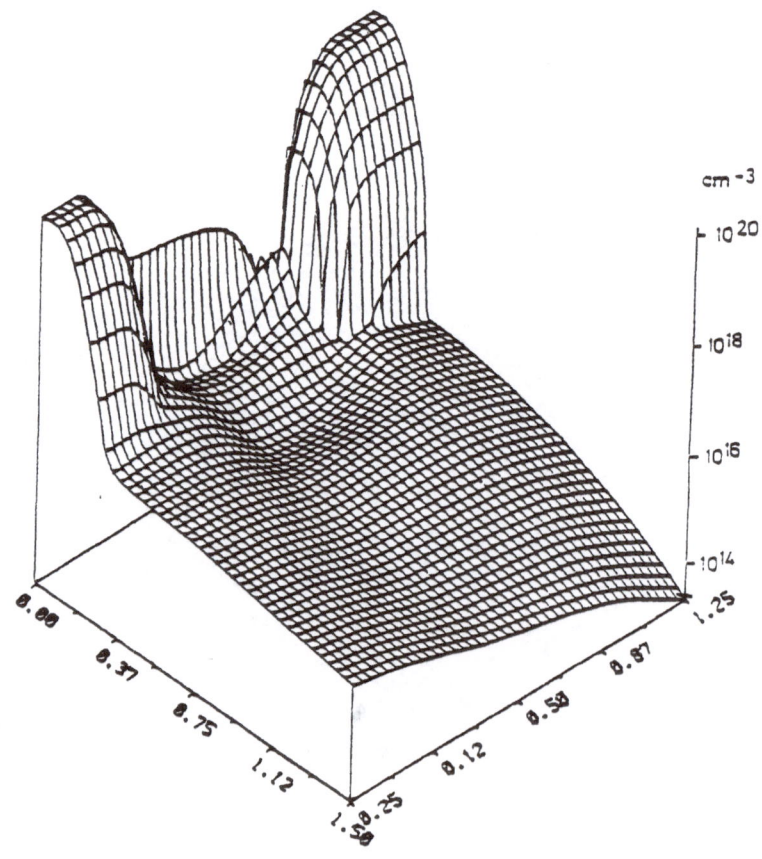

Fig. 4.3-3 Concentration of electrons (U_{GS}=2V, U_{DS}=5.6V)

Fig. 4.3-2 and Fig. 4.3-3 show the electron distribution for both operating points in a logarithmic scale. At the first operating point, Fig. 4.3-2, the transistor is turned off; there is no inversion layer between the source and drain regions which can be found, as expected, in Fig. 4.3-3 at the second operating point. It should be noted that in Fig. 4.3-3 the electron density does not drop below the intrinsic number in contrast to Fig. 4.3-2. The reason for this is source barrier lowering brought about by the increased hole density.

The corresponding hole densities are given in Fig. 4.3-4 and Fig. 4.3-5, respectively. One should bear in mind that all the holes outside the undisturbed bulk region are generated by impact ionization. In agreement with the electron densities the hole density is also much larger for U_{GS}=2V. The large hole density near the source partially compensates the acceptor doping. Thus the potential barrier at the source is lowered and high electron injection from the source region ensues.

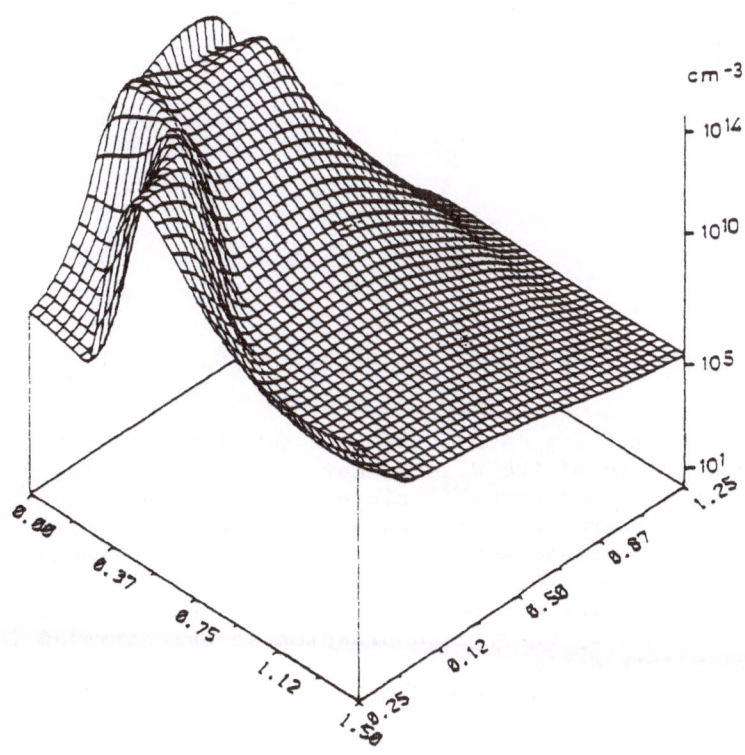

Fig. 4.3-4 Concentration of holes (U_{GS}=0V, U_{DS}=8V)

554

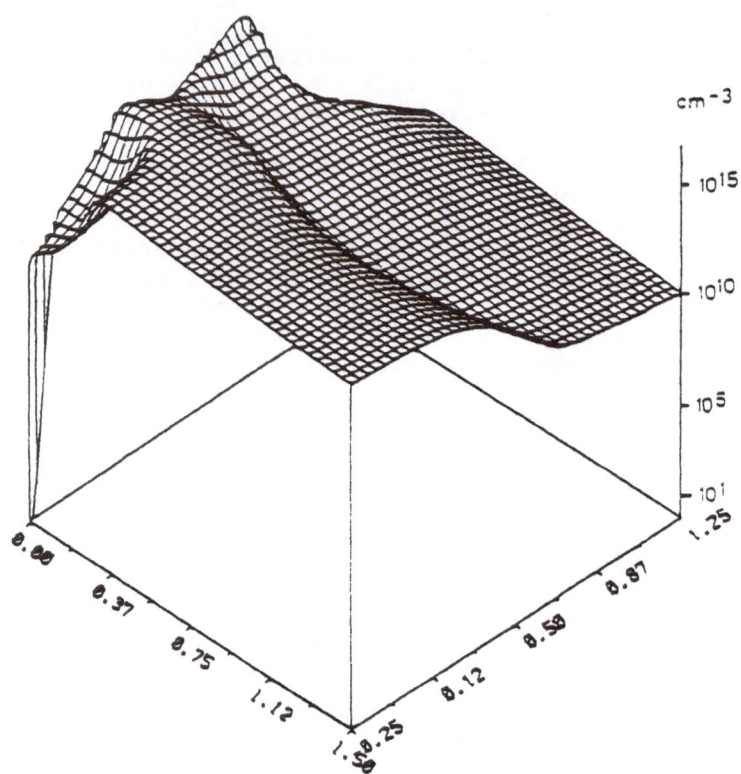

Fig. 4.3-5 Concentration of holes (U_{GS}=2V, U_{DS}=5.6V)

Looking at Fig. 4.3-1 again, we find that the negative resistance branch of the U_{GS}=0V characteristic for large current levels leads into a vertical slope, i.e. the decrease of U_{DS} is stopped. The corresponding drain voltage is called "sustain voltage"; it increases weakly with increasing U_{GS} because a large gate bias smoothes the electric field distribution thus lowering its peak value. The existence of a nearly unique sustain voltage can be explained by heavy recombination as demonstrated in /134/. A good many interesting results concerning avalanche breakdown phenomena have been published in /33/, /57/, /86/, /114/, /133/.

4.4 A Simple n-MOS Inverter with Depletion Load

Simple inverters are eminently important for the design of an integrated logic circuit. They consist of a driver transistor and a load which is usually another transistor as these devices can be fabricated more easily than resistors. The driver transistor should be normally off in order to turn it off easily. With the input voltage logically high, this device should turn on and should exhibit high conductivity between source and drain; the voltage drop across this transistor will, therefore, be small and the output voltage will be very low. The load transistor has to combine two features: when the driver transistor is turned on, nearly the total voltage should drop across the load and the current must be moderate. On the other hand, when the driver transistor is turned off, it should be able to supply the output without a significant voltage drop. These features can be achieved by a depletion (normally on) transistor with a smaller channel width than the driver transistor.

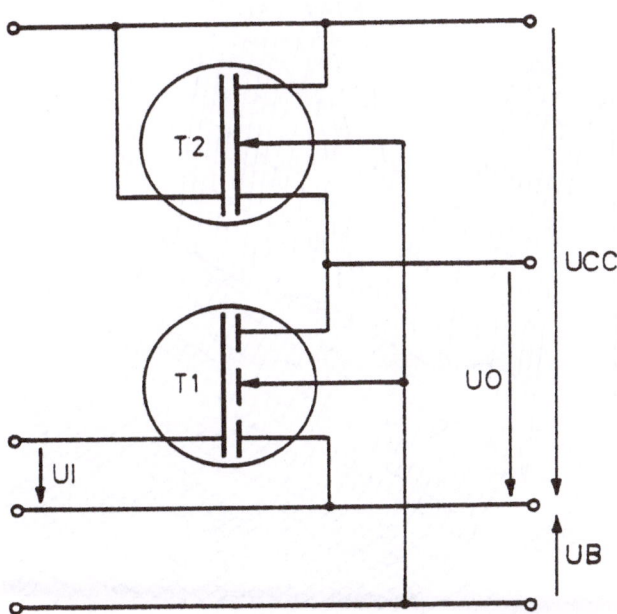

Fig. 4.4-1: A simple n-MOS inverter

Fig. 4.4-1 shows a typical n-MOS inverter with depletion load. In the following we shall analyze a depletion transistor which can be used as depletion load device together with device 3 of chapter 4.1.

556

To realize a depletion transistor, donor impurities have to be implanted into the channel. With this technique a conducting channel between source and drain is created. However, if the dose and depth of the implantation are moderate, the device can still be controlled via the gate voltage. In this way the threshold voltage can be shifted towards a negative value.

Fig. 4.4-2 shows the doping profile of the load device indicating the shallow implantation at the surface of the channel. The implantation has been carried out with Antimony, a dose of $10^{12} cm^{-2}$, and an energy of 180 keV. The other device data are identical with the third transistor of chapter 4.1. The shallow Boron implantation, however, has been omitted.

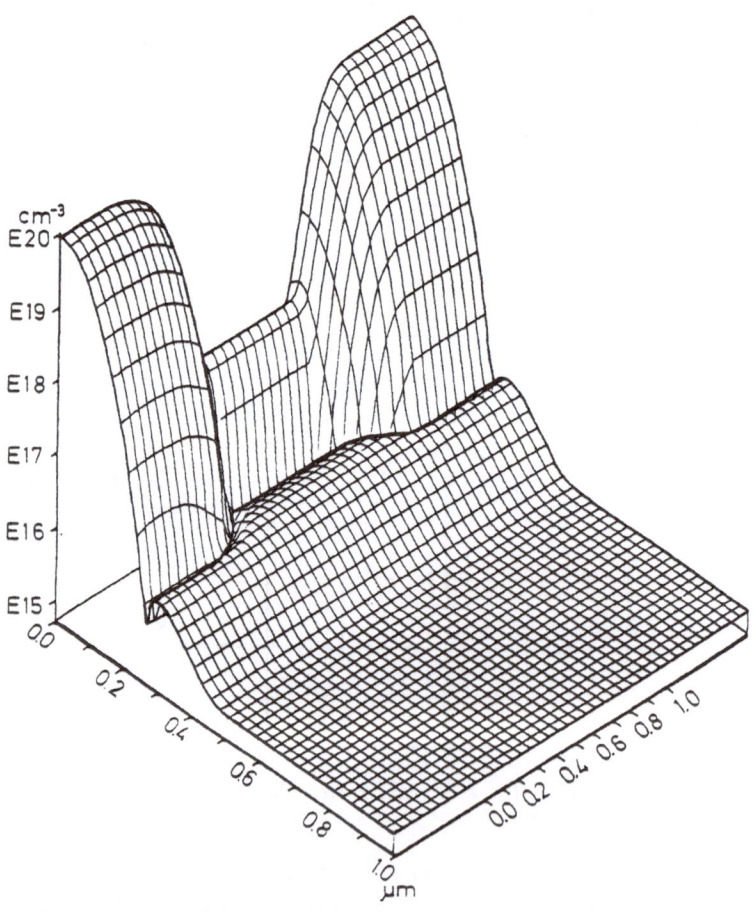

Fig. 4.4-2: Doping profile of depletion transistor

To get a more principal understanding of depletion—mode devices, some internal distributions will be discussed for the same operating point as has been chosen in chapter 4.1 (UD=3V, UG=0V, UB=0V).

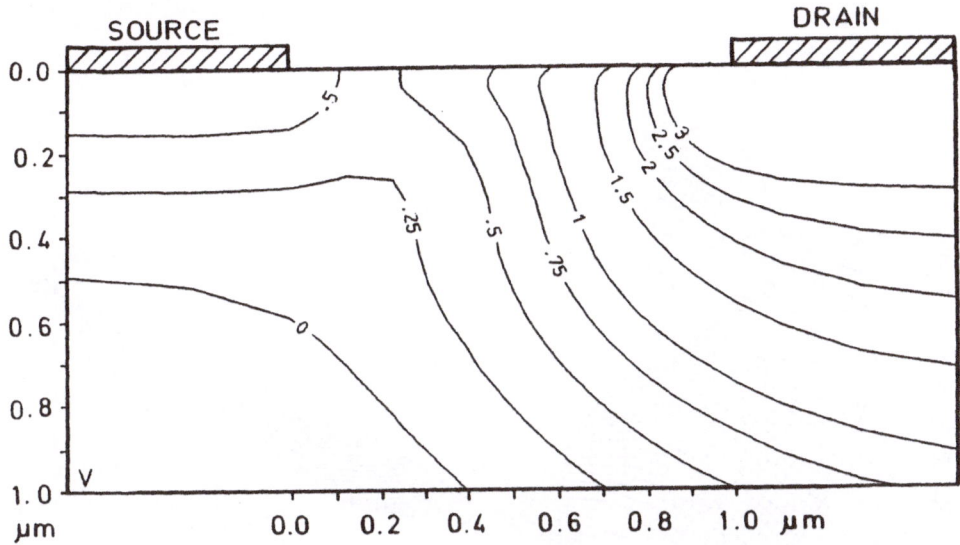

Fig. 4.4-3: Potential distribution in depletion device

The potential distribution is given in Fig. 4.4-3. No effective barrier between source and channel can be seen in contrast to the pictures of chapter 4.1. A source-channel diode does not exist and the built-in potential at the n⁺n junction remains small.

Fig. 4.4-4 shows the electron distribution in the load device. An electron channel with its maximum electron density at a depth of about 100nm can be seen. The onset of pinch-off is due to the fact that donor channel implantation was low enough to be easily depleted. Similarly, the channel could be depleted by a negative gate voltage.

In Fig. 4.4-5 the distribution of the lateral current density is presented. The channel is rather wide because of the donor implantation profile; that can be also deduced from the electron density. Thus the load device is able to conduct much more current at the same peak current density than an enhancement device. It is also to note that no indications of punch-through exist.

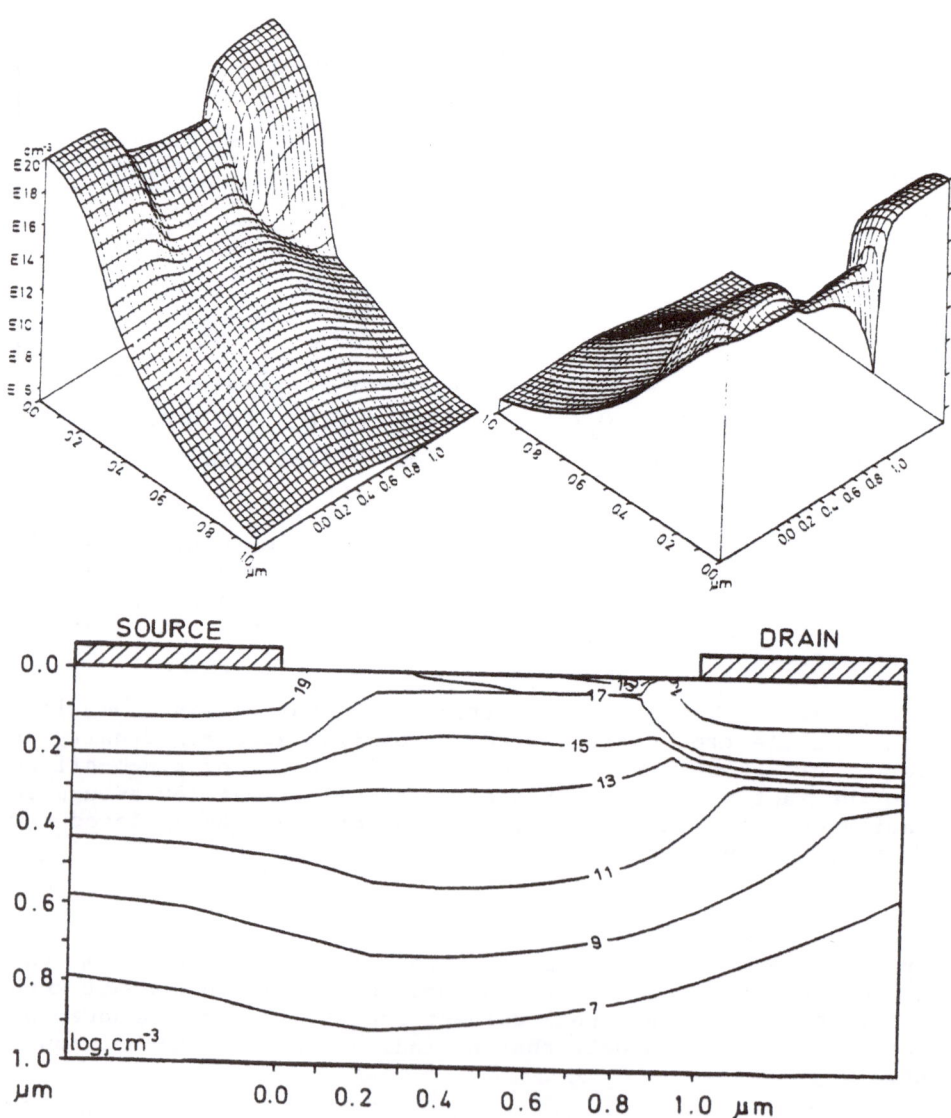

Fig. 4.4-4: Electron distribution in depletion device

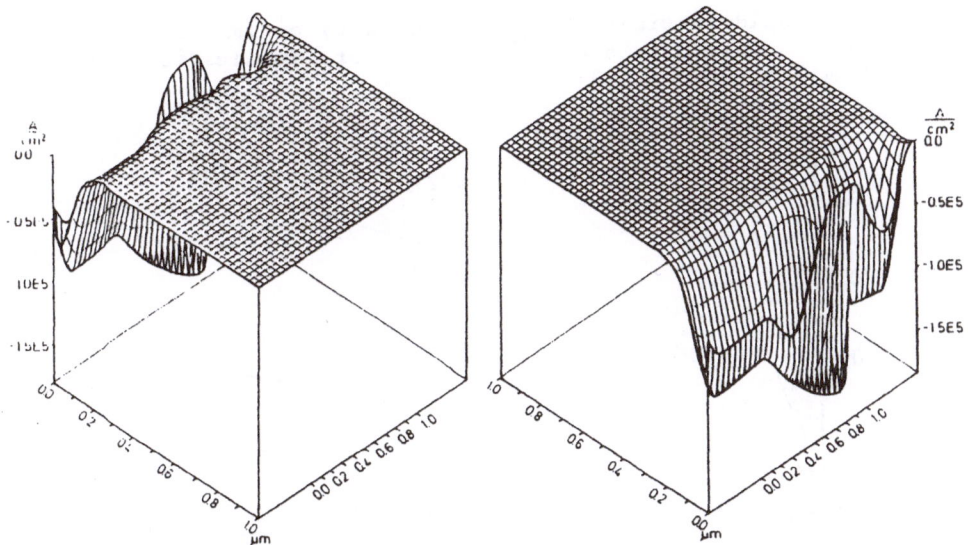

Fig. 4.4-5: Lateral current density in depletion device

The transfer function of an n-MOS inverter consisting of two transistors – as just discussed – has been calculated. The channel widths for the enhancement and depletion transistors have been chosen to be 20 μm and 2.5 μm, respectively to get a proper ß ratio /72/. Substrate voltage has been set to -2 V to increase the threshold voltage of the enhancement transistor and to enhance the signal to noise ratio. Fig. 4.4-6 shows the output characteristic of the driver transistor in solid and the load characteristic in dashed lines. As drain and gate voltages are identical for the load transistor and the threshold voltage is negative, this device always operates in the triode region. In the given circuit the load characteristic is linear which is due to the compensation of substrate and drain voltage influence on threshold voltage.

The transfer characteristic is given in Fig. 4.4-7. The low-level is 0.2 Volt, the high level is 3 Volt and is identical with the supply voltage because the load transistor does not produce a voltage drop without current flow. The noise margin at the low-level is very low (0.2 Volt) which is a well known problem with miniaturized logic circuits. The reason for this unfortunate phenomenon is to find in the low threshold voltage of the driver transistor. The noise margin at the high-level is 1.36 Volts and is certainly sufficient. Voltage amplification is -2.5 and should be acceptable. However, the performance of the inverter could definitely be improved by optimizing the device parameters. This offers a broad field of application for the present model.

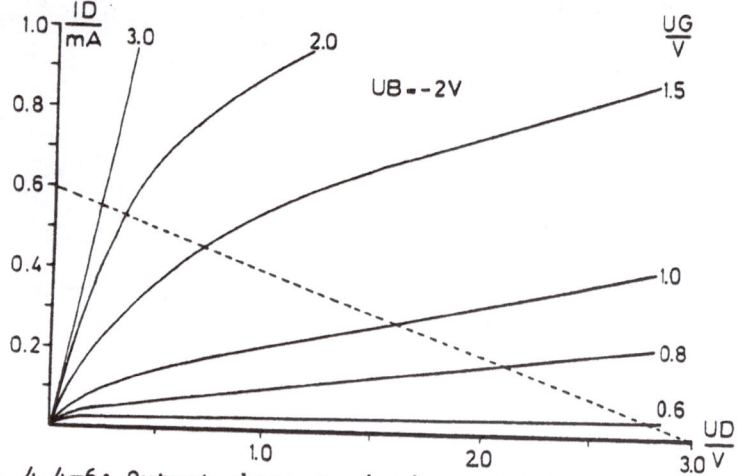

Fig. 4.4-6: Output characteristics of drive and load device

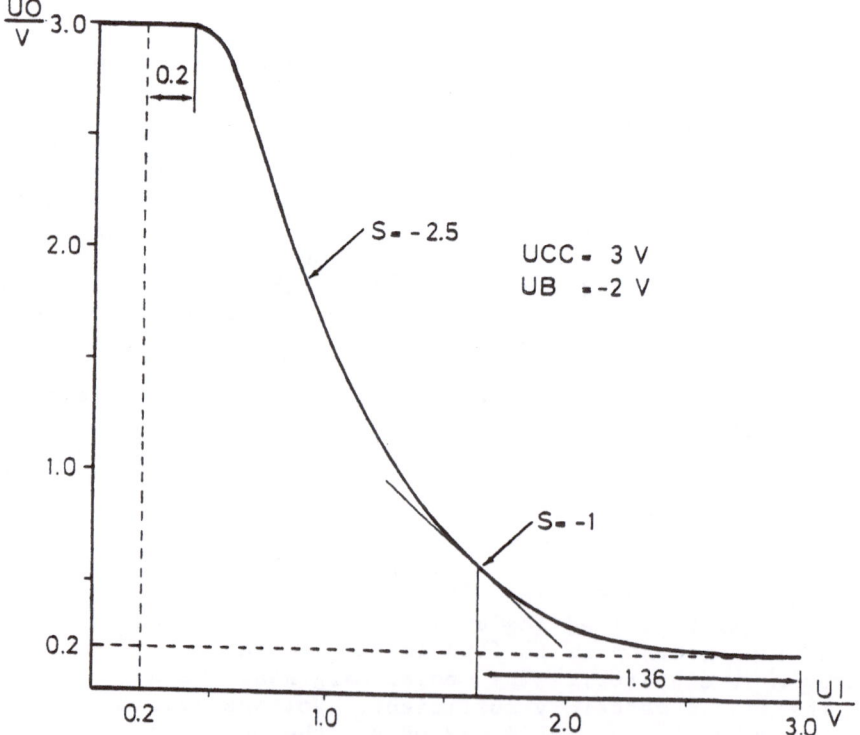

Fig. 4.4-7: Transfer characteristic of n-MOS inverter

4.5 A CMOS Inverter

In this chapter we shall discuss a simulation of a CMOS inverter which is the basic cell of any integrated circuit in CMOS technology. Of primary interest in this technology are power consumption, propagation delay and impact ionization, the latter mainly because of the disadvantageous "Latch-Up" which is induced by substrate current. The device data for the inverter which will be discussed here are summarized in Fig. 4.5-1.

	n-channel	p-channel	
r_j	0.36	0.36	μm
L	2.6	2.6	μm
L_{eff}	2.2	2.2	μm
W	20	20	μm
t_{ox}	30	30	nm
Gate	n-poly Si	p-poly Si	
Doping	p-well $2.4 \cdot 10^{15}$	substrate $6 \cdot 10^{14}$	cm^{-3}
		channel implantation	
Dose	$2 \cdot 10^{12}$ (B)	$1.3 \cdot 10^{12}$ (As)	cm^{-3}
Energy	40	180	keV

Fig. 4.5-1: Process data of CMOS devices

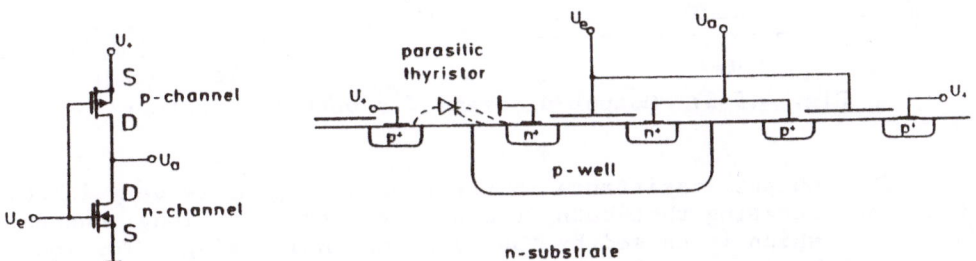

Fig. 4.5-2: The basic diagram of the inverter

The basic diagram of the CMOS Inverter and its principal technological realisation are given in Fig. 4.5-2. Using the process data of Fig. 4.5-1, threshold voltages of about 1.2V (n-channel transistor) and -1.2V (p-channel transistor) are achieved. As these values are rather large, good signal-to-noise behavior can be expected in contrast to the inverter discussed in the previous chapter.

The output diagrams for both independent transistors have been calculated within the range $0 < U_{DS} < 5V$ for $U_{GS}=2$, 3, 4, and 5 Volts and are given Fig. 4.5-3. Although the threshold voltages of both transistors are equal (their absolute values), the current in the n-channel transistor is much larger than in the p-channel transistor (for given drain and gate bias) which has to be contributed to the lower mobility of holes. The difference in the current values is less pronounced for high drain voltages because the saturation velocities of both carrier types only differ slightly.

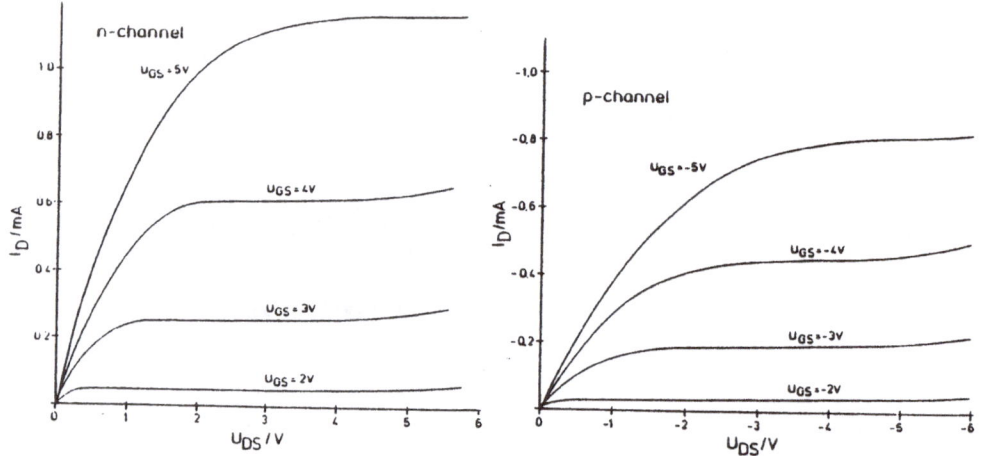

Fig. 4.5-3: Output diagrams for both transistors

The output resistance in the pentode regime is very large, thus demonstrating that both transistors exhibit long channel behavior which is caused by the large channel doping. The long-channel behavior can also be seen in the potential distribution in Fig. 4.5-4 for both transistors for a subthreshold case ($U_{GS}=\pm0.5V$, $U_{DS}=\pm5V$). The surface potential is constant along the channel up to the pinch-off region. This is a typical subthreshold behavior of long channel devices. The pinch-off region extends more towards source in the depth than at the surface as the doping level decreases with increasing distance from the surface. In the p-MOST the pinch-off region is also more extended than in the n-MOST which is due to the different doping levels.

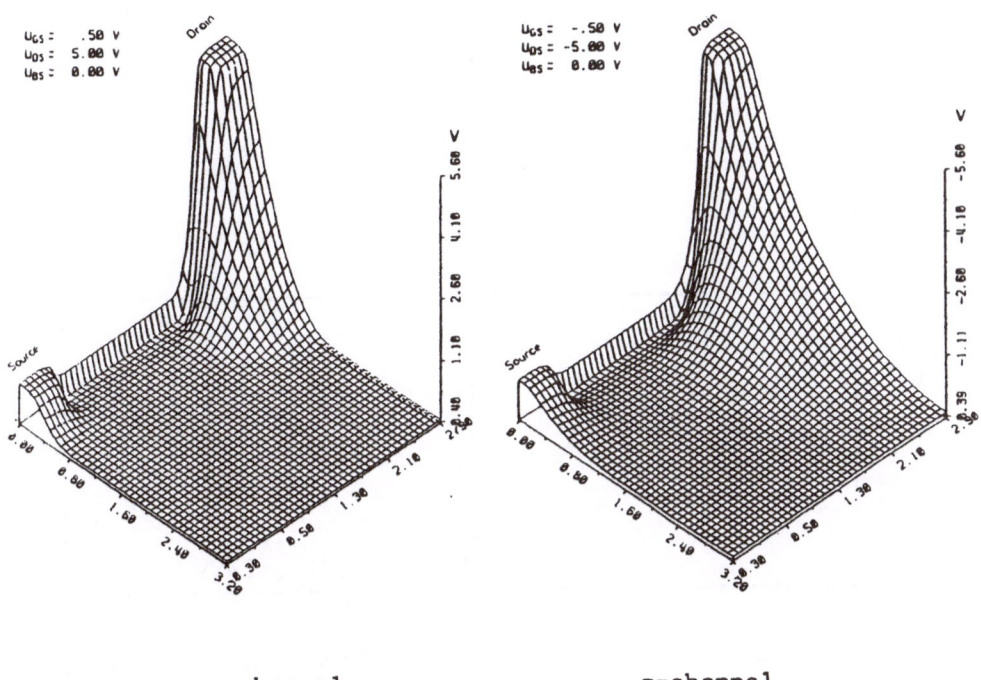

n-channel   p-channel

Fig. 4.5-4: Potential distribution

The transfer characteristic for the given transistor pair has been calculated and is shown in Fig. 4.5-5. Because of different carrier mobilities in both transistors this characteristic is shifted by about 90mV from the symmetry-line. As expected, noise immunity is very good. Specifying $U_a < 0.5V$ for the L-level and $U_a > 4.5V$ for the H-level it can be deduced from Fig. 4.5-5 that these levels are obtained at $U_e > 2.5V$ and $U_e < 2.22V$, respectively. Voltage amplification is very large in the active region ($v_u = -50$) because both transistors operate in the pentode regime with very large output resistance.

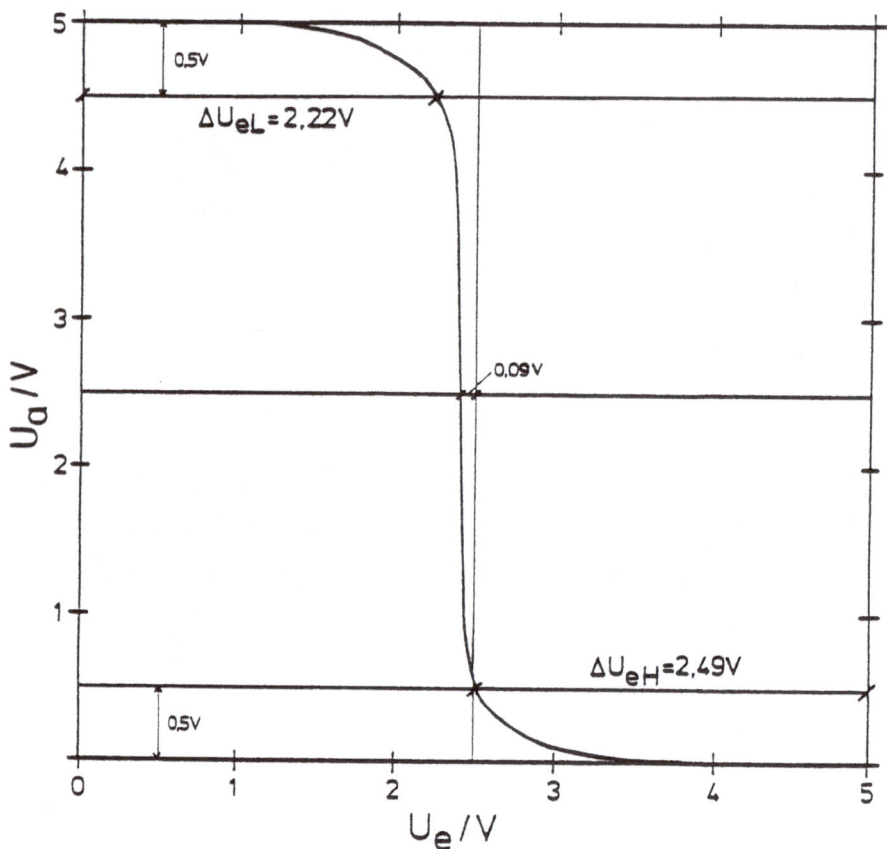

Fig. 4.5-5: Transfer characteristic

Fig. 4.5-6 shows the supply current as function of the input voltage. The very low stand-by current is the main power of the CMOS technology and has to be contributed to the excellent subthreshold behavior of the simulated transistors. However, the benefit of little power consumption can only be utilized in the static or low-frequency operation as parasitic capacitances have to be charged and discharged during any logic sweep. Another component contributing to increased power consumption in the dynamic operation is the current flux through the transistors during the change of the input voltage.

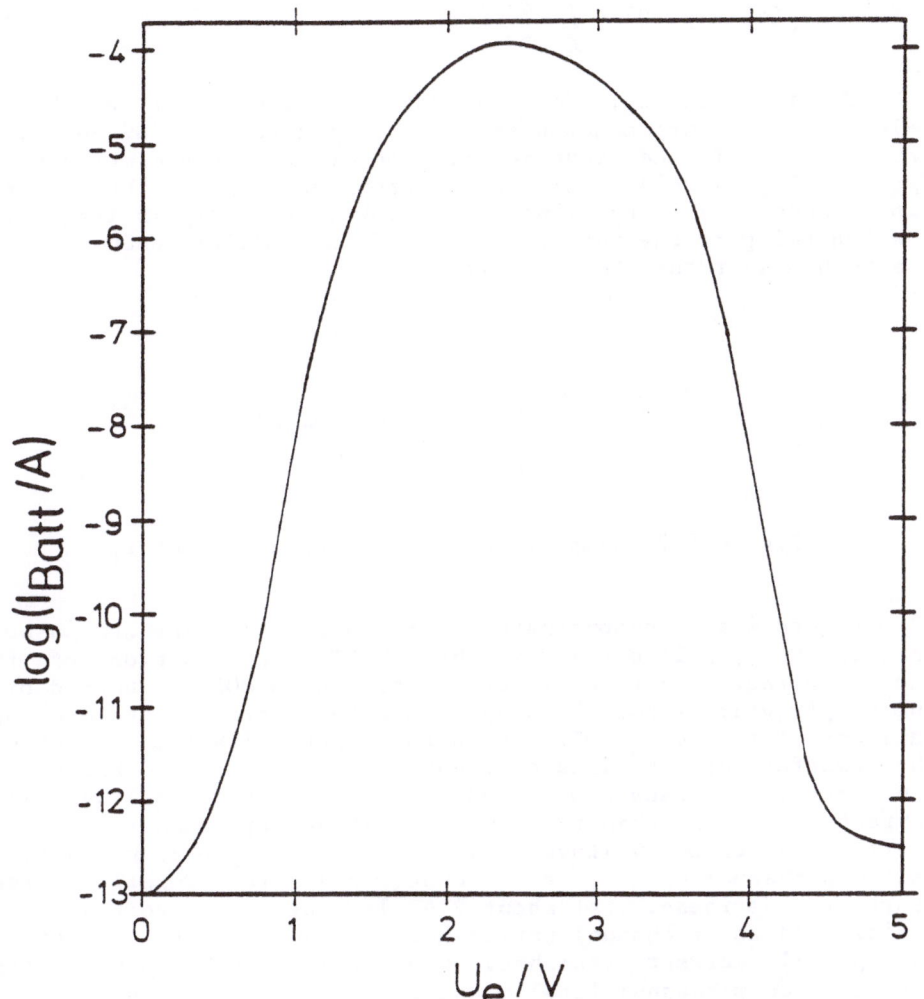

Fig. 4.5-6: Supply current versus input voltage

Now let us estimate the propagation delay for the chosen inverter. As our analysis is only static, no exact calculation can be presented here. However, the transistors are usually so fast that switching speed is determined mainly by parasitic capacitances rather than by the rise times of the node currents of the transistors. So let us neglect any transient behavior of the transistors themselves. If we apply an ideal L-H voltage jump on input, the p-channel transistor is turned off instantly and can, therefore, be ignored (cf. Fig. 4.5-7). The $u_a(t)$ characteristic is given by the integral equation

$$u_a(t) = u_a(0) - \int_0^t \frac{i_d(t)}{C_a} \, dt \qquad\qquad (4.5.1)$$

The fall time t_f, defined as the time after which the L-level at the output has been reached, can be found by numerical integration of the inverse of the output characteristic for $U_{GS}=5V$ (Fig. 4.5-5). t_f is proportional to the output capacitance and is $t_f=4.3$ns for $C_a=1$pF. The delay is larger for the L-H swing at the output because of the smaller conductance of the p-channel transistor ($t_r=7.6$ns).

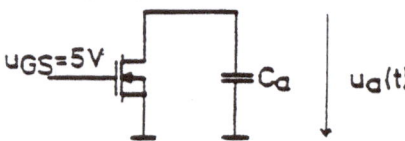

Fig. 4.5-7: Simplification for estimation of t_f

Figure 4.5-8 demonstrates the substrate currents in both transistors (p-well current in the n-MOST) as function of the input voltage. For low input voltage the n-MOST is turned off, the supply voltage totally drops along this transistor and U_{DS} vanishes for the p-MOST. When the input voltage is increased, the transfer current rises exponentially (subthreshold regime of the n-channel transistor) and the same applies to the p-well current. However, when the input voltage approaches about 2 Volts, the output voltage, which is the drain-to-source voltage of the n-channel device, and consequently the p-well current start to decrease. At about 2.4Volts the output voltage drops rapidly and the p-channel device now yields substrate current and the p-well current vanishes. Increasing the input voltage further, the p-channel finally comes into the subthreshold region ($U_e > 3.8V$) and the substrate current again decreases. The p-MOST exhibits larger leakage current than the n-MOST as its doping level is in general smaller. However, as dark space phenomena /123/ are not accounted for in the model, the substrate current level may be considerably smaller than computed.

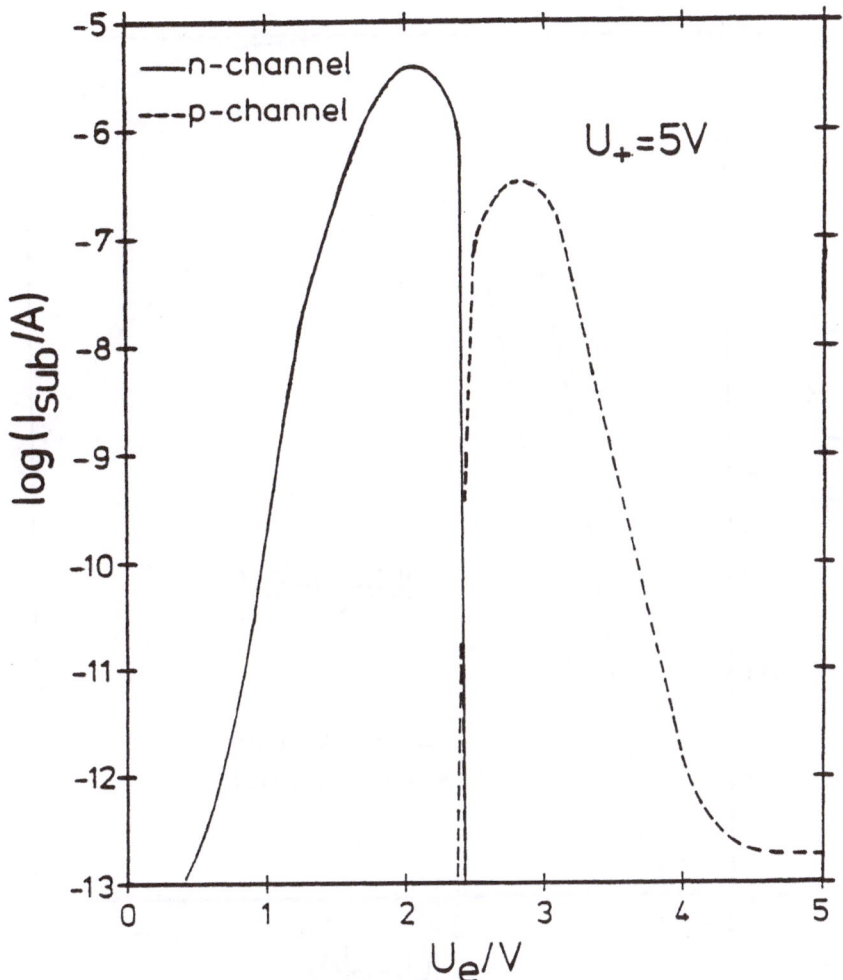

Fig. 4.5-8: Substrate currents

Fig. 4.5-9 shows substrate currents for a fixed drain voltage ($U_{DSn}=U_{SDp}=5V$). These operating points are not part of the transfer characteristic and cannot occur in the steady state case. However, they may occur during a logic sweep. As holes exhibit less ionization activity than electrons, substrate currents in the p-MOST are smaller by at least one decade compared to the substrate currents in the n-MOST.

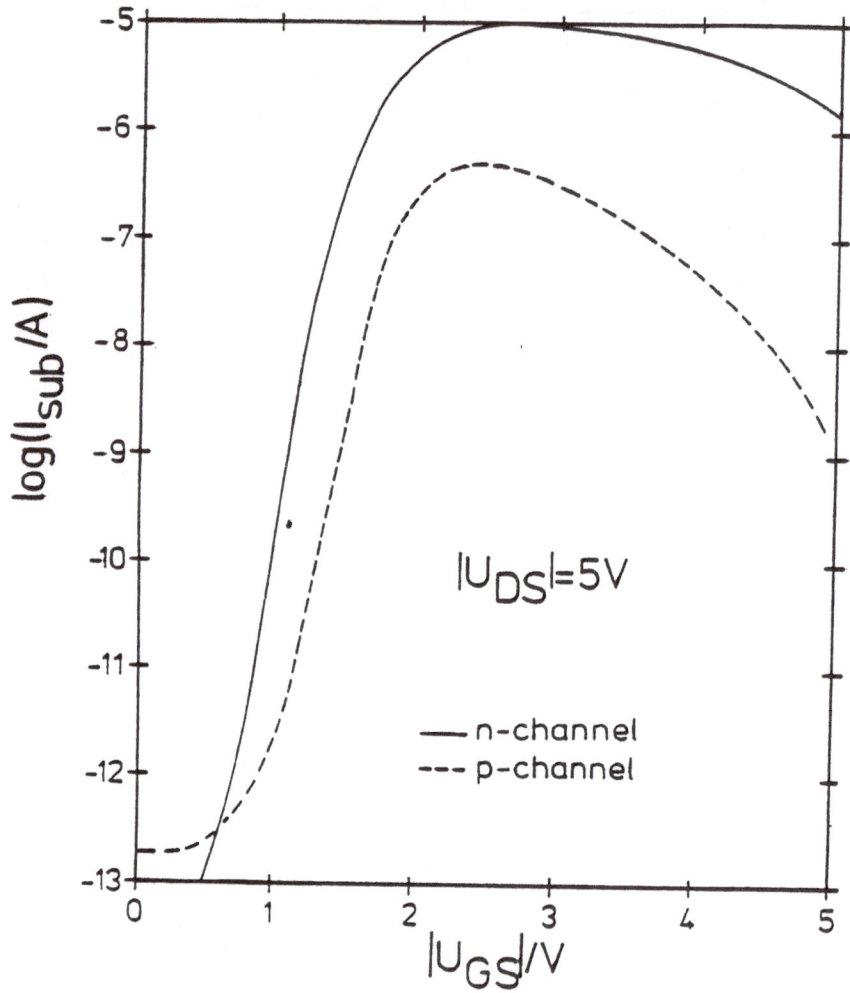

Fig. 4.5-9: Substrate currents for fixed drain voltage

Substrate current is important in CMOS technology as it initiates current injection from source into substrate. Thus the parasitic n^+pnp^+ thyristor (which consists of n-MOST source, p-well, n-substrate, and p-MOST source of a nearby inverter, cf. Fig. 4.5-1) may be triggered. The so called "latch-up" may result in the burn out of the entire circuit. Critical values of the substrate current for triggering the parasitic thyristor are about $10^{-5}A$-$10^{-4}A$ /125/. Contacts to the the p-well and the substrate result in improved latch-up performance /125/.

5. CONCLUSION

In this paper we tried to sketch the state of the art in modeling MOS transistors with numerical methods. The underlying physics has been discussed and the importance of increasingly sophisticated numerical methods has been briefly outlined. It has become evident that only progress in basic semiconductor physics will lead to the development of models which are capable of simulating device behaviour more reliably and which will match the technological advances of the recent device miniaturisation. One highly important objective of a model, its ability to predict the performance of a new device prior to having built the actual device, can only be reached if the physical parameters of the basic equations are analyzed even more thoroughly. This possibly implies a complete re-evaluation of some commonly accepted assumptions and approximations and it also seems to be the only way to get rid of the enormous amount of fitting parameters and the heuristic formulae which just simulate more or less precisely some complex physical phenomena. The inherent physics has to be analyzed very carefully before one can begin to synthesize a model which is able to simulate reality better. The power of a numerical model to predict device behaviour has been demonstrated using our MOS-transistor simulation program MINIMOS.

However, still much effort in analysis and simulation will have to be spent to make device miniaturisation and integration keep pace with the speed of recent days.

ACKNOWLEDGMENT

This work has been supported by the "Fond zur Förderung der wissenschaftlichen Forschung" (Projekt Nr. S22/11). Essential help of Siemens AG, Munich, in providing MOS devices is gratefully acknowledged. Critically reading our manuscript by Dr. J.Machek is very appreciated. We are in dept to Dipl.Ing. E.Langer for his excellent help in preparing this manuscript. Last but not least the authors wish to thank Dipl.Ing. D. Schornböck and the whole staff of the computer center for the excellent computer access.

REFERENCES

/1/ Adachi T., Yoshii A., Sudo T., "Two-Dimensional Semiconductor Analysis Using Finite-Element Method", IEEE Trans. Electron Devices, Vol.ED-26, pp.1026-1031, 1979.

/2/ Adler M.S., "A Method for Achieving and Choosing Variable Density Grids in Finite Difference Formulations and the Importance of Degeneracy and Band Gap Narrowing in Device Modeling", Proc. NASECODE I Conf., pp.3-30, 1979.

/3/ Adler M.S., "A Method for Terminating Mesh Lines in Finite Difference Formulations of the Semiconductor device Equations", Solid-State Electron., Vol.23, pp.845-853, 1980.

/4/ Agajanian A.H., "A Bibliography on Semiconductor Device Modeling", Solid-State Electron., Vol.18, pp.917-929, 1965.

/5/ Anderson C.L., Crowell C.R., "Threshold Energies for Electron-Hole Pair Generation by Impact Ionization in Semiconductors", Physical Review, Vol.B5, pp.2267-2272, 1972.

/6/ Antoniadis D.A., Hansen S., Dutton R.W., "SUPREM II - a Program for IC Process Modeling and Simulation", Stanford Technical Report No.5019-2, 1978.

/7/ Arora N.D., Hauser J.R., Roulston D.J., "Electron and Hole Mobilities in Silicon as a Function of Concentration and Temperature", IEEE Trans. Electron Devices, Vol.ED-29, pp.292-295, 1982.

/8/ Baccarani G., Mazzone A.M., "On the Diffusion Current in Heavily Doped Silicon", Solid-State Electron., Vol.18, pp.469-470, 1975.

/9/ Bank R.E., Rose D.J., "Parameter Selection for Newton-Like Methods Applicable to Nonlinear Partial Differential Equations.", SIAM J.Numer.Anal., Vol.17, pp.806-822, 1980.

/10/ Baraff G.A., "Distribution Functions and Ionization Rates for Hot Electrons in Semiconductors", Physical Review, Vol.128, pp.2507-2517, 1962.

/11/ Baraff G.A., "Maximum Anisotropy Approximation for Calculating Electron Distributions; Application to High Field Transport in Semiconductors", Physical Review, Vol.133, pp.A26-33, 1964.

/12/ Barnes J.J., Shimohigashi K., Dutton R.W., "Short-Channel MOSFET's in the Punchthrough Current Mode", IEEE Trans. Electron Devices, Vol.ED-26, pp.446-453, 1979.

/13/ Barnes J.J., Lomax R.J., "Transient 2-Dimensional Simulation of a Submicrometre Gate-Length M.E.S.F.E.T.", Solid-State Communications, Vol.13, pp.93-95, 1973.

/14/ Bell D.A., "Improved Formulation of Gummel's Algorithm for solving the 2-Dimensional Current-flow Equations in Semiconductor Devices", Electron. Lett., Vol.8, No.22, pp.536-538, 1972.

/15/ Benedek P., Silvester P., "Capacitance of Parallel Rectangular Plates Separated by a Dielectric Sheet", IEEE Trans. Microwave Theory and Techniques, Vol.MTT-20, pp. 504-510, 1972.

/16/ Birkhoff G., "The Numerical Solution of Elliptic Equations", SIAM, Philadelphia, 1971.

/17/ Blakey P.A., "Comments on "Effects of Carrier Diffusion on the Small-Signal Behavior of IMPATT Diodes"", IEEE Trans. Electron Devices, Vol.ED-27, p.299, 1980.

/18/ Blatt F.J., "Physics of Electronic Conduction in Solids", McGraw-Hill, New York, 1968.

/19/ Brown G.W., Lindsay B.W., "The Numerical Solution of Poisson's Equation for Two-Dimensional Semiconductor Devices", Solid-State Electron., Vol.19, pp.991-992, 1976.

/20/ Buturla E.M., Cottrell P.E., Grossman B.M., Salsburg K.A., "Finite-Element Analysis of Semiconductor Devices: The FIELDAY Program", IBM J. Res. Dev., Vol.25, pp.218-231, 1981.

/21/ Buturla E.M., Cotrell P.E., "Simulation of Semiconductor Transport Using Coupled and Decoupled Solution Techniques", Solid-State Electron., Vol.23, pp.331-334, 1980.

/22/ Buturla E.M., Cotrell P.E., Grossman B.M., Salsburg K.A., Lawlor M.B., McMullen C.T. "Three-Dimensional Finite Element Simulation of Semiconductor Devices", Proc. International Solid-State Circuits Conf., pp.76-77, 1980.

/23/ Canali C., Majni G., Minder R., Ottaviani G., "Electron and Hole Drift Velocity Measurements in Silicon and Their Empirical Relation to Electric Field and Temperature", IEEE Trans. Electron Devices, Vol.ED-22, pp.1045-1047, 1975.

/24/ Caughey D.M., Thomas R.E., "Carrier Mobilities in Silicon Empirically Related to Doping and Field", Proc. IEEE, Vol.52, pp.2192-2193, 1967.

/25/ Chamberlain S.G., Husain A., "Three-Dimensional Simulation of VLSI MOSFET's", Proc. International Electron Devices Meeting, pp.592-595, 1981.

/26/ Chwang R., Chung-Whei Kao, Crowell C.R., "Normalized Theory of Impact Ionization and Velocity Saturation in Nonpolar Semiconductors via a Markov Chain Approach", Solid-State Electron., Vol.22, pp.599-620, 1979.

/27/ Chynoweth A.G., "Ionization Rates for Electrons and Holes in Silicon", Physical Review, Vol.109, pp.1537-1540, 1958.

/28/ Chynoweth A.G., "Uniform Silicon p-n Junctions. 2. Ionization Rates for Electrons", J. Appl. Phys., Vol.31, pp.1161-1165, 1960.

/29/ Coe D.J., Brockman H.E., Nicholas K.H., "A Comparison of Simple and Numerical Two-Dimensional Models for the Threshold Voltage of Short-Channel MOSTs", Solid-State Electron., Vol.20, pp.993-998, 1977.

572

/30/ Coen R.W., Muller R.S., "Velocity of Surface Carriers in Inversion Layers on Silicon", Solid-State Electron., Vol.23, pp.35-40, 1980.

/31/ Conradt R., "Auger-Rekombination in Halbleitern", in: Festkörperprobleme XII, pp.449-464, Vieweg, Braunschweig, 1972.

/32/ Cotrell P.E., Buturla E.M., "Two-Dimensional Static and Transient Simulation of Mobile Carrier Transport in a Semiconductor", Proc. NASECODE I Conf., pp.31-64, 1979.

/33/ Crowell C.R., Sze S.M., "Temperature Dependence of Avalanche Multiplication in Semiconductors", Appl. Phys. Lett., Vol.9, pp.242-244, 1966.

/34/ Curtis A.R., Reid J.K., "The Choice of Step Lengths When Using Differences to Approximate Jacobian Matrices", J. Inst. Maths Applics, Vol.13, pp.121-126, 1974.

/35/ D'Avanzo D.C., "Modeling and Characterization of Short-Channel Double Diffused MOS Transistors", Stanford Technical Report No. G-201-6, 1980.

/36/ Dalal V.L., "Avalanche Multiplication in Bulk n-Si", Appl. Phys. Lett., Vol.15, pp.379-381, 1969.

/37/ Dang L.M., Konaka M., "A Two-Dimensional Computer Analysis of Triode-Like Characteristics of Short-Channel MOSFET's", IEEE Trans. Electron Devices, Vol.ED-27, pp.1533-1539, 1980.

/38/ De La Moneda F.H., "Threshold Voltage from Numerical Solution of the Two-Dimensional MOS Transistor", IEEE Trans. Circuit Theory, Vol.CT-20, pp.666-673, 1973.

/39/ De Mari A., "An Accurate Numerical Steady-State One-Dimensional Solution of the P-N Junction", Solid-State Electron., Vol.11, pp.33-58, 1968.

/40/ De Mari A., "An Accurate Numerical One-Dimensional Solution of the P-N Junction under Arbitrary Transient Conditions", Solid-State Electron., Vol.11, pp.1021-2053, 1968.

/41/ Debye P.P., Conwell E.M., "Electrical Properties of N-Type Germanium", Physical Review, Vol.93, pp.693-706, 1954.

/42/ Demoulin E., Greenfield J.A., Dutton R.W., Chatterjee P.K., Tasch A.F., "Process Statistics of Submicron MOSFET's", Proc. International Electron Devices Meeting, pp.34-37, 1979.

/43/ Dennard R.H., Gaensslen F.H., Yu H.N., Rideout V.L., Bassous E., Le Blanc A.R., "Design of Ion-Implanted MOSFET's with Very Small Physical Dimensions", IEEE J. Solid-State Circuits, Vol.SC-9, pp.256-268, 1974.

/44/ Deuflhard P., "A Modified Newton Method for the Solution of Ill-Conditioned Systems of Nonlinear Equations with Application to Multiple Shooting", Numer. Math., Vol.22, pp.289-315, 1974.

/45/ Deuflhard P., Heindl G., **"Affine Invariant Convergence Theorems for Newton's Method and Extensions to Related Methods"**, SIAM J.Numer.Anal., Vol.16, pp.1-10, 1979.

/46/ Dongarra J.J., Moler C.B., Bunch J.R., Stewart G.W., **"LINPACK"**, SIAM, Philadelphia, 1979.

/47/ Dorkel J.M., Leturcq Ph., **"Carrier Mobilities in Silicon Semi-Empirically Related to Temperature, Doping and Injection Level"**, Solid-State Electron., Vol.24, pp.821-825, 1981.

/48/ Duff I.S., **"A Survey of Sparse Matrix Research"**, Proc. IEEE, Vol.65, pp.500-535, 1977.

/49/ Duff I.S., **"Practical Comparison of Codes for the Solution of Sparse Linear Systems"**, A.E.R.E. Harwell, Oxfordshire, 1979.

/50/ Dupont T., Kendall R.D., Rachford H.H., **"An Approximate Factorization Procedure for Solving Self-Adjoint Elliptic Difference Equations"**, SIAM, J. Num. Anal., Vol.5, pp.559-573, 1968.

/51/ Dutton R.W., Hansen S.E., **"Process Modeling of Integrated Circuit Device Technology"**, Proc. IEEE, Vol.69, pp.1305-1320, 1981.

/52/ Dziewior J., Schmid W., **"Auger Coefficients for highly doped and highly excited Silicon"**, Appl. Phys. Lett., Vol.31, pp.346-348, 1977.

/53/ Engl W.L., Manck O., Wieder A.W., **"Device Modeling"**, in: Process and Device Modeling for Integrated Circuit Design, pp.3-17, Noordhoff, Leyden, 1977.

/54/ Engl W.L., Dirks H., **"Functional Device Simulation by Merging Numerical Building Blocks"**, Proc. NASECODE II Conf., pp.34-62, 1981.

/55/ Engl W.L., Dirks H., **"Models of Physical Parameters"**, in: Introduction to the Numerical Analysis of Semiconductor Devices and Integrated Circuits, pp.42-46, Boole Press, Dublin, 1981.

/56/ Engl W.L., Dirks H., **"Numerical Device Simulation Guided by Physical Approaches"**, Proc. NASECODE I Conf., pp.65-93, 1979.

/57/ Fair R.B., Sun R.C., **"Threshold-Voltage Instability in MOSFET's Due to Channel Hot-Hole Emission"**, IEEE Trans. Electron Devices, Vol.ED-28, pp.83-94, 1981.

/58/ Forsythe G.E., Wasow W.R., **"Finite Difference Methods for Partial Differential Equations"**, Wiley, New York, 1960.

/59/ Fox L., **"Finite-Difference Methods in Elliptic Boundary-Value Problems"**, in: The State of the Art in Numerical Analysis, pp.799-881, Academic Press, London, 1977.

/60/ Frey J., **"Physics Problems in VLSI Devices"**, in: Introduction to the Numerical Analysis of Semiconductor Devices and Integrated Circuits, pp.47-50, Boole Press, Dublin, 1981.

574

/61/ Frey J., "Transport Physics for VLSI", in: Introduction to the Numerical Analysis of Semiconductor Devices and Integrated Circuits, pp.51-57, Boole Press, Dublin, 1981.

/62/ Gaensslen F.H., Jaeger R.C., "Temperature Dependent Threshold Behaviour of Depletion Mode MOSFET's", Solid-State Electron., Vol.22, pp.423-430, 1979.

/63/ Gansner M.,Ilegems M.,Schwob P.,Dutoit M., "Modelisation des Structures Microelectroniques de Petites Dimensions", Proc. Journees d'Electronique, pp.93-105, 1980.

/64/ Gibbons J., Johnson W.S., Mylroie S.W., "Projected Range Statistics", Halstead Press, Strandsberg, 1975.

/65/ Grant W.N., "Electron and Hole Ionization Rates in Epitaxial Silicon at High Electric Fields", Solid-State Electron., Vol.16., pp.1189-1203, 1973.

/66/ Greenfield J.A., Dutton R.W., "Nonplanar VLSI Device Analysis Using the Solution of Poisson's Equation", IEEE Trans. Electron Devices, Vol.ED-27, pp.1520-1532, 1980.

/67/ Greenfield J.A., Hansen S.E., Dutton R.W., "Two-Dimensional Analysis for Device Modeling", Stanford Technical Report No. G-201-7, 1980.

/68/ Grimes R.G., Kincaid D.R., Young D.R., "ITPACK 2A - A Fortran Implementation of Adaptive Accelerated Iterative Methods for Solving Large Sparse Linear Systems", University of Texas, Austin, Report. No. CNA-164, 1980.

/69/ Gummel H.K., "A Self-Consistent Iterative Scheme for One-Dimensional Steady State Transistor Calculations", IEEE Trans. Electron Devices, Vol.ED-11, pp.455-465, 1964.

/70/ Hachtel G.D., Mack M.H., O'Brien R.R., "Semiconductor Analysis Using Finite Elements-Part 2: IGFET and BJT Case Studies", IBM J. Res. Dev., Vol.25, pp.246-260, 1981.

/71/ Hachtel G.D., Mack H.H., O'Brien R.R., Speelpennig B., "Semiconductor Analysis Using Finite Elements-Part 1: Computational Aspects", IBM J. Res. Dev., Vol.25, pp.232-245, 1981.

/72/ Hamilton D.J., Howard W.G., "Basic Integrated Circuit Engineering", McGraw-Hill Kogakusha, Tokyo, 1975.

/73/ Hart J.F., Cheney E.W., Lawson C.L., Maehly H.J., "Computer Approximations", Wiley, New York, 1968.

/74/ Haug A., "Strahlungslose Rekombination in Halbleitern (Theorie)", in: Festkörperprobleme XII, Vieweg, Braunschweig, 1972.

/75/ Hauser J.R., "Threshold Energy for Avalanche Multiplication in Semiconductors", J. Appl. Phys., Vol.37, pp.507-509, 1966.

/76/ Heimeier H.H., "A Two-Dimensional Numerical Analysis of a Silicon N-P-N Transistor", IEEE Trans. Electron Devices, Vol.ED-20, pp.708-714, 1973.

/77/ Hess K., "Ballistic Electron Transport in Semiconductors", IEEE Trans. Electron Devices, Vol.ED-28, pp.937-940, 1981.

/78/ Heywang W., Pötzl H.W., "Bandstruktur und Stromtransport", Springer, Berlin, 1976.

/79/ Jacoboni C., Canali C., Ottaviani G., Quaranta A.A., "A Review of Some Charge Transport Properties of Silicon", Solid-State Electron., vol.20, pp.77-89, 1977.

/80/ Jaggi R., Weibel H., "High-Field Electron Drift Velocities and Current Densities in Silicon", Helv.Phys.Acta., Vol.42, pp.631-632, 1969.

/81/ Johnson L.W., Riess R.D., "An Error Analysis for Numerical Differentiation", J. Inst. Maths Applics, Vol.11, pp.115-120, 1973.

/82/ Kahng D., Atalla M.M., "Silicon-Silicon Dioxide Field Induced Surface Devices", IRE-AIEE, Solid-State Device Res. Conf., 1960.

/83/ Kani K., "A Survey of Semiconductor Device Analysis in Japan", Proc. NASECODE I Conf., pp.104-119, 1979.

/84/ Kennedy D.P., Murley P.C., "Steady State Mathematical Theory for the Insulated Gate Field Effect Transistor", IBM J. Res. Dev., Vol.17, pp.2-12, 1973.

/85/ Kennedy D.P., O'Brien R.R., "Two-Dimensional Mathematical Analysis of a Planar Type Junction Field-Effect Transistor", IBM J. Res. Dev., Vol.13, pp.662-674, 1969.

/86/ Kotani N., Kawazu S., "A Numerical Analysis of Avalanche Breakdown in Short-Channel MOSFETS", Solid-State Electron., Vol.24, pp.681-687, 1981.

/87/ Kotani N., Kawazu S., "Computer Analysis of Punch-Through in MOSFET's", Solid-State Electron., Vol.22, pp.63-70, 1979.

/88/ Lee C.A., Logan R.A., Batdorf R.L., Kleimack J.J., Wiegmann W., "Ionization Rates of Holes and Electrons in Silicon", Physical Review, Vol.134, pp.A761-773, 1964.

/89/ Lee H., Sansbury J.D., Dutton R.W., Moll J.L., "Modeling and Measurement of Surface Impurity Profiles of Laterally Diffused Regions", IEEE J. Solid-State Circuits, Vol.SC-13, pp.455-461, 1978.

/90/ Lee H., "Two-Dimensional Impurity Diffusion Studies: Process Models and Test Structures for Low-Concentration Boron Diffusion", Stanford Technical Report No.G201-8, 1980.

/91/ Leguerre R., "Approximate Values of the Multiplication Coefficient in One-sided Abrupt Junctions", Solid-State Electron., Vol.19, pp.875-881, 1976.

/92/ Li S.S., Thurber W.R., "The Dopant Density and Temperature Dependence of Electron Mobility and Resistivity in n-Type Silicon", Solid-State Electron., Vol.20, pp.609-616, 1977.

/93/ Liu S., Hoefflinger B., Pederson D.O., "Interactive Two-Dimensional Design of Barrier Controlled MOS Transistors", IEEE Trans. Electron Devices, Vol.ED-27, pp.1550-1558, 1980.

/94/ Loeb H.W., Andrew R., Love W., "Application of 2-Dimensional Solutions of the Shockley-Poisson Equation to Inversion-Layer M.O.S.T. Devices", Electron. Lett., Vol.4, pp.352-354, 1968.

/95/ Longo H.E., "Ladungsträgermultiplikation bei MIS-Transistoren", Zeitschrift f. Angew. Physik, Vol.29, p.166, 1970.

/96/ Manck O., Engl W.L., "Two-Dimensional Computer Simulation for Switching a Bipolar Transistor out of Saturation", IEEE Trans. Electron Devices, Vol.ED-24, pp.339-347, 1975.

/97/ Marhsak A.H., Shrivastava R., "Law of the Junction for Degenerate Material with Position-Dependent Band Gap and Electron Affinity", Solid-State Electron., Vol.22, pp.567-571, 1979.

/98/ Marsal D., "Die Numerische Lösung partieller Differential-gleichungen", Bibliographisches Institut, Mannheim, 1976.

/99/ Masuda H., Nakai M., Kubo M., "Characteristics and Limitation of Scaled-Down MOSFET's due to Two-Dimensional Field-Effect", IEEE Trans. Electron Devices, Vol.ED-26, pp.980-986, 1979.

/100/ Matsumoto H., Sawada K., Asai S., Hirayama M., Nagasawa K., "Effect of Long Term Stress on IGFET Degradations Due to Hot Electron Trapping", Trans. Electron Devices, Vol.ED-28, pp.923-928, 1981.

/101/ McKay K.G., "Avalanche Breakdown in Silicon", Physical Review, Vol.94, pp.877-884, 1954.

/102/ McKay K.G., McAfee K.B., "Electron Multiplication in Silicon and Germanium", Physical Review, Vol.91, pp.1079-1084, 1953

/103/ Mertens R.P., Van Meerbergen J.L., Nijs J.F., Van Overstraeten R.J., "Measurement of the Minority-Carrier Transport Parameters in Heavily Doped Silicon", IEEE Trans. Electron Devices, Vol.ED-27, pp.949-955, 1980.

/104/ Meyer G.H., "On Solving Nonlinear Equations with a One-Parameter Operator Imbedding", SIAM J.Numer.Anal., Vol.5, pp.739-752, 1968.

/105/ Miller S.L., "Avalanche Breakdown in Germanium", Physical Review, Vol.99, pp.1234-1241, 1955.

/106/ Miller S.L., "Ionization Rates for Holes and Electrons in Silicon", Physical Review, Vol.105, pp.1246-1249, 1957.

/107/ Mock M.S., "A Time-Dependent Numerical Model of the Insulated-Gate Field-Effect Transistor", Solid-State Electron., Vol.24, pp.959-966, 1981.

/108/ Mock M.S., "A Two-Dimensional Mathematical Model of the Insulated-Gate Field-Effect Transistor", Solid-State Electron., Vol.16, pp.601-609, 1973.

/109/ Mock M.S., "Convergence and Accuracy in Stationary Numerical Models", in: Introduction to the Numerical Analysis of Semiconductor Devices and Integrated Circuits, pp.58-62, Boole Press, Dublin, 1981.

/110/ Moglestue C., "A Monte-Carlo Particle Model Study of the Influence of the Doping Profiles on the Characteristics of Field-Effect Transistors", Proc. NASESCODE II Conf., pp.244-249, 1981.

/111/ Moglestue C., Beard S.J., "A Particle Model Simulation of Field Effect Transistors", Proc. NASECODE I Conf., pp.232-236, 1979.

/112/ Moll J.L., Overstraeten R. van, "Charge Multiplication in Silicon p-n Junctions", Solid-State Electron., Vol.6, pp.147-157, 1963.

/113/ Moll J.J, Meyer N.I., "Secondary Multiplication in Silicon", Solid-State Electron., Vol.3, pp.155-158, 1961.

/114/ Müller W., Risch L., Schütz A., "Analysis of Short Channel MOS Transistors in the Avalanche Multiplication Regime" IEEE Trans. Electron Devices, to be published.

/115/ Nakagawa A., "One-Dimensional Device Model of the npn Bipolar Transistor Including Heavy Doping Effects under Fermi Statistics", Solid-State Electron., Vol.22, pp.943-949, 1979.

/116/ Nussbaum A., "Inconsistencies in the Original Form of the Fletcher Boundary Conditions", Solid-State Electron., Vol.21, pp.1178-1179, 1978.

/117/ Ochi S., Okabe T., Yoshida I., Yamaguchi K., Nagata K., "Computer Analysis of Breakdown Mechanism in Planar MOSFET's", IEEE Trans. Electron Devices, Vol.ED-27, pp.399-400, 1980.

/118/ Ogawa T., "Avalanche Breakdown and Multiplication in Silicon pin Junctions", Jap. J. Appl. Phys., Vol.4, pp.473-484, 1965.

/119/ Oh S.Y., Ward D.E., Dutton R.W., "Transient Analysis of MOS Transistors", IEEE Trans. Electron Devices, Vol.ED-27, pp.1571-1578, 1980.

/120/ Oka H., Nishiuchi K., Nakamura T., Ishikawa H., "Computer Analysis of a Short-Channel BC MOSFET", IEEE Trans. Electron Devices, Vol.ED-27, pp.1514-1520, 1980.

/121/ Oka H., Nishiuchi K., Nakamura T., Ishikawa H., "Two-Dimensional Numerical Analysis of Normally-Off Type Buried Channel MOSFET's", Proc. International Electron Devices Meeting, pp.30-33, 1979.

/122/ Okuto Y., Crowell C.R., "Energy Conservation Considerations in the Characterization of Impact Ionization in Semiconductors", Physical Review, Vol.B6, pp.3076-3081, 1972.

/123/ Okuto Y., Crowell C.R., "Ionization Coefficients in Semiconductors: A Nonlocalized Property", Physical Review, Vol.B10, pp.4284-4296, 1974.

/124/ Ortega J.M., Rheinboldt W.C., "Iterative Solution of Nonlinear Equations in Several Variables", Academic Press, New York, 1970.

/125/ Peyne R.S., Grant W.N., Bertram W.J., "Elimination of Latch-Up in Bulk CMOS", Proc. International Electron Devices Meeting, pp.248-251, 1980.

/126/ Rheinboldt W.C., "Methods for Solving Systems of Nonlinear Equations", SIAM, Philadelphia, 1974.

/127/ Rokus A., "Zur numerischen Lösung linearer pentagonaler Gleichungssysteme hohen Ranges", Diplomarbeit, Technische Universität Wien, 1980.

/128/ Runge H., "Distribution of Implanted Ions under Arbitrarily Shaped Mask Edges", Phys. Status Solidi, Vol.(a)39, pp.595-599, 1977.

/129/ Ryssel H., Ruge I., "Ionenimplantation", Teubner, Stuttgart, 1978.

/130/ Sabnis A.G., Clemens J.T., "Characterization of the Electron Mobility in the inverted (100) SI-Surface", Proc. International Electron Devices Meeting, pp.18-21, 1979.

/131/ Sayle W.E., Lauritzen P.E., "Avalanche Ionization Rates Measured in Silicon and Germanium at Low Electric Fields", IEEE Trans. Electron Devices, Vol.ED-18, pp.58-66, 1971.

/132/ Scharfetter D.L., Gummel H.K., "Large-Signal Analysis of a Silicon Read Diode Oscillator", IEEE Trans. Electron Devices, Vol.ED-16, pp.64-77, 1969.

/133/ Schütz A., Selberherr S., Pötzl H.W., "A Two-Dimensional Model of the Avalanche Effect in MOS Transistors", Solid-State Electron., Vol.25, pp.177-183, 1982.

/134/ Schütz A., Selberherr S., Pötzl H.W., "Analysis of Breakdown Phenomena in MOSFET's", IEEE Trans. Computer-Aided-Design of Integrated Circuits, Vol.CAD-1, pp.77-85 1982.

/135/ Schütz A., Selberherr S., Pötzl H.W., "Numerical Analysis of Breakdown Phenomena in MOSFET's", Proc. NASECODE II Conf., pp.270-274, 1981.

/136/ Seeger K., "Semiconductor Physics", Springer, Wien, 1973.

/137/ Selberherr S., Schütz A., Pötzl H., "Investigation of Parameter Sensitivity of Short Channel MOSFETS", Solid-State Electron., Vol.25, pp.85-90, 1982.

/138/ Selberherr S., Fichtner W., Pötzl H.W., "MINIMOS - a Program Package to Facilitate MOS Device Design and Analysis", Proc. NASECODE I Conf., pp.275-279, 1979.

/139/ Selberherr S., Schütz A., Pötzl H., "MINIMOS - Zweidimensionale Modellierung von MOS-Transistoren", Elektronikschau, Vol.9, pp.18-23, Vol.10, pp.54-58, 1980.

/140/ Selberherr S., Schütz A., Pötzl H.W., **"MINIMOS - a Two-Dimensional MOS Transistor Analyzer"**, IEEE Trans. Electron Devices, Vol.ED-27, pp.1540-1550, 1980.

/141/ Selberherr S., Guerrero E., **"Simple and Accurate Representation of Implantation Parameters by Low Order Polynomals"**, Solid-State Electron., Vol.24, pp.591-593, 1981.

/142/ Selberherr S., Schütz A., Pötzl H., **"Two-Dimensional MOS Transistor Modelling"**, European Electronics, Vol.1, pp.20-30, 1981.

/143/ Selberherr S., **"Zweidimensionale Modellierung von MOS-Transistoren"**, Dissertation, Technische Universität Wien, 1981.

/144/ Sherman A.H., **"On Newton-Iterative Methods for the Solution of Systems of Nonlinear Equations"**, SIAM J.Numer.Anal., Vol.15, pp.755-771, 1978.

/145/ Shockley W., **"Problems Related to p-n Junctions in Silicon"**, Solid-State Electron., Vol.2, pp.35-67, 1961.

/146/ Slotboom J.W., **"The pn-Product in Silicon"**, Solid-State Electron., Vol.20, pp.279-283, 1977.

/147/ Smith G.D., **"Numerical Solution of Partial Differential Equations: Finite Difference Methods"**, Clarendon Press, Oxford, 1978.

/148/ Smith R.A., **"Semiconductors"**, Cambridge University Press, Cambridge, 1978.

/149/ Spirito P., **"Avalanche Multiplication Factors in Ge and Si Abrupt Junctions"**, IEEE Trans. Electron Devices, Vol.ED-21, pp.226-231, 1974.

/150/ Stoer J., Bulirsch R., **"Einführung in die Numerische Mathematik II"**, Springer, Berlin, 1978.

/151/ Stone H.L., **"Iterative Solution of Implicit Approximations of Multidimensional Partial Differential Equations"**, SIAM J.Numer.Anal., Vol.5, pp.530-558, 1968.

/152/ Sun E., Moll J., Berger J., Alders B.,, **"Breakdown Mechanism in Short-Channel MOS Transistors"**, Proc. International Electron Devices Meeting, pp.478-482, 1978.

/153/ Sun S.C., Plummer J.D., **"Electron Mobility in Inversion and Accumulation Layers on Thermally Oxidized Silicon Surfaces"**, IEEE Trans. Electron Devices, Vol.ED-27, pp.1497-1508, 1980.

/154/ Sutherland A.D., **"A Two-Dimensional Computer Model for the Steady-State Operation of MOSFET's"**, US. Army Electronics Command Res. and Dev. Techn. Rep., ECOM-75-1344-F, 1977.

/155/ Sutherland A.D., **"An Algorithm for Treating Interface Surface Charge in the Two-Dimensional Discretization of Poisson's Equation for the Numerical Analysis of Semiconductor Devices such as MOSFET's"**, Solid-State Electron., Vol.23, pp.1085-1087, 1980.

/156/ Sutherland A.D., "On the Use of Overrelaxation in Conjuncton with Gummel's Algorithm to Speed the Convergence in a Two-dimensional Computer Model for MOSFET's", IEEE Trans. Electron Devices, Vol.ED-27, pp.1297-1298, 1980.

/157/ Sze S.M., Gibbons G., "Avalanche Breakdown Voltages of Abrupt and Linearly Graded p-n Junctions in Ge, Si, GaAs, and GaP", Appl. Phys. Lett., Vol.8, pp.111-113, 1966.

/158/ Sze S.M., "Physics of Semiconductor Devices", Wiley, New York, 1969.

/159/ Takacs D.,Schwabe U.,Bürker U., "The Influence of Temperature on the Tolerances of MOS-Transistors in a One-Micrometer Technology", Proc. International Electron Devices Meeting, pp.569-573, 1980.

/160/ Tang J.Y., Shichijo H., Hess K., Iafrate G.J., "Band-Structure Dependent Impact Ionization in Silicon and Gallium Arsenide", Journal de Physique, pp.C7:63-69, Montpellier, 1981.

/161/ Temple V.A.K., Adler M.S., "Calculation of the Diffusion Curvature Related Avalanche Breakdown in High-Voltage Planar p-n Junctions", IEEE Trans. Electron Devices, Vol.ED-22, pp.910-916, 1975.

/162/ Thornber K.K., "Applications of Scaling to Problems in High-Field Electronic Transport", J. Appl. Phys., Vol.52, pp.279-290, 1981.

/163/ Thornber K.K., "Relation of Drift Velocity to Low-Field Mobility and High-Field Saturation Velocity", J. Appl. Phys., Vol.51, pp.2127-2136, 1980.

/164/ Tielert R., "Two-Dimensional Numerical Simulation of Impurity Redistribution in VLSI Processes", IEEE Trans. Electron Devices, Vol.ED-27, pp.1479-1483, 1980.

/165/ Toyabe T., Yamaguchi K., Asai S., Mock M., "A Numerical Model of Avalanche Breakdown in MOSFET's", IEEE Trans. Electron Devices, Vol.ED-25, pp.825-832, 1978.

/166/ Toyabe T., Asai S., "Analytical Model of Threshold Voltage and Breakdown Voltage of Short-Channel MOSFET's derived from Two-Dimensional Analysis", IEEE Trans. Electron Devices, Vol.ED-26, pp.453-461, 1979.

/167/ Troutman R.R., "Low-Level Avalanche Multiplication in IGFET's", IEEE Trans. Electron Devices, Vol.ED-23, pp.419-425, 1976.

/168/ Troutman R.R., "VLSI Limitations from Drain-Induced Barrier Lowering.", IEEE Trans. Electron Devices, Vol.ED-26, pp.461-469, 1979.

/169/ Van de Wiele F., Van Overstraeten R., De Man H., "Graphical Method for the Determination of Junction Parameters and of Multiplication Parameters", Solid-State Electron., Vol.13, pp.25-36, 1970.

/170/ Van Overstraeten R., De Man H., **"Measurement of the Ionization Rates in Diffused Silicon p-n Junctions"**, Solid-State Electron., Vol.13, pp.583-608, pp.1970.

/171/ Van Overstraeten R.J., De Man H.J., Mertens R.P., **"Transport Equations in Heavy Doped Silicon"**, IEEE Trans. Electron Devices, Vol.ED-20, pp.290-298, 1973.

/172/ Van Roosbroeck W.V., **"Theory of Flow of Electrons and Holes in Germanium and Other Semiconductors"**, Bell Syst. Techn. J., Vol.29, pp.560-607, 1950.

/173/ Van Vliet K.M., **"On Fletcher's Boundary Conditions"**, Solid-State Electron., Vol.22, pp.443-444, 1979.

/174/ Van Vliet K.M., **"The Shockley-Like Equations for the Carrier Densities and the Current Flows in Materials with a Nonuniform Composition"**, Solid-State Electron., Vol.23, pp.49-53, 1980.

/175/ Vandorpe D., Borel J., Merckel G., Saintot P. **"An Accurate Two-Dimensional Numerical Analysis of the MOS Transistor"**, Solid-State Electron., Vol.15, pp.547-557, 1972.

/176/ Vandorpe D., Xuong N.H., **"Mathematical 2-Dimensional Model of Semiconductor Devices"**, Electron. Lett., Vol.7, pp.47-50, 1971.

/177/ Varga R.S., **"Matrix Iterative Analysis"**, Prentice-Hall, Englewood Cliffs, 1962.

/178/ Wachspress E.L., **"Iterative Solution of Elliptic Systems"**, Prentice-Hall, Englewood Cliffs, 1966.

/179/ Wada T., Dang R.L.M., **"Modification of ICCG Method for Application to Semiconductor Device Simulators"**, Electron. Lett., Vol.18, pp.265-266, 1982.

/180/ Wolf H.F., **"Semiconductors"**, Wiley, New York, 1971.

/181/ Wolff P.A., **"Theory of Multiplication in Silicon and Germanium"**, Physical Review, Vol.95, pp.1415-1420, 1954.

/182/ Yamaguchi K., **"Field-Dependant Mobility Model for Two-Dimensional Numerical Analysis of MOSFET's"**, IEEE Trans. Electron Devices, Vol.ED-26, pp.1068-1074, 1979.

/183/ Yamaguchi K., Toyabe T., Kodera H., **"Two-Dimensional Analysis of Triode-Like Operation of Junction Gate FET's"**, IEEE Trans. Electron Devices, Vol.ED-22, pp.1047-1049, 1975.

/184/ Yokoyama K.Y., Yoshii A., Horiguchi S., **"Threshold Sensitivity Minimization of Short-Channel MOSFET's by Computer Simulation"**, IEEE Trans. Electron Devices, Vol.ED-27, pp.1509-1514, 1980.

/185/ Yoshii A., Horiguchi S., Sudo T., **"A Numerical Analysis for Very Small Semiconductor Devices"**, Proc. International Solid-State Circuits Conf., pp.80-81, 1980.

/186/ Yoshii A., Kitazawa H., Tomizawa M., Horiguchi S., Sudo T., **"A Three-Dimensional Analysis of Semiconductor Devices"**, IEEE Trans. Electron Devices, Vol.ED-29, pp.184-189, 1982.

FIELDAY - FINITE ELEMENT DEVICE ANALYSIS

K. A. Salsburg

International Business Machines, Federal Systems Division
Manassas, Virginia
P. E. Cottrell
E. M. Buturla

International Business Machines, General Technology Division
Essex Junction, Vermont

Abstract

Two- and three-dimensional mobile carrier transport in a semi-conductor is simulated in the FIELDAY program using the finite element method. A wide variety of physical effects important in bipolar and field effect transistors can be modeled. The finite element method transforms the continuum description of mobile carrier transport in a semiconductor device to a simulation model at a discrete number of points. Coupled and decoupled algorithms offer two methods of linearizing the differential equations. Direct techniques are used to solve the resulting matrix equations. Pre- and post-processors enable users to rapidly generate new models and analyze results. Specific examples illustrate the flexibility and accuracy of FIELDAY. Projections of future advancements in the program are discussed.

SECTION 1 INTRODUCTION

Very Large Scale Integrated (VLSI) circuits require computer simulation at the process, device, circuit, and system levels. In this paper, the FInite ELement Device AnalYsis (FIELDAY) program, used extensively throughout IBM to simulate semiconductor devices, is described. FIELDAY is a general purpose program which numerically solves the semiconductor transport equations in one, two, or three dimensions for steady state or transient operating conditions.* FIELDAY can simulate an arbitrary metal-insulator-semiconductor structure and can model a wide variety of physical effects which are important for the accurate simulation of bipolar and field effect devices. Applications range from analysis of short- and narrow-channel effects on the threshold of IGFET's to the transient simulation of heavily-doped bipolar transistors.

The motivation for this CAD tool is easy to understand when other options for obtaining the same information are considered. There are two choices: the better is to fabricate and characterize devices; the other is to use simpler models which are based either on one-dimensional approximations or on extrapolations of data from devices "similar" to those of interest. Both of these approaches have advantages and limitations when compared to FIELDAY.

Fabrication and testing of actual devices is the best way to discover all of the implications and limitations of any new technology. While this is required of any technology considered for development, it is probably the most expensive and time-consuming option. Devices fabricated for a new VLSI technology may cost over one million dollars to fabricate and may take six months to complete.

Device simulation is a cost-effective method of determining whether a new technology is worth developing. Once that decision has been made, simulation can substitute for many costly matrix experiments that normally are required to optimize new process and device structures. This is important in VLSI technologies because of the statistical nature of device design. Function must be guaranteed for the large number of devices in modern electronic systems. It is not possible to produce enough experimental hardware to test even the most critical combinations for possible parameters and structures. Large numbers of devices and significant variations in structure from device-to-device make a statistical design imperative.

Another advantage that simulation offers is certainty of structure and physical parameters of the device, i.e., device

---

*E.M. Buturla, P.E. Cottrell, B.M. Grossman, K.A. Salsburg, "Finite-Element Analysis of Semiconductor Devices: The FIELDAY Program," IBM Journal of Research and Development, Vol. 25, No. 4, July 1981, pp 218-231.

design information can be derived before a new fabrication process has been stabilized. The device can be optimized early in the product development cycle. A final advantage is that the internal operation of any device can be examined easily through multidimensional simulation; experimental techniques can approximate this for only a few parameters.

Simpler models and the extrapolation of data from existing device structures cannot predict accurately how new device structures will behave. This can be determined only by a model based on fundamental physical assumptions. The modeling difficulty is compounded by the near unity aspect ratios of devices used in VLSI technologies; e.g., a minimum size n-channel IGFET has a length, width, and depth of similar dimensions. This means that three-dimensional modeling is required to accurately describe this device. The transient behavior of bipolar devices must be described, mainly because of the three-dimensional effects associated with small emitters.

This paper describes the FIELDAY algorithm and the associated interactive pre-processor and post-processor programs which provide a comprehensive device design package. The following sections describe the physical model and the numerical algorithm and outline the interactive pre-and post-processing capabilities. Also presented are specific examples of applications of the FIELDAY program, including the analysis of short and narrow IGFETs, a Vertical Metal Oxide Silicon (VMOS) transistor and bipolar transistors, and expected enhancements to the program.

SECTION 2 PHYSICAL MODEL

The Semiconductor Transport Equations

Two specie mobile charge carrier flow in a semiconductor can be described by the following set of equations [1]:

$$\nabla^2\psi + \frac{q}{\epsilon}\,(p - n + N_D - N_A + N_Q) = 0 \tag{1}$$

$$J_n = q\{D_n\nabla n - \mu_n n\nabla(\psi + \Delta V_c)\} \tag{2}$$

$$J_p = -q\{D_p\nabla p + \mu_p p\nabla(\psi - \Delta V_v)\}\,. \tag{3}$$

The principle of charge conservation requires that

$$\nabla \cdot J_n - q\left(\frac{\partial n}{\partial t} + R_n\right) = 0, \text{ and} \tag{4}$$

$$\nabla \cdot J_p + q\left(\frac{\partial p}{\partial t} + R_p\right) = 0. \tag{5}$$

The three unknown quantities are the space-charge potential (Ψ), and the electron (n) and hole (p) mobile charge densities at each instant of time. N_D and N_A are the donor and acceptor impurity densities, N_Q is the density of fixed charged particles, the constant q is the magnitude of the electronic charge, ε is the permittivity, and J_n and J_p are the electron and hole current densities. R_n and R_p are the electron and hole recombination rate densities, μ_n and μ_p are the electron and hole mobilities, ΔV_c and ΔV_v are changes in the conduction and valence band edges [2] of heavily-doped semiconductors, and D_n and D_p are the electron and hole diffusion coefficients.

Poisson's equation relates the space-charge potential of mobile and fixed charges with mobile charge densities given by Boltzmann statistics. The flow of mobile charge carriers is described by the equations of electron and hole current continuity. The electron and hole mobilities that appear in these equations are functions of electric field strength $|\nabla \Psi|$, and impurity density [3]. The diffusion coefficients D_n and D_p are related to the electron and hole mobilities by the Einstein relationship.

In the FIELDAY model, the recombination-generation mechanisms include carrier generation due to avalanche multiplication, photo-generation, and Auger and Shockley-Read-Hall recombination [4-6]. These recombination-generation mechanisms couple the two current continuity equations and introduce strong nonlinearities, particularly for the case of avalanche multiplication.

The presence of avalanche-generated charge carriers is included in the semiconductor transport equations through the generation rate (G):

$$\nabla \cdot J_n = -G \tag{6}$$

$$\nabla \cdot J_p = G \tag{7}$$

The rate at which charge carriers are generated by impact ionization within a differential volume element is equal to the product of the probability of ionization per unit distance traveled by the impacting electron or hole and the magnitude of the electron or hole current density flowing through the volume element. The electron-hole pairs generated per unit volume per unit time (G), is given by Equation (8);

$$G = \alpha_n J_n + \alpha_p J_p \, , \qquad (8)$$

where

α_n = probability of ionization per unit distance of impacting electron travel, and

α_p = probability of ionization per unit distance of impacting hole travel.

The probabilities α_n and α_p are functions of temperature (T), ionization energy (ε_i), mean free path for optical phonon generation ($\lambda_{n,p}$), phonon energy (ε_r), and electric-field strength in the direction of charge-carrier flux ($E \cdot J_{n,p}$). The function for ionization probability [7] is given by Equation (9);

$$\alpha_{n,p} = \frac{1}{\lambda_{n,p}} \exp\left(a_0 + \frac{a_1}{|E \cdot J_{n,p}|} + \frac{a_2}{|E \cdot J_{n,p}|2}\right) \qquad (9a)$$

when $(E \cdot J_{n,p} > 0)$;

$$\alpha_{n,p} = 0, \ (E \cdot J_{n,p} \le 0) \qquad (9b)$$

$$a_0 = -757r^2 + 75.5r - 1.92 \qquad (10a)$$

$$a_1 = (46r^2 - 11.9r + 1.75 \times 10^{-2})\chi_{n,p} \qquad (10b)$$

$$a_2 = (11.5r^2 - 1.17r + 3.9 \times 10^{-4})\chi_{n,p}^2 \qquad (10c)$$

$$r = \frac{\epsilon_r}{\epsilon_i} \tanh \frac{\epsilon_r}{2kT} \tag{11}$$

$$\chi_{n,p} = \frac{\epsilon_i}{q\lambda_{n,p}} \tag{12}$$

$$\lambda_{n,p} = (\lambda_0)_{n,p} \tanh \frac{\epsilon_r}{2kT} \tag{13}$$

Note that $\alpha_{n,p}$ is zero when $E \cdot J_{n,p}$ is less than or equal to zero (Equation 9b), because the electric field is either decelerating or doing no work at all on the impacting specie when the angle between the direction of the electric field and the impacting specie is greater than, or equal to, 90 degrees.

Auger recombination is a three-particle process where two electrons and a hole interact leaving one electron plus a photon; or two holes and an electron interact leaving one hole plus a photon. For $n_p > n_i^2$ (nonequilibrium), these events occur at a frequency that can be expressed by:

$$R_{n,p} = \alpha_n n^2 p + \alpha_p' p^2 n \tag{14}$$

where $R_{n,p}$ is the electron and hole recombination rate density. More generally:

$$R_{n,p} = (\alpha_n n^{B_n} + \alpha_p p^{B_p})(np - n_{i_{eff}}^2) \tag{15}$$

where $n_{i_{eff}}$ is the equilibrium np product. For band gap narrowing:

$$n_{i_{eff}}^2 = n_{i_0}^2 \exp\left\{(\Delta E_c + \Delta E_v) / (kT/q)\right\} \tag{16}$$

where n_{i_0} is the intrinsic concentration of a lightly-doped semi-conductor and ΔE_c and ΔE_v are the changes in the conduction and valence band edges due to heavy doping. Equation (15) degenerates

to Equation 14 for $n_p \gg n_{i_{eff}}^2$. This ratio is relatively independent of temperature. Values of α_n and α_p are of the order of
10^{-31} cm^6 sec^{-1} and B_n and B_p typically have values of 1.0. This
means that for heavily-doped emitters ($N_D \sim 10^{20}$) minority carrier
lifetimes of 10^{-9} seconds are expected:

$$\tau_p \approx 1/\alpha_n n^2 \approx 10^{-9} \text{ sec.} \tag{17}$$

Boundary Conditions

The three semiconductor transport equations with three unknowns require three boundary conditions. Boundary conditions
are specified at contacts and along the entire edge of the semiconductor device model. At contacts, the space-charge potential
and the electron and hole densities are required. At ohmic contacts, thermal equilibrium and space-charge neutrality determine
charge carrier densities. Carrier concentrations at Schottky contacts are either set to fixed values [8] or modulated by a thermionic recombination velocity [9].

The capability of specifying current boundary conditions and
modeling simple circuit elements is available also in the FIELDAY
program. In order to specify current as a contact boundary condition, a relationship between current (I), and potential (Ψ), in
the neighborhood of a contact must be found. The previously-given
sum of Equations (2) and (3) integrated over the area of contact
gives this relationship. The current flowing into a contact of
area A is given by the following equation:

$$I(\psi) = \int_A q \left\{ D_n \nabla n - D_p \nabla p - \mu_n n \nabla(\psi + \Delta V_c) \right.$$

$$\left. - \mu_p p \nabla(\psi - \Delta V_v) \right\} \cdot d\hat{s} - \int_A \epsilon \frac{\partial (\nabla \psi)}{\partial t} \cdot d\hat{s} \tag{18}$$

If a capacitor, resistor, and current source are connected to a
contact, $I(\Psi)$ in Equation (18) is given by:

$$I(\psi) = i + c \frac{\partial (V_{applied} - \psi)}{\partial t} + V_{applied}$$

$$\tag{19}$$

$$- (\psi \pm \ln |N_D - N_A| / n_i \mp \Delta V_v) / R_{\substack{p \\ n}}^c$$

Here $V_{applied}$ is the voltage applied to the load and i is
the strength of the current source (positive value of i for current flowing into the contact).

At the noncontact boundaries of the semiconductor, the normal component of the electron and hole current densities and the
electric-field strength are all equal to zero.   In insulator
regions, only the space-charge potential and its normal derivative are considered.

Bipolar, Unipolar and No Current Flow

Many devices can be simulated accurately without modeling
the current flow of one or both mobile carriers. Hole current
can be ignored in the simulation of an n-channel IGFET under most
operating conditions. In this case, the equation for hole current
continuity is not used. If the flow of both carrier types can be
ignored, as in the simulation of the capacitance of a reverse
biased p-n junction, only Poisson's equation is solved.

The FIELDAY model may operate in one of three modes. The
first assumes bipolar current flow and determination of three
unknown quantities, Ψ, n and p. The second assumes unipolar current flow with two unknowns, Ψ and n, while the third assumes no
current flow and uses only one unknown, Ψ. In insulator regions,
there are no mobile charge carriers and only Poisson's equation
is needed.

SECTION 3   NUMERICAL APPROACH

The solution of the governing nonlinear, coupled partial differential equations by classic techniques, i.e., integration and
application of boundary conditions, is impossible for any except
the most basic problems. Instead, an approximation technique is
neccessary to transform the continuum problem to a discrete one.
The unknown variables are determined at a large but finite number
of points in both space and time so that an accurate solution of
the equations is obtained. In FIELDAY, the finite element method
transforms Poisson's equation, and a hybrid finite-difference/
finite-element technique transforms the current continuity equations [10]. Euler's method approximates the rate of change of
mobile charge density with time. These equations then are linearized by one of two methods. The first decouples the three discrete equations and solves them iteratively [11]. The second,
more involved, approach solves the equations simultaneously using
Newton's method [12,13]. Either approach results in large, sparse
matrix equations which must be solved numerous times to obtain
the final solution.

Poisson's Equation

The finite-element method transforms Poisson's equation from a continuous to a discrete form. Convenient piecewise approximations over arbitrary regions or elements are constructed. Using appropriate energy conservation principles, an elemental expression is obtained which relates the unknown (ψ) to elemental properties and charge density.

The following functional is used to approximate Poisson's equation [14]:

$$I = \int_V \frac{\epsilon}{2} (\nabla \psi)^2 dv - \int_V q\psi(p - n + N_D - N_A + N_Q)\, dv \qquad (20)$$

where V is the volume of the domain. The functional represents the energy of the system and may be expressed as the sum of a large but finite number (λ) of energies for all of the elements:

$$I = \sum_{e=1}^{\lambda} I_e \qquad (21)$$

Equation (20) is applied then to a single element over which a linear variation of (ψ) with respect to position is assumed. At equilibrium, energy is a minimum, and the first variation with respect to (Ψ) is zero. By taking the first variation and integrating over space, the following expression is obtained:

$$[A]\{\psi\} = [B]\{p - n + N_D - N_A + N_Q\}\ . \qquad (22)$$

The matrix [A] is symmetric and its terms are a function of the element geometry and permittivity. The number of nodes in an element is the order of [A]. For two-dimensional triangular and three-dimensional right-prismatic elements, the order is 3 and 6. The vectors $\{\psi\}$ and $\{p - n + N_D - N_A + N_Q\}$ represent the space-charge potential and charged-particle densities at each node of the finite element.

The matrix [B] is the element volume distribution matrix, which relates the portion of area or volume to a particular node of an element. The form of the distribution matrix is important. The traditional procedure, common in structural analysis and usually denoted as the lumped method, assumes equal portions of area or volume associated with each node. The consistent method assumes the same approximation for charge density as for the potential, in this case a linear variation [15]. Figure 1 shows a triangular mesh for an oxide-silicon interface structure. In a metal-oxide-silicon field effect transistor (MOSFET), an accurate, physically-realistic solution of this region is of importance since the device-current

flow is controlled by the interface potential. With the simplifying assumption of spatially constant charge density, the volumes
associated with nodes 1 and 2 in Figure 1a are 4/3 A and 2/3 A,
respectively, when calculated using the above methods. Since the
LaPlacian matrix has similar entries relating nodes 1 and 2 to
their respective neighbors, even a uniform charge distribution
will produce a solution with differences in potential along the
oxide-silicon interface. Thus, the solution does not reflect
the symmetry inherent to the problems. Indeed, any weighting
scheme that depends on how a regular array of nodes is connected
by triangles will result in anomalous solutions.

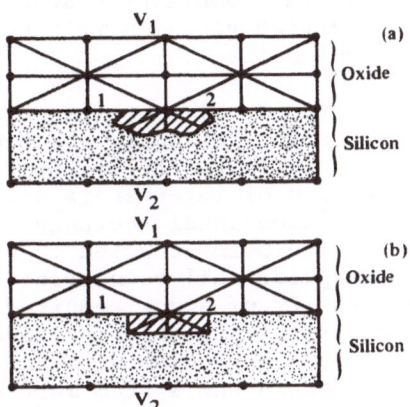

Figure 1.   Source distribution approximations at the oxide
            semiconductor interface.   (a) Equal areas method.
            (b) Closest node method.   The area of each tri-
            angle is A.

These difficulties can be minimized by choosing a weighting
scheme that does not depend on the way the nodes are connected.
This is achieved by associating an incremental area within a triangle with the node of the triangle that is closest to it. Thus,
the area associated with each node is defined by the perpendicular
bisectors of the sides of the triangle as shown in Figure 2. This
method is independent of the way the triangles are used, and the
deviations introduced by infrequently-used obtuse triangles are
small. Figure 1b shows the application of this method to the
oxide-semiconductor interface problem. The area weight of nodes
1 and 2 is A and the solution of Poisson's equation now will
reflect the problem's planar symmetry. For a three-dimensional
prismatic element, the volume associated with each node is the
appropriate portion of the triangular face times one-half the
altitude of the element.

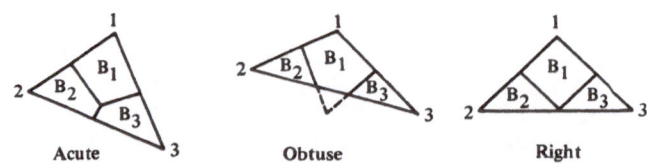

Figure 2. Source function distribution for acute, right, and
obtuse triangles.

The elemental matrix, Equation (22), is applied to each element of the domain and all contributions are combined into a global
matrix. The order of the global matrix is the total number of
nodes approximating the domain. Boundary conditions are applied
and the solution of the modified global matrix equation will yield
the value of ψ at each nodal point.

An alternative to the finite element method commonly used to
obtain approximate solutions for Poisson's equation is the finite
difference method, which approximates derivatives by using differences. The physical domain of interest is divided conceptually
into a large number of two-dimensional rectangular cells. Space-
charge potential in the device being studied is calculated at the
intersections of the rectangular cell boundaries called nodal
points (Figure 3). The density of nodes required in any region
of the device is related directly to the magnitude of the potential gradient in that region. A problem with the finite difference method is that all nodes in the system must be at the corners
of neighboring cells. This constraint forces an increase in the
density of nodes in areas otherwise of little interest, solely
because they are horizontally or vertically aligned with regions
containing higher potential variations. Since the solution's computer time requirements are related to the number of nodes, a more

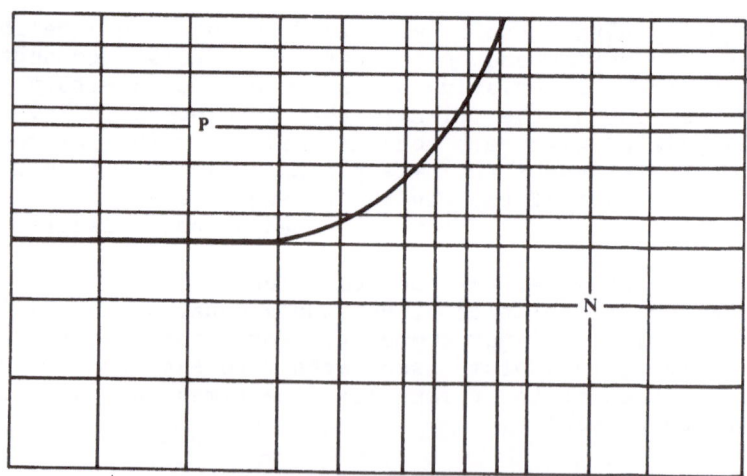

Figure 3. Finite difference mesh.

desirable system would be one which allows changes in the density of nodes across the device as potential variations dictate, yet remain independent of the geometry of neighboring nodes. The finite element method allows continuous and independent variation in cell density resulting in fewer nodes being required for a given problem [16].

Current Continuity Equations

The current continuity equations are solved using a hybrid scheme. The finite element division of space is used along with a difference approximation to describe current flow between nodes. The electron continuity equation is transformed using Gauss' theorem:

$$\int_V \nabla \cdot J_n dv = \int_S J_n \cdot dS = \int_V q \left(\frac{\partial n}{\partial t} + R_n \right) dv \tag{23}$$

where the surface, S, surrounds volume, V. For each element, the surface and volume are defined consistent with the distribution scheme for Poisson's equation. The discrete approximation of current is derived by assuming constant current density within the element. With this approximation, the total current flowing from node i to j along side k, as shown in Figure 4, is

$$I_k = q\mu_n \frac{kT}{q} \frac{d_k}{l_k} \left\{ n_j Z(\Delta_k) - n_i Z(-\Delta_k) \right\} \tag{24}$$

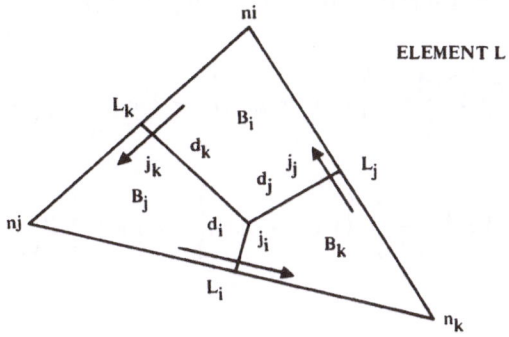

Figure 4. Current continuity in triangular element.

where d_k and l_k are flux cross-section and length of side k, and μ_n is the electron mobility for the element. The Bernoulli function is defined by

$$Z(\Delta_k) = \frac{\Delta_k}{e^{\Delta_k} - 1} \tag{25}$$

where Δ_k is the potential difference along the kth side. Evaluation of the Bernoulli function about the point $\Delta_k = 0$ requires special attention.

With the use of Euler's method to approximate the rate of change of mobile carrier density with respect to time, the discrete elemental electron current continuity equation becomes

$$\left([C] + [B]\frac{1}{\Delta t} \right) \left\{ n(t + \Delta t) \right\} = [B] \left\{ \frac{n(t)}{\Delta t} - R_n \right\} \tag{26}$$

Note that [C] and R_n are evaluated at time t. The matrix [C] is nonsymmetric and its coefficients are of the form $\mu(d/l)Z(\Delta)$.

The global form of the electron current continuity matrix equation is assembled on an element-by-element basis. The total number of unknowns may be less than the Poisson matrix equation, since current flows only in the semiconductor. A similar development is followed for the hole current continuity equation. The following is the discrete representation of the governing differential equations:

$$\{F_1\} = [A]\{\psi\} - [B]\{p - n + N_D - N_A + N_Q\}, \tag{27}$$

$$\{F_2\} = \left([C] + [B]\frac{1}{\Delta t} \right) \left\{ n(t + \Delta t) \right\} - [B] \left\{ \frac{n(t)}{\Delta t} - R_n \right\}, \text{ and} \tag{28}$$

$$\{F_3\} = \left([D] + [B]\frac{1}{\Delta t} \right) \left\{ p(t + \Delta t) \right\} - [B] \left\{ \frac{p(t)}{\Delta t} - R_p \right\} \tag{29}$$

Linearization Scheme

Since the discrete equations are nonlinear and coupled, a linearization scheme is required to solve them. Two algorithms have been implemented in FIELDAY. Both require an initial guess of the solution followed by an adjustment of the guess according to certain criteria until Equations (27) through (29) are satisfied to an acceptable degree of accuracy. The first method, described by Gummel, decouples the three equations and solves them serially. The second technique uses Newton's method to linearize the three equations and solves them simultaneously. Each approach has its own merits.

With the decoupled approach, the coupled differential equations are solved serially. In the "inner loop," the linearized Poisson Equation (1), is solved repeatedly until a consistent set of ψ, n and p is obtained. This is accomplished by assuming that the quasi-Fermi potentials are known functions of position. Equation (1) is linearized with Newton's method. This results in the following normalized expression

$$([A] + [B]\{n^k + p^k\})(\delta^{k+1}) = -\{F^k\} \tag{30}$$

where

$$\delta^{k+1} = (\psi^{k+1} - \psi^k)/(kT/q)$$

and F^k is the residual of the homogeneous equation evaluated with the results of the kth iteration. With fixed quasi-Fermi potentials the expressions

$$n^{k+1} = n^k \exp(\delta^{k+1})$$
$$p^{k+1} = p^k \exp(\delta^{k+1}) \tag{31}$$

aid in the evaluation of n and p by avoiding exponential overflows. Equation (30) is solved many times to find self-consistent values of ψ^k, n^k and p^k. The resulting values then are used to obtain new values of n and p consistent with current continuity. This entails evaluating the coefficients of Equation (27) with known space-charge potentials, ψ, and solving for the mobile carrier densities. This, in essence, updates the quasi-Fermi potentials. Poisson's Equation (27) then is evaluated to test for global convergence. The inner-outer loop cycle is repeated until self-consistent values of ψ, n and p are obtained. This process is outlined in Figure 5.

The decoupled approach is attractive since portions of the program can be written and tested independently. The disadvantage of the approach is possible slow convergence since, in many applications, the equations are strongly coupled.

The coupled linearization technique uses Newton's method to linearize Equations (1)-(3) simultaneously. This method is

596

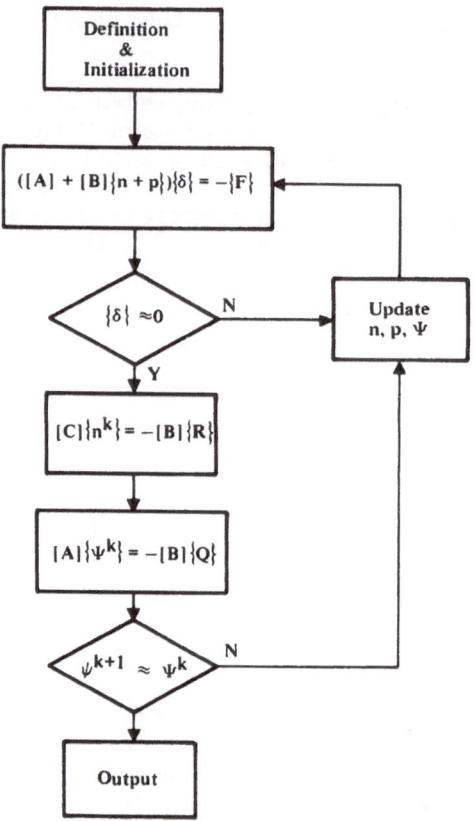

Figure 5. Decoupled algorithm (Gummel's algorithm).

attractive since quadratic convergence is anticipated as the
solution is approached. We derive

$$
\begin{bmatrix}
\partial F_1/\partial \psi & \partial F_1/\partial n & \partial F_1/\partial p \\
\partial F_2/\partial \psi & \partial F_2/\partial n & \partial F_2/\partial p \\
\partial F_3/\partial \psi & \partial F_3/\partial n & \partial F_2/\partial p
\end{bmatrix}
\begin{Bmatrix}
\Delta \psi \\
\Delta n \\
\Delta p
\end{Bmatrix}
=
\begin{bmatrix}
-F_1 \\
-F_2 \\
-F_3
\end{bmatrix}
\tag{32}
$$

where the F_i and dF_i/dx are evaluated for the kth iteration and
$\Delta x = x^{k+1} - x^k$. Equation (32) represents the basic finite element
equation for a single node. In a two-dimensional, first-order triangular element, the F and the coefficients will have different
values at each of the three nodes. Therefore, the elemental version is a ninth-order system of linear equations. These elemental

equations are summed into a sparse global matrix representing the entire device model. They are solved repetitively for the incremental values of n, p and ψ until convergence is obtained.

Although mathematically more attractive than the decoupled approach, there are difficulties from a practical standpoint. The main implementation problem is the evaluation of the derivative terms in Equation (32), since the field or doping dependent mobilities, recombination-generation, and the Bernoilli function (Equation (25)), must be differentiated with respect to the variables ψ, n and p.

Matrix Solution Method

With either linearization scheme, matrix equations must be solved. The matrices are large and their order is usually between 100 and 5000. They also are sparse since the number of nonzero terms usually is less than five percent. The solution of the matrix equations accounts for most of the computation; therefore, a judicious choice of technique is important.

A direct-matrix-solution technique was chosen rather than an iterative technique since the solution time for iterative methods depends on the numerical conditioning of the matrix and may fail to converge in some situations. Direct methods require much more storage, but the solution time for different problems of the same order will not vary drastically and always will yield a set of results. The approach in FIELDAY is to use a symbolic and numeric factorization procedure for direct solution of the appropriate matrix Equation [17].

In the coupled approach, the Waterloo Sparse Matrix Package (SPARSPAK) [18] is used to solve the matrix equation. The SPARSPAK package is primarily the work of J. Alan George and Joseph Liu and provides a number of direct-solution techniques. SPARSPAK requires the nonzero terms of the matrix, its row and column pointers, and a single work vector; it then internally allocates storage it needs from this vector. Among the techniques available in SPARSPAK are nested dissection and quotient-tree implicit block factorization ordering [19,20]. The nested dissection algorithm is used to order the coefficient matrix so that an explicit LU factor may be determined efficiently. This ordering results in a low multiplication count, which is a good measure of the amount of computation. The quotient-tree implicit block factorization technique produces an ordering that gives the matrix a block-like structure. This structure permits the LU factorization to be carried out implicitly without having to store much of the "fill" for the factored matrix. Significant storage savings are achieved then at the expense of an increase in computation.

The decoupled approach requires the solution of three matrix equations of the order N-by-N. The coupled approach requires the solution of a single 3N-by-3N equation. Since the amount of computation is proportional to the square of the order of the matrix, the coupled approach requires more computations per iteration. However, the decoupled approach also may require more iterations. The question of which approach should be used has been addressed in a previous work, where it was shown that the decoupled approach was more efficient for "weakly" nonlinear problems and the coupled approach more efficient for "highly" nonlinear and transient problems [21].

Three-dimensional problems require more computation resources than two-dimensional problems. For a mesh of the same degree of accuracy, the three-dimensional solution requires on order N^3 more CPU time and storage where N is the number of planes in the three-dimensional model. Since N typically may be equal to or greater than 10, the three-dimensional analyses are significantly more expensive.

Modified Newton's Method

A simple modification of the Newton technique can reduce dramatically the amount of time spent solving the matrix equations. In the modified method, the Jacobian matrix is computed and factored only every mth iteration. The factor m is determined internally within FIELDAY and is adjusted as the solution proceeds. Significant savings can result; e.g., in a series of six test problems, total computation time was cut in half. In fact, the savings was greatest for larger problems since a greater part of their execution time was spent solving the matrix equations. There are additional savings that can result when using the modified Newton approach for a sequence of similar problems. This occurs for transient or steady-state analyses with similar boundary conditions. In those cases, the Jacobian Matrix from the preceding problem can be reused, which again saves in the calculation effort.

SECTION 4 PRE- AND POST-PROCESSING

Significant time and cost benefits are achieved by using interactive graphics. A package of interactive programs has been designed to be used on the IBM 3277 Display Station Graphics Attachment. These programs are used to generate complex finite element models and to analyze results. For preprocessing, the model generation programs fall into three categories: model definition, impurity concentration designation, and input verification.

Model definition consists of generating a finite element mesh, assigning material properties to each element, and designating contact nodes. The FIELDAY user has several options available for model definition. POINTS is a semiautomatic generation scheme. A rectangular grid work is defined with uniform spacings, which are changed as required. The model is synthesized from rectangular regions, each with a constant permittivity and then scaled to the problem dimensions. The user also supplies information indicating the element material type and contact position. A bipolar mesh generated with POINTS is shown in Figure 6. To create this mesh by hand takes about one month; to generate it with POINTS and its interactive graphics capability requires no more than a few hours of the user's time. POINTS generates, displays, and stores the mesh for later use by FIELDAY. During mesh display, the contact nodes are indicated and interactive windowing and node and element numbering may be utilized.

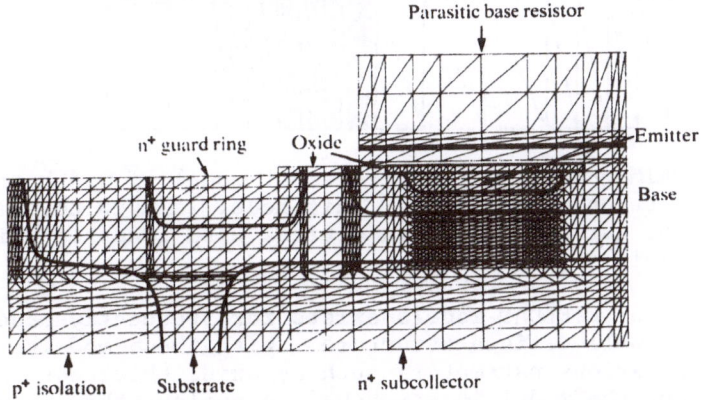

Figure 6. POINTS-generated finite-element mesh used to simulate a bipolar transistor. The parasitic base resistor models the flow of base current around the end of the emitter in the third dimension, which is not simulated.

Another option for model definition is the (TRIangular Mesh generator) TRIM. Here, the user specifies the boundary of a region and selects a mathematically-regular grid, such as a rectangular mesh. TRIM generates a conformal map of that mesh onto the user's model. The same mesh may be mapped onto geometrically-similar models; thus, the user need not respecify the mesh generation information but only a few details relative to his model. Contact designation, mesh storage, and display features are similar to those available for POINTS. Figure 7 shows a mesh, with an unusual semirecessed oxide shape, generated by TRIM. Note the varying density of elements over the model. This increases the accuracy of solution at points where the fields rapidly are changing.

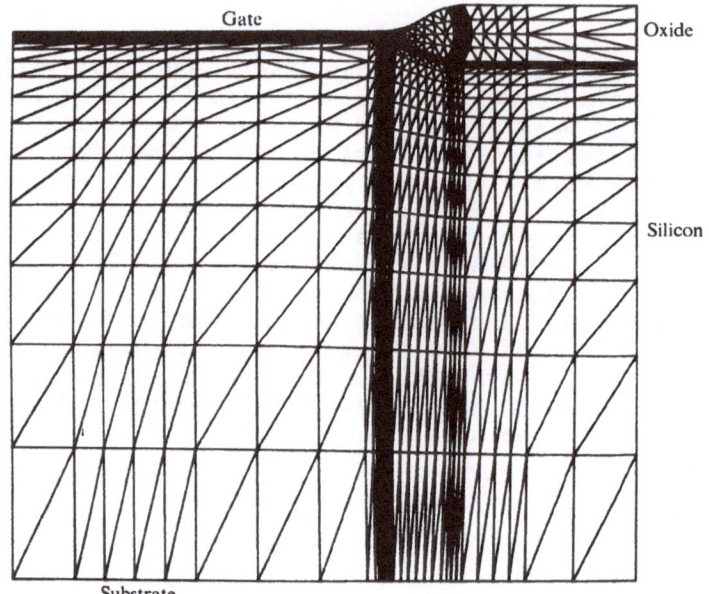

Figure 7.   TRIM-generated finite-element model of width cross-
section of an IGFET. This model is used to simulate
the narrow channel in devices with recessed-oxide
field regions.

A library of FIELDAY models exists for frequently-modeled
devices with a well-defined structure. With these models, the
user supplies various parameters such as oxide thickness and junction depth, and the model is stretched to reflect the given parameters. Again, mesh storage and display are possible. For three-
dimensional simulations, a two-dimensional model is created using
one of the above techniques and is replicated in the third dimension to produce a mesh of right-triangular prisms. The user then
can delete elements, change their material properties, or reassign
contacts. Figure 8 shows a short and narrow IGFET and the mesh
used for simulation.

Impurity concentration is designated by assigning an
electrically-active impurity ion density to each node of the
finite-element mesh. This is accomplished by specifying measured values, employing a process simulator, or describing a profile as the sum of analytic expressions. The pre-processing
program, DOPING, allows viewing of the impurity concentration
profile prior to FIELDAY execution to ensure that the device being
modeled is the desired one. The impurity concentration can be
displayed with contour plots, line graphs, or perspective plots.
Figure 9 shows a perspective plot of the doping profile for a
bipolar transistor.

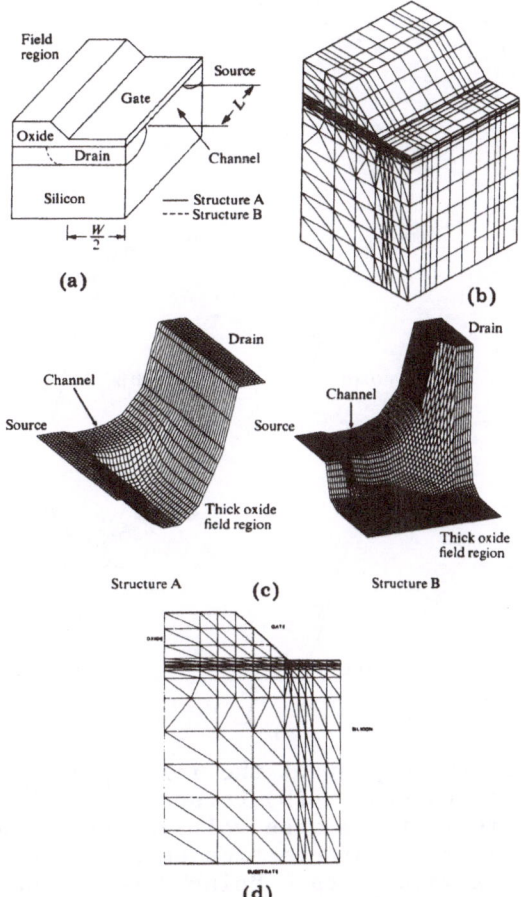

Figure 8. (a) Structure of a short and narrow IGFET device.
Oxide thicknesses are 50 and 800 nm for the active
and field devices, respectively. The source and
drain are abrupt, cylindrical junctions 0.5 μm deep.
The channel doping may be described by $C = C_o \exp$
$[-(Y - R)^2/2S^2]^2 + C_b$, where $C_o = 2.5 \times 10^{16}$ cm^{-3},
$S = 0.2$ μm, $C_b = 1.0 \times 10^{15}$ cm^{-3}, and $R = 0.0$ μm.

(b) The finite-element mesh used to simulate a 1.5-
by-1.5 μm short and narrow IGFET containing 13 planes
of 164 nodes each.

(c) Modeled surface potential for Structures A and B
at a drain-to-source bias of 5.0 V and a source-to-
substrate bias of 1.0 V. The device length and width
is 1.5 μm.

(d) Finite element mesh of width cross-section.

602

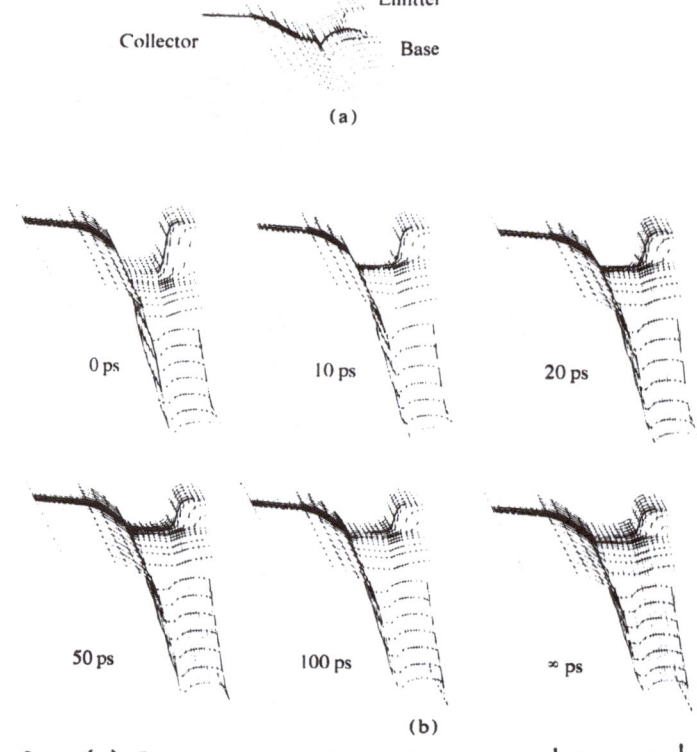

(a)

(b)

Figure 9. (a) Perspective plot of the log $\left| N_A - N_D \right|$ for a
bipolar transistor. (b) Perspective plots of
log (n) following a base voltage step. The
gradual increase of electron density under the
emitter shows that base "pinch" resistance is
limiting the device's switching speed.

Input verification consists of mesh checking and input consistency checking. During mesh checking, the finite element mesh
is examined for errors and poorly-shaped elements. The areas of
the mesh in which problems occur are highlighted. The user simultaneously can watch to determine how to modify the mesh in the
event of errors. The types of errors that can arise include overlapping elements and dangling nodes. Poorly-shaped elements are
obtuse triangles or elements with large aspect ratios. Other
input is examined for completeness and consistency.

For post-processing, a program called FEMPLOT permits rapid
interpretation of the FIELDAY analysis with the ability to inter-
actively view the results. FEMPLOT will display nodal values of
potential, electron density or hole density with contour plots,
line graphs, or perspective plots. The elemental values of electric field strength and current density also may be displayed.
Contour plots are a means of displaying nodal data values. Lines

of equal value are drawn through points of equal value interpolated along the sides of the elements. Perspective graphs are a means of displaying all the nodal data from a two-dimensional surface of a model. Figure 9 shows a series of perspective plots of the log of the electron concentration at various times during the transient response. The gradual increase of electron density under the emitter shows that base "pinch" resistance is limiting the device's switching speed.

Model definition using graphics replaces the time-consuming task of meticulously defining every node and element in the finite element mesh, which formerly took 65-70% of the total analysis time. FEMPLOT minimizes the time the FIELDAY user spends searching through results on a node-by-node or element-by-element basis. Device designers are able to optimize designs by rapidly viewing simulation results and noting activity within the device that cannot be measured experimentally. This capability results in fewer design projects terminating after considerable resource expenditure. Therefore, the use of interactive graphics results in higher engineering productivity by reducing development time, lowering development costs, and making more resources available for product optimization.

Device Simulation Results

In this section, several applications of the FIELDAY program are described which illustrate its features, accuracy, and flexibility. First, a model of the effect of short and narrow channels on IGFET thresholds is presented. Separate two-dimensional models simulate the short-channel effect for wide devices and the narrow-channel effect for long devices. The accuracy of these simulations is demonstrated by the close agreement between the model and experimental data. A three-dimensional model of an IGFET also is presented. Comparisons of two- and three-dimensional results show the need for this approach for short and narrow devices. A VMOS transistor was optimized using the program. In this case, simulation showed that a superior device could be designed. Simulation of the transient response of a bipolar transistor is described, and again excellent agreement between data and model is demonstrated. Three-dimensional transient capability of the program is illustrated by simulation of reverse recovery of an ellipsoidal junction.

Short- and Narrow-Channel Effects in IGFETs

The FIELDAY program frequently has been used to model short- and narrow-channel effects in IGFETs [13, 22-26]. The short channel effect is the decrease in the threshold voltage due to potential barrier lowering caused by the proximity of the large drain potential. A related phenomenon is the reduction of the sensitivity of the threshold voltage to the substrate-source bias for short-channel lengths. Narrow-channels have the opposite effect. The threshold increases as the edges of the high-threshold field region move toward the center of the lower-threshold active device. Since channel length and width vary from device-to-device across a VLSI chip, and to a large extent from chip-to-chip, a statistical analysis of the effect of these parameters on threshold is required. Variation in channel length may contribute more than 50% to the total threshold tolerance of a well-designed IGFET.

One evaluated device has the structure shown in Figure 10. All the structural parameters used as input data for the model were determined by physical measurements independent of the experimental device characteristics. The background concentration was assumed constant and the source and drain profiles were considered Gaussian along the radii of the cylindrical, lateral diffusion regions. Figure 11 clearly shows the effect of varying channel length and drain bias. For comparison purposes, the modeling results have been shifted by a constant value in gate voltage to compensate for the effect of fixed charge, work function differences, and boron depletion near the oxide-silicon interface. Figure 12 shows more complete characteristics for the 1.2 micron channel length device with a drain potential of 4 volts. For all biases and channel lengths, the modeling results match the experimental results to well within the limits of error imposed by inaccuracies in measurement of the physical device parameters.

The short- and narrow-channel effects are modeled independently for long and narrow, or for short and wide, devices by

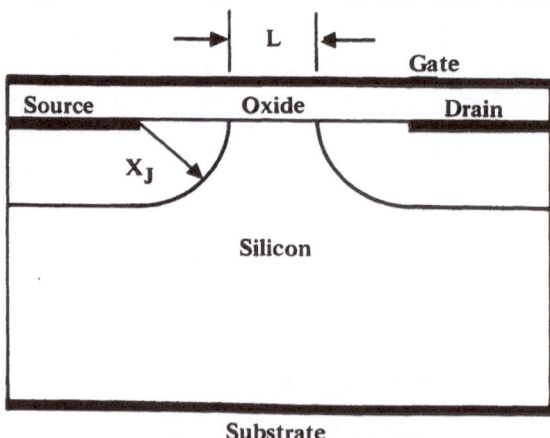

Figure 10. IGFET device structure: $X_J = 1.85 \mu m$, $T_{ox} = 0.07 \mu m$, $C_o = 1.0 \times 10^{19}$ cm^{-3}, and $C_B = 6.5 \times 10^{15}$ cm^{-3}.

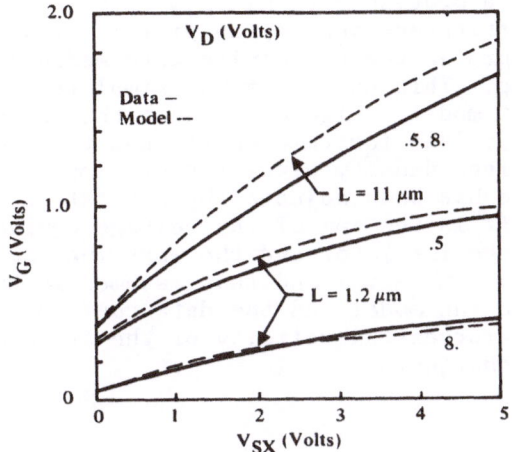

Figure 11. Gate voltage required for 10^{-8} W/L amperes of source-drain current versus source to substrate bias.

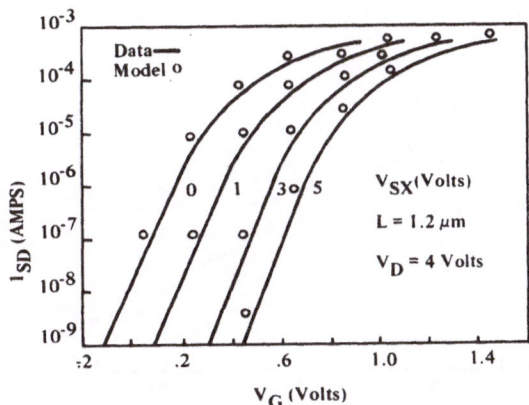

Figure 12. Source-drain current versus gate bias for varying source-substrate bias.

simulation of mid-channel two-dimensional cross-sections. Short and narrow devices must be simulated by a three-dimensional model. Here, application of the model to devices as short as 0.7 microns and as narrow as 1.5 microns are discussed.

The effect of short-channels was modeled and measured on wide
devices designed for a process with minimum feature sizes of 1.0
microns. The oxide thickness and doping profile of a capacitor
with the same structure as the FET device were measured by a pulsed
capacitance technique. This capacitor was simulated with a one-
dimensional transient model. Figure 13 shows the agreement between
measured capacitance values for this device and those derived from
the displacement current density predicted by simulation. The
simulated capacitance has been adjusted by a constant value of
gate bias. This shift is the sum of the voltage equivalent of
the charge found in the insulator and the work function difference
between the actual gate material and that assumed in the model.
The agreement between the model and the data confirms the accuracy
of the model and reinforces the validity of the transient capacitance measurement technique.

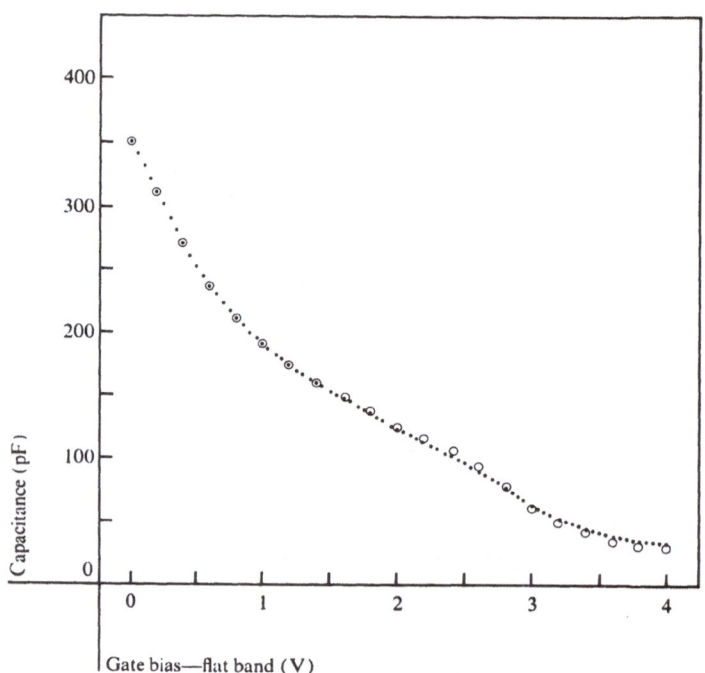

Figure 13. Measured (O) and simulated (●) capacitance versus
gate-to-substrate bias. The simulated results have
been shifted by -0.56 V to account for work function and oxide charge differences. The channel
doping parameters are C_o = 5.24 x 10^{16} cm^{-3}, C_b =
1.2 x 10^{15} cm^{-3}, S = 15.3 nm, and R = -24.0 nm.
The oxide thickness is 27.8 nm.

Devices with the previous structure were simulated for channel lengths of 10.0, 2.0, 1.3, 1.0, and 0.7 microns with a two-dimensional model having 1620 nodes. The inversion charge of the capacitor as a function of gate voltage also was modeled. Figure 14 shows the modeled and actual threshold as a function of source-to-substrate bias for a 10.0-micron-long device. The effect of decreasing channel length on threshold is illustrated in Figure 15.

The narrow-channel effect was investigated for the Silicon and Aluminum Metal Oxide Semiconductor (SAMOS) transistor [24, 26, 27]. The threshold of this device was determined by modeling the width cross-section of the transistor. The threshold was calculated by linearly extrapolating the variation of inversion with

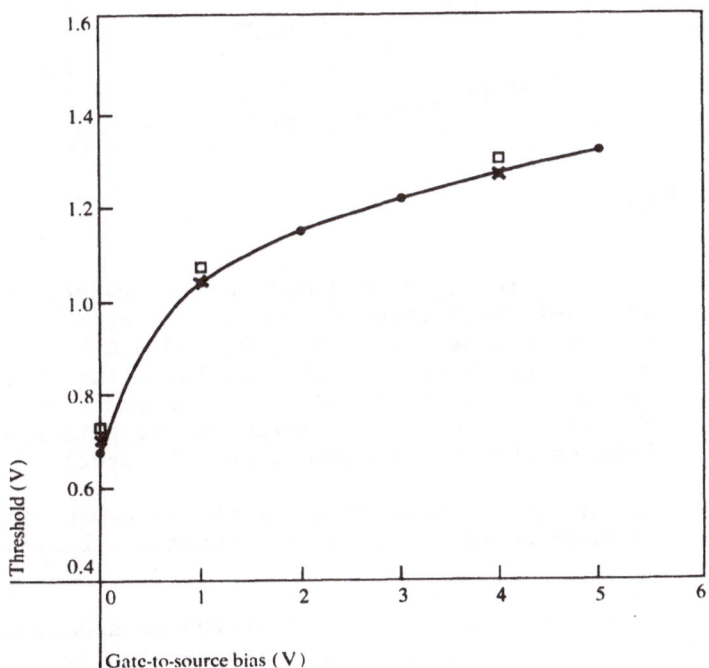

Figure 14.   Measured (□) and modeled (● capacitor model, x IGFET model) long-channel threshold. The simulated threshold has been adjusted by -0.56 V to account for work function and effective oxide charge difference between the actual devices and the model. The threshold of the 10.0-μm device is defined at 40 nA of normalized source current. The threshold of the IGFET as modeled by the capacitor is defined at 10^{10} electrons of inversion charge per square cm. The vertical and lateral junction depths are 0.25 and 0.15 μm, respectively.

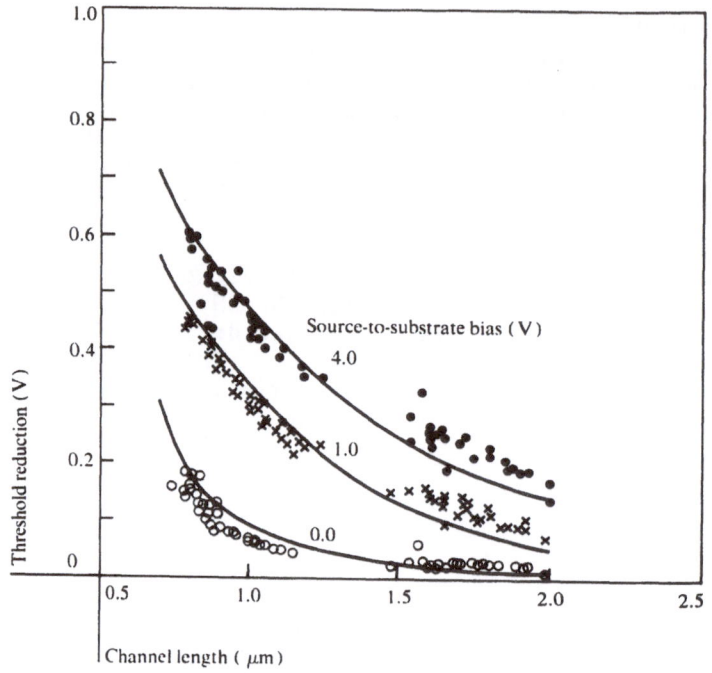

Figure 15.   Measured (●, x, 0) and modeled (-) reduction in
              threshold with channel length at source-to-
              substrate biases of 0.0, 1.0, and 4.0 V and a
              drain bias of 4.0 V. The simulated threshold
              has been adjusted by -0.56 V to account for work
              function and effective oxide charge difference
              between the actual devices and the model.

gate bias to zero charge. Figure 16 shows the agreement between
the empirical and modeled threshold at two substrate biases for
long devices.

    A three-dimensional simulation [22] of the threshold of a
short and narrow device was made on two devices with structures
defined in Figure 8. The difference between the two structures
is the shape of the diffused source and drain. In Structure A,
these diffusions extend under the field oxide, while in Structure
B they are terminated at the edge of the thin-oxide region. Figure 8b shows the 2132-node finite-element mesh used to model these
devices. One plane of the three-dimensional finite element mesh
used in analysis is shown in Figure 8d. Simulated subthreshold
characteristics were used to define the threshold of a device 1.5
microns long and wide. The sensitivity of this threshold to

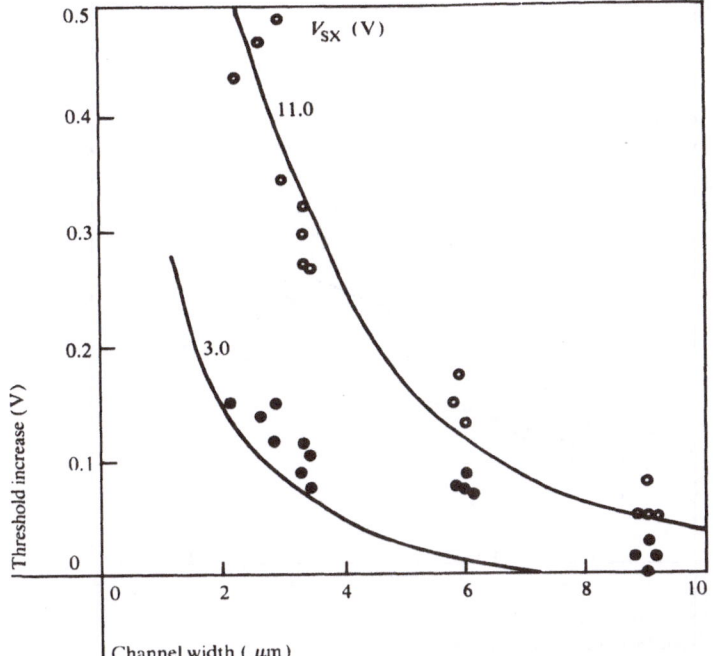

Figure 16.   Measured (O, ●) and modeled (-) increase in threshold with decreasing device width for a long-channel SAMOS transistor.

increasing source-to-substrate bias is shown in Figure 17. The threshold as predicted also is shown by a composite of separate, two-dimensional short- and narrow-channel models. For the composite model, the threshold is defined as the algebraic sum of the threshold changes predicted by the independent short- and narrow-channel models and the threshold of a long and wide device. In this case, the width and length of the cross-sections were taken at mid-channel. The composite model is inadequate because it fails to predict the effect of relatively minor differences between Structures A and B. In addition, the result of the composite model produces a threshold dependence on source-to-substrate bias that bears little functional resemblance to the actual characteristics.

The different behavior of Structures A and B can be explained easily. The extension of the diffusions raises the surface potential in the field-oxide region near the active device. This reduces the impact of the narrow-channel effect and results in a lower threshold for the device with Structure A. Figure 18 shows equipotential contour plots of the faces of the model of Structure A.

610

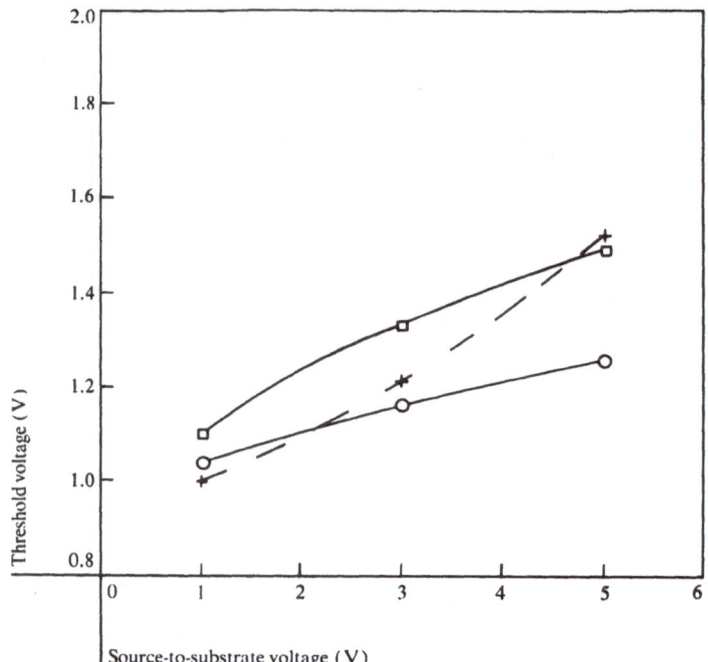

Figure 17.   Three-dimensional (0,□) and composite (+) model
threshold versus source-to-substrate bias for a
device 1.5 μm long and wide with Structures A (0)
and B (□).   The device structure is described in
Figure 8.   The drain-to-source bias is 5.0 V.

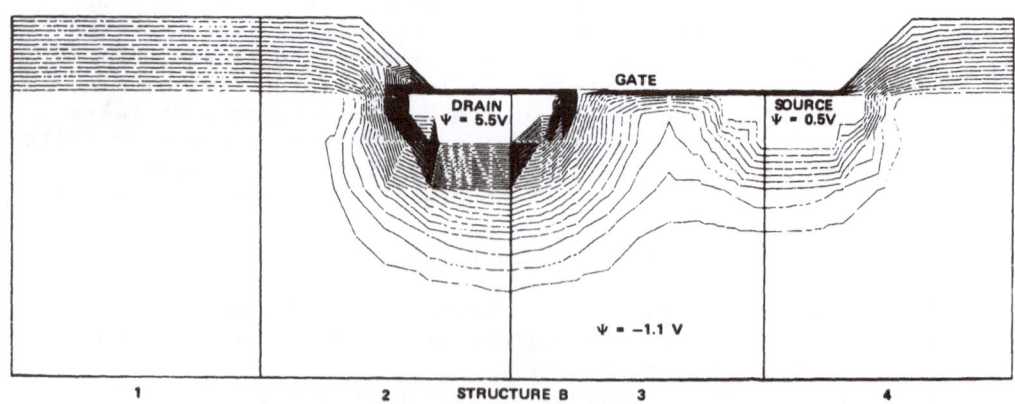

Figure 18.   Equipotential contours for a 1.5 by 1.5 μm device
at V_{DS} = 5.0 and V_{SX} = 1.0 V.

igure 8c shows a perspective plot of the surface potential for
tructures A and B. Here, the influence of the extensions of the
ource and drain under the field oxide can be readily seen.

Separate analysis of the short- and narrow-channel effects
f a device over a range of sizes and operating conditions takes
everal weeks of work and approximately 20 CPU hours on an IBM
ystem/370 Model 168. This expense results in about 800 values
f current as a function of bias and device size. These can be
educed to 80 values of threshold. It is worth the cost and
ffort, since it could take up to six months to obtain similar
esults empirically, at over 100 times the cost.

imulation of VMOS field-effect transistors

VMOS transistors are being investigated for use in one-device
andom access memories [28, 29]. A VMOS memory cell uses devices
hich have the shape of inverted pyramids to charge and discharge
he buried-diffusion storage capacitor. A typical cell structure
s shown in Figure 19. The threshold of this transistor must be
igh enough to prevent discharge of the capacitor when the surface
iffusion is grounded and low enough to adequately charge the
apacitor when the surface diffusion and the gate are biased
ositively.

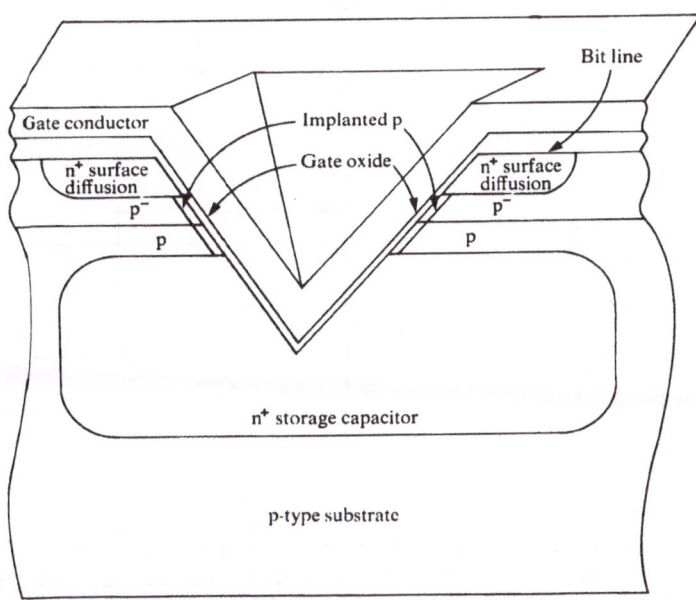

Figure 19. VMOS dynamic memory cell structure.

Meeting these criteria with a VMOS device is difficult because of its asymmetrical structure. The p-type region above the buried n+ diffusion raises the threshold of the device when charging the capacitor. This reduces the stored charge. The doping level in this p-type region cannot be reduced because the decreased capacitance of the buried diffusion also would reduce the stored charge. The presence of this layer has an additional effect. It causes the threshold of the transistor to decrease rapidly with increasing voltage on the buried diffusion. This requires an additional increase in the nominal threshold so that the capacitor will not discharge, when the surface diffusion is grounded, and further will degrade the stored charge. These effects can be described by a single figure of merit defined as the sum of a holding and a charging loss. As shown in Figure 20a, the charging loss is the increase in threshold with increasing surface-diffusion-to-storage-capacitor bias. The holding loss, shown in Figure 20b, is the decrease in threshold with increasing storage-to-surface-diffusion

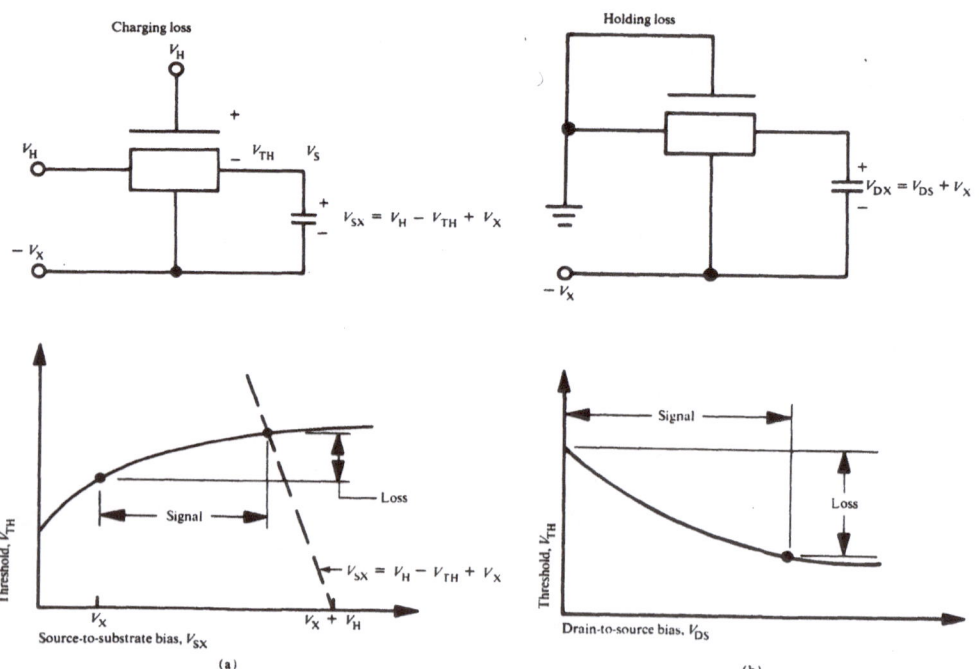

Figure 20.   Schematics of (a) charging loss and (b) holding loss of a one-device memory cell. The charging loss is caused by the increase of threshold with source-to-substrate bias. The holding loss is caused by the reduction in threshold with drain-to-source bias.

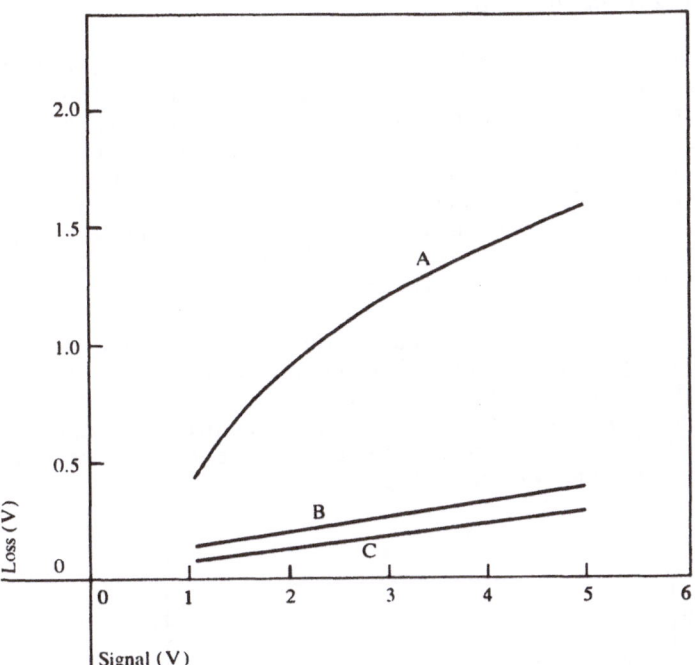

Figure 21.   Total holding plus charging loss versus stored
voltage for a conventional VMOS transistor
(Curve A), an implanted VMOS transistor (Curve
B), and an implanted planar transistor (Curve C).

bias. This loss represents the inefficiency of the transistor
caused by dependence of the threshold on source and drain bias.

Various transistor designs were investigated using the FIELDAY
program [30]. It was found that the transistor characteristics
were very sensitive to the variation of the doping level and position of the p-type region. A steep profile, along with an additional implanted p-type region under the gate oxide, produced a
nearly optimum device. The loss-versus-signal characteristics of
this implanted VMOS device, a conventional VMOS transistor, and a
planar device with similar oxide thickness and channel length are
shown in Figure 21. With a 5-V power supply, the conventional
VMOS device can store only 75% of the charge of a planar device,
while the implanted device can store 90%. Thus, a 20% increase
in efficiency was predicted through simulation. This may be
translated directly into increased memory performance or reduced
chip area and cost.

Transient Simulation of Bipolar Transistors

In most cases, the performance and function of IGFET devices can be predicted from their steady-state behavior. The intrinsic speed of these devices exceeds the speed at which practical IGFET-integrated circuits can operate. This is because IGFETs are majority-carrier devices with no significant minority-carrier injection. On the other hand, bipolar devices are minority-carrier devices and their performance in an integrated circuit depends on the transient response of individual devices. Thus, transient simulation of bipolar devices is most important.

An npn bipolar transistor was simulated in two dimensions with the FIELDAY program [31]. The response of the collector current to a rapidly increasing base voltage was measured and modeled. Figure 22 shows the predicted and actual transient response of the collector voltage. Good correlation is shown between experimental and modeled characteristics.

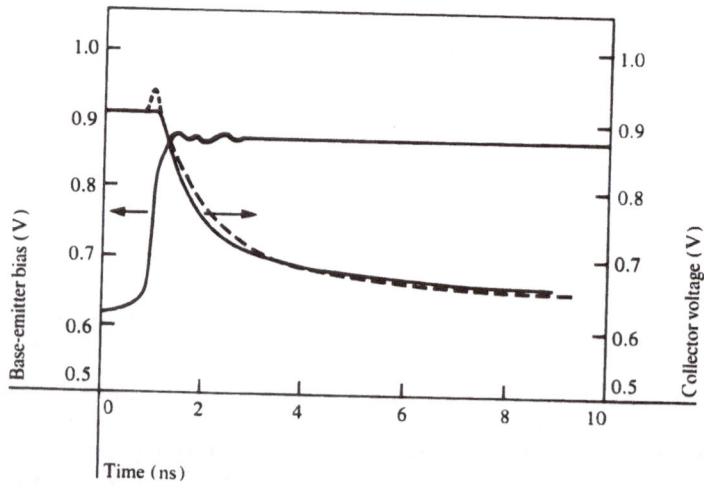

Figure 22. Simulated (—) and measured (- -) transient response of a bipolar transistor to an abrupt change in base-emitter bias. The base voltage in the simulation has been decreased by 0.028 V to make the modeled and simulated steady-state collector currents equivalent.

The transient three-dimensional capability of FIELDAY is demonstrated with the simulation of the reverse recovery of an ellipsoidal junction [32]. A structure similar to that shown in Figure 23 may be found at the four corners of every integrated

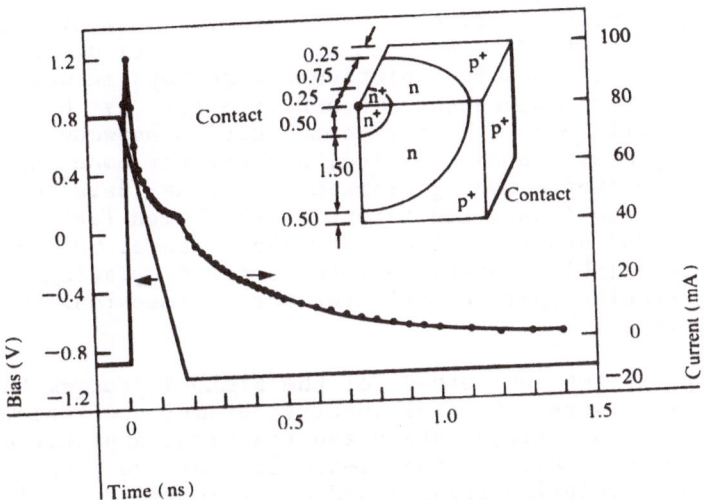

Figure 23.   Reverse recovery current of an ellipsoidal junction.
The bias is changed from +0.8 V to -1.0 V in 10-ps
steps.   The abrupt change in current at 0.0 and 0.18 ns
is the displacement current which is proportional to
the rate of change in bias across the junction.   Dimensions in μm.

bipolar transistor.   The reverse recovery of this structure will
play an important role in the performance of bipolar devices as
the area of emitters is reduced.   The results are shown in Figure
23.   In first order, the recovery time is in agreement with that
predicted by classical theory:

$$I = W_b^2/2D = 0.45 \text{ ns.}$$

Although detailed transient analysis of three-dimensional
structures presently is costly, this example illustrates the feasi-
bility of such a capability and serves to guide further development
of algorithms which would allow routine use.

SECTION 6   SUMMARY AND CONCLUSIONS

Future improvements will be made to the FIELDAY program in
the areas of enhanced physics, speed, and user friendliness.   It
is well-known that certain forms of ionized radiation, when incident on a semiconductor circuit, can cause soft-errors in that
circuit.   Implementing the α-particle physics in FIELDAY will
enable the designer to evaluate the effect of process changes,
such as deep-buried implants, on soft-error rate sensitivity.

Improved mobility models, particularly at the surface, would improve the FET threshold results considerably. Computation times will continue to be decreased by a mixture of new techniques in the linear equation solution and perhaps in new hardware such as an array processor. A new data base has been designed which will permit better communication between the pre-processing programs, process, device, and circuit modeling programs, and the post-processing programs. The new data base will contain all the run inputs and outputs. The human factors of the use of the FIELDAY system will be improved. Easier mesh generation will be available. Suitable methods for displaying and interpreting results, particularly for three-dimensional models, must be discovered.

The capabilities and methods of the FIELDAY program have been described in this work. Several specific examples of one-, two-, and three-dimensional steady-state and transient applications have been presented to illustrate the flexibility of the program. Close correlation of simulation results and experimental data illustrate the accuracy of the model and the credibility of its underlying assumptions and computation methods. Pre-processors speed the creation of new models through interactive mesh generation. Post-processing programs allow rapid examination of the internal operation of devices and subsequent improvement of design.

These capabilities form a comprehensive CAD tool which allows prediction of the characteristics of new devices and rapid response to problems affecting device function and reliability. In its predictive role, FIELDAY can be used to evaluate new device concepts and to optimize device design prior to fabrication. This is important because of the time and cost involved in evaluating new ideas for integrated devices.

FIELDAY, and similar programs, fill a gap in the development of integrated circuits. This gap is between the generation and design of new device concepts and the simulation of circuits using those devices. This role is important for any CAD tool because significant changes and improvements in design often can be made only at the early stages of development. In addition, these tools can replace costly matrix experiments and allow device design and process stabilization to occur simultaneously. Careful simulation can avoid disastrous and costly mistakes that often plague new product development.

In a responsive role, FIELDAY can offer rapid and definitive analysis of device phenomena that limit circuit function or affect the reliability of a product. In this case, a hypothesis can be proposed and tested without fabricating devices, thus reducing the time required to solve this type of problem. Simulation

allows examination of the internal operation of devices, and the
resulting insights often spark innovative solutions.

FOOTNOTES

1. Cottrell, P. E. and E. M. Buturla. "Application of the Finite
Element Method to Semiconductor Transport," presented at the Asilomar Conference on Circuits and Systems, Pacific Grove, CA, (1976).
2. Slotboom, J. W. and H. C. DeGraaff. "Measurements of Band
Gap Narrowing in Si Bipolar Transistors," *Solid State Electron.* *19*,
(1976) 857-862.
3. Caughey, D. M. and R. E. Thomas. "Carrier Mobilities in Silicon Empirically Related to Doping and Field," *Proc. IEEE 55*, (1967)
2192-2193.
4. Dziewior J. and W. Schmid. " Auger Coefficients for Highly
Doped and Excited Silicon," *Appl. Phys. Lett. 31*, (1977) 346-348.
5. Hall, R. N. "Electron-Hole Recombination in Germanium," *Phys.*
Rev. 87, (1952) 387.
6. Shockley, W. and W. T. Read. "Statistics of the Recombination
of Holes and Electrons," *Phys. Rev. 87* (1952) 835-842.
7. Crowell, C. R. and S. M. Sze. "Temperature Dependence of Avalanche Multiplication in Semiconductors," *Appl. Phys. Lett. 9*
(1966) 242-244.
8. Schottky, W. *Naturwissenschaften 26* (1938) 843.
9. Crowell, C. R. and S. M. Sze. "Current Transport in Metal-
Semiconductor Barriers," *Solid-State Electron. 9* (1966) 1035.
10. Buturla, E. M. and P. E. Cottrell. "Two-Dimensional Finite
Element Analysis of Semiconductor Phenomena," presented at the
International Conference on Numerical Methods in Electric and Magnetic Field Problems, Santa Margherita, Italy, 1976.
11. Gummel, H. K. "A Self-Consistent Iterative Scheme for One-
Dimensional Steady State Transistor Calculations," *IEEE Trans.*
Electron Devices ED-11, (1964) 445-465.
12. Hachtel, G. D., M. Mack, and R. R. O'Brien. "Semiconductor
Device Analysis Via Finite Elements," presented at the Eighth
Asilomar Conference on Circuits and Systems, Pacific Grove, CA,
(1974).
13. Cottrell, P. E. and E. M. Buturla. "Two-dimensional Static
and Transient Simulation of Mobile Carrier Transport in a Semiconductor," *Proceedings of the NASECODE I Conference*, Boole Press,
Dublin, Ireland (1979) 31-64.
14. O. C. Zienkiewicz and Y. K. Cheung, "Finite Elements in the
Solution of Field Problems," *The Engineer*, Sept. 1965, pp 507-510.
15. Archer, J. S. "Consistent Matrix Formulations for Structural
Analysis Using Influence Coefficient Techniques," presented at
the First American Institute of Aeronautics and Astronautics Annual
Meeting, June 1964, Paper No. 64-488.

618

16. Buturla, E. M. and P. E. Cottrell. "Two-Dimensional Finite Element Analysis of Semiconductor Steady-State Transport Equations," presented at the International Conference on Computational Methods in Nonlinear Mechanics, Austin, TX, 1974.

17. International Business Machines Program Product, *Subroutine Library-Mathematics User's Guide*, Order No. SH12-5300-1, available through IBM branch offices.

18. George, J. A. and J. Liu. *Users Guide for SPARSPAK: Waterloo Sparse Linear Equations Package*, Department of Computer Science, University of Waterloo, Waterloo, Ontario, Canada (1979).

19. George, J. A. and J. Liu. "Algorithms for Matrix Partitioning and the Numerical Solution of Finite Element Systems," *SIAM Journal of Numerical Analysis*, Vol. 15, (1978) 297.

20. George, J. A. and J. Liu. "An Automatic Nested Dissection Algorithm for Irregular Finite Element Problems," *SIAM Journal of Numerical Analysis*, Vol. 15 (1978) 1053.

21. Buturla, E. M. and P. E. Cottrell. "Simulation of Semiconductor Transport Equations Using Coupled and Decoupled Solution Techniques," *Solid-State Electron. 23*, (1980) 331-334.

22. Buturla, E. M., P. E. Cottrell, B. M. Grossman, M. B. Lawlor, C. T. McMullen, and K. A. Salsburg. "Three Dimensional Simulation of Semiconductor Devices," *IEEE International Solid State Circuits Conference Digest of Technical Papers*, San Francisco, (Feb. 1980) 76-77.

23. Cottrell, P. E. and E. M. Buturla. "Steady State Analysis of Field Effect Transistors via the Finite Element Method," *Digest of the IEEE International Electron Devices Meeting*, Washington, DC, (Dec. 1975), 51-54.

24. Noble, W. P. and P. E. Cottrell. "Narrow Channel Effects in Insulated Gate Field Effect Transistors," *Digest of the IEEE International Electron Devices Meeting*, Washington, DC, (Dec. 1976), 582-586.

25. Gaensslen, F. H. "Geometry Effects of Small MOSFET Devices," *IBM J. Res. Develop. 23*, (1979) 682-688.

26. Kotecha, H. and W. P. Noble. "Interaction of IGFET Field Design with Narrow Channel Device Operation," *Digest of the IEEE International Electron Devices Meeting*, Washington, DC, (Dec. 1980), 724-727.

27. Larsen, Richard A. "A Silicon and Aluminum Dynamic Memory Technology," *IBM J. Res. Develop. 24*, (1980) 268-282.

28. Barnes, J.J. and D.N. Shabde, and F.B. Jenne. "The Buried-Source VMOS Dynamic RAM Device," *Digest of the IEEE International Electron Devices Meeting*, Washington, DC, (Dec. 1977), 272-275.

29. Hoffman K. and R. Losehand. "VMOS Technology Applied to Dynamic RAMs," *IEEE J. Solid State Circuits SC-13*, (1978) 617-622.

30. Cottrell, P.E. and E.M. Buturla. "Threshold and Subthreshold Characteristics of VMOS FETs via a Two-dimensional Finite Element Model," presented at the Device Research Conference, June 1978, abstract published in IEEE Trans. Electron Devices ED-25,(1978) 1346.

31. Eardley, D. IBM General Technology Division laboratory, East Fishkill, NY, private communication.

32. Buturla, E.M., P.E. Cottrell, B.M. Grossman, C.T. McMullen, and K.A. Salsburg. "Three-Dimensional Transient Finite Element Analysis of the Semiconductor Transport Equations," *Proceedings of the NASECODE II Conference,* Boole Press, Dublin, Ireland, (June 1981).